T0199835

Green Chemistry
for Environmental
Sustainability

Sustainability: Contributions through Science and Technology

Series Editor: Michael C. Cann, Ph.D.
Professor of Chemistry and Co-Director of Environmental Science
University of Scranton, Pennsylvania

Preface to the Series

Sustainability is rapidly moving from the wings to center stage. Overconsumption of non-renewable and renewable resources, as well as the concomitant production of waste has brought the world to a crossroads. Green chemistry, along with other green sciences technologies, must play a leading role in bringing about a sustainable society. The **Sustainability: Contributions through Science and Technology** series focuses on the role science can play in developing technologies that lessen our environmental impact. This highly interdisciplinary series discusses significant and timely topics ranging from energy research to the implementation of sustainable technologies. Our intention is for scientists from a variety of disciplines to provide contributions that recognize how the development of green technologies affects the triple bottom line (society, economic, and environment). The series will be of interest to academics, researchers, professionals, business leaders, policy makers, and students, as well as individuals who want to know the basics of the science and technology of sustainability.

Michael C. Cann

Published Titles

Green Chemistry for Environmental Sustainability
Edited by Sanjay Kumar Sharma, Ackmez Mudhoo, 2010

Forthcoming Title

Microwave Heating as a Tool for Sustainable Chemistry
Edited by Nicholas E. Leadbeater, 2010

Sustainability: Contributions through Science and Technology

Series Editor: Michael Cann

Green Chemistry for Environmental Sustainability

Edited by
Sanjay Kumar Sharma
Ackmez Mudhoo

CRC Press
Taylor & Francis Group
Boca Raton London New York

CRC Press is an imprint of the
Taylor & Francis Group, an **informa** business

CRC Press
Taylor & Francis Group
6000 Broken Sound Parkway NW, Suite 300
Boca Raton, FL 33487-2742

First issued in paperback 2019

© 2011 by Taylor and Francis Group, LLC
CRC Press is an imprint of Taylor & Francis Group, an Informa business

No claim to original U.S. Government works

ISBN-13: 978-1-4398-2473-3 (hbk)
ISBN-13: 978-0-367-26243-3 (pbk)

This book contains information obtained from authentic and highly regarded sources. Reasonable efforts have been made to publish reliable data and information, but the author and publisher cannot assume responsibility for the validity of all materials or the consequences of their use. The authors and publishers have attempted to trace the copyright holders of all material reproduced in this publication and apologize to copyright holders if permission to publish in this form has not been obtained. If any copyright material has not been acknowledged please write and let us know so we may rectify in any future reprint.

Except as permitted under U.S. Copyright Law, no part of this book may be reprinted, reproduced, transmitted, or utilized in any form by any electronic, mechanical, or other means, now known or hereafter invented, including photocopying, microfilming, and recording, or in any information storage or retrieval system, without written permission from the publishers.

For permission to photocopy or use material electronically from this work, please access www.copyright.com (http://www.copyright.com/) or contact the Copyright Clearance Center, Inc. (CCC), 222 Rosewood Drive, Danvers, MA 01923, 978-750-8400. CCC is a not-for-profit organization that provides licenses and registration for a variety of users. For organizations that have been granted a photocopy license by the CCC, a separate system of payment has been arranged.

Trademark Notice: Product or corporate names may be trademarks or registered trademarks, and are used only for identification and explanation without intent to infringe.

Library of Congress Cataloging-in-Publication Data

Green chemistry for environmental sustainability / editors, Sanjay Kumar Sharma, Ackmez Mudhoo.
 p. cm. -- (Sustainability, contributions through science technology)
 Includes bibliographical references and index.
 ISBN 978-1-4398-2473-3 (hardcover : alk. paper)
 1. Environmental chemistry--Industrial applications. I. Sharma, Sanjay Kumar. II. Mudhoo, Ackmez. III. Title. IV. Series.

TP155.2.E58G744 2011
660--dc22

2010019527

**Visit the Taylor & Francis Web site at
http://www.taylorandfrancis.com**

**and the CRC Press Web site at
http://www.crcpress.com**

For my loving wife, Pratima, on our tenth wedding anniversary

—Sanjay K. Sharma

With affection for Aryanna, mum, and dad

—Ackmez Mudhoo

Contents

Preface

The thinking behind writing this book rests much on the following mighty saying of Albert Einstein (1879–1955).

We shall require a substantially new manner of thinking if mankind is to survive.

The last five decades have been a time of unprecedented change in the general way of living in the world in many walks of life. Aside from all the political, economic, and social developments that have taken place, many of the key changes in society have stemmed from the numerous advances in science and technology. These various technological and societal developments are highly interactive and have consequences that may be either desirable or undesirable from the viewpoint of energy consumption, ecological, and environmental degradation.

Some of these advances in modern society for the betterment in the overall living conditions of humans, and to some extent for animals, have brought with them negative consequences in complex mechanisms that cause collateral environmental damage, are virtually irreversible, and have yet to be faced head-on. As well as air pollution and global warming, which are thought to arise from the ever-expanding use of energy, there are other problems of a global nature that may be grossly categorized as the following: environmental pollution of natural waters and soils, bioaccumulation of heavy metals and other harmful molecules in living biota including humans, unequal distribution of energy, and ecological disruption in several biospheres. Additionally, inequalities in the geographic distribution of natural resources, especially petroleum, make some countries wealthy and others relatively poor.

Following the establishment of the 12 Principles of Green Chemistry (Anastas and Warner, 1998), there has been a gradual and constructive growth in our understanding of what green chemistry means. Green chemistry is a relatively young science in its own respect. Interest in this subject, however, is growing rapidly and, although no concerted agreement has been reached as yet about the exact content and limits of this interdisciplinary field, there appears to be increasing interest in umpteen environmental topics, which are based on chemistry embodied in this subject. To the pleasant surprise of all, this increased understanding of the principles that are the backbone of green chemistry has spurred many outstanding efforts to implement chemical processes and innovative technologies that are incrementally moving society toward more sustainable practices and products that embody and foster environmental stewardship and environmental protection. Environmental stewardship is the responsibility for environmental quality shared by all those whose actions affect the environment. In 2005, the Environmental Protection Agency laid out a vision for environmental stewardship recognizing it as a means to a more sustainable future.

However, the green chemistry community has always strived to convince many people in the chemical community that green chemistry is different from the historical pollution prevention initiatives of the 1980s that green chemistry burgeoned from. The proposal and risk of changing the way chemistry is done and applied, the development of new chemistries and chemical synthetic pathways, and the establishment of novel and benign chemical processes that are significantly more efficient using nonpetrochemical and renewable feedstocks have been challenging notions for a world that has been surrounded by the products of petroleum for more than a century. Green chemistry has hence brought a relatively prompt and positive shift in the paradigm as it concerns the overall use and management of natural resources and raw materials for the development of society with a promise to cause far less pronounced harm to the environment at large.

By adopting green chemistry principles and methodologies, researchers all over the globe are devising new processes to help protect and ultimately save the environment from further damage. The writing of this book was undertaken because it was earnestly intended to be a work that encompassed some of the various relevant aspects linked and linking green chemistry practice to environmental sustainability. In this book, research on and the application of green chemistry and engineering in addressing current issues that are both of environmental and social nature have been presented. The book covers sustainable development and environmental sustainability through chapters contributed in green chemistry and engineering research, the design and synthesis of environmentally benign chemical processes, green approaches to minimize and/or remediate environmental pollution, the development of biomaterials, biofuel and bioenergy production, biocatalysis, and policies and ethics in green chemistry. We hope the book provides an insightful text on the chemical and biochemical technologies that are being studied, optimized and eventually developed to promote environmental sustainability. We also feel it provides up-to-date information on some selected fields where the Principles of Green Chemistry are being embraced for safeguarding and improving the quality of the environment.

REFERENCE

Anastas, P.T. and Warner, J.C. 1998. *Green Chemistry: Theory and Practice*. New York: Oxford University Press.

Sanjay K. Sharma
Ackmez Mudhoo

Acknowledgments

This undertaking has provided us with a unique opportunity to renew some old friendships and hopefully to weave some new ones in the pursuit to gather the expertise required for compiling this edited book on "Green Chemistry" and "Environmental Sustainability." The primary acknowledgment, without any reservation, goes to our pool of esteemed contributors for the way they have graciously responded with characteristic good humor and indulgence to our relatively modest deadlines. We hope they feel that the final result does ample justice to their painstaking efforts and mental strain exercised in writing and getting their respective chapter(s) ready for this book. We are also appreciative and grateful to other colleagues and fellow researchers who volunteered their help in reviewing the scientific contents of manuscripts received for this book.

Professor S.K. Sharma especially expresses his heartfelt gratitude to his respected parents, Dr. M.P. Sharma and Smt. Parmeshwari Devi. He also wishes to extend his warmest regards to Professor R.K. Bansal who has been a source of inspiration to him, and to Dr. V.K. Agrawal, Chairman of the Institute of Engineering and Technology, Alwar (India), for his encouraging words. Ackmez Mudhoo equally expresses his appreciation for the faith his parents, Azad A. Mudhoo and Ruxana B. Mudhoo, his brother, Assad, and sister-in-law, Tina, have placed in him throughout the writing of this book. Ackmez Mudhoo is also grateful to Professor Romeela Mohee (University of Mauritius, Mauritius), Dr. Vinod K. Garg and Meetali (both from Guru Jambheshwar University of Science and Technology, Hisar, Haryana, India) and Dr. Zhi-Qing Lin (Southern Illinois University, Edwardsville, Illinois) for their encouragement.

Sanjay K. Sharma
Ackmez Mudhoo

Editors

Professor Sanjay K. Sharma is a very well-known author of many books and hundreds of articles over the last 20 years. He is presently a professor and head of the Department of Chemistry and Environmental Engineering at the Institute of Engineering and Technology (IET), Alwar, Rajasthan, India. Dr. Sharma did his postgraduate (1995) and PhD (1999) work from the University of Rajasthan, Jaipur. His field of work was synthetic organophosphorus chemistry and computational chemistry for which he attained his PhD. In 1999, he joined IET and started working in the field of environmental chemistry and established a Green Chemistry Research Laboratory. His work in the field of green corrosion inhibitors is very well recognized and praised by the international research community. He is a member of the American Chemical Society and the Green Chemistry Network (Royal Society of Chemists, United Kingdom) and is also a lifetime member of various international professional societies including the International Society of Analytical Scientists, Indian Council of Chemists, International Congress of Chemistry and the Environment and Indian Chemical Society. Dr. Sharma has 6 textbooks and over 40 research papers of national and international repute to his credit. Dr. Sharma is also serving as editor-in-chief for two international research journals *RASAYAN Journal of Chemistry* and *International Journal of Wastewater Treatment and Green Chemistry* and is a reviewer in many other international journals.

Ackmez Mudhoo obtained his Bachelor's degree in Engineering (Hons.) in chemical and environmental engineering from the University of Mauritius in 2004. He then successfully completed his master of philosophy (M.Phil.) degree through research in the department of chemical and environmental engineering, University of Mauritius in 2007. His main research interests encompass the design of composting systems and analysis of composting processes and the biological treatment of solid wastes and wastewater. Ackmez has 24 international journal publications and 4 conference papers to his credit, and an additional 5 research and review papers in the pipeline. Ackmez is also serving as peer reviewer for *Waste Management, International Journal of Environment and Waste Management, Journal of Hazardous Materials, Journal of Environmental Informatics, Environmental Engineering Science, RASAYAN Journal of Chemistry and Water*

Research, and is the editor-in-chief for two international research journals *International Journal of Process Wastes Treatment* and *International Journal of Wastewater Treatment and Green Chemistry*. He worked as a consultant chemical process engineer for China International Water & Electric Corp. (CWE, Mauritius) from February 2006 to March 2008. Ackmez is presently a lecturer in the Department of Chemical and Environmental Engineering at the University of Mauritius.

Contributors

Vikash Babu
Department of Biotechnology
Indian Institute of Technology
   Roorkee
Roorkee, Uttarakhand, India

Roberto Ballini
Chemistry Division
School of Science and Technolgy
University of Camerino
Camerino, Italy

George D. Bennett
Department of Chemistry
Millikin University
Decatur, Illinois

Bijan Choudhury
Department of Biotechnology
Indian Institute of Technology
   Roorkee
Roorkee, Uttarakhand, India

Carl Dalhammar
International Institute for
   Industrial Environmental
   Economics (IIIEE)
Lund University
Lund, Sweden

Nicola D'Antona
Istituto Chimica Biomolecolare
Consiglio Nazionale delle
   Ricerche (CNR)
Catania, Italy

Om Parkash Dhankher
Department of Plant, Soil and Insect
   Sciences
University of Massachusetts
Amherst, Massachusetts

Serena Gabrielli
Chemistry Division
School of Science and Technolgy
University of Camerino
Camerino, Italy

Pritee Goyal
Department of Applied Science
G.L. Bajaj Institute of Technology
   and Management
Greater Noida, India

Meenu Gupta
Department of Biotechnology
Indian Institute of Technology Roorkee
Roorkee, Uttarakhand, India

Nazrul Haq
Department of Applied Chemistry
Aligarh Muslim University
Aligarh, Uttar Pradesh, India

Samir Kumar Khanal
Department of Molecular Biosciences
   and Bioengineering
University of Hawaii at Mānoa
Honolulu, Hawaii

Telemachus C. Koliopoulos
Environmental Consultancy
Department of Surveying Engineering
Technological Educational Institute
   of Athens
Athens, Greece

Georgia Koliopoulou
Department of Experimental Physiology
University of Athens
Athens, Greece

Buddhi P. Lamsal
Department of Food Science and
   Human Nutrition
Iowa State University
Ames, Iowa

Fei Li
Suzhou Key Lab for Selenium
   and Human Health
Suzhou Institute for Advanced Study,
   USTC
Suzhou, Jiangsu, China

Zhi-Qing Lin
Environmental Sciences Program and
   Department of Biological Sciences
Southern Illinois University
Edwardsville, Illinois

Ali Mohammad
Department of Applied Chemistry
Aligarh Muslim University
Aligarh, Uttar Pradesh, India

Romeela Mohee
Department of Chemical and
   Environmental Engineering
University of Mauritius
Reduit, Mauritius

Raffaele Morrone
Istituto Chimica Biomolecolare
Consiglio Nazionale delle Ricerche
   (CNR)
Catania, Italy

Ackmez Mudhoo
Department of Chemical and
   Environmental Engineering
University of Mauritius
Reduit, Mauritius

Saoharit Nitayavardhana
Department of Molecular Biosciences
   and Bioengineering
University of Hawaii at Mānoa
Honolulu, Hawaii

Alessandro Palmieri
Chemistry Division
School of Science and Technolgy
University of Camerino
Camerino, Italy

Catalina Pisoschi
Department of Biochemistry
University of Medicine and Pharmacy
Craiova, Romania

Ileana Prejbeanu
Department of Environmental Health
University of Medicine and Pharmacy
Craiova, Romania

Oana Purcaru
Department of Biochemistry
University of Medicine and Pharmacy
Craiova, Romania

Abdul Rauf
Department of Chemistry
Aligarh Muslim University
Aligarh, Uttar Pradesh, India

Rashmi Sanghi
Facility for Ecological and Analytical
   Testing
Indian Institute of Technology Kanpur
Kanpur, Uttar Pradesh, India

Sanjay K. Sharma
Department of Chemistry and
   Environmental Engineering
Institute of Engineering and
   Technology
Alwar, Rajasthan, India

Shweta Sharma
Department of Chemistry
Aligarh Muslim University
Aligarh, Uttar Pradesh, India

Shilpi
Department of Biotechnology
Indian Institute of Technology Roorkee
Roorkee, Uttarakhand, India

Prachand Shrestha
Energy Biosciences Institute
University of California
Berkeley, California

Shalini Srivastava
Department of Chemistry
Dayalbagh Educational Institute
Agra, Uttar Pradesh, India

Camelia Stanciulescu
Department of Biochemistry
University of Medicine and Pharmacy
Craiova, Romania

Daniela Tache
Department of Biochemistry
University of Medicine and Pharmacy
Craiova, Romania

Devin Takara
Department of Molecular Biosciences
    and Bioengineering
University of Hawaii at Mānoa
Honolulu, Hawaii

Preeti Verma
Facility for Ecological and Analytical
    Testing
Indian Institute of Technology Kanpur
Kanpur, Uttar Pradesh, India

Xuebin Yin
Suzhou Key Lab for Selenium
    and Human Health
Suzhou Institute for Advanced Study,
    USTC
Suzhou, Jiangsu, China

Liu Ying
Suzhou Key Lab for Selenium
    and Human Health
Suzhou Institute for Advanced Study,
    USTC
Suzhou, Jiangsu, China

Wei Zhang
Department of Chemistry
University of Massachusetts Boston
Boston, Massachusetts

1 Green Chemistry and Engineering

A Versatile Research Perspective

Sanjay K. Sharma, Ackmez Mudhoo, and Wei Zhang

CONTENTS

1.1 INTRODUCTION

Green chemistry (environmentally benign chemistry) involves the utilization of a set of principles that reduces or eliminates the use or generation of hazardous substances in the design, manufacture, and application of chemical products (Kidwai and Mohan, 2005). In practice, green chemistry covers a much broader range of issues than the definition suggests (Lancaster, 2000). In addition to using and producing better chemicals with less waste, green chemistry also involves reducing other associated environmental impacts, including a reduction in the amount of energy used in chemical processes (Kidwai and Mohan, 2005). Green chemistry is not different from traditional chemistry inasmuch as it embraces the same creativity and innovation that has always been central to classical chemistry. However, there is a crucial difference in that, historically, synthetic chemists have not been seen to rank the environment very high in their priorities (Kidwai and Mohan, 2005). However with an increased awareness for environmental protection, environmental pollution prevention, safer industrial ecology, and cleaner production technologies worldwide, there is a heightened interest and almost a grand challenge for chemists to develop new products, processes, and services that achieve the necessary social, economical, and environmental objectives. Since the types of chemicals and the types of transformations are very varied in the chemical industry and chemical research world, so are the green chemistry solutions that have been proposed. Anastas and Warner (1998) brilliantly developed "The twelve principles of green chemistry," which are valuable benchmark guidelines for practicing chemists and engineers in developing and assessing how green a synthesis, compound, process, or technology is.

The 12 principles of green chemistry (Anastas and Warner, 1998) are as follows:

Prevention: It is better to prevent waste than to treat or clean up waste after it has been created.

Atom Economy: Synthetic methods should be designed to maximize incorporation of all materials used in the process into the final product.

Less Hazardous Chemical Syntheses: Wherever practicable, synthetic methods should be designed to use and generate substances that possess little or no toxicity to human health and the environment.

Designing Safer Chemicals: Chemical products should be designed to effect their desired function while minimizing their toxicity.

Safer Solvents and Auxiliaries: Use of auxiliary substances (solvents, separation agents, etc.) should be made unnecessary wherever possible and innocuous when used.

Design for Energy Efficiency: Energy requirements of chemical processes should be recognized for their environmental and economic impacts and should be minimized. If possible, synthetic methods should be conducted at ambient temperature and pressure.

Use of Renewable Feedstocks: A raw material or feedstock should be renewable rather than depleting whenever technically and economically practicable.

Reduce Derivatives: Unnecessary derivatization (use of blocking groups, protection/deprotection, and temporary modification of physical/chemical processes) should be minimized or avoided if possible, because such steps require additional reagents and can generate waste.

Catalysis: Catalytic reagents (as selective as possible) are superior to stoichiometric reagents.

Design for Degradation: Chemical products should be designed so that at the end of their function they break down into innocuous degradation products and do not persist in the environment.

Real-Time Analysis for Pollution Prevention: Analytical methodologies need to be further developed to allow for real-time, in-process monitoring and control prior to the formation of hazardous substances.

Inherently Safer Chemistry for Accident Prevention: Substances and the form of a substance used in a chemical process should be chosen to minimize the potential for chemical accidents, including releases, explosions, and fires.

In course of time, our knowledge on the temporal and spatial scales of environmental problems has increased (Mihelcic et al., 2003). Conventional pollutants were first studied and managed on a regional scale. The scale of issues in time and space increased further when bioaccumulative persistent toxic pollutants and global climate change chemicals of concern were considered (Mihelcic et al., 2003). Many models and tools were developed for environmental impact assessments that have been successful in predicting impacts for selected chemicals in selected environmental settings. These models have joined air and water quality aspects to point and non-point sources and have been very useful for the development of emission control and compliance strategies (Mihelcic et al., 2003). Unfortunately, these models were aimed primarily at evaluating the quantity of pollutants that could be discharged into the environment with acceptable impact. These efforts did not focus on pollution prevention nor did they, in many cases, develop mechanistic understanding of how pollutants are initially formed and released. Also, while individuals involved in this strategy of waste management were very adept at calculating the carrying capacity of the natural world, they assumed that pollutant generation and release were a normal part of commerce. In addition, the attractiveness of end-of-pipe approaches (Polshettiwar and Varma, 2008) to waste management decreased, and strategies such as environmentally conscious manufacturing, eco-efficient production, or pollution prevention gained prominence. These "green" approaches to the design and development of processes and products have dexterously served as the basis for green chemistry and green engineering (Mihelcic et al., 2003).

Green chemistry and green engineering bring about changes in the hazard of a product at the most fundamental level, that is, the molecular level (Lankey and Anastas, 2002). The types of hazards that are of concern for their impact on human health and the environment can be viewed simply as physical or chemical properties of the substances being used. In simpler terms, the molecular structure of a chemical substance is a determining factor in its properties, such as potential health hazards. Practitioners of green chemistry and engineering focus on modifying these intrinsic properties to reduce or eliminate the hazardous nature of these substances (Lankey and Anastas, 2002). It is for this reason that green chemistry and green engineering are capable of accomplishing their goals throughout the life cycle of a product or process; inherent properties do not change merely by moving through the various life-cycle stages. Life-cycle innovation, therefore, is recognizing that, by designing the fundamental properties of the substances used in particular stages of the life cycle, green chemistry and green engineering have the power to impact on the entire life cycle of a product or process (Anastas and Lankey, 2000), and make it more environmentally safe (Polshettiwar and Varma, 2008) and sustainable. All the more, the roles of chemists and engineers are more than ever intertwined and complementary for the progress of green engineering. The decisions made by chemists in designing chemical products and processes have a direct impact on the options available to chemical and process engineers (Kirchhoff, 2003). The physical and chemical properties of a material strongly influence the type of reactor that must be used in a given process. The task of the engineer is simplified when chemists design products and processes that reduce or eliminate the use and generation of hazardous substances (Kirchhoff, 2003). Green chemistry therefore indispensably provides a foundation on which it can be built.

1.2 GREEN ENGINEERING AND SUSTAINABILITY

Green engineering (Anastas et al., 2000) focuses on how to achieve sustainability through science and technology (Fiksel, 1998; Skerlos et al., 2001). The twelve principles of green engineering are strongly concomitant with those of green chemistry provide a structured framework for scientists and engineers to engage in when designing new materials, products, processes, and systems that are benign to human health and the environment (Anastas and Zimmerman, 2003). Engineers use these principles as guidelines to help ensure that designs for products, processes, or systems have the fundamental components, conditions, and circumstances necessary to be more sustainable. Furthermore, the breadth of the principles' applicability is important (Anastas and Zimmerman, 2003). With regard to design architecture (i.e., molecular architecture required to construct chemical compounds, product architecture to create an automobile, or urban architecture to build a city (Anastas and Zimmerman, 2003), the same green engineering principles must be applicable, effective, and appropriate. Otherwise, these would not be principles but a mere compilation of a list of useful techniques that have been successfully demonstrated under specific conditions.

Indeed, sustainability will be one of the main drivers for innovation in order to allow the technical industries to care for the well-being of consumers in a safe and healthy environment (Höfer and Bigorra, 2007). In this particular context, sustainability is defined as the design of human and industrial systems to ensure that humankind's use of natural resources and cycles does not lead to a diminished quality of life owing to either losses in future economic opportunities or adverse impacts on social conditions, human health, and the environment (Thomas, 2003). These requirements reflect that social conditions, economic opportunity, and environmental quality are essential if society's development goals are to reconcile with international environmental limitations. Accordingly, fundamental research, education, and knowledge transfer are needed to meet this vision.

Hence, the "design for sustainability" concept will be important during the design and development of sustainability over the full product and process life cycle. In the chemical industry, "design for sustainability" is more than what is often manifested via sustained development of green chemical routes, process intensification, and process redesign with extensive research and development programs at all levels of chemistry and chemical engineering. All the more, sustainability has been widely endorsed as the overarching goal of environmental policy (Thomas, 2003). The hope or long-term vision is that a strong, just, and wealthy society can be consistent with a clean environment, healthy ecosystems, and a beautiful planet. There has, however, been considerable doubt expressed about the potential for scientific research in this area. Thomas (2003) argues and maintains that many key issues of sustainability can be expressed as tractable research questions that build on existing research. Although research in sustainability is inherently broad and multifaceted, it falls naturally into three basic areas: use of materials and energy, land use, and human development. These three areas are interrelated, and in various contexts two or perhaps all three might be considered together.

1.3 RESEARCH AREAS IN GREEN CHEMICAL ENGINEERING

Over the past decade, green chemistry has convincingly demonstrated how fundamental scientific methodologies may be devised and applied to protect human health and the environment in an economically beneficial manner (Anastas and Kirchhoff, 2002). However, significant progress is underway in several key research areas, such as biosynthesis, biochemical engineering, biocatalysis, photocatalysis, heterogeneous catalysis, design of safer chemicals and environmentally benign solvents, sonochemistry, microwave (MV)-assisted polymerization, green corrosion inhibition (Sharma et al., 2009a, 2009b, 2009c), and the development of renewable feedstocks. Research to reduce the environmental impacts of engineered systems spans a wide range of topics, disciplines, and levels of focus. Similar to all engineering design approaches, green engineering research will constantly progress. Areas in the research and development of green chemistry have been identified as follows by Tundo et al. (2000):

Use of alternative feedstocks: Use of feedstocks that are both renewable, rather than depleting, and less toxic to human health and the environment.

Use of innocuous reagents: Use of reagents that are inherently less hazardous and are catalytic whenever feasible.

Employing natural processes: Use of biosynthesis, biocatalysis, and biotech-based chemical transformations for efficiency and selectivity.

Use of alternative solvents: Design and utilization of solvents that have reduced potential for detriment to the environment and serve as alternatives to currently used volatile organic solvents, chlorinated solvents, and solvents that damage the natural environment.

Design of safer chemicals: Use of molecular structure design—and consideration of the principles of toxicity and mechanism of action—to minimize

the intrinsic toxicity of the product while maintaining its efficacy of
function.

Developing alternative reaction conditions: Design of reaction conditions that
increase the selectivity of the product and allow for dematerialization of the
product separation process.

Minimizing energy consumption: Design of chemical transformations that
reduce the required energy input in terms of both mechanical and thermal
inputs and the associated environmental impacts of excessive energy usage.

Numerous outstanding and "breakthrough" research findings and examples (over
5000 published research articles) of green chemistry and green engineering may be
found in the literature. While it is not feasible to comprehensively reproduce the
essence of each of these research articles in the present exposé, the case studies,
examples, descriptions, and discussions provided in the following sections hopefully
constitute an insightful and representative selection of green chemistry research and
green engineering technology where significant advancement has been made and is
still in progress.

1.3.1 Biosyntheses and Biocatalysts

Sustainability, industrial ecology, eco-efficiency, and green chemistry direct the
development of the next generation of materials, products, and processes. Biodegradable plastics and bio-based polymers that are based on annually renewable agricultural and biomass feedstock can form the basis for a portfolio of sustainable,
eco-efficient products that can compete and capture markets currently dominated by
products based exclusively on petroleum feedstock (Mohanty et al., 2002). Natural/
biofiber composites (biocomposites) are emerging as a viable alternative to glass
fiber-reinforced composites especially in automotive and building product applications. The combination of biofibers such as kenaf, hemp, flax, jute, and henequen
with polymer matrices from both nonrenewable and renewable resources to produce
composite materials that are competitive with synthetic composites requires special
attention. Natural fiber-reinforced polypropylene composites have attained commercial attraction in automotive industries. Natural fiber (polypropylene or natural fiber)
polyester composites are not sufficiently eco-friendly because of the petroleum-
based source and the nonbiodegradable nature of the polymer matrix. Using natural
fibers with polymers based on renewable resources will resolve many environmental
issues (Mohanty et al., 2002). By embedding biofibers with renewable resource-
based biopolymers such as cellulosic plastics, polylactides, starch plastics, polyhy-
droxyalkanoates (bacterial polyesters), and soy-based plastics, the so-called green
bio-composites are continuously being developed.

With advances in biotechnology, biocatalysis, and biosynthesis (Kidwai and
Mohan, 2005; Ran et al., 2008) there have been significant accomplishments in the
use of biologically based feedstocks/raw materials as a viable alternative to petroleum feedstocks for a number of chemical processes (Frost and Draths, 1995).
Indeed, as a result of recent advances in genomics, proteomics, and pathway engineering, biocatalysis is emerging as the best among green technologies. Enzymes are

highly efficient with excellent regioselectivity and stereoselectivity (Ran et al., 2008). By conducting reactions in water under ambient reaction conditions, both the use of organic solvents and energy input are minimized. The selection of feedstocks has a major effect, not only on the efficacy of the synthetic pathway, but also on the environmental and health effects of the process. More than 98% of all organic chemicals are derived from petroleum (Szmant, 1989). Achieving a sustainable chemical industry dictates switching from depleting finite sources to renewable feedstocks (Anastas and Kirchhoff, 2002). Research in this area has focused on both the macro- and molecular level. The carbohydrate economy provides a rich source of feedstocks for synthesizing commodity (Lynd et al., 1999) and specialty chemicals. Agricultural wastes have been converted into useful chemical intermediates such as levulinic acid, alcohols, ketones, and carboxylic acids (Anastas and Kirchhoff, 2002). Chitosan, a biopolymer with a wide range of potential applications, is being currently explored for use in the oil-drilling industry (Kumar et al., 2000). At the molecular level, genetic engineering produces valuable chemical products via nontraditional pathways (Anastas and Kirchhoff, 2002).

1.3.1.1 Laccases

Biocatalysts in organic synthesis have been less used in the past, probably because they were not commercially available (Riva, 2006). The search for new, efficient, and environmentally benign processes for the textile and pulp and paper industries has increased the interest in these essentially "green" catalysts, which work with air and produce water as the only by-product, making them more generally available to the scientific community (Riva, 2006). Consequently, a significant number of reports that focus on the biochemical properties of these proteins and/or on their applications in technological and bioremediation processes in addition to their use in chemical reactions have been published in the past decade. Laccases are a group of oxidative enzymes that belong to the multinuclear copper-containing oxidases. They are glycoproteins, which are ubiquitous in nature—they have been reported in higher plants, and in virtually every fungus that has been examined for them (Thurston, 1994). They catalyze the monoelectronic oxidation of substrates at the expense of molecular oxygen (Riva, 2006). Interest in these essentially "eco-friendly" enzymes (they work with air and produce water as the only by-product) has grown significantly over recent years: their uses extend from the textile to the pulp and paper industries, and from food applications to bioremediation processes. Laccases also have uses in organic synthesis, where their typical substrates are phenols and amines, and the reaction products are dimers and oligomers derived from the coupling of reactive radical intermediates (Riva, 2006).

The reactions catalyzed by laccases proceed by the monoelectronic oxidation of a suitable substrate molecule (phenols and aromatic or aliphatic amines) to the corresponding reactive radical (Riva, 2006). The redox process takes place with the assistance of a cluster of four copper atoms that form the catalytic core of the enzyme; they also confer the typical blue color to these enzymes because of the intense electronic absorption of the Cu–Cu linkages (Piontek et al., 2002). The overall outcome of the catalytic cycle is the reduction of one molecule of oxygen to two molecules of water and the concomitant oxidation of four substrate molecules to produce four radicals (Claus, 2004). These reactive intermediates can then produce dimers, oligomers, and polymers. Several applications of laccases have been proposed, which use one of the protocols described. Specifically, the direct substrate oxidation of phenol derivatives has been investigated in bioremediation efforts to decontaminate industrial wastewaters. The polymeric polyphenolic derivatives that result from the laccase-catalyzed oxidative couplings are usually insoluble and can be separated easily by filtration or sedimentation (Riva, 2006). With regard to the synthetic applications of laccase-mediator systems, investigations of the laccase-mediated delignification process have shown that redox mediators drive these enzymes toward the oxidation of nonphenolic substituents, particularly of benzyl alcohol groups, in several lignin model compounds, for example, the oxidation of adlerol to the corresponding ketone, adlerone (Barreca et al., 2003).

1.3.1.1.1 Bio-Based Energy

Waste contains three primary constituents, cellulose, hemicellulose, and lignin, and can contain other compounds (e.g., extractives) (Okonko et al., 2009). Cellulose and hemicellulose are carbohydrates that can be broken down by enzymes, acids, or other compounds to simple sugars, and then fermented to produce ethanol renewable electricity, fuels, and biomass-based products (van Wyk, 2001). When the amount of organic agricultural waste (such as corn stalks, leaves, and wheat straw from wheat-processing facilities, and sawdust and other residues from wood mills) is also considered, this component of solid waste could be a principal resource for biodevelopment (van Wyk, 2001). Materials of organic origin are known as biomass (a term that describes energy materials that emanate from biological sources) and are of major importance to sustainable development because they are renewable as opposed to nonorganic materials and fossil carbohydrates (van Wyk, 2001). Sugar will be the key feedstock of the future, as it can be used not only to ferment ethanol for transportation fuel, but also for a whole set of new, basic building blocks. Molecules such as lactic acid, succinic acid, propylene glycol, or 3-hydroxy propionic acid produced at 20 cents per pound can catalyze the innovation of new chemical product families, similar to the innovation boost based on the cracker chemicals in the middle of the twenty-first century (Okonko et al., 2009). Indeed, the combination of bio-based feedstock, bio-processes, and new products offers the potential to revolutionize structures of the chemical industry.

Omoyinmi et al. (2004) demonstrated that animal protein production varied from 0.91 to 1.41 g kg^{-1} of waste in the earthworm, from 1.15 to 1.4 g kg^{-1} of waste in the garden snail, and from 0.9 to 1.6 g kg^{-1} of waste in the palm grub. It was also shown that the short life cycle and production of a large number of offsprings could be harnessed for the raising of feed for fish/livestock and in some cases for human consumption. This culture of invertebrates offered economic benefits to the farmer and improved on the environmental quality by transforming wastes into beneficial products (Okonko et al., 2009). In 2004, Kareem and Akpan reported that the use of agricultural by-products as a substrate for enzyme production was cheap and could facilitate the large-scale production of industrial enzymes in the tropics. Eight isolates of *Rhizopus* sp. were obtained from the environment and were grown on solid media for the production of pectinase enzymes. Three media formulated from agricultural materials were the following: medium A (ricebran + cassava starch, 10:2 w/w), medium B (cassava starch + soybean, 1:2 w/w), and medium C (ricebran + soybean + casein hydrolysate, 10:20.5 w/w). The result obtained by Kareem and Akpan (2004) showed that medium A gave the highest pectinase activity of 1533.3 μm L^{-1}. The three solid media supported profuse mycelia growth of *Rhizopus* species and enhanced its pectinase-producing potential.

Lignocellulose consists of lignin, hemicellulose, and cellulose (Sun and Cheng, 2002). Hydrolysis methods have also been used to degrade lignocellulose to alkali and acid (Nguyen et al., 1999). However, many enzyme processes are preferred over acid or alkaline processes since they are specific biocatalysts that can operate under much milder reaction conditions, do not produce undesirable products, and are environmentally friendly (Polshettiwar and Varma, 2008). The future of fermentation technology will be greatly enhanced by lignocellulose conversion. To develop the carbohydrate potential of biowaste materials, its cellulose content has to be converted into sugars such as glucose that can be used as starting compounds in the biosynthesis of many bioproducts. This conversion process could either be acid or enzyme catalyzed; the concentrated acid process for producing sugars was reported as early as 1883 and there has been a simultaneous saccharification and fermentation of cellulose into different useful products. This treatment disrupts the hydrogen bonding between cellulose chains, and once it has been decrystallized it becomes extremely susceptible to hydrolysis. Wastes from biomass can also provide raw materials for a variety of bio-based products (Okonko et al., 2009). For example, plastics from biomass are being produced using polylactic acid from corn. According to Block (1999), executive order and proposed bill will boost bio-based products and bioenergy in nations that see the need for it. The demand for ethanol has the most significant market where ethanol is used either as a chemical feedstockor as an octane enhancer, or petrol additive. Brazil produces ethanol from the fermentation of cane juice, whereas in the United States corn is used (Okonko et al., 2009). In the United States, fuel ethanol has been used in gasohol or oxygenated fuels since the 1980s. These gasoline fuels contain up to 10% ethanol by volume (Sun and Cheng, 2002). The production of ethanol from sugars or starch impacts negatively on the economics of the process, thus making ethanol more expensive compared with fossil fuels. However, the huge amounts of residual plant biomass considered as waste can potentially be converted into various different value-added products, including biofuels, chemicals, and cheap energy sources for fermentation, improved animal feeds, and human nutrients. Biodiesel production is a completely renewable resource. Biodiesel product is made from soya and canola, which is a self-sustaining fuel. Best of all, it provides a market for excess soybean oil production (Okonko et al., 2009). Biodiesel is a substitute for fuels that produce a lot of soot and carbons. These poisonous elements are associated with regular diesel fuel emissions (especially buses). However, biodiesel has been around for decades as a supplement that is added to conventional diesel fuel to improve the lubricity of diesel engines. A biodiesel fuel consists mainly of methyl esters in soybean oil. Many car manufacturers are exploring the possibilities of developing vehicles that can accommodate a biodiesel product by creating a diesel car that is friendly to the use of vegetable oil blended with diesel fuel. In addition to displacing North America's reliance on imported petroleum, the use of biodiesel product has been shown to reduce air pollution and greenhouse gases. On par with the biodiesel industry, a recent article in *The Chemical Engineer* magazine (February 2009, Issue 812, pp. 26–27, *Particulates matter*, UK) described the use of a simulated moving bed reactor (SMBR) developed by Dr. Viviana Silva and Professor Alírio Rodrigues (Faculdade de Engenharia da Universidade do Porto, Department of Chemical Engineering, Laboratory of Separation and Reaction Engineering (LSRE), Portugal) to produce green diesel additives. According to the inventors, the synthesis of diethylacetal via the SMBR-based technology for the production and separation of acetal, a green diesel additive, by means of acid ion-exchange resins (Amberlyst 15 wet) as catalysts/adsorbent is the first-known process of its kind. On the basis of data gathered by the researchers of LSRE from a pilot SMBR unit, the conversion of acetaldehyde can reach a maximum of 98–100% with purities of both extract and raffinate being above 99%. The ethanol from the extract can be dehydrated in a commercial membrane process at 99% purity and recycled to the SMBR. This represents a 50% reduction in energy costs when compared to traditional distillation processes. The acetal from the raffinate is separated by distillation, obtaining pure acetal and an azeotropic mixture of ethanol–acetal recyclable to the SMBR, thereby avoiding the use of solvents to break the azeotrope and associated solvent distillation costs.

1.3.1.1.2 Biopolymers

Polymers from renewable resources have worldwide interest and are attracting an increasing amount of attention for predominantly two major reasons: environmental concerns and the fact that petroleum resources are finite (Yu and Chen, 2009). Generally, polymers from renewable resources can be classified into three groups: natural polymers, synthetic polymers from bioderived monomers, and polymers from microbial fermentation. The development of synthetic polymers using bio-derived monomers provides a new direction for the production of biodegradable polymers from renewable resources (Yu and Chen, 2009). One of the most promising polymers in this regard is poly(lactic acid) (PLA) because it is made from agricultural products and is readily biodegradable. Lactide is a cyclic dimer prepared by the controlled depolymerization of lactic acid, which in turn can be obtained by the fermentation of corn, sugar cane, or sugar beet (Yu and Chen, 2009). Although PLA is not a new polymer, better manufacturing practices have improved the economics of producing monomers from agricultural feedstocks and, as such, PLA is at the forefront of the emerging biodegradable plastics industries.

PLA belongs to the family of aliphatic polyesters commonly made from α-hydroxy acids, which include polyglycolic acid and polymandelic acid (Garlotta, 2001). PLA is commercially interesting because of its good strength properties, film transparency, biodegradability, and availability. PLA is also manufactured by biotechnological processes from renewable resources (Yu and Chen, 2009). Although many sources of biomass can be used, corn has the advantage of providing the required high-purity lactic acid. PLA can be synthesized from lactic acid in two ways: a direct polycondensation reaction and a ring-opening polymerization of a lactide monomer. The technique of ring-opening polymerization has the advantage of providing a product with a higher molecular weight (Yu and Chen, 2009).

In nature, a special group of polyesters is produced by a wide variety of microorganisms for internal carbon and energy storage as part of their survival mechanism. Poly(b-hydroxybutyrate) (PHB) was first mentioned in the scientific literature as early as 1901 and detailed studies began in 1925. Over the next 30 years, PHB inclusion bodies were studied primarily as an academic curiosity. The energy crisis of the 1970s was an incentive to seek naturally occurring substitutes for synthetic plastics, which then sped up the research and commercialization of PHB (Yu and Chen, 2009). This biopolymer has received much research attention in recent years, with a large number of publications concerned with biosynthesis, microstructure, mechanical and thermal properties, and biodegradation through to genetic engineering (Yu and Chen, 2009).

1.3.1.2 Microbial Biosynthesis

Microbial biosynthesis of natural products is an emerging area of metabolic engineering and industrial biotechnology that offers significant advantages over conventional chemical methods or extraction from biomass. Lower energy requirements, lower CO_2 emissions, less toxic waste in the form of solvents and metal catalysts, simpler purification schemes, renewable feed stocks such as corn or soybeans, and the general ability of enzymes to perform chiral synthesis are among the improvements to be gained by microbial synthesis. Verhoef et al. (2007) constructed *Pseudomonas putida* strain S12palB1 that produces *p*-hydroxybenzoate from renewable carbon sources via the central metabolite l-tyrosine. *P. putida* S12palB1 was based on the platform strain *P. putida* S12TPL3, which has an optimized carbon flux toward l-tyrosine. Phenylalanine ammonia lyase (Pal) was introduced for the conversion of l-tyrosine into *p*-coumarate, which is further converted into *p*-hydroxybenzoate by endogenous enzymes. *p*-Hydroxybenzoate hydroxylase (PobA) was inactivated to prevent the degradation of *p*-hydroxybenzoate. These modifications resulted in stable accumulation of *p*-hydroxybenzoate at a yield of 11% (C-mol C-mol^{-1}) on glucose or on glycerol in shake-flask cultures. In a glycerol-limited fed-batch fermentation, a final *p*-hydroxybenzoate concentration of 12.9 mM (1.8 g L^{-1}) was obtained, at a yield of 8.5% (C-mol C-mol^{-1}). A twofold increase in the specific *p*-hydroxybenzoate production rate (q_p) was observed when l-tyrosine was supplied to a steady-state C-limited chemostat culture of *P. putida* S12palB1. This implied that l-tyrosine availability was the bottleneck for *p*-hydroxybenzoate production under these conditions. When *p*-coumarate was added instead, q_p increased by a factor 4.7, indicating that Pal activity is the limiting factor when sufficient l-tyrosine is available. Thus, two major leads for further improvement of the *p*-hydroxybenzoate production by *P. putida* S12palB1 were identified by Verhoef et al. (2007). Chandran et al. (2003) recently examined the impact of increased availability of phosphoenolpyruvate during shikimic acid biosynthesis in *Escherichia coli* K-12 constructs carrying plasmid-localized *aroF*^{FBR} and *tktA*inserts encoding, respectively, feedback-insensitive 3-deoxy-D-*arabino*-heptulosonic acid 7-phosphate synthase and transketolase. Strategies for increasing the availability of phosphoenolpyruvate were based on amplified expression of *E. coli ppsA*-encoded phosphoenolpyruvate synthase or heterologous expression of the *Zymomonas mobilis glf*-encoded glucose facilitator. The highest titers and yields of shikimic acid biosynthesized from glucose in 1 L fermentor runs were achieved using *E. coli* SP1.l*pts*/pSC6.090B, which expressed both *Z. mobilis glf*-encoded glucose facilitator protein and *Z. mobilis glk*-encoded glucose kinase in a host deficient phosphoenolpyruvate:carbohydrate phosphotransferase system. At 10 L scale with yeast extract supplementation, *E. coli* SP1. l*pts*/pSC6.090B synthesized 87 g L^{-1} of shikimic acid in 36% (mol/mol) yield with a maximum productivity of 5.2 g L^{-1} h^{-1} for shikimic acid synthesized during the exponential phase of growth. Butanol production relies on petroleum-based syntheses since the challenges in improving natural biocatalysts hinder the efficient synthesis starting from renewable biomass. An alternative biocatalyst was developed to allow butanol synthesis using *Escherichia coli* by Leonard and Prather (2008). The engineering of the biocatalyst entailed the simultaneous introduction of the multigene butanol biosynthetic pathway of solventogenic *Clostridium*. To improve the functionality of the engineered pathway, the expression of a synthetic gene encoding for a rate-limiting step enzyme was equally explored. Furthermore, to elicit high productivity, the native metabolism of the butanol-producer strain was reprogramed to direct carbon flow and cofactor regeneration. Leonard and Prather (2008) hence harnessed the synergistic combination of the *de novo* pathway constructions for the synthesis of butanol. Leonard and Prather (2008) also report that the manipulability of such synthetic biocatalyst can circumvent impediment of the biochemical synthesis of butanol from renewable feedstock. Meijnen et al. (2008) engineered the solvent-tolerant bacterium *Pseudomonas putida* S12 to utilize xylose as a substrate by expressing xylose isomerase (XylA) and xylulokinase (XylB) from *Escherichia coli*. The initial yield on xylose was low (9% [g CDW g^{-1} substrate], where CDW is cell dry weight), and the growth rate was poor (0.01 h^{-1}). The main cause of the low yield was the oxidation of xylose into the dead end product xylonate by endogenous glucose dehydrogenase (Gcd). Subjecting the XylAB-expressing *P. putida* S12 to laboratory evolution yielded a strain that efficiently utilized xylose (yield, 52% [g CDW g^{-1} xylose]) at a considerably improved growth rate (0.35 h^{-1}). Meijnen et al. (2008) attributed the high yield in part to Gcd inactivity, whereas the improved growth rate may be connected to alterations in the primary metabolism. Surprisingly, without any further engineering, the evolved D-xylose-utilizing strain metabolized L-arabinose as efficiently as D-xylose. Furthermore, despite the loss of Gcd activity, the ability to utilize glucose was not affected. Thus, a *P. putida* S12-derived strain was obtained that efficiently utilizes the three main sugars present in lignocellulosic hydrolysate: glucose, xylose, and arabinose. According to Meijnen et al. (2008), the strain they have developed will form the basis for a platform host for the efficient production of biochemicals from renewable feedstock.

1.3.1.2.1 Biocatalysts Research

The limited number of suitably well-characterized biocatalysts continues to limit progress in the application of biological routes in the synthesis of compounds for novel pharmaceuticals, materials, or performance chemicals (Bommarus and Polizzi, 2006). In this situation, the discovery of novel biocatalysts or novel functionalities or substrates on existing ones is an important task. Therefore, the demand for new and useful biocatalysts is steadily and rapidly increasing (Drepper et al., 2006) and a range of novel biocatalysts obtainable through one of the three techniques: environmental sampling or screening, protein engineering on existing enzymes, and extension of the catalytic profile of existing catalysts, is being synthesized and tested. Cytochromes P450 are ubiquitously distributed enzymes, which were discovered about 50 years ago and which possess high complexity and display a broad field of activity. They are hemoproteins encoded by a superfamily of genes converting a broad variety of substrates and catalyzing a variety of interesting chemical reactions (Bernhardt, 2006). This enzyme family is involved in biotransformation of drugs, bioconversion of xenobiotics, metabolism of chemical carcinogens, biosynthesis of physiologically important compounds such as steroids, fatty acids, eicosanoids, fat-soluble vitamins, and bile acids, the conversion of alkanes, terpenes, and aromatic compounds, as well as the degradation of herbicides and insecticides. There is also a broad versatility of reactions catalyzed by cytochromes P450 such as carbon hydroxylation, heteroatom oxygenation, dealkylation, epoxidation, aromatic hydroxylation, reduction, and dehalogenation (Bernhardt, 2006). More than 5000 different P450 genes have been cloned up to date (Refer to http://drnelson.utmem.edu/CytochromeP450.html for more details). Members of the same gene family are defined as usually having ≥40% sequence identity to a P450 protein from any other family. Mammalian sequences within the same subfamily are always >55% identical. The numbers of individual P450 enzymes in different species differ significantly, showing the highest numbers observed so far in plants (Bernhardt, 2006). The structure–function relationships of cytochromes P450 are far from being well understood and their catalytic power has so far hardly been used for biotechnological processes. Nevertheless, the set of interesting reactions being catalyzed by these systems and the availability of new genetic engineering techniques allowing to heterologously express them and to improve and change their activity, stability, and selectivity as well as the increasing interest of the industry in life sciences make them promising candidates for biotechnological application in the future (Bernhardt, 2006).

Heme-thiolate haloperoxidases are undoubtedly the most versatile biocatalysts of the hemeprotein family and they share catalytic properties with at least three further classes of heme-containing oxidoreductases, namely, classic plant and fungal peroxidases, cytochrome P450 monooxygenases, and catalases (Hofrichter and Ullrich, 2006). For a long time, only one enzyme of this type—the chloroperoxidase (CPO) of the ascomycete *Caldariomyces fumago*—has been known. The enzyme is commercially available as a fine chemical and catalyzes the unspecific chlorination, bromination, and iodation (but no fluorination) of a variety of electrophilic organic substrates via hypohalous acid as actual halogenating agent. In the absence of halide, CPO resembles cytochrome P450s and epoxidizes and hydroxylates activated substrates such as organic sulfides and olefins; aromatic rings, however, are not susceptible to CPO-catalyzed oxygen transfer. Recently, a second fungal haloperoxidase of the heme-thiolate type has been discovered by Hofrichter and Ullrich (2006) in the agaric mushroom *Agrocybe aegerita*. The UV–vis adsorption spectrum of the isolated enzyme showed little similarity to that of CPO but is almost identical to a resting state P450. The *Agrocybe aegerita* peroxidase (AaP) has strong brominating as well as weak chlorinating and iodating activities, and biocatalyzes both benzylic and aromatic hydroxylations of toluene and naphthalene (Hofrichter and Ullrich, 2006).

AaP and related fungal peroxidases could become promising biocatalysts in biotechnological applications because they seemingly fill the gap between CPO and P450 enzymes and act as "self-sufficient" peroxygenases. From the environmental point of view, the existence of a halogenating mushroom enzyme is interesting because it could be linked to the multitude of halogenated compounds known to be found in these organisms. *Candida antarctica* lipase B (CAL-B) is a very effective catalyst for the production of amines and amides using different enzymatic procedures (Gotor-Fernández et al., 2006). The simplicity of use, low cost, commercial availability, and recycling possibility make this lipase an ideal tool for the synthesis and resolution of a wide range of nitrogenated compounds that can be used for the production of pharmaceuticals and interesting manufactures in the industrial sector. Nonsteroidal anti-inflammatory drugs (NSAIDs) are used for the treatment of inflammation and pain in various rheumatic and musculoskeletal disorders (Morrone et al., 2008). One of the most important classes of NSAIDs is the "profens," derivatives of 2-arylpropionic acid. These compounds have been used in clinical science for more than 30 years and, although clinical studies have proved that the *S*-enantiomer acts as the active anti-inflammatory agent, and are mainly marketed nowadays as racemic mixture. For this reason, several investigations have been carried out to obtain enantiomerically pure (*S*)-2-arylpropanoic acids. Among the well established procedures, the biocatalytic separation of the racemate has often furnished satisfactory results and the use of lipases in the resolution of chiral acids has provided good results both by hydrolysis of the preformed esters and by direct esterification in nonaqueous medium. Morrone et al. (2008) prepared enantiopure *S*-fenoprofen through the kinetic resolution of *rac*-fenoprofen catalyzed by lipase B from *C. antarctica* (Novozym 435), using an irreversible process of esterification that exploits orthoformates as alcohol donor. The results of Morrone et al. (2008) showed that even in the presence of an irreversible methodology it was possible to obtain the resolution of *rac*-fenoprofen with biocatalysts at a low enantioselectivity. Furthermore, easy recovery of the (*S*)-fenoprofen from reaction mixture, possibility to recycle the ester after racemization, and absence of by-products given the soft conditions of reaction made the procedure of Morrone et al. (2008) suitable for industrial purposes. The bottleneck to obtain an optimum performance of an enzyme is often reliant on devising an effective method for its immobilization (Sheldon, 2007). Sheldon (2007) devised a novel, versatile, and effective methodology for enzyme immobilization as cross-linked enzyme aggregates (CLEAs). This method is elegantly straightforward and involves the precipitation of the enzyme from aqueous buffer followed by cross-linking of the resulting physical aggregates of enzyme molecules. Sheldon (2007) demonstrated this method to be applicable to a wide variety of enzymes, including in addition to a wide variety of hydrolases, lyases such as nitrile hydratases and oxynitrilases, and oxidoreductases such as laccase and galactose oxidase. CLEAs are stable, recyclable biocatalysts exhibiting high catalytic productivities. Because the methodology is essentially a combination of purification and immobilization into a single step, the enzyme does not need to be of high purity. The technique is also applicable to the preparation of combi-CLEAs, containing two or more enzymes, for use in one-pot, multistep syntheses (Sheldon, 2007). Initially, Cao et al. (2000, 2001) studied the synthesis of CLEAs from penicillin G amidase (EC 3.5.1.11), an industrially important enzyme used in the synthesis of semisynthetic penicillin and cephalosporin antibiotics. The free enzyme had limited thermal stability and a low tolerance to organic solvents, which thence made it an ideal candidate for stabilization as a CLEA (Sheldon, 2007). Indeed, penicillin G amidase CLEAs, prepared by precipitation with ammonium sulfate, proved to be effective catalysts for the synthesis of ampicillin (Cao et al., 2000, 2001). Outstandingly, the productivity of the CLEA was even higher than that of the free enzyme from which it was made and substantially higher than that of the cross-linked enzyme crystals (CLECs) whose productivity as a commercial catalyst was much lower.

1.3.2 CATALYSIS

Some of the major advances in chemistry, especially industrial chemistry, over the past generation have been in the area of catalysts. Through the use of catalyst, chemists and chemical engineers have found ways of removing the need for large quantities of reagents that would otherwise have been needed for chemical transformation and would eventually have contributed to the waste stream (Kidwai and Mohan, 2005). The area of catalysis is sometimes referred to as the "foundational pillar" of green chemistry (Anastas et al., 2001; Anastas and Kirchhoff, 2002). Catalytic reactions (Dijksman et al., 2001; Dias et al., 2001) often reduce energy requirements and decrease separations due to increased selectivity and they may allow the use of renewable feedstocks. Some interesting examples in the emerging areas of catalysis are discussed below.

1.3.2.1 Nanocatalysis

Recently, the use of nanoparticulate materials in catalysis has attracted considerable attention because of their improved efficiency under mild and environmentally benign conditions in the context of green chemistry. Because of enormously large and highly reactive surface area, nanoparticles (NPs) exhibit some unique properties in comparison with bulk materials. Among many other NPs, silica-based NPs have been well studied because of the following reasons: (1) silica NPs are easy to synthesize at room temperature, (2) NP size can be easily tuned, (3) easy adjustment of synthesis parameters leads to NPs with narrow size distribution (monodispersed NPs), (4) silica NPs are stable in organic solvents, and (5) they are environment-friendly materials (Banerjee et al., 2009). Owing to these attractive features, silica NPs found widespread applications in the synthesis of core–shell hybrid nanomaterial for catalysis of organic reactions. Dominguez-Quintero et al. (2003) synthesized nanostructured palladium materials supported on silica for the catalytic hydrogenation of benzene, 2-hexanone, and cyclohexanone (Dominguez-Quintero et al., 2003). Astruc (2007) reported the synthesis of palladium NPs (PdNPs) as efficient green homogeneous and heterogeneous carbon–carbon coupling precatalysts. The most studied reaction is the Heck coupling of halogenoarenes with olefins that usually proceeds at high temperature (120–160°C). Under such conditions, the Pd^{II} precursor is reduced to Pd^0, forming PdNPs from which Pd atom leaching, subsequent to oxidative addition of the aryl halide onto the PdNP surface, is the source of very active molecular catalysts. According to Astruc (2007), many other C–C coupling reactions can also be efficiently and "greenly" catalyzed by species produced from preformed PdNPs. For catalysis of these reactions, leaching of active Pd atoms from the PdNPs may also provide a viable molecular mechanistic scheme. Hou et al. (2007) also found that polyethylene glycol (PEG)-modified silica surfaces were able to effectively stabilize and immobilize palladium NPs for their use as selective oxidation catalysts in combination with supercritical CO_2 ($scCO_2$) as reaction medium under mild conditions. These catalysts show high activity and excellent stability under continuous-flow operation. Saxena et al. (2007) reported the use of highly monodispersed, easily recyclable, and cheap nickel NPs (Ni-NPs) as an efficient green catalyst for chemo-selective oxidative coupling of thiols. The Ni-NPs act as a green catalyst that can selectively catalyze oxidative coupling of thiols to disulfides without producing any overoxidized products at room temperature and with excellent yield under air atmosphere.

Recently, Murugadoss and Chattopadhyay (2008) also reported on the catalytic activity of a new metal NP–polymer composite consisting of silver (Ag) NPs and environment-friendly (green) chitosan. The polymer not only acted as the reducing agent for the metal ions, but also stabilized the product NPs by anchoring them (Murugadoss and Chattopadhyay, 2008). The majority of the particles produced in this way showed a size of <5 nm. The catalytic activity of the composite was investigated photometrically by monitoring the reduction of 4-nitrophenol (4NP) in the presence of excess $NaBH_4$ in water, under both heterogeneous and microheterogeneous conditions. The Ag NPs in the composite retained their catalytic activities even after using them for 10 cycles. The observations of Murugadoss and Chattopadhyay (2008) also suggest that the catalytic efficiency under microheterogeneous conditions is much higher than that under heterogeneous conditions. Thus, Murugadoss and Chattopadhyay (2008) confidently concluded that the composite they have developed represents an ideal case of an environmentally friendly and stable catalyst, which works under heterogeneous as well as microheterogeneous conditions with the advantage of nanoscopic particles as the catalyst. Shi et al. (2009) also recently designed a green, environmentally benign, and efficient synthesis of sulfonamides catalyzed by nano-Ru/Fe_3O_4. Despite the importance of sulfonamide derivatives as intermediates in drug synthesis, till now such a transformation as proposed by Shi et al. (2009) is rarely known. For the first time a dehydrogenation–condensation–hydrogenation sequence of alcohols and sulfonamides has been realized in the presence of a nanostructured catalyst. The magnetic property of the catalyst system allows for convenient isolation of the product and efficient recycling of the catalyst. A variety of coupling reactions of benzylic alcohols and sulfonamides including various heterocycles were successfully realized, often with >80% isolated yield. Advantageously, only one equivalent of the primary alcohol is consumed in the process.

1.3.2.2 Solvent-Free Catalysis

With increasing global environmental concerns, the design of "solvent-free" green processes has definitely gained significant attention from synthetic organic chemists. As a result, many reactions are being designed to proceed cleanly and efficiently in the solid state or under solvent-free conditions. Less chemical pollution, lower cost, and an easier workup procedure are the main reasons for the recent increase in the popularity of solvent-free reactions. Thioethers play important roles in biological and chemical processes (Cremlyn, 1996) and also serve as useful building blocks for various organosulfur compounds. Therefore, synthesis of thioethers in "green" and simple ways has much importance. A new route for the synthesis of linear and vinyl thioethers has been recently demonstrated by Banerjee et al. (2009) using bare silica NPs as catalysts at room temperature under solvent-free conditions. The catalyst can be reused up to 6 times without loss of catalytic activity. Banerjee et al. (2009) reported a simple and efficient protocol for the synthesis of linear and vinyl thioethers via anti-Markovnikov addition of thiols to alkenes and alkynes using a recyclable native silica NP catalyst at room temperature.

Reddy et al. (2006) carried out the Knoevenagel condensation of various aliphatic, aromatic, and heterocyclic aldehydes with malononitrile in the liquid phase in one step by employing a sulfate-ion-promoted zirconia solid acid catalyst. This catalyst facilitated the reaction under solvent-free conditions at moderate temperatures providing excellent yields of the products. The sulfate-ion-promoted zirconia catalyst was synthesized by immersing a finely powdered hydrous $Zr(OH)_4$ into the 1 mol L^{-1} H_2SO_4 solution and subsequent drying and calcination at 923 K. The $Zr(OH)_4$ was prepared from an aqueous $ZrOCl_2 \cdot 8H_2O$ solution by hydrolysis with dilute aqueous ammonia. Recently, Kumar and Maurya (2008) reported an organocatalyzed protocol for one-pot synthesis of 1,4-dihydropyridines via three-component coupling of cinnamaldehyde, aniline, and β-keto esters under solvent-free conditions at ambient temperature. Kumar and Maurya (2008) found that the reaction is generally very fast and the products are obtained in high yield.

The condensation of indoles with carbonyl compounds is of importance as this reaction provides a direct and interesting route toward bis(indolyl)methanes (Penieres-Carrillo et al., 2003). Various biological and pharmaceutical activities have been reported for this class of compounds. Bis(indolyl)methanes are useful for promoting beneficial estrogen metabolism and are also effective in the prevention of cancer due to their ability to modulate certain cancer-causing estrogen metabolites. All the more, these compounds may normalize abnormal cell growth associated with cervical dysplasia. As bis(indolyl)methanes are so important compounds in pharmaceutical chemistry, their synthesis have received increasing attention. Hasaninejad et al. (2007) synthesized an efficient solvent-free procedure for the preparation of bis(indolyl)methanes via the condensation of indoles with aldehydes as well as ketones in the presence of catalytic amount of phosphorus pentoxide/silica gel (P_2O_5/SiO_2) at room temperature. The advantages of this method are generality, high yields, short reaction times, ease of product isolation, low cost, and ecologically friendly.

Protection of the hydroxy groups plays an essential role in multistep organic synthesis including natural products and biologically active compounds (Hajipour et al., 2009). Among the protecting groups for protection of alcohols, tetrahydropyranylation is one of the most frequently used processes for protecting hydroxyl groups, because tetrahydropyrane (THP) ethers are less expensive, easy to deprotect, and stable enough under strong basic media such as Grignard and alkyl lithium reagents, oxidative conditions, and alkylating and acylating reagents (Hajipour et al.,

2009). A green, efficient, and large-scale method for tetrahydropyranylation of alcohols in the presence of a catalytic amount of pyridinium chloride at room temperature under solvent-free conditions has been lately reported by Hajipour et al. (2009). The latter developed a simple, large-scale, and efficient method for tetrahydropyranylation of various alcohols using pyridinium chloride. The mild reaction conditions, short reaction times, good to high yields, low cost, easy preparation, easy handling of pyridinium chloride as ionic liquid and acid catalyst, straightforward isolation of product and catalyst, and reusability of catalyst are the "green chemistry" advantageous traits of this method.

1.3.2.3 Novel Catalysts

Heteropolyacids (HPAs) are well-defined molecular clusters that are remarkable for their molecular and electronic structural diversity (Heravi et al., 2007) and their quite diverse significance in many areas, for example, catalysis, medicine, and materials science (Misono et al., 2000). The applications of HPAs, in the field of catalysis are growing continuously. These compounds possess unique properties such as Brönsted acidity, possibility to modify their acid–base and redox properties by changing their chemical composition (substituted HPAs), ability to accept and release electrons, high proton mobility, easy workup procedures, easy filtration, and minimization of cost and waste generation due to reuse and recycling of these catalysts (Heravi et al., 2006a). Because of their stronger acidity, they generally exhibit higher catalytic activity than conventional catalysts such as mineral acids, ion exchange resins, mixed oxides, and zeolites (Bamoharram et al., 2006a). In the context of green chemistry, the substitution of harmful liquid acids by solid reusable HPAs as catalysts in organic synthesis (Heravi et al., 2007) is the most promising application of these acids (Bamoharram et al., 2006b). Benzoxadiazepines have been found to possess marked biological effects as central nervous system stimulants. They have also been reported as antibacterial and anti-inflammatory agents, pesticides, and insecticides (Tandon and Kumar, 2004). Few methods are reported in the literature for the synthesis of benzoxadiazepines. During the course of their (Heravi et al., 2007) studies toward the development of HPAs as efficient heterogeneous catalysts (Heravi et al., 2006b), Heravi et al. (2007) reported the synthesis of 3,1,5-benzoxadiazepines derivatives by cyclization of o-PDA and acyl chlorides in the presence of a catalytic amount of various types of HPAs, including $H_{14}[NaP_5W_{30}O_{110}]$, $H_5[PMo_{10}V_2O_{40}]$, and $H_6[P_2W_{18}O_{62}]$.

The Collins group from the Carnegie Mellon University, New York, USA has designed a series of catalysts that activate hydrogen peroxide to bleach wood pulp (Kirchhoff, 2003). Termed as tetra amido macrocyclic ligand (TAML) activators, these catalysts are selective, effective over a wide pH range, consume less energy, and eliminate the problem of chlorinated by-products. This technology has applications in the pulp and paper industry, which is gradually shifting toward better reagents for pulp delignification and bleaching (Kirchhoff, 2003). Chlorine dioxide has largely replaced chlorine as a bleaching agent because it produces fewer chlorinated organic by-products, many of which are linked to cancer and endocrine disruption. Bleaching with chlorine dioxide, however, uses stoichiometric amounts of reagents, is energy intensive, and generates trace amounts of chlorinated organics. Combining minute quantities of TAML activators with hydrogen peroxide offers a totally chlorine-free approach for pulp bleaching (Kirchhoff, 2003). Furthermore, this technology is effective in bleaching the highly colored effluent streams associated with paper mills.

Olefin metathesis has become a tool for synthetic organic and polymer chemists. Well-defined, functional group-tolerant catalysts have allowed these advances (Grubbs, 2007). Olefin metathesis or transalkylidenation is an organic reaction that entails redistribution of alkylene fragments by the scission of carbon–carbon double bonds in olefins. Since its discovery, olefin metathesis has gained widespread use in research and industry for making products ranging from medicines and polymers to enhanced fuels. Its advantages include the creation of fewer side products and hazardous wastes. Yves Chauvin, Robert H. Grubbs, and Richard R. Schrock actually shared the 2005 Nobel Prize in Chemistry for "the development of the metathesis method in organic synthesis." Recently, Michrowska et al. (2006) developed a novel green catalyst, bearing a polar quaternary ammonium group, which is very stable and can be easily prepared from commercially available reagents. The catalyst can be efficiently used for olefin metathesis not only in traditional but also in aqueous media. Various ring-closing-, cross-, and enyne-metathesis reactions were conducted in water–methanol mixtures in air. Michrowska et al. (2006) found that the electron-withdrawing quaternary ammonium group not only activates the catalyst chemically, but also at the same time allows its efficient separation after reaction. The application of the catalyst of Michrowska et al. (2006) leads to organic products of high purity.

Benzodiazepines are pharmacologically very important compounds and widely used as anticonvulsant, antianxiety, analgesic, sedatives, antidepressive, and hypnotic activities (Gholap and Tambe, 2008). In addition, 1,5-benzodizepines are used as starting materials for the synthesis of various heterocyclic ring compounds. Owing to the widespread applications of benzodiazepines, the synthesis involving inexpensive and environmentally safe catalytic systems is a great challenge to pharmaceutical chemists. Several methods were documented for the synthesis of 1,5-benzodiazepines, which include the condensation of o-phenylenediamines (OPD) with α,β-unsaturated compounds, α-haloketones, or with ketones catalyzed by $BF_3 \cdot OEt_2$, $NaBH_4$, and polyphosphoric acid on SiO_2, $MgO–POCl_3$, $Yb(OTf)_3$, $Sc(OTf)_3$ ionic liquids, acetic acid-MWI, NBS, $Ga(OTf)_3$, and $ZrCl_4$ (Gholap and Tambe, 2008). However, despite their potential utility, many of these methods suffer from drawbacks such as relatively low yield, high reaction temperature, use of expensive reagents, and long reaction time (Gholap and Tambe, 2008). Therefore, the development of mild and efficient methods would extend the scope of this conversion. Use of a catalytic quantity of green solid acids that could be removed from the reaction mixture easily, avoiding expensive and toxic reagents, could be useful for this purpose. To this end, Gholap and Tambe (2008) devised a method for the synthesis of 2,3-dihydro-1H-1,5-benzodiazepines from the cyclocondensation of OPD and enolizable ketones using boric acid as a green catalyst in water. The corresponding 2,3-dihydro-1H-1,5-benzodiazepines were obtained in excellent yield at room temperature (25°C). Short reaction time, eco-friendly, and environment-friendly catalytic system are the most advantageous features of the method by Gholap and Tambe (2008). In conclusion, the latters' procedure is simple and general for various enolizable ketones involving water as a green solvent.

1.3.3　Green Solvents

Solvents are auxiliary materials used in chemical synthesis. They are not an integral part of the compounds undergoing reaction, yet they play an important role in chemical production and synthesis (Li and Trost, 2008). By far, the largest amount of "auxiliary waste" in most chemical productions is associated with solvent usage. In a classical chemical process, solvents are used extensively for dissolving reactants, extracting and washing products, separating mixtures, cleaning reaction apparatuses, and dispersing products for practical applications. Although the invention of various exotic organic solvents has resulted in some remarkable advances in chemistry, the legacy of such solvents has led to various environmental and health concerns (Li and Trost, 2008). Consequently, as part of green chemistry efforts, various cleaner solvents have been evaluated as replacements. The idea of "green" solvents expresses the goal to minimize the environmental impact resulting from the use of solvents in chemical production (Capello et al., 2007). The following discussion will focus on three types of green solvents: ester solvents, supercritical carbon dioxide (scCO$_2$), and ionic liquids, which have been receiving heightening interest and application in green chemistry research. A new category of solvents—near critical water—is introduced in the last section.

1.3.3.1　Ester Solvents

Ester solvents are actually the largest group of green solvents (Höfer and Bigorra, 2007). Speciality solvents such as glycerol carbonate can be used as nonreactive diluents in epoxy or polyurethane systems. Ethyl lactate has been reported as a photo-resist carrier solvent and a clean-up solvent in microelectronics and semiconductor manufacturing, and 2-ethylhexyl lactate can be used as degreaser and as a green solvent in agrochemical formulations, for example, for the protection of paddy rice crops (Taranta and Buckpesch, 2005). Ester solvents have also been applied for use in industrial cleaners. The Dow Haltermann Custom Processing (DHCP)-oxygenated solvents product range covers a series of ester solvents that can be applied in a wide range of formulated chemical products including inks, coatings, lubricants, and industrial cleaners. All DHCP products are designed to provide high technical performance at low environmental impact and low occupational risk. ESTASOL™ oxygenated solvents, COASOL™ coalescing aid, 2-ethylhexyl acetate, and di-*n*-butyl ether are strong polar solvents for use as alternatives to chlorinated solvents, aromatics, and ketones in cleaning and a wide range of functional fluids.

1.3.3.2　Supercritical Carbon Dioxide

Supercritical carbon dioxide refers to carbon dioxide that is in a fluid state while also being at or above both its critical temperature and pressure, yielding rather uncommon properties. Supercritical carbon dioxide is seen as a promising green solvent because it is nontoxic, and a by-product of other industrial processes. Furthermore, separation of the reaction components from the starting material is much simpler than with traditional organic solvents. Supercritical carbon dioxide is well established for use as a processing solvent in polymer applications such as polymer modification, formation of polymer composites, polymer blending, microcellular foaming, particle production, and polymerization (Nalawade et al., 2006). Its gas-like diffusivity and liquid-like density in the supercritical phase allow replacing conventional, often noxious, solvents with supercritical CO_2 (Nalawade et al., 2006). Wang et al. (2007) report the smooth aerobic oxidation of styrene catalyzed by $PdCl_2$/CuCl in the supercritical carbon dioxide and poly(ethylene glycol) biphasic system. A high conversion of styrene and yield of acetophenone were obtained in the presence of a relatively low catalyst loading. This environmentally benign biphasic catalytic system involving $scCO_2$ can be applied to the Wacker oxidation of various alkenes (Wang et al., 2007). Salgin (2007) investigated the extraction of jojoba seed oil using both supercritical CO_2 and supercritical CO_2 + ethanol mixtures. The recovery of jojoba seed oil was performed in a green and hi-tech separation process at operating pressures of 25, 35, and 45 MPa, operating temperatures of 343 and 363 K, supercritical fluid flow rates of 3.33×10^{-8}, 6.67×10^{-8}, and 13.33×10^{-8} m^3 s^{-1}, entrainer concentrations of 2, 4, and 8 vol%, and average particle diameters of 4.1×10^{-4}, 6.1×10^{-4}, 8.6×10^{-4}, and 1.2×10^{-3} m. It was found that a green chemical modifier such as ethanol could enhance the solubilities, initial extraction rate, and extraction yield of jojoba seed oil from the seed matrix as compared to supercritical CO_2. The supercritical fluid extraction involved short extraction time and minimal usage of small amounts of entrainer to the CO_2. Salgin (2007) reported an estimated 80% extraction of the total jojoba seed oil during the constant rate period at a pressure of 35 and 45 MPa. Villarroya et al. (2008) report the horseradish peroxidase (HRP)-mediated inverse emulsion polymerization of water-soluble acrylamide in $scCO_2$. The enzymatic polymerization takes place within water droplets formed in $scCO_2$. These are stabilized either as reversed micelles using perfluoropolyether ammonium carboxylate or in the absence of a stabilizer using very high shear. The viability of water-in-CO_2 (W/C) emulsion as a reaction medium for *in situ* enzyme-mediated polymerization (Villarroya et al., 2008) has been tested for the first time.

There is significant interest in enzymes as they have proven to be powerful and environment-friendly natural catalysts for the polymerization of water-soluble monomers that can function under milder reaction conditions than those used in traditional free radical polymerization techniques. Hence, the combination of $scCO_2$ and water as reaction medium is a significant advancement made by Villarroya et al. (2008) in natural polymerization process. Short-chain esters of carboxylic acids and monoterpene alcohols are very important compounds in food, cosmetic, and pharmaceutical industries. Lipase-catalyzed esterification of β-citronellol and lauric acid was performed in organic solvents and in supercritical carbon dioxide by Habulin et al. (2008). The optimal conditions found in organic solvents at atmospheric pressure comprise an equimolar ratio of substrates, a temperature of 60°C, an agitation speed of 500 rpm, and 12% (w/w of substrates) enzyme preparation of lipase B from *C. antarctica*. Habulin et al. (2008) noted that the highest conversion in $scCO_2$ was obtained at 60°C and 10 MPa with ethyl methylketone serving as a cosolvent. In food chemistry research, supercritical pasteurization is receiving increasing attention as an alternative technology for foodstuff pasteurization, but often the possible effects on the perceptible quality are not sufficiently considered. To address this latter issue, besides standard microbial analysis, Gasperi et al. (2009) recently investigated the impact of CO_2/N_2O supercritical pasteurization (100 bar, 36°C, and 10 min treatment time) on the quality traits of fresh apple juice, much linked to consumer perception. Discriminative sensory analysis and basic chemical characterization (total solids, sugars, organic acids, and polyphenols) could not clearly demonstrate any induced modification of the treated juice, while headspace analysis of volatile compounds (both by gas chromatography–mass spectrometry and proton transfer reaction mass spectrometry) led Gasperi et al. (2009) to deduce a general depletion of the volatile compounds that must be considered in the development of a stabilization method based on supercritical gases.

1.3.3.3 Ionic Liquids

A very peculiar class of solvents is ionic liquids. Ionic liquids define a class of fluids rather than a small group of individual examples. The most commonly studied systems contain phosphonium, imidazolinium (Burrell et al., 2007), or tricaprylmethyl ammonium cations, with varying heteroatom functionality. One regularly suggested advantage of ionic liquids, which positions them as solvents for green chemistry, is the intrinsic lack of vapor pressure. Ionic liquids have brought about particular scientific interest for extraction or separation technologies, and phase transfer catalysis. Ionic liquids are opening up a promising new field of nonaqueous enzymology (Yang and Pan, 2005). As compared with those observed in conventional organic solvents, enzymes in ionic liquids have presented enhanced activity, stability, and selectivity. Advantages of using ionic liquids over the use of normal organic solvents as reaction medium for biocatalysis also include their high ability of dissolving a wide variety of substrates, especially those highly polar ones, and their widely tunable solvent properties through appropriate modification of cations and anions. Research has been conducted (and is in progress) to harness the solvent properties of ionic liquids, their effects on enzyme performance such as activity, stability, and selectivity, and their applications in biocatalysis (Yang and Pan, 2005). The Knoevenagel condensation of aromatic aldehydes with malononitrile and ethyl cyanoacetate using amino-functionalized ionic liquid, 1-aminoethyl-3-methylimidazolium hexafluorophosphate as catalyst, was successfully performed in aqueous media by Cai et al. (2006), and it was found that the catalyst could be recycled and reused at least 6 times without apparent loss of activity. Arce et al. (2007) successfully demonstrated the ability of the ionic liquid 1-ethyl-3-methylimidazolium ethylsulfate ([emim][EtSO$_4$]) to act as an extraction solvent for liquid–liquid extraction and as an azeotrope breaker for extractive distillation to separate the azeotropic mixture ethyl *tert*-butyl ether (ETBE) + ethanol, thus purifying the tertiary ether, which is the most used additive to improve the octane index of gasoline.

Lou et al. (2006) successfully conducted the hydrolysis of immobilized *C. antarctica* lipase B (Novozym 435)-catalyzed enantioselective of D,L-phenylglycine methyl ester to enatiopure D-phenylglycine in the systems with ionic liquids. The Novozym 435 exhibited excellent activity and enantioselectivity in the system containing the ionic liquid BMIM·BF$_4$ compared to several typical organic solvents tested. Lou et al. (2006) found that cations and, particularly, anions of ionic liquids have a significant effect on the reaction, and the ionic liquid BMIM·BF$_4$, which showed to be the most suitable for the reaction, gave the highest initial rate and enantioselectivity among various ionic liquids examined. In the study of Chiappe et al. (2004), soluble epoxide hydrolase (sEH) was shown to catalyze the hydrolysis of epoxides using the ionic liquids [bmim][PF_6], [bmim][N(Tf)$_2$], and [bmim][BF_4] where bmim = 1-butyl-3-methylimidazolium, PF_6 = hexafluorophosphate, (Tf)$_2$ = bis(trifluoromethylsulfonyl)imide, and BF_4 = tetrafluoroborate) comprise the reaction media. Reaction rates were generally comparable with those observed in buffer solution, and when the cress enzyme was used, the hydrolysis of *trans*-β-methylstyrene oxide gave, through a stereoconvergent process, the corresponding optically active (1*S*,2*R*)-*erythro*-1-phenylpropane-1,2-diol. To demonstrate the effectiveness of ionic liquids as green solvents, Kumar and Malhotra (2008) synthesized nucleoside-based antiviral drugs using imidazolium-based ionic liquids as reaction medium. The ionic liquids were proved to be better solvents for all the nucleoside in terms of solubility and reaction medium as compared to conventional molecular solvents. In the medical analysis and tissue-processing areas, asymmetry of cations and the type of anions play a key role in the properties of ionic liquids as fixatives for tissue preservation. Pernak et al. (2005) have proven that 1-methyl-3-octyloxymethylimidazolium tetrafluoroborate is a very good fixative, with similar effects as formalin. The study of Pernak et al. (2005) showed that 1-methyl-3-octyloxymethylimidazolium tetrafluoroborate is applicable for both histological and immunohistochemical purposes. After treatment with 1-methyl-3-octyloxymethylimidazolium tetrafluoroborate, tissue sections are more intensely stained. With respect to expression patterns and staining intensity, immunohistochemical staining is comparable in tissues fixed in formalin and the selected ionic liquids. Hence, Pernak et al. (2005) demonstrated the properties of 1-methyl-3-octyloxymethylimidazolium tetrafluoroborate for tissue preservation in histopathological procedures, eliminating the requirement of formalin, which is toxic to cells (Nilsson et al., 1998). In the field of bioenergy and biomass utilization, Ha et al. (2007) and Nguyen et al. (2008) recently accomplished interesting results with ionic liquids. Ha et al. (2007) successfully demonstrated the production of biodiesel in ionic liquids through immobilized *C. antarctica* lipase-catalyzed methanolysis of soybean oil. Among the tested 23 ionic liquids, the highest fatty acid methyl esters (FAMEs) production after 12 h at 50°C was achieved in [Emim][TfO]. The production yield of 80% was eight times higher compared to the conventional solvent-free system. It was around 15% higher than the FAMEs production system using *tert*-butanol as an additive. The results of Ha et al. (2007) for a high production yield in ionic liquids showed that ionic liquids are potential green and efficient reaction media for biodiesel production. The application of thermophilic bacteria to the production of hydrogen (H_2) from a cellulosic biomass by Nguyen et al. (2008) has captured the attention of engineering researchers and chemists, all alike. *Thermotoga neapolitana* is considered as a potential hydrogen producer because of its ability to be used for the direct H_2 fermentation from raw cellulose, but its H_2 yield is not high. In their study, Nguyen et al. (2008) tested three chemical pretreatment methods with an ionic liquid, acid, and alkali, respectively, on cellulose to enhance the cellulose conversion of this strain into H_2. Batch cultivations were carried out to investigate the influence of the chemical pretreatment methods on H_2 production under nonsterile conditions. The highest H_2 concentration in the headspace obtained from the ionic liquid-pretreated cellulose was 36.1% (v/v) with ethanol as an antisolvent. This was clearly higher than that from the acid- and alkali-pretreated celluloses (22–24%).

Cultivation with cellulose pretreated with the ionic liquid at a concentration of 10% combining with N_2 sparging showed a maximum cumulative H_2 yield of 1280 mL H_2 L^{-1} culture. This value is approximately 10-fold higher than that of raw cellulose (122 mL H_2 L^{-1} culture). The H_2 yield obtained from fermentation with ionic liquid-pretreated cellulose was 2.2 mol H_2 mol^{-1} glucose equivalents. In this respect, the results of Nguyen et al. (2008) convincingly showed that the pretreatment of cellulose using an ionic liquid has considerable potential for improving the direct conversion of cellulosic substrates into H_2 by bacteria.

1.3.3.4 "Near-Critical" Water

Under normal conditions, oil and water do not mix. But "near-critical" water which is very hot but still liquid water at temperatures of 250°C–300°C and pressures of 6.89 MPa can act as a good solvent for both salts and nonpolar organic compounds, including oils. This makes ordinary water an ideal reaction solvent for certain chemical processes. Researchers at the Georgia Institute of Technology have been studying a wide range of chemical processes in search of applications where the special properties of this near-critical water might provide both economic and environmental advantages. Their work could lead to the replacement of traditional organic solvents in certain specialty chemical processes. Dr. Charles Eckert, director of Georgia Tech's Specialty Separations Center and a professor in the School of Chemical Engineering said: "Our goal is to do the technical work to see where we can use this as a replacement process and to couple that with an economic analysis to see where this can be used profitably." Hence, the study of near-critical water for chemical processing is part of a larger Georgia Tech initiative aimed at applying sustainable technology to broad areas of manufacturing. As further explained by Dr. Eckert, "Sustainable development to us means producing something that has both economic and environmental advantage, then doing the technology transfer necessary to get someone to use it." Hence, the promise is that "near-critical" water offers outstanding performance as a benign solvent for both organic and ionic compounds (Eckert et al., 1999). Conventionally, Friedel–Crafts acylations require the presence of Lewis acids such as aluminum chloride, ferric chloride, and zinc chloride. However in near-critical water, phenol could be converted with acetic acid to produce 20-hydroxyacetophenone, 40-hydroxyacetophenone, and phenyl acetate without the addition of an acid catalyst (Kidwai and Mohan, 2005). Sakaki et al. (1996) examined the noncatalytic decomposition characteristics of cellulose in near-critical water by heating a sealed reactor in which the cellulose and water were charged in a salt bath kept at 305°C, 355°C, or 405°C. As part of the improved green characteristics of the reaction, it was observed that the cellulose was rapidly decomposed to water solubles (WSs), and the WS was further decomposed after the WS yield reached nearly 80%. The heating time giving the maximum WS yield was shortened to under 15 s by increasing the treatment temperature to over 355°C. Chandler et al. (1997) reported the successful alkylation reactions in near-critical water in the absence of added acid catalysts. Water has a strong tendency to ionize as the temperature is increased, allowing water in the near-critical region (250–350°C) to act as an effective acid catalyst. This behavior of water was actually harnessed in the study of Chandler et al. (1997). By simultaneously employing near-critical water as the reaction solvent and catalyst, the need for environmentally hazardous organic solvents and acid catalysts was eliminated, and the high hydronium ion content of near-critical water proved to be sufficient to synthesize a variety of substituted phenols with tertiary, secondary, and even primary alcohols. Still, Nolen et al. (2003) investigated the Claisen–Schmidt condensation of benzaldehyde with 2-butanone and used their reactions to demonstrate the ability to conduct conventionally acid- or base-catalyzed reactions homogeneously using near-critical water without the addition of a catalyst.

1.3.4 Efficient Chemical Syntheses

Green chemistry for chemical synthesis addresses our future challenges in working with chemical processes and products by inventing novel reactions that can maximize the desired products and minimize the by-products, designing new synthetic schemes and apparatuses that can simplify operations in chemical productions, and seeking greener solvents that are inherently, environmentally, and ecologically benign (Li and Trost, 2008). For example, presently, the main feedstock of chemical products comes from nonrenewable petroleum that is being depleted rapidly both for chemical and energy needs. However, nature provides a vast amount of biomass in the renewable forms of carbohydrates, amino acids, and triglycerides to obtain organic products, but a major obstacle to using renewable biomass as feedstock is the need for novel chemistry to transform the large amounts of biomass selectively and efficiently, in its natural state, without extensive functionalization, defunctionalization, or protection (Li and Trost, 2008). Hence, the need arises for conducting research to eventually design new and more feedstock- and energy-efficient chemical pathways for product synthesis.

1.3.4.1 Direct Conversion of C–H Bonds

Direct transformation of the C–H bonds of organic molecules into desired structures without extra chemical transformations represents another class of major desirable reactions (Chatani et al., 2001; Jia et al., 2001). In nature, a variety of organic compounds can be oxidized easily by molecular oxygen or other oxygen donors in the cells of bacteria, fungi, plants, insects, fish, and mammals (Shapiro and Caspi, 1998). It is worth noting the important advances in biomimetic approaches to such oxidations (Costas et al., 2004). Hydroxylation of linear alkanes or methane to generate terminal alcohols is very useful in the synthesis of chemicals and fuels (Olah et al., 2006). For example, (NH)-indoles and tetrahydroisoquinolines were converted directly into alkaloids by using such a coupling (Li and Li, 2005). More recently, an elegant cross-coupling of two aryl C–H bonds to form arene–arene coupling products was reported by Fagnou, Sanford, and others (Hull and Sanford, 2007).

1.3.4.2 Synthesis without Protections

Because of the nature of classical chemical reactivity, organic synthesis extensively utilizes protection–deprotection of functional groups, which increases the number of steps in synthesizing the desired target compounds (Li and Trost, 2008). Novel chemistry is needed to perform organic synthesis without protection and deprotection. Recently, progress has been made on this subject. Baran et al. (2007) reported a total synthesis of a natural product without any protecting groups.

1.3.4.3　Other Examples of Novel and "Green" Syntheses

Thioureas have attracted much attention by virtue of their bioactivities as pharmaceuticals and pesticides. Various thiourea derivatives and their metal complexes exhibit analgesic, anti-inflammatory, antimicrobial, anticancer, and antifungal activities. Moreover, thioureas are important as building blocks in the synthesis of heterocycles. Ziyaei Halimehjani et al. (2009) recently reported a highly efficient and simple synthesis of unsymmetrical thioureas based on the reaction of readily synthesized dithiocarbamates with amines, without using any catalyst under solvent-free conditions. The short reaction time, high yields, and solvent-free conditions are advantages of this method. An excellent example of synthetic redesign can be found in Pfizer's production of sertraline (Kirchhoff, 2003), the active ingredient in the antidepressant Zoloft. The new manufacturing process has doubled product yield while eliminating 140 tons of $TiCl_4$, 440 tons of solid TiO_2 waste, 150 tons of 35% HCl, and 100 tons of 50% NaOH on an annual basis. Key to improving the manufacture of sertraline was changing from a mixture of four organic solvents, used in the first step, to a single solvent, ethanol. Solvent usage decreased from 60,000 L for every 1000 kg of product in the discovery phase to 6000 L for every 1000 kg of product in the combined process. The new commercial process has provided improved safety and material handling and decreased energy and water usage, while doubling the overall yield (Kirchhoff, 2003). By mimicking natural oxidations by certain enzymes that have iron at their active sites, two research groups made significant progress toward environment-friendly reactions that oxidize organic compounds using relatively nontoxic metal catalysts. The catalyst ligands used with the iron are inexpensive and simple to synthesize, unlike the complex porphyrin ligands called hemes that are found in some enzymes. And the new reactions used hydrogen peroxide as the oxidizing agent, which resulted in harmless water as the by-product.

Another excellent example of "green and sustainable chemistry" contributing to society and mankind has been realized recently by experts from the Asahi Kasei Corporation. Experts from Asahi Kasei Corporation have reviewed in their paper published in *Polymer Journal* (Fukuoka et al., 2007) the world's first process that has succeeded in developing and leading to the industrialization by Asahi Kasei Corporation for producing an aromatic polycarbonate (PC) using CO_2 as the starting material. The carbonate group of PC links directly to the residual aromatic groups of the bisphenol. Until Asahi Kasei's new process had been revealed, all of carbonate groups of PC in the world were derived from CO as starting material. Furthermore, more than about 90% of PC has been produced by the so-called "phosgene process," and the PC contains Cl impurities. It needed to use not only highly toxic and corrosive phosgene made from CO and Cl_2 as a monomer, but also very large amounts of CH_2Cl_2 and water, and needed to clean a large amount of wastewater. The new process from Asahi Kasei Corporation by contrast has enabled the high-yield production of the two products, Cl-free and high-quality PC and high-purity monoethylene glycol (MEG), starting from ethylene oxide (EO), by-product of CO_2 and bisphenol-A. The PC produced by the new process has many excellent properties compared with conventional PCs. The new process not only overcomes drawbacks in the conventional processes, but also achieves resource and energy conservation. The reduction of CO_2 emissions (0.173 tCO_{2e} per ton PC) is also achieved in the new process, because all CO_2 is utilized as the component consisting of main chains of the products. To date, the newly constructed commercial plant of Chimei–Asahi Corp. (Taiwan), a joint venture between Asahi Kasei Corporation and Chi Mei Corp., has been successfully operating achieving fullproduction since June 2002. The initial capacity (PC: 50,000 tons year^{-1}) has now increased to 150,000 tons year^{-1}.

1.3.5 MW-ASSISTED SYNTHESES

MW chemistry is the science of applying MW irradiation to chemical reactions. MWs act as high-frequency electric fields and will generally heat any material containing mobile electric charges, such as polar molecules in a solvent or conducting ions in a solid. Polar solvents are heated as their component molecules are forced to rotate with the field and lose energy in collisions. The use of MW irradiation as an alternative heat source is becoming more and more popular in chemistry. MV ovens mainly owe their popularity to the often observed enhanced reaction rates (Hoogenboom and Schubert, 2007). Nevertheless, these enhanced reaction rates can be often explained by the increased reaction temperatures that are allowed by the use of closed (pressurized) reactors. The use of such closed reaction vials has opened a completely unexplored area of high-temperature chemistry under MW irradiation. The closed reaction vessels are also exploited to replace high-boiling solvents by low-boiling solvents, which simplify product isolation. In addition, the direct heating of molecules under MW irradiation leads to very fast and homogeneous heating that has resulted in the reduction of side reactions, cleaner products, and higher yields (Hoogenboom and Schubert, 2007).

1.3.5.1 MW Chemistry and Polymers

The use of MW irradiation in polymer chemistry is an emerging field of research and has been mainly focused on step-growth polymerizations, ring-opening polymerization, and (controlled) radical polymerizations. Linear aromatic polyamides, polyimides, and poly(amideimide)s exhibit excellent thermal, mechanical, and chemical stabilities. As a result, these materials are often used in high-performance applications. However, the rigid structure of these materials makes them hardly soluble in organic solvents and, therefore, the use of high-temperature MW-assisted polymerization procedures has proven to be advantageous. 4,4′-Azobenzoyl chloride has been shown to react with eight different 5,5′-disubstituted hydantoin derivatives and it was seen that the hydantoin moieties improved the solubility of the resulting polyamides while the high thermal stability of the material was retained. The authors claimed higher yields and efficiencies for the MW-assisted polymerization procedure in o-cresol compared to a standard polymerization method with conventional heating in N,N-dimethylacetamide (DMAc) or bulk. Most of the recent investigations on MW-assisted polycondensations have been performed on the synthesis of poly(amide-imides). The reported polymerizations were all performed using domestic MW ovens (Hoogenboom and Schubert, 2007). The monomers were placed and ground in a porcelain dish. After the addition of a small amount of o-cresol, the mixture was ground again followed by MW heating. The MW-assisted synthesis of a series of optically active poly(amide-imides), in which the chirality resulted from the incorporation of amino acids. N,N'-(pyromellitoyl)-bis(amino acid chloride)s were reacted with eight hydantoin derivatives. The field of MW-assisted polymerizations will continue to expand rapidly in the coming years inspired by the large number of beneficial effects of MW irradiation that were observed in recent years. Several questions such as specific MW absorption in copolymerizations as well as upscaling issues still need to be dealt with to further harness this technique. However, it is expected that the use of MW irradiation will evolve from a research topic into a common research tool in polymer science (Hoogenboom and Schubert, 2007). Whether MW-assisted polymerization procedures may be incorporated into future commercial processes depend much on the feasibility of a process that should equally have a clear economical advantage over thermal heating.

1.3.5.2 MW-Assisted Chemistry in Medicine

MW-assisted organic synthesis has revolutionized organic synthesis. Small molecules can be built in a fraction of the time required by classical thermal methods. As a result, this technique has rapidly gained acceptance as a valuable tool for accelerating drug discovery and development processes (Mavandadi and Pilotti, 2006). The imidazole moiety is an important heterocyclic nucleus due to their widespread biological activities and use in synthetic chemistry. The substituted imidazoles are well known as inhibitors of P38MAP kinase and therapeutic agents. Imidazole chemistry, because of its use in ionic liquids and in N-heterocyclic carbenes (NHCs), gave a new dimension to the area of organometallics and "green chemistry." In addition, the imidazole ring system is one of the most important substructures found in a large number of natural products and pharmacologically active compounds such as the hypnotic agent etomidate, and the proton pump inhibitor omeprazole. Owing to their great importance, many synthesis strategies have been developed. In 1882, Radziszewski and Japp reported the first synthesis of the imidazole from 1,2-dicarbonyl compound, various aldehydes, and ammonia, to obtain the 2,4,5-triphenyl imidazoles. Another method is the four-component one-pot condensation of glyoxals, aldehydes, amines, and ammonium acetate in refluxing acetic acid, which is the most desirable convenient method. Very recently, literature survey reveals several methods for synthesis of 2,4,5-triaryl imidazoles using $ZrCl_4$, zeolite HY/silica gel, $NaHSO_3$, sulfanilic acid, iodine, and ionic liquid. However, many of these methodologies exhibit one or more disadvantages, such as low yields, high-temperature requirement, prolonged reaction time, highly acidic conditions, requirement of excess of catalysts, and use of solvents. Therefore, there is a strong demand for a simple, highly efficient, environmentally benign, and versatile method for the one-pot synthesis of 2,4,5-triarylimidazole derivatives. Realizing the need, Shelke et al. (2008) devised a simple and highly efficient method for a three-component condensation of benzil/benzoin, aldehydes, and ammonium acetate under MW irradiation in the presence of glyoxylic acid (CHOCOOH) in solvent-free condition to afford the corresponding 2,4,5-triarylimidazole derivatives in high yields. The remarkable advantages offered by their method are inexpensive and readily available catalyst, simple procedure, much faster (1–3 min) reactions, and high yield of products. Among the various classes of nitrogen-containing heterocyclic compounds, quinoxalines are important components of several pharmacologically active compounds (Mohsenzadeh et al., 2007). Although rarely described in nature, synthetic quinoxaline ring is a part of a number of antibiotics that inhibit the growth of Gram-positive bacteria and are also active against various transplantable tumors. From the synthesis point of view, despite remarkable efforts over the last decade, the development of effective methods for the synthesis of quinoxaline is still an important challenge. By far, the most common method relies on the condensation of an aryl 1,2-diamine with a 1,2-dicarbonyl compound in refluxing ethanol or acetic acid for 2–12 h. With the thought that it is nowadays of prime interest to connect research in chemistry and environmental protection, Mohsenzadeh et al. (2007) used two benign, simple, rapid, solvent-free and versatile MW energy transfer systems for the synthesis of quinoxalines, (2,3-diphenylquinoxaline and its derivatives, 2,3-diphenyl-4a,5,6,7,8,8a-hexahydroquinoxaline, and 2-phenylquinoxaline and also pyridine-2,3-diamine and 2,3-dihydro-5,6-diphenylpyrazine). The first approach of Mohsenzadeh et al. (2007) consisted in the use of mineral supports on the reaction of 1,2-dicarbonyl (benzil) or a-hydroxyketone (acyloin) with 1,2-diamines. Among mineral supports, the efficiency of acidic alumina as catalyst was clearly proved (80–86%). Moreover, it had an oxidative role in the tandem oxidation of acyloin to the corresponding 1,2-dicarbonyl. The second approach by Mohsenzadeh et al. (2007) was the use of polar paste system on the reaction of 1,2-dicarbonyl (benzil) with 1,2-diamines. The excellent products yield (90–97%) and simple washing with water and filtration in the absence of organic solvent made it an impressive alternative green synthetic route.

Although both methods implemented in the study of Mohsenzadeh et al. (2007) had been designed to avoid additional purification steps, the latter was found to be greener, cleaner, and with an easier workup. In genetic engineering research, Cavallaro et al. (2009) very recently prepared new copolymers, useful for gene delivery, based on α,β-poly-(N-2-hydroxyethyl)-D, L-aspartamide (PHEA) as a polymeric backbone and bearing an oligoamine such as diethylenetriamine in the side chain using MW-assisted heating in order to reduce solvent volume and make the reaction faster. While concluding that the syntheses were successful, the results of Cavallaro et al. (2009) support the use of these copolymers as gene delivery systems of the future and the use of MWs makes the proposed synthetic method advantageous as time and solvents are both saved.

Diels–Alder reactions form a well-studied group of MW-assisted reactions (Keglevich et al., 2007). The cycloaddition of 1,2-dihydrophosphinine oxides and dimethyl acetylenedicarboxylate (DMAD) or N-phenylmaleimide (NPMI) results in the formation of 2-phosphabicyclo octadiene and bicyclooctene, respectively. Under traditional heating in boiling toluene, the reaction was rather slow. On MW irradiation at 110°C in the absence of any solvent, the cycloaddition became 30 times faster and the side reactions deriving mainly from the polymerization of the starting materials were suppressed. Hence, the yields were excellent (Keglevich and Dudás, 2006).

1.3.5.3 MW Irradiation Technology for Waste Remediation

Waste treatments to induce stabilization or recovery of waste materials represent an important part of modern research efforts. MW irradiation technology has proved to be a powerful tool to convey energy exactly where it is needed by the process and also as an energy-efficient alternative to current heating technologies employed in the processing and treatment of waste (Appleton et al., 2005), as well as to allow operation in peculiar environments, even in remote-controlled modality (Corradi et al., 2007). In an extensive review by Appleton et al. (2005), it has been concluded that there is indeed significant potential for MW technology to be employed as an alternative heating source in the treatment of waste streams and environmental remediation. Jones et al. (2002) assessed the areas in environmental engineering where MW heating applications were made possible. They have subsequently short-listed contaminated soil remediation, waste processing, minerals processing, and activated carbon regeneration as the broad and potential areas of research interest, and identified as contaminated soil vitrification, volatile organic compounds (VOCs) treatment and recovery, waste sludge processing, mineral ore grinding, and carbon in pulp gold recovery as specific areas of potential commercial development. However, several major limitations thwart these technologies from being widely employed on large scales. These include the absence of sufficient data to quantify the dielectric properties of the treated waste streams (Jones et al., 2002; Appleton et al., 2005), and technical difficulties encountered when upgrading successful laboratory or pilot-scale processes to the industrial scale. This has meant that commercialization of MW heating processes for environmental engineering applications has so far been slow. In fact, commercialization is only deemed viable when MW heating offers additional process-specific advantages over conventional methods of heating (Jones et al., 2002).

A team from the Westinghouse Savannah River Technology Center (WSRC—a DOE Laboratory), and the University of Florida (UF), has been active for over a decade in the development of MW technology for specialized waste management applications. This interaction has resulted in the development of unique equipment and uses of MW energy for a variety of important applications for remediation of hazardous and radioactive wastes (Wicks et al., 2000). Some of the many wastes under study using MW treatments have included remediation of discarded electronic circuitry and reclamation of the precious metals within, incinerator ashes, medical and infectious wastes, industrial wastes/sludges, rubber products including tires, asbestos, groundwaters, VOCs, shipboard wastes, contaminated soils and sediments, and radioactive wastes and sludges (high-, low- and intermediate-level wastes, transuranic, and mixed wastes). MW energy interacts with matter in a way different from all other thermal treatment processes. As a result of these unique features, the advantages of using MW energy for treating a vast array of hazardous wastes can include many potential advantages. The advantages ultimately realized will depend both on the type and characteristics of the wastes to be treated. Some of the most important features of MW treatment of hazardous wastes are significant waste volume reduction, rapid heating, high-temperature capabilities, selective heating, enhanced chemical reactivity, ability to treat wastes *in situ*, treatment or immobilization of hazardous components to meet regulatory requirements for storage, transportation or disposal, rapid and flexible process that can also be made remote, portability of equipment and process, improved safety, including reductions in personnel exposure of potentially hazardous chemicals or materials for processing and disposition, and energy savings and cleaner energy source compared to some more conventional systems (Wicks et al., 2000).

A new technology—WSRC/UF Tandem Microwave System—has been developed by WSRC and UF and applied to a variety of waste management applications, including disposition of electronic circuitry. This technology consists of a tandem MW unit that makes use of direct and hybrid MW energy, which is designed to not only treat primary waste, but also simultaneously, off-gases, when necessary. Several studies involving MW treatment of emissions from processing of electronic circuitry have been carried out in the past and it has been noted that the emissions of key components (such as benzene, toluene, ethylbenzene, styrene, naphthalene, and 1,3,5-trimethylnenzene) were generally reduced by factors of 10–1000. Although only laboratory-scale studies have been mostly performed, a preliminary assessment showed that the overall concept and process can be mocked-up to a much larger scale. Over 40 different types of materials and mixes of materials have been studied thus far in the tandem MW system, for destruction and remediation of radioactive transuranic wastes. Volume reductions of 60–90% were generally achieved and main organic constituents, sources for gas generation during storage, were eliminated by the MW treatment. The products produced were either ashes or vitrified materials, depending on the application (Wicks et al., 2000).

Electron beam (EB) and MW treatments are two of the most emerging biological decontamination techniques because in many cases they provide distinct advantages over conventional processes in terms of product properties, process time saving, increased process yield, and environmental compatibility. Both EB (which is an ionizing radiation) and MW (which is a nonionizing radiation) sterilization techniques are based on the radiation ability to alter physical, chemical, and biological properties of materials. Martin et al. (2005) applied separate EB irradiation, separate MW heating, and combined (successive and simultaneous) EB irradiation and MW heating to the reduction of viable cells of *Escherichia coli*, *Salmonella typhi*, *Staphylococcus intermedius*, *Pseudomonas aeruginosa*, and *Trichinella spiralis*. The results of their study concerning the disinfection by separate and combined EB and MW irradiation of sewage sludge performed by a food industry wastewater treatment station (vegetable oil plant) have demonstrated that the simultaneous EB and MW irradiation produced the biggest reduction of microorganisms. According to Martin et al. (2006), it seemed that MW irradiation could cause the modification of the microorganisms sensitivity to EB irradiation. Thus, the application of combined EB and MW irradiation led to greater lethal effects than the EB irradiation alone. Also, the tests demonstrated that irradiation time and the upper limit of EB required absorbed dose, which ensures a good decontamination effect, could be reduced by a factor at least of two by additional use of MW energy to EB energy. Martin et al. (2006) more importantly brought forward that the combined EB and MW irradiation could become a new disinfection/sterilization method, commercially viable alternative to classical thermal or chemical destruction. Also, ionizing irradiation costs could be much decreased and the application of low-intensity radiation sources, which are less expensive and cleaner, could be extended for the sanitation/sterilization of a wide variety of materials including food items, medical objects, hospital waste, wastewater, and sewage sludge. The technology of sludge irradiation by MW followed by composting could be feasibly developed to produce disinfected compost for agriculture (Martin et al., 2006).

The treatment of industrial wastewater is always among the most severe obstacles faced by all countries. Research carried out so far in this respect and with regard to the applicability of MW technology has shown that MW technology may be effective in treating wastewater at the various stages of the treatment process (Bo et al., 2002). Ramon et al. (2003) demonstrated the use of an MW heating system, employed in the chemical digestion step, for the determination of chemical oxygen demand in wastewater. The results were first compared with those provided by standard methods using reference substances. The problems arising from abrupt heating of the sample and the potential thermal decomposition of potassium dichromate were also examined and two different approaches to sample digestion involving a gradually increasing irradiation time were tested. By optimizing the operating conditions, Ramon et al. (2003) had the digestion time reduced to 8–60 times with respect to the standard method (the reference digestion time is 5 min). Particularly in difficult digestions, the proposed approach of using MW heating is firmly believed to provide a substantially improved degradation with respect to conventional procedures. The procedure was applied to wastewater from various industries by Ramon et al. (2003) and was found to ensure thorough digestion of all samples and to provide favorable results in all cases tested. In their study, Zhihui et al. (2005) investigated the synergistic effects of several MW-assisted advanced oxidation processes (MW/AOPs) for the degradation of 4-chlorophenol (4-CP). The efficiencies of the degradation of 4-CP in dilute aqueous solution for a variety of AOPs with or without MW irradiation were compared. The results of Zhihui et al. (2005) showed that the synergistic effects between MW and H_2O_2, UV/H_2O_2, and TiO_2 photocatalytic oxidation (PCO) resulted in a high degradation efficiency for 4-CP, thus supporting the potential of MW/AOPs for treatment of industrial wastewater. Regarding organic contaminant degradation during wastewater treatment processes, Bo et al. (2006) equally studied the MW-assisted oxidation process for degradation of p-nitrophenol (PNP) from aqueous solution (a synthetic wastewater). Their process consisted of a granular activated carbon (GAC) fixed-bed reactor, an MW source, solution and air supply system, and a heat exchanger. The process was operated in continuous flow mode, air was applied for oxygen supply, and the GAC acted as an MW energy absorption material as well as the catalyst for PNP degradation. MW power, air flow, GAC dose, and influent flow proved to be major factors that influenced PNP degradation and the subsequent results showed that PNP was degraded effectively by this new process. Under a given condition of PNP concentration 1330 mg L^{-1}, MW power 500 W, influent flow 6.4 mL min^{-1}, and air flow 100 mL min^{-1}, a PNP removal of 90%, was achieved. Zhihui et al. (2005) concluded that the biodegradability of the solution was improved apparently after treatment by MW-assisted oxidation process, which benefits further treatment of the solution using the biochemical method. Recently, Gromboni et al. (2007) evaluated the coupling of MW radiation with photo-Fenton reaction for wastewater treatment. This strategy was performed in a focused MW digestion oven with open vessels and ultraviolet radiation was obtained from MW lamps (MWLs) activated by MW radiation. The operational conditions were established considering the extent of degradation of chlorfenviphos and cypermethrin used for bovine ticks (*Boophilus michoplus*) control. Gromboni et al. (2007) found that the combination of MWLs and Fenton reaction degraded pesticide residues efficiently by more than 98% in about 4 min.

1.3.6 Sonochemical Reactions

Ultrasonic irradiation is widely used in chemistry and elsewhere. Imaging techniques using echolocation, such as SONAR systems for target detection or echography in health care, medical imaging and diagnostics, biological cell-disruption, nondestructive testing of materials, and thermoplastic welding represent perhaps the best-known use of ultrasound (Cravotto and Cintas, 2006). Chemical applications extend to such varied areas as organic and organometallic chemistry, materials science, aerogels, food chemistry, and medicinal research (Cravotto and Cintas, 2006). Sonochemistry is a branch of chemical research dealing with the chemical effects and applications of ultrasonic waves, that is, sound with frequencies above 20 kHz that lie beyond the upper limit of human hearing. Although the range of ultrasonic frequencies can be extended up to 100 MHz, it is customary to divide ultrasound into two distinct regions: conventional power ultrasound, up to 100 kHz, which especially affects chemical reactivity in liquids (although higher frequencies can also achieve so), and diagnostic ultrasound (2–10 MHz) with applications in both medicine and materials processing (Cravotto and Cintas, 2006). Sonochemistry harmoniously shares with green/sustainable chemistry aims such as the use of less hazardous chemicals and solvents, reduced energy consumption, and increased product selectivity. In this regard, ultrasound and MW heating are in many instances complementary techniques for driving chemical reactions (Cravotto and Cintas, 2006).

The analysis of numerous experiments has revealed that ultrasound has no effect on chemical pathways and reaction rates have often been comparable to those of non-irradiated processes. Thus, in many heterogeneous reactions the application of ultrasound has the same effect as a high-speed agitator or a homogenizer in which fluids do not cavitate (Cravotto and Cintas, 2006). Enhanced yields and rates can be observed owing to the mechanical effects of the shock waves. Chemical effects of ultrasound will occur only if an elemental reaction is the sonication-sensitive step or when the high-energy species released after cavitational collapse do indeed participate as reaction intermediates. At this point it is important to point out that most organic chemists are only interested in using ultrasound as a convenient tool to enhance the yields and rates of many chemical reactions. The high temperatures and pressures developed locally by cavitation prove advantageous in a multitude of common reactions, in which conventional conditions or high-speed stirring cannot achieve the same results. Whether these effects fall under the heading of true or false sonochemistry, they all contribute to the current trend to adopt this emerging green technology.

1.3.6.1 Sonochemistry and Metal Complexes

The reactivity of transition metal complexes can often be enhanced by ultrasound under homogeneous conditions. Chromium aryl(alkoxy)carbenes react with propargylic alcohols under sonication (20 kHz) to afford b-lactones in good yields. These Dötz cyclizations also take place under thermal activation, but sonication works faster and, in addition, favors the formation of less heavily substituted b-lactones. Palladium fluoride complexes, which have become important and useful organometallics, can be prepared by a new ultrasound-promoted ligand exchange between $[(Ph_3P)_2Pd(Ar)I]$ and AgF in aromatic solvents. No I–F exchange occurs without sonication. The process can also be conducted in the presence of a catalytic amount (5–10 mol%) of the corresponding aryl iodide, a variant that is beneficial for the purity of the product. By the ultrasonic procedure the first dinuclear organopalladium m-fluorides and their mononuclear analogs stabilized by trialkylphosphine ligands have been synthesized. On the basis of recent research using ultrasound, Nandurkar et al. (2008) report a simple and convenient methodology for complex formation of a wide variety of transition metals/alkaline earth metals with 1,3-diketones under sonication. Their method showed a significant rate enhancement for metal complex formation in the presence of ultrasound as compared to their silent counterpart, thereby providing higher yields. Similarly, Shingote et al. (2008) also successfully performed the asymmetric transfer hydrogenation of ketones using Ru(II)arene/amino alcohol catalyst system that proceeds with significant rate enhancement by ultrasound promotion. Comparison of the silent reactions carried out at 25°C with reactions under sonochemical activation at 25°C clearly showed an enhancement in the catalytic activity by 5–10 times without significantly affecting the enantioselectivity. More recently, Raouafi et al. (2009) carried out the nucleophilic aromatic substitution under ultrasound irradiation of a dichlorobenzene iron η^6-complex with various secondary amines. The reaction time at moderate temperatures was considerably shortened compared to silent reaction conditions at room temperature (several days) or at solvent refluxing temperature (12–48 h).

1.3.6.1.1 Ultrasound and Materials Science

Although the search for new polymers with improved properties continues to attract great research interest, the economic drivers in the polymer industry demand continual improvement of existing materials. This has led to a large effort aimed at modifying existing polymers (Suslick and Price, 1999). The effects of ultrasound on polymers can be both physical and chemical. Irradiation of liquids with ultrasound can cause solely physical changes from acoustic streaming, such as rapid mixing and bulk heating. Although cavitation is not always necessary for these effects, they almost always accompany cavitation (Suslick and Price, 1999). Examples of physical changes induced by ultrasound in polymer systems include the dispersal of fillers and other components into base polymers (as in the formulation of paints), the encapsulation of inorganic particles with polymers, modification of particle size in polymer powders, and, perhaps most important, the welding and cutting of thermoplastics (Price, 1996). In contrast, chemical changes can also be created during ultrasonic irradiation, invariably as a result of cavitation, and these effects have been used to benefit many areas of polymer chemistry. Although not nearly as commonly used as organic materials, polymers with backbones consisting of inorganic elements have some special properties that are making them commercially significant. For example, the polysiloxanes (silicones) and polyphosphazenes remain flexible to very low temperatures. Poly(organosilanes) have a conjugated silicon backbone that confers interesting photo- and electroactive properties. It has been reported by Price et al. (1995) that ultrasound produces a significant acceleration in the cationic polymerization of cyclic siloxanes to give the commercially very important silicone resins. Polymers produced under sonication had narrower polydispersities but higher molecular weights than those produced under normal conditions. The acceleration of the polymerization was caused by more efficient dispersion of the acid catalyst throughout the monomer, leading to a more homogeneous reaction and hence a lower distribution of chain lengths. There is also the issue of ultrasonic degradation of long chains, which must be considered during the late stages of such polymerizations.

1.3.6.1.2   Impediments to Ultrasound Use at Industrial Level

Nevertheless, to date, industrial interest in sonochemistry has somewhat lagged behind research work investigating new applications and the fundamentals of how ultrasound influences chemical systems (Cravotto and Cintas, 2006). A major factor is the difficulty that underlies scale-up. Although a good deal of work has been carried out in developing scaled-up systems, and many development modules and prototypes have been produced, large-scale and industrial processing is restricted by ultrasonic path lengths and the limitations in the power delivery systems. A more widespread adoption of sonochemistry has almost also certainly been hindered by an inability to work at throughputs that are realistic for industrial production. In a recent interesting review, Ruecroft et al. (2005) have presented an overview of the application of power ultrasound to crystallization of organic molecules, and the equipment developed in recent years in which sonocrystallization and sonochemistry may be carried out at industrial scale. Ruecroft et al. (2005) point out explicitly that the availability, rather than nonavailability, of *robust* large-scale equipment remains critical to establishing the viability of this technology for industrial use in the near future. Large-scale equipment is available for the application of power ultrasound to wastewater treatment (in particular, anaerobic digestion) and to a range of activities in the food processing industry (e.g., emulsification, microbial fermentation, cell disintegration, homogenizing). Technologies for the scale-up of ultrasonic processing have been developed in many different ways, which may be classified generally as (1) probes, that is, tips or other small-area devices delivering very high local intensities, in a flow cell or large volume, (2) opposing parallel transducers arranged around a duct, through which the process solution or suspension flows. Probe systems have been developed that employ a high-intensity probe operating in a flow cell or immersed in a large volume. The probe is in direct contact with the process fluid, but in such systems, the acoustic energy is not particularly well focused other than at the tip. Systems where opposing parallel plate transducers are used to insonate a fluid flowing through a duct contain the acoustic energy to a greater extent than a probe system; however, the energy is still not particularly well focused to a central volume (Ruecroft et al., 2005). Devices where an annular duct is employed through which the process fluid flows and around which transducers are placed may be considered to be similar to an opposing parallel plate arrangement where limited focusing of ultrasound occurs. An approach to the design of scaled-up units for industrial application is to avoid resonance and standing waves and to develop systems that avoid coherent wave relationships. There are a number of advantages in using noncoherent ultrasound; the more even distribution of the ultrasound through the working fluid is a key benefit. In addition one can design equipment with greater flexibility in terms of dimensions, frequency, and configuration (Ruecroft et al., 2005). The reader is invited to read through Ruecroft et al. (2005) for additional details of the latest industrial-scale ultrasound equipment.

1.3.6.1.3 Industrial Applications of Ultrasound

Ultrasonics has been used extensively in a variety of application namely to clean mechanical parts, weld plastic materials, upgrade ores, and create slurries in the mining industry. Additional applications of ultrasonic irradiation or sonication have been rather slow to develop, but several are emerging now. We now present a few of these applications. The role ultrasonics can play in influencing crystallizations (sonocrystallization) is of particular interest to the chemical and pharmaceutical industries. Sonication can impact both nucleation and growth. It can induce nucleation in the metastable zone without seeding, while reducing the extent of primary nucleation. The nonseeding aspect is an advantage for contained, sterile environments. Cavitational collapse may create nucleation similar to heterogeneous nucleation from trace particle impurities in a liquid or a surface imperfection. The practice is claimed to allow precise production of high-purity organic and inorganic crystalline substances, including intermediates and active pharmaceutical ingredients (APIs). The technology has been validated in a good manufacturing practice (GMP) environment. SulphCo, Sparks, Nev., promote ultrasonics for upgrading heavy sour crude oils into lighter sweeter material (sonocracking). Its technology reportedly increases gravity and reduces sulfur and nitrogen levels and viscosity, thus providing more product per barrel. A typical Middle East crude oil contains 40–45% residuum and 0–5% asphaltenes. The sonocracking process converts a portion of these undesirable components into lighter, more desirable fractions. Sulfur and residuum content is claimed to be reduced by up to 80%. A test unit with a throughput of 2000 billion barrels per day of petroleum products was recently installed at OIL–SC in South Korea. It has demonstrated a 5-degree rise in gravity on Arab Medium crude oil. Sivakumar and Rao (2001) reviewed the use of ultrasound as a tool to improve the cleanliness of leather production for each stage of leather processing, showing that ultrasound has the added advantages of improving the process efficiency, reducing the process time, and improving the quality of leather produced. The rate of diffusion of chemicals such as dyes through the pores of the leather matrix can be enhanced by the use of power ultrasound. Thus, the optimum amount of chemicals required can be employed to improve the percent exhaustion of chemicals, thereby minimizing the significant amount of unspent chemicals in the effluent and reducing pollution load to a large extent. Dyeing experiments carried out in the presence of ultrasound (150 W and 33 kHz) have led to an increase of about 40% in dye exhaustion and a decrease in dyeing time by about 55% compared with the process in the absence of ultrasound under stationary conditions (Sivakumar and Rao, 2001).

Compared with conventional drumming conditions, ultrasound increases dye exhaustion by about 28% and reduces dyeing time by 25%. Ultrasound can also function as a physical activator and reduce the use of conventional chemicals in some of the processing stages, such as soaking for cleaning the raw skin/hide, liming for loosening hair, and degreasing for removing natural fat. Therefore, ultrasound can be employed effectively in leather processing to enjoy multifaceted benefits (Sivakumar and Rao, 2001). Albu et al. (2004) used ultrasound to increase the extraction efficiency of carnosic acid from the herb *Rosmarinus officinalis* using butanone, ethyl acetate, and ethanol as solvents for the food and pharmaceutical industry. Both dried and fresh leaves of the herb were collected, and when sonication was performed at the same temperature, an improvement in the yields of carnosic acid for all three solvents and shortened extraction time was observed. Sonication also reduced the solvent effect so that ethanol, which is a poor solvent under conventional conditions, reached a similar level of extraction efficiency to the other two when sonicated. The extraction of dried herb with ethanol proved to be more efficient than that of fresh material, probably due to the water present in the latter. In the textile sector, Basto et al. (2007) evaluated the potential of using ultrasound to enhance the bleaching efficiency of laccase enzyme on cotton fabrics. Ultrasound of low intensity (7 W) and relatively short reaction time (30 min) seemed to act in a synergistic way with the enzyme in the oxidation/removal of the natural coloring matter of cotton. The increased bleaching effect could be attributed to improved diffusion of the enzyme from the liquid phase to the fibers surface and throughout the textile structure. In the food industry, Zheng and Sun (2006) also thoroughly assessed the application of power ultrasound to food freezing as a relatively new subject but with recent research advances showing its promising potential. Resulting from acoustic effects, the application of power ultrasound is beneficial to many food-freezing processes. If ultrasound is applied to the process of freeze preservation of fresh foodstuffs, it shortens the freezing process, and leads to a product of better quality. If it is applied to freeze concentration and freeze-drying processes, it can be used to control crystal size distribution in the frozen product. Furthermore, power ultrasound can also bring several benefits to the process of partial freezing of ice cream inside a scraped surface freezer, for example, reducing crystal size and preventing incrustation on freezing surface. Therefore, ultrasonic freezing process could have promising applications in freezing of high-value food (ingredients) and pharmaceutical products (Zheng and Sun, 2006). However, for the future development of this technology, several problems still remain to be explored. More fundamental research is still needed in order to identify factors that affect the ability of power ultrasound in performing the above functions. New food drying processes utilizing ultrasound pretreatment for dehydration are also becoming increasingly popular in the food industry. Fernandes and Rodrigues (2007), Fernandes et al. (2008) successfully demonstrated the drying of banana and pineapple. The study by Fernandes and Rodrigues (2007) showed that water diffusivity increases after application of ultrasound and that the overall drying time for banana was reduced by 11%, which represented an economy of energy since air drying is energy cost intensive. During the ultrasonic treatment the bananas lost sugar; so ultrasonic pretreatment can be an interesting process to produce dried fruits with low sugar content, too. The use of ultrasound as a pretreatment prior to air drying was compared to the use of osmotic dehydration as a pretreatment prior to air drying in the study by Fernandes and Rodrigues (2007). Their results showed that the use of ultrasonic pretreatment is interesting when large amounts of water needs to be removed from the fruit, a case in which the combined processing time (pretreatment and air drying) is shortened. Biodiesel is commonly produced in batch reactors using heat and mechanical mixing as energy input. The subsequent slow reaction kinetics and poor mass transfer during biodiesel production by transesterification render the overall process performance, plant capacity, biodiesel yield, and low quality. The ultrasonic reactors recently developed by Hielscher Ultrasonics GmbH (Germany) have addressed this issue of slow kinetics and actually improved the transesterification kinetics significantly and lowered excess methanol, and less catalyst is required for biodiesel processing (Figure 1.1). The ultrasonic cavitational mixing serves as an effective alternative means to achieve a better mixing in the commercial biodiesel processing and it also provides the necessary activation energy for the industrial biodiesel transesterification. All the more, the ultrasonic reactors

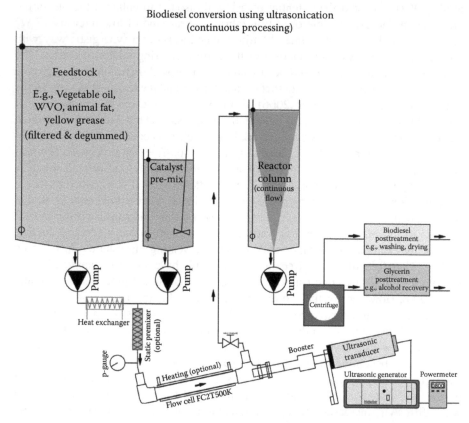

FIGURE 1.1 Inline ultrasonic mixing of oil and methanol for biodiesel production. (Reprinted from Thomas Hielscher, General Manager, Hielscher Ultrasonincs GmbH (Thomas@hielscher.com). Available from http://www.hielscher.com/ultrasonics/biodiesel_ultrasonic_mixing_reactors.htm. With permission.)

from Hielscher Ultrasonics GmbH allow for the continuous inline processing with biodiesel yields in excess of 99%. The additional "green chemistry and green engineering" features of this new technology are the following: the ultrasonic reactors reduce the processing time from the conventional 1–4 h batch processing to less than 30 s; more importantly, ultrasonication reduces the separation time from 5 to 10 h (using conventional agitation) to less than 60 min; ultrasonication also helps to decrease the amount of catalyst required by up to 50% due to the increased chemical activity in the presence of cavitation; when using ultrasonication the amount of excess methanol required is reduced and the resulting glycerin is of higher purity.

1.3.6.1.4   Fluorous Synthesis

Fluorous synthesis was introduced by Horvath and Rabai in 1994 in the developed fluorous biphasic catalysis (Horvath and Rabai, 1994). The character of temperature-dependent miscibility of fluorous and organic phases was utilized for homogeneous catalysis at high temperature and biphasic separation at lower temperature. Since the mid-1990s, Gladysz, Curran, Zhang, and many others extended the utility of fluorous synthesis for solution-phase high-throughput synthesis, separation of biomolecules, and noncovalent immobilizations for biochemistry applications (Gladysz et al., 2004). Fluorous chemistry has demonstrated a broad combinatorial capability and also shown a great potential for green chemistry applications (Curran and Lu, 2001; Horvath, 2008; Zhang, 2009a).

The early development of fluorous technology was focused on biphasic catalysis using "heavy fluorous" catalysts. It offers advantages of catalyst recycling and reduction of the amount of organic solvents for product separations. However, the disadvantages are the high cost and the persistent nature of fluorous solvents. The development of "light fluorous" chemistry significantly improves the green chemistry aspect of fluorous chemistry. In light fluorous chemistry, much short perfluorocarbon chains are used as phase tags to facilitate separations (Zhang, 2003). Fluorous solid-phase extraction (F-SPE) can replace fluorous liquid extraction for product purification (Zhang and Curran, 2006). Thus, no fluorous solvents are required in the reaction and separation processes.

1.3.6.1.5   Fluorous Catalysis

Catalysis always plays a critical role in organic synthesis and is an important topic of green chemistry research. Significant efforts have been made in the development of catalyst recycling techniques, environment-friendly solvent systems, and metal-free organocatalysis. Fluorous catalysis without using fluorous solvents has been achieved by using fluorous silica gel and PTFE as the noncovalent support for recovery of fluorous catalysts from organic solvents. Figure 1.2 shows a Suzuki reaction using a fluorous Pd catalyst **1**, which was immobilized on fluorous silica gel through fluorous affinity interactions (Tzschucke et al., 2002). The reaction was carried out in a common organic solvent. The fluorous catalyst absorbed on the fluorous silica gel was recovered and reused three times without significant loss of reactivity.

Organocatalysis attracts a high level of current interest because of its novel activation mechanisms and environment-friendly nature. Since it requires relative high catalyst loading (5–30 mol%), the development of recovery technique for

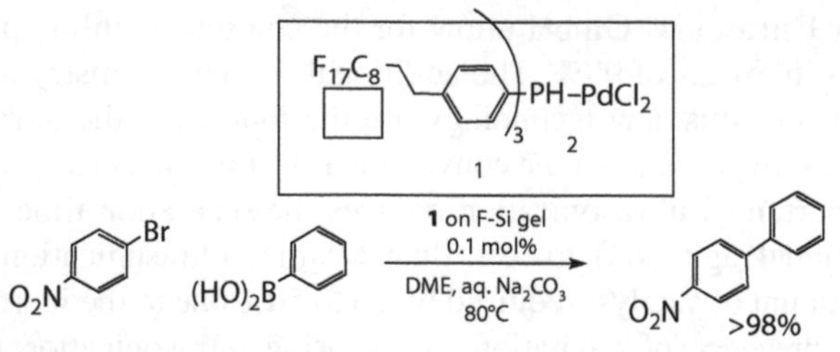

FIGURE 1.2 Reaction and recovery of fluorous Pd catalyst.

organocatalysis is highly demanding. Fluorous chemistry has demonstrated catalyst recovery capability while retaining the homogeneous catalyst reactivity. An example of fluorous imidazolidinone (MacMillan) catalyst **3** for the asymmetric Diels–Alder reactions is shown in Figure 1.3. Compared with the original imidazolidinone catalyst **2**, fluorous catalyst **3** generated **4**-endo product in a better yield (86%) and enantioselectivity (93% *ee*). Fluorous catalyst **3** was recovered by F-SPE in 84% yield with 99% purity. It can be directly used for the next round of reaction. On the other hand, regular imidazolidinone catalyst **2** was recovered by acid–base workup in 65% yield with only 74% purity. It cannot be used for the next round of reaction without further purification.

Use of fluorous pyrrolidine sulfonamides **5** for enantioselective aldol reactions and Michael additions are shown in Figure 1.4. The aldol reaction of cyclohexanone and 4-nitrobenzaldehyde under 10 mol% of the catalyst gave product **6** in 90% yield with a 70% ee and a 2:1 *syn/anti* ratio. The catalyst can be recovered by F-SPE and reused up to 6 times without significant loss of catalytic activity.

1.3.6.1.6   Fluorous Linker-Facilitated Synthesis

An important part of fluorous synthesis is related to the development of fluorous linkers for chromatography-free purifications. In discovery and medicinal chemistry labs, a major task is to synthesize small-scale but large numbers of compounds;

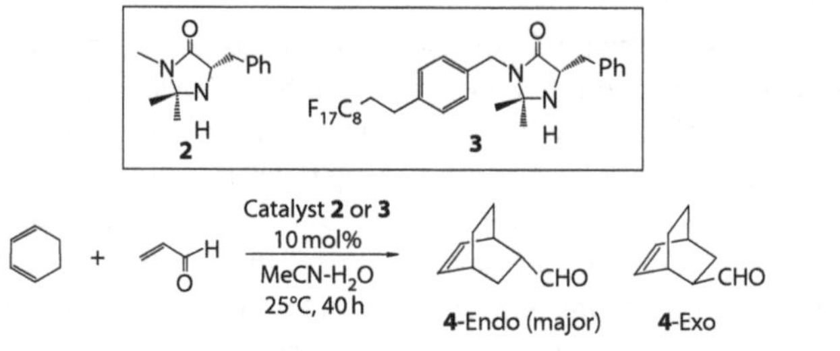

Catalyst	Yield	Endo: exo	ee% (endo)	Catalyst recovery	Purity of recovered cat.
2	82%	90.3:9.7	88.4	65%	74%
3	86%	93.4:6.6	93.4	84%	99%

FIGURE 1.3 Imidazolidinone-catalyzed Diels–Alder reactions.

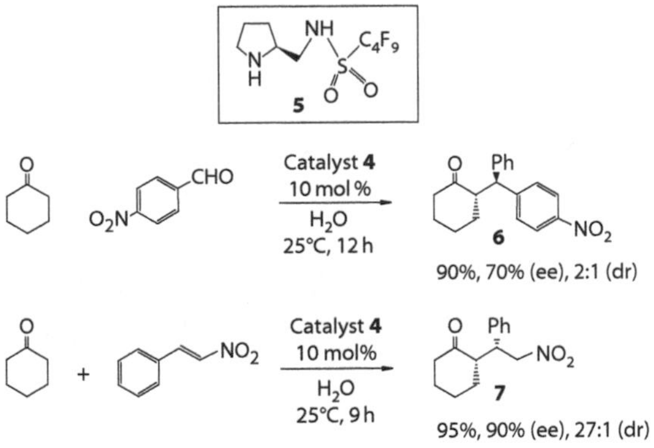

FIGURE 1.4 Fluorous pyrrolidine sulfonamide-catalyzed Michael addition and aldol reaction.

chromatography is the popular separation technique, but it is the most time-consuming step and generates large amounts of chemical wastes. Fluorous linkers could play double roles in multistep synthesis, not only for functional group protection, but also as phase tags for F-SPE separations (Zhang, 2004). At the linker cleavage step, displacement reactions are employed to detach the fluorous linker and introduce the new groups. This two-in-one strategy maximizes the separation advantage of the fluorous tag and minimizes the effort for tagging and detagging steps. The example shown in Figure 1.5 demonstrates the utility of perflurooctanesufonyl linker in the synthesis of proline-fused heterocyclic compounds (Zhang and Chen, 2005). Fluorous benzaldehydes were used for one-pot, three-component [3 + 2] cycloaddition to form intermediates 5. The fluorous linker was removed by Suzuki coupling reactions and generated the biaryl functional group. Numerous examples of using protective and displaceable fluorous linkers can be found in literature (Zhang, 2009b).

An example of using fluorous linker-attached amino acid for the synthesis of N-alkylated dihydropteridinones 14 is shown in Figure 1.6 (Nagashima and Zhang, 2004). The 4,6-dichloro-5-nitropyrimidine reacted with fluorous amino esters to form fluorous-attached compounds 10. They were then reacted with secondary amines to form 11. Reduction of the nitro group yielded compounds 12 followed by MW-assisted cyclization that yielded compounds 13. Compounds 13 were N-alkylated to give products 14. Numerous examples of using protective and displaceable fluorous linkers can be found in the literature (Zhang, 2009b).

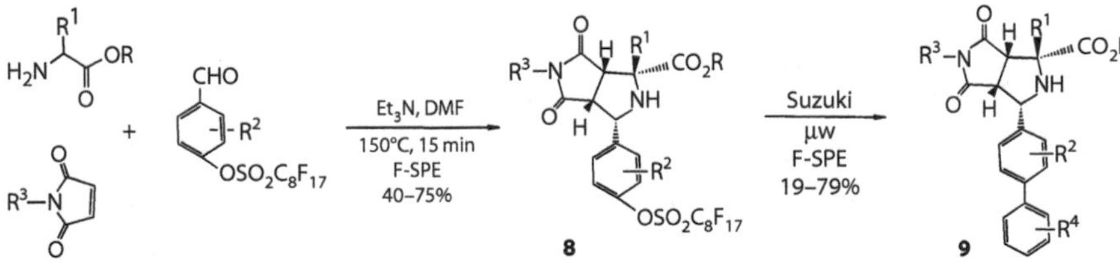

FIGURE 1.5 Synthesis of diaryl-substituted proline derivatives 6.

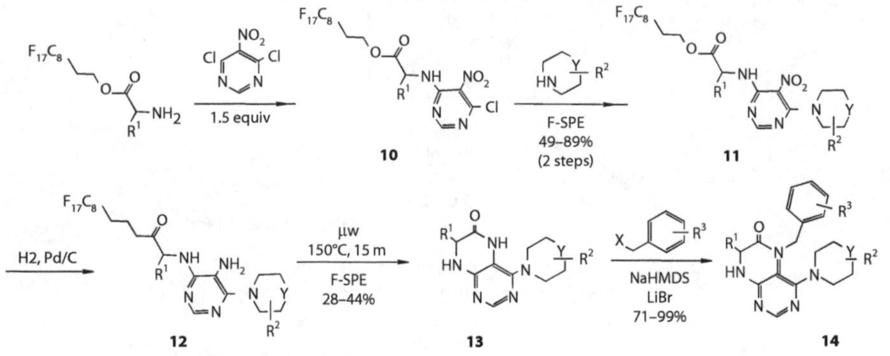

FIGURE 1.6   Fluorous amino acid-facilitated synthesis of *N*-alkylated dihydropteridinones.

1.3.6.1.7 Fluorous Multicomponent Reactions

Atom economy is an important principle of green chemistry. A multicomponent reaction (MCR) generates multiple bonds in a single reaction process, which is a highly efficient way to construct complicated molecules. Conducting post-MCR modification reactions further increases the molecular complexity and molecular diversity. It is a highly efficient and a green chemistry approach to make complex molecules. F-MCR employs a fluorous input as the limiting agent. After the MCR, the fluorous component is separated from the reaction mixture by F-SPE and then used for post-MCR modifications. The fluorous tag can be removed by intermolecular or intramolecular displacement reactions (Zhang, 2007). Figure 1.7 highlights the synthesis of benzodiazepine-quinazolinone alkaloids starting from a Ugi reaction (Zhang et al., 2007). A fluorous benzyl amine (F-BnNH$_2$) was used as the limiting agent for Ugi reactions to form products **7**. This compound was readily separated by F-SPE and then subjected to several steps of postcondensation reactions and finally to form products **8**. In this example, combinatorial techniques including MCR, fluorous linker, and MW reaction were employed for natural product synthesis. Another example of the Ugi/de-Boc/cyclization method for the synthesis of quinoxalinones **18** is shown in Figure 1.8 (Zhang and Tempest, 2004). Ugi reactions of an F-Boc protected-aniline were completed under MW irradiation. Condensation products **17** were easily isolated from the reaction mixture containing excess aldehydes and unreacted acids by F-SPE. MW-assisted TFA deprotection of F-Boc group promoted the cyclization to form quinoxalinones **18**.

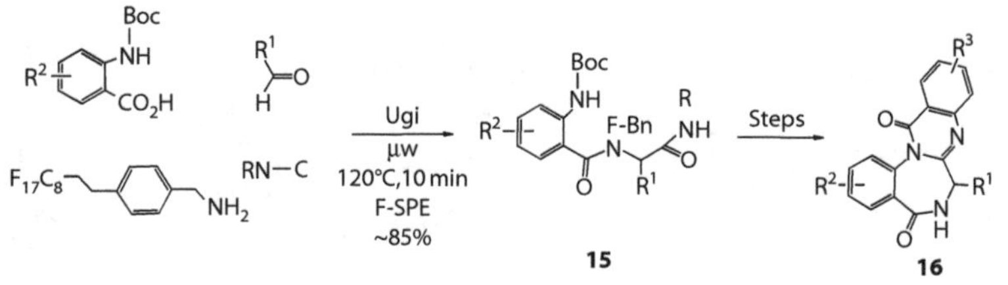

FIGURE 1.7   Synthesis of benzodiazepine-quinazolinones **8**.

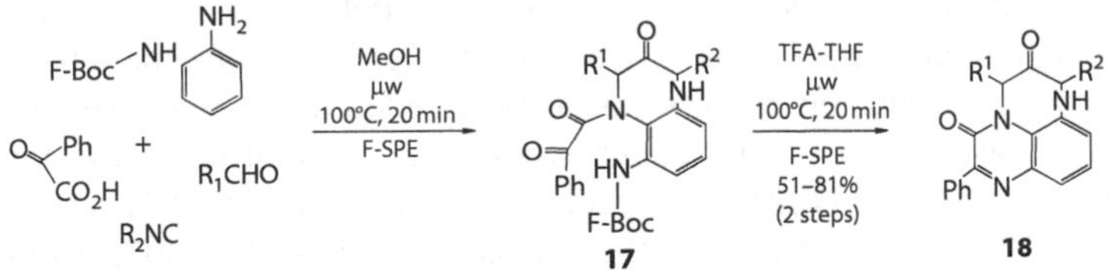

FIGURE 1.8 F-MCR facilitated synthesis of quinoxalinones **18**.

1.4 CONCLUDING REMARKS

The forthcoming challenges in resource, environmental, economical, and societal sustainabilities require more efficient and benign scientific technologies for working with chemical processes and products. Green chemistry addresses such challenges by opening a wide and multifaceted research scope eventually allowing the invention of novel reactions that can maximize the desired products and minimize the by-products, the design of new synthetic schemes and apparatuses that can simplify operations in chemical productions, and seek greener solvents that are inherently, environmentally, and ecologically benign (Li and Trost, 2008). Tangibly green and environmentally safe research findings from green chemistry are then translated into industrial-scale processes with the help of green engineering principles. Because of practical, logistical, economic, inertial, and institutional reasons (Anastas and Zimmerman, 2003), it will be essential in the near future to optimize unsustainable products, production processes and synthetic pathways, and systems that are currently in place. This is a vital short-term measure, and the green engineering principles once again provide a useful framework for accomplishing this optimization. Having both researchers from academia and specialized research institutions and those who support or manage research understand the basic concepts of green chemical engineering will definitely help bridge the gaps between broad goals and models, basic and applied research, and successful implementation of environmentally friendly and economically sustainable technologies (Anastas and Lankey, 2002). Therefore, marrying the principles of the sustainability concept as broadly promoted by the green engineering principles with established cost, quality, and performance standards will be the continual endeavor for economies in general and for the chemical industry in particular. It is hence a ubiquitous need to steer research and development efforts in such a direction that will constitute a powerful tool for fostering sustainable innovation.

Green chemistry and green engineering are no instantaneous panacea for remediating the pressing environmental concerns and impacts to our modern era, but putting their twelve mighty principles into practice will patiently and surely help to pave the way to a world where the grass is greener.

ACKNOWLEDGMENTS

First, we express our appreciation to researchers worldwide, who have put in commendable effort and made rapid progress in the various areas of green chemistry and green engineering, leading to a cleaner environment. We nevertheless extend our apologies to all those people whose work and research findings could not be cited and discussed in our exposé. Finally, we wish to explicitly thank the following people whose valuable data and discussions, as reported in their respective publications and reports, have been of considerable significance in writing this chapter: Riva (2006), Yu and Chen (2009), Morrone et al. (2008), Sheldon (2007), Hou et al. (2007), Shi et al. (2009), Banerjee et al. (2009), Hasaninejad et al. (2007), Heravi et al. (2007), Kirchhoff (2003), Michrowska et al. (2006), Gholap and Tambe (2008), Wang et al. (2007), Cai et al. (2006), Arce et al. (2007), Kidwai and Mohan (2005), Li and Li (2005), Li and Trost (2008), Bora et al. (2002), Hoogenboom and Schubert (2007), Shelke et al. (2008), Keglevich et al. (2007), Cravotto and Cintas (2006), and Thomas Hielscher (General Manager of Hielscher Ultrasonics GmbH). We are also grateful to our other colleagues and the reviewers whose criticisms have benefited this chapter.

REFERENCES

Albu, S., Joyce, E., Paniwnyk, L., Lorimer, J.P., and Mason, T.J. 2004. Potential for the use of ultrasound in the extraction of antioxidants from *Rosmarinus officinalis* for the food and pharmaceutical industry. *Ultrasonics Sonochemistry*, 11(3–4):261–65.

Anastas, P.T., Heine, L., and Williamson, T.C. 2000. *Green Engineering*. American Chemical Society: Washington, DC.

Anastas, P.T. and Kirchhoff, M.M. 2002. Origins, current status, and future challenges of green chemistry. *Accounts of Chemical Research*, 35:686–94.

Anastas, P.T., Kirchhoff, M.M., and Williamson, T.C. 2001. Catalysis as a foundational pillar of green chemistry. *Applied Catalysis A*, 221:3–13.

Anastas, P.T. and Lankey, R.L. 2000. Life–cycle assessment and green chemistry: The Yin and Yang of industrial ecology. *Green Chemistry*, 6:289–95.

Anastas, P.T. and Warner, J.C. 1998. *Green Chemistry, Theory and Practice*. Oxford University Press: Oxford, UK.

Anastas, P.T. and Zimmerman, J.B. 2003. Design through the 12 Principles of Green Engineering. *Environmental Science and Technology*, 37:94–101.

Appleton, T.J., Colder, R.I., Kingman, S.W., Lowndes, I.S., and Read, A.G. 2005. Microwave technology for energy-efficient processing of waste. *Applied Energy*, 81:85–113.

Arce, A., Rodríguez, H., and Soto, A. 2007. Use of a green and cheap ionic liquid to purify gasoline octane boosters. *Green Chemistry*, 9:247–53.

Astruc, D. 2007. Palladium nanoparticles as efficient green homogeneous and heterogeneous carbon–carbon coupling precatalysts: A unifying view. *Inorganic Chemistry*, 46:1884–94.

Bamoharram, F.F., Heravi, M.M., Roshani, M., Gharib, A., and Jahangir, M. 2006a. A catalytic method for synthesis of γ-butyrolactone, ε-caprolactone and 2-cumaranone in the presence of Preyssler's anion, $[NaP_5W_{30}O_{110}]^{14-}$, as a green and reusable catalyst. *Journal of Molecular Catalysis A: Chemistry*, 252:90–95.

Bamoharram, F.F., Heravi, M.M., Roshani, M., and Akbarpour, M. 2006b. Catalytic performance of Preyssler heteropolyacid as a green and recyclable catalyst in oxidation of primary aromatic amines. *Journal of Molecular Catalysis A: Chemistry*, 255:193–98.

Banerjee, S., Das, J., and Santra, S. 2009. Native silica nanoparticle catalyzed anti-Markovnikov addition of thiols to inactivated alkenes and alkynes: A new route to linear and vinyl thioethers. *Tetrahedron Letters*, 50:124–27.

Baran, P.S., Maimone, T.J., and Richter, J.M. 2007. Total synthesis of marine natural products without using protecting groups. *Nature*, 446:404–08.

Barreca, A.M., Fabbrini, M., Galli, C., Gentili, P., and Ljunggren, S. 2003. Laccase-mediated oxidation of lignin model for improved delignification procedures. *Journal of Molecular Catalysis B: Enzymatic*, 26:105–10.

Basto, C., Tzanov, T., and Cavaco-Paulo, A. 2007. Combined ultrasound-laccase assisted bleaching of cotton. *Ultrasonics Sonochemistry*, 14:350–54.

Bernhardt, R. 2006. Cytochromes P450 as versatile biocatalysts. *Journal of Biotechnology*, 124:128–45.

Block, D. 1999. Executive order and proposed bill will boost biobased products and bioenergy. *Biocycle Magazine*, 40:55–57.

Bo, L.L., Li, M.W., Quan, X., Chen, S., Xue, D.M., and Li, C.B. 2002. Treatment of high concentration organic wastewater by microwave catalysis. *Microwave and Millimeter Wave Technology*, 2002. *Proceedings, ICMMT 2002*, pp. 289–92.

Bo, L., Quan, X., Chen, S., Zhao, H., and Zhao, Y. 2006. Degradation of *p*-nitrophenol in aqueous solution by microwave assisted oxidation process through a granular activated carbon fixed bed. *Water Research*, 40:3061–68.

Bommarus, A.S. and Polizzi, K.M. 2006. Novel biocatalysts: Recent developments. *Chemical Engineering Science*, 61:1004–16.

Bora, U., Chaudhuri, M.K., and Dehury, S.K. 2002. Green chemistry in Indian context—Challenges, mandates and chances of success. *Current Science*, 82:1427–36.

Burrell, A.K., Del Sesto, R.E., Baker, S.N., McCleskey, T.M., and Baker, G.A. 2007. The large scale synthesis of pure imidazolium and pyrrolidinium ionic liquids. *Green Chemistry*, 9:449–54.

Cai, Y., Peng, Y., and Song, G. 2006. Amino-functionalized ionic liquid as an efficient and recyclable catalyst for Knoevenagel reactions in water. *Catalysis Letters*, 109:61–64.

Cao, L., van Langen, L.M., Rantwijk, F., and Sheldon, R.A. 2001. Cross-linked aggregates of penicillin acylase: Robust catalysts for the synthesis of β-lactam antibiotics. *Journal of Molecular Catalysis B: Enzymatic*, 11:665–70.

Cao, L., van Rantwijk, F., and Sheldon, R.A. 2000. Cross-linked enzyme aggregates: A simple and effective method for the immobilization of penicillin acylase. *Organic Letters*, 2:1361–64.

Cavallaro, G., Licciardi, M., Scirè, S., and Giammona, G. 2009. Microwave-assisted synthesis of PHEA-oligoamine copolymers as potential gene delivery systems. *Nanomedicine*, 4:291–303.

Capello, C., Fischer, U., and Hungerbühler, K. 2007. What is a green solvent? A comprehensive framework for the environmental assessment of solvents. *Green Chemistry*, 9:927–34.

Chandler, K., Deng, F., Dillow, A.K., Liotta, C.L., and Eckert, C.A. 1997. Alkylation reactions in near-critical water in the absence of acid catalysts. *Industrial & Engineering Chemistry Research*, 36:5175–79.

Chandran, S.S., Yi, J., Draths, K.M., von Daeniken, R., Weber, W., and Frost, J.W. 2003. Phosphoenolpyruvate availability and the biosynthesis of shikimic acid. *Biotechnology Progress*, 19:808–14.

Chatani, N. et al. 2001. $Ru_3(CO)_{12}$-catalyzed coupling reaction of sp3 C–H bonds adjacent to a nitrogen atom in alkylamines with alkenes. *Journal of the American Chemical Society*, 123:10935–41.

Chiappe, C., Leandri, E., Lucchesi, S., Pieraccini, D., Hammock, B.D., and Morisseau, C. 2004. Biocatalysis in ionic liquids: The stereoconvergent hydrolysis of *trans*-β-methylstyrene oxide catalyzed by soluble epoxide hydrolase. *Journal of Molecular Catalysis B: Enzymatic*, 27:243–48.

Claus, H. 2004. Laccases: Structure, reactions, distribution. *Micron*, 35:93–96.

Corradi, A., Lusvarghi, L., Rivasi, M.R., Siligardi, S., Veronesi, P., Marucci, G., Annibali, M., and Ragazzo, G. 2007. Waste treatment under microwave irradiation. *Advances in*

Microwave and Radio Frequency Processing, Part IV. Monika Willert-Porada, ed. Springer: Berlin, Heidelberg, Germany, pp. 341–48.

Costas, M., Mehn, M.P., Jensen, M.P., and Que, L. 2004. Dioxygen activation at mononuclear nonheme iron active sites: Enzymes, models, and intermediates. *Chemical Reviews*, 104:939–86.

Cravotto, G. and Cintas, P. 2006. Power ultrasound in organic synthesis: Moving cavitational chemistry from academia to innovative and large-scale applications. *Chemical Society Reviews*, 35:180–96.

Cremlyn, R.J. 1996. *An Introduction to Organo–Sulfur Chemistry.* John Wiley & Sons: New York.

Curran, D.P. and Lu, Z. 2001. Fluorous techniques for the synthesis and separation of organic molecules. *Green Chemistry*, 3:G3–G7.

Dias, E.L., Brookhart, M., and White, P.S. 2001. Rhodium(I)-catalyzed homologation of aromatic aldehydes with trimethylsilyldiazomethane. *Journal of the American Chemical Society*, 123:2442–43.

Dijksman, A., Marino-Gonzàlez, A., Payeras, A.M., Arends, W., and Sheldon, R.A. 2001. Efficient and selective aerobic oxidation of alcohols into aldehydes and ketones using Ruthenium/TEMPO as the catalytic system. *Journal of the American Chemical Society*, 123:6826–33.

Dominguez-Quintero, O., Martinez, S., Henriquez, Y., D'Ornelas, L., Krentzien, H., and Osuna, J.J. 2003. *Molecular Catalysis, A,* 197:185–91.

Drepper, T., Eggert, T., Hummel, W., Leggewie, C., Pohl, M., Rosenau, F., Wilhelm, S., and Karl-Erich Jaeger, E. 2006. Novel biocatalysts for white biotechnology. *Biotechnology Journal*, 1:777–86.

Eckert, C.A., Glaeser, R., and Brown, J.S. 1999. Tuning of chemical reactions with expanded solids. *Proceedings of the Fifth Conference of Supercritical Fluids and their Applications*, Verona, Italy.

Fernandes, F.A.N., Linhares Jr., F.E., and Rodrigues, S. 2008. Ultrasound as pre-treatment for drying of pineapple. *Ultrasonics Sonochemistry*, 15:1049–54.

Fernandes, F.A.N. and Rodrigues, S. 2007. Ultrasound as pre-treatment for drying of fruits: Dehydration of banana. *Journal of Food Engineering*, 82:261–67.

Fiksel, J. 1998. *Design for Environment: Creating Eco-Efficient Products and Processes.* McGraw-Hill: New York.

Frost, J.W. and Draths, K.M. 1995. Sweetening chemical manufacture. *Chemistry in Britain*, 31:206–10.

Fukuoka, S., Tojo, M., Hachiya, H., Aminaka, M., and Hasegawa, K. 2007. Green and sustainable chemistry in practice: Development and industrialization of a novel process for polycarbonate production from CO_2 without using phosgene. *Polymer Journal*, 39:91–114.

Garlotta, D.A. 2001. Literature review of poly(lactic acid). *Journal of Polymers and the Environment*, 9:63–84.

Gasperi, F., Aprea, E., Biasioli, F., Carlin, S., Endrizzi, I., Pirretti, G., and Spilimbergo, S. 2009. Effects of supercritical CO_2 and N_2O pasteurisation on the quality of fresh apple juice. *Food Chemistry*, 115:129–36.

Gholap, S.S. and Tambe, G.B. 2008. Boric acid catalyzed convenient synthesis of 2,3-dihydro-1h-1,5-benzodiazepines in water under mild conditions. *RASAYAN Journal of Chemistry*, 1:862–64.

Gladysz, J.A., Curran, D.P., and Horvath, I.T. (eds), 2004. *Handbook of Fluorous Chemistry.* Wiley–VCH: Weinheim.

Gotor-Fernández, V., Busto, E., and Gotor, V. 2006. *Candida antarctica* lipase B: An ideal biocatalyst for the preparation of nitrogenated organic compounds. *Advanced Synthesis & Catalysis*, 348:797–812.

Gromboni, C.F., Kamogawa, M.Y., Ferreira, A.G., Nóbrega, J.A., and Nogueira, A.R.A. 2007. Microwave-assisted photo-Fenton decomposition of chlorfenvinphos and cypermethrin in residual water. *Journal of Photochemistry and Photobiology A: Chemistry*, 185:32–37.

Grubbs, R.H. 2007. Olefin metathesis. *Tetrahedron*, 60:7117–7140.

Ha, S.H., Lan, M.G., Lee, S.H., Hwang, S.M., and Koo, Y.-M. 2007. Lipase-catalyzed biodiesel production from soybean oil in ionic liquids. *Enzyme and Microbial Technology*, 41:480–83.

Habulin, M., Šabeder, S., Sampedro, M.A., and Knez, Z. 2008. Enzymatic synthesis of citronellol laurate in organic media and in supercritical carbon dioxide. *Biochemical Engineering Journal*, 42:6–12.

Hajipour, A.R., Kargosha, M., and Ruoho, A.E. 2009. Tetrahydropyranylation of alcohols under solvent–free conditions. *Synthetic Communications*, 39:1084–91.

Hasaninejad, A., Zare, A., Sharghi, H., Niknam, K., and Shekouhy, M. 2007. P_2O_5/SiO_2 as an efficient, mild, and heterogeneous catalytic system for the condensation of indoles with carbonyl compounds under solvent-free conditions. *ARKIVOC*, (xiv):39–50.

Heravi, M.M., Bakhtiari, K., and Bamoharram, F.F. 2006a. 12-Molybdophosphoric acid: A recyclable catalyst for the synthesis of Biginelli-type 3,4-dihydropyrimidine-2(1H)-ones. *Catalysis Communications*, 7:373–76.

Heravi, M.M., Behbahani, F.K., and Bamoharram, F.F. 2006b. H14[NaP5W30O110]: A heteropolyacid catalyzed acetylation of alcohols and phenols in acetic anhydride. *Journal of Molecular Catalysis A: Chemical*, 253:16–19.

Heravi, M.M., Sadjadi, S., Oskooie, H.A., Shoar, R.H., and Bamoharram, F.F. 2007. Heteropolyacids as green and reusable catalysts for the synthesis of 3,1,5-benzoxadiazepines. *Molecules*, 12:255–62.

Höfer, R. and Bigorra, J. 2007. Green chemistry—A sustainable solution for industrial specialties applications. *Green Chemistry*, 9:203–12.

Hofrichter, M. and Ullrich, R. 2006. Heme-thiolate haloperoxidases: versatile biocatalysts with biotechnological and environmental significance. *Applied Microbiology and Biotechnology*, 71:276–88.

Hoogenboom, R. and Schubert, U.S. 2007. Microwave-assisted polymer synthesis: Recent developments in a rapidly expanding field of research. *Macromolecular Rapid Communications*, 28:368–86.

Horvath, I.T. 2008. Solvents from nature. *Green Chemistry*, 10:1024–28.

Horvath, I.T. and Rabai, T. 1994. Facile catalyst separation without water: Fluorous biphase hydroformylation of olefins. *Science*, 266:72–76.

Hou, Z., Theyssen, N., and Leitner, W. 2007. Palladium nanoparticles stabilised on PEG-modified silica as catalysts for the aerobic alcohol oxidation in supercritical carbon dioxide. *Green Chemistry*, 9:127–32.

Hull, K.L. and Sanford, M.S. 2007. Catalytic and highly regioselective cross-coupling of aromatic C–H substrates. *Journal of the American Chemical Society*, 129:11904–05.

Jia, C., Kitamura, T., and Fujiwara, Y. 2001. Catalytic functionalization of arenes and alkanes via C–H bond activation. *Accounts of Chemical Research*, 34:633–39.

Jones, D.A., Lelyveld, T.P., Mavrofidis, S.D., Kingman, S.W., and Miles, N.J. 2002. Microwave heating applications in environmental engineering—a review. *Resources, Conservation and Recycling*, 34:75–90.

Kareem, S.O. and Akpan, I. 2004. Pectinase production by *Rhizopus* sp. cultured on cheap agricultural materials. In *Book of Abstract no. 56 of the 1st International Conference on Science and National Development*, UNAAB, from 25 to 28th October, 2004, p. 53.

Keglevich, G. and Dudás, E. 2006. Microwave promoted efficient synthesis of 2-phosphabicyclo[2.2.2]octadiene-and octene 2-oxides under solventless conditions in Diels–Alder reaction. *Synthesis Communications*, 37:3191.

Keglevich, G., Kerényi, A., Sipos, M., Ujj, V., Makó, A., Csontos, I., Novák, T., Bakó, P., and Greiner, I. 2007. Green chemical approaches and tools in the development of environmentally friendly synthetic methods. *Periodica Polytechnica Chemical Engineering*, 51:53–56.

Kidwai, M. and Mohan, R. 2005. Green chemistry: An innovative technology. *Foundations of Chemistry*, 7:269–87.

Kirchhoff, M.M. 2003. Promoting green engineering through green chemistry. *Environmental Science and Technology*, 37:5349–53.

Kumar, G., Bristow, J.F., Smith, P.J., and Payne, G.F. 2000. Enzymatic gelation of the natural polymer chitosan. *Polymer*, 41:2157–2168.

Kumar, V. and Malhotra, S.V. 2008. Synthesis of nucleoside-based antiviral drugs in ionic liquids. *Bioorganic & Medicinal Chemistry Letters*, 18:5640–42.

Kumar, A. and Maurya, R.A. 2008. Organocatalysed three-component domino synthesis of 1,4-dihydropyridines under solvent free conditions. *Tetrahedron*, 64:3477–82.

Lancaster, M. 2000. Green Chemistry. *Education and Chemistry*, 37:40–46.

Lankey, R.L. and Anastas, P.T. 2002. Life-cycle approaches for assessing green chemistry technologies. *Industrial & Engineering Chemistry Research*, 41:4498–4502.

Leonard, E. and Prather, K.J. 2008. *De novo* biocatalyst design: An alternative strategy for the petroleum-free synthesis of biobutanol. Chemical Engineering, MIT, 77 Massachusetts ave., Bldg 66–425, Boston, MA 02139 (*The 30th Symposium on Biotechnology for Fuels and Chemicals*, May 2008).

Li, Z. and Li, C.-J. 2005. CuBr-catalyzed direct indolation of tetrahydroisoquinolines via cross-dehydrogenative coupling between sp3 C–H and sp2 C–H bonds. *Journal of the American Chemical Society*, 127:6968–69.

Li, C.-J. and Trost, B.M. 2008. Green chemistry for chemical synthesis. *Proceedings of the National Academy of Sciences USA*, 105:13197–202.

Lou, W.-Y., Zong, M.-H., Liu, Y.-Y., and Wang, J.-F. 2006. Efficient enantioselective hydrolysis of D,L-phenylglycine methyl ester catalyzed by immobilized *Candida antarctica* lipase B in ionic liquid containing systems. *Journal of Biotechnology*, 125:64–74.

Lynd, L.R., Wyman, C.E., and Gerngross, T.U. 1999. Biocommodity engineering. *Biotechnology Progress*, 15:777–93.

Martin, D., Craciun, G., Manaila, E., Ighigeanu, D., Togoe, I., Oproiu, C., Margaritescu, I., and Iacob, N. 2006. Waste treatment by microwave and electron beam irradiation. *Proceedings of the 2nd Environmental Physics Conference*, 18–22 February 2006, Alexandria, Egypt, pp. 91–100.

Martin, D.I., Margaritescu, I., Cirstea, E., Togoe, I., Ighigeanu, D., Nemtanu, M.R., Oproiu, C., and Iacob, N. 2005. Application of accelerated electron beam and microwave irradiation to biological waste treatment. *Vacuum*, 77:501–06.

Mavandadi, F. and Pilotti, A. 2006. The impact of microwave-assisted organic synthesis in drug discovery. *Drug Discovery Today*, 11:165–74.

Meijnen, J.-P., de Winde, J.H., and Ruijssenaars, H.J. 2008. Engineering *Pseudomonas putida* S12 for efficient utilization of D-xylose and L-arabinose. *Applied and Environmental Microbiology*, 74:5031–37.

Mihelcic, J.R., Crittenden, J.C., Small, M.J., Shonnard, D.R., Hokanson, D.R., Huichen, Q., Sorby, S.H., James, V., Sutherland, J.W., and Schnoor, J.L. 2003. Sustainability science and engineering: The emergence of a new metadiscipline. *Environmental Science and Technology*, 37:5314–24.

Michrowska, A., Gulajski, L., Kaczmarska, Z., Mennecke, L., Kirschning, A., and Grela, K. 2006. A green catalyst for green chemistry: Synthesis and application of an olefin metathesis catalyst bearing a quaternary ammonium group. *Green Chemistry*, 8:685–88.

Misono, M., Ono, L., Koyano, G., and Aoshima, A. 2000. Heteropolyacids. Versatile green catalysts usable in a variety of reaction media. *Pure and Applied Chemistry*, 72:1305–11.

Mohanty, A.K., Misra, M., and Drzal, L.T. 2002. Sustainable bio-composites from renewable resources: Opportunities and challenges in the green materials world. *Journal of Polymers and the Environment*, 10:19–26.

Mohsenzadeh, F., Aghapoor, K., and Darabi, H.R. 2007. Benign approaches for the microwave-assisted synthesis of quinoxalines. *Journal of the Brazilian Chemical Society*, 18(2):297–303.

Morrone, R., D'Antona, N., and Nicolosi, G. 2008. Convenient preparation of (s)-fenoprofen by biocatalysed irreversible esterification. *RASAYAN Journal of Chemistry*, 1:732–37.

Murugadoss, A. and Chattopadhyay, A. 2008. A "green" chitosan–silver nanoparticle composite as a heterogeneous as well as micro-heterogeneous catalyst. *Nanotechnology*, 19:9.

Nagashima, T. and Zhang, W. 2004. Fluorous parallel synthesis of an *N*-alkylated dihydropteridinone library. *Journal of Combinatorial Chemistry*, 6:942–49.

Nalawade, S.P., Picchioni, F., and Janssen, L.P.B.M. 2006. Supercritical carbon dioxide as a green solvent for processing polymer melts: Processing aspects and applications. *Progress in Polymer Science*, 31:19–43.

Nandurkar, N.S., Patil, D.S., and Bhanage, B.M. 2008. Ultrasound assisted synthesis of metal-1,3-diketonates. *Inorganic Chemistry Communications*, 11:733–36.

Nguyen, T.-A. D., Han, S.J., Kim, J.P., Kim, M.S., Oh, Y.W., and Sim, S.J. 2008. Hydrogen production by the hyperthermophillic eubacterium, *Thermotoga neapolitana*, using cellulose pretreated by ionic liquid. *International Journal of Hydrogen Energy*, 33:5161–68.

Nguyen, Q.A., Keller, F.A., Tucker, M.P., Lombard, C.K., Jenkins, B.M., Yomogida, D.E., and Tiangco, V.M. 1999. Bioconversion of mixed solids waste to ethanol. *Applied Biochemistry and Biotechnology*, 77:455–471.

Nilsson, J.A., Zheng, X., Sundqvist, K., Liu, Y., Atzori, L., Elfwing, A., Arvidson, K., and Grafström, R.C. 1998. Toxicity of formaldehyde to human oral fibroblasts and epithelial cells: Influences of culture conditions and role of thiol status. *Journal of Dental Research*, 77:1896–1903.

Nolen, S.A., Liotta, C.L., Eckert, C.A., and Gläser, R. 2003. The catalytic opportunities of near-critical water: A benign medium for conventionally acid and base catalyzed condensations for organic synthesis. *Green Chemistry*, 5:663–69.

Okonko, I.O., et al. 2009. Utilization of food wastes for sustainable development. *EJEAFChe*, 8:263–86.

Olah, G.A., Goeppert, A., and Prakash, G.K.S. 2006. *Beyond Oil and Gas: The Methanol Economy*. Wiley–VCH: Weinheim (Germany).

Omoyinmi, G.A.K., Adebisi, A.A., and Fagade, S.O. 2004. Farm organic wastes in the production of animal protein. In *Book of Abstract no. 46 of the 1st International Conference on Science and National Development*, organized by Col. of Nat. Sci, UNAAB, from 25 to 28th October, 2004, p. 43.

Penieres-Carrillo, G., Garcia-Estrada, J.-G., Gutierrez-Ramirez, J.-L., and Alvarez-Toledano, 2003. Infrared-assisted eco-friendly selective synthesis of diindolylmethanes. *Green Chemistry*, 5:337–39.

Pernak, A., Iwanik, K., Majewski, P., Grzymisławski, M., and Pernak, J. 2005. Ionic liquids as an alternative to formalin in histopathological diagnosis. *Acta Histochemica*, 107:149–56.

Piontek, K. et al. 2002. Crystal structure of a laccase from the fungus *Trametes versicolor* at 1.90A° resolution containing a full complement of coppers. *Journal of Biological Chemistry*, 277:37663–69.

Polshettiwar, V. and Varma, R.S. 2008. Aqueous microwave chemistry: A clean and green synthetic tool for rapid drug discovery. *Chemical Society Reviews*, 37:1546–57.

Price, G.J. 1996. In *Chemistry under Extreme or Non-Classical Conditions*, R. van Eldik, C.C. Hubbard, eds. Wiley & Sons: New York.

Price, G.J., Wallace, E.N., and Patel, A.M. 1995. In *Silicon Containing Polymers*, R.G. Jones, ed., p. 147. Royal Society of Chemistry: Cambridge, UK.

Ramon, R., Valero, F., and del Valle, M. 2003. Rapid determination of chemical oxygen demand using a focused microwave heating system featuring temperature control. *Analytica Chimica Acta*, 491:99–109.

Ran, N., Zhao, L., Chen, Z., and Tao, J. 2008. Recent applications of biocatalysis in developing green chemistry for chemical synthesis at the industrial scale. *Green Chemistry*, 10:361–72.

Raouafi, N., Belhadj, N., Boujlel, K., Ourari, A., Amatore, C., Maisonhaute, E., and Schöllhorn, B. 2009. Ultrasound-promoted aromatic nucleophilic substitution of dichlorobenzene iron(II) complexes. *Tetrahedron Letters*, 50:1720–22.

Reddy, B.M., Patil, M.K., Rao, K.N., and Reddy, G.K. 2006. An easy-to-use heterogeneous promoted zirconia catalyst for Knoevenagel condensation in liquid phase under solvent-free conditions. *Tetrahedron Letters*, 47:4409–13.

Riva, S. 2006. Laccases: Blue enzymes for green chemistry. *TRENDS in Biotechnology*, 24:219–26.

Ruecroft, G., Hipkiss, D., Ly, T., Maxted, N., and Cains, P.W. 2005. Sonocrystallization: The use of ultrasound for improved industrial crystallization. *Organic Process Research & Development*, 9:923–32.

Sakaki, T., Shibata, M., Miki, T., Hirosue, H., and Hayashi, N. 1996. Decomposition of cellulose in near-critical water and fermentability of the products. *Energy Fuels*, 10:684–88.

Salgin, U. 2007. Extraction of jojoba seed oil using supercritical CO_2 + ethanol mixture in green and high-tech separation process. *The Journal of Supercritical Fluids*, 39:330–37.

Saxena, A., Kumar, A., and Mozumd, S. 2007. Ni-nanoparticles: An efficient green catalyst for chemo-selective oxidative coupling of thiols. *Journal of Molecular Catalysis A: Chemical*, 269:35–40.

Shapiro, S. and Caspi, E. 1998. The steric course of enzymic hydroxylation at primary carbon atoms. *Tetrahedron*, 54:5005–40.

Sharma, S.K., Mudhoo, A., and Khamis, E. 2009a. Adsorption studies, modeling and use of green inhibitors in corrosion inhibition: An overview of recent research. *The Journal of Corrosion Science and Engineering*, 11:1–24. (Available at http://www.jcse.org/view-paper.php?vol=11&pap=14.)

Sharma, S.K., Mudhoo, A., Jain, G., and Sharma, J. 2009b. Inhibitory effects of *Ocimum tenuiflorum* (Tulsi) on the corrosion of zinc in sulphuric acid: A green approach. *RASAYAN Journal of Chemistry*, 2:332–339.

Sharma, S.K., Mudhoo, A., Jain, G., and Khamis, E. 2009c. Corrosion inhibition of neem (*Azadirachta indica*) leaves extract as a green corrosion inhibitor for zinc in H_2SO_4. *Green Chemistry Letters and Reviews*, 2:47–51.

Sheldon, R.A. 2007. Cross-linked enzyme aggregates (CLEAs): Stable and recyclable biocatalysts. *Biochemical Society Transactions*, 35:1583–87.

Shelke, K., Kakade, G., Shingate, B., and Shingare, M. 2008. Microwave-induced one-pot synthesis of 2,4,5-triarylimidazoles using glyoxylic acid as a catalyst under solvent-free conditions. *RASAYAN Journal of Chemistry*, 1:489–94.

Shi, F., Tse, M.K., Zhou, S., Pohl, M.-M., Radnik, J., Hobner, S., Jahnisch, K., Brockner, A., and Beller, M. 2009. Green and efficient synthesis of sulfonamides catalyzed by nano-Ru/Fe$_3$O$_4$. *Journal of American Chemical Society*, 131:1775–79.

Shingote, S.K., Kelkar, A.A., Borole, Y.L., Joshi, P.D., and Chaudhari, R.V. 2008. Ultrasound promoted asymmetric transfer hydrogenation of ketones using Ru(II)arene/amino alcohol catalyst system. *Ultrasonics Sonochemistry*, 15:289–93.

Sivakumar, V. and Rao, P.G. 2001. Application of power ultrasound in leather processing: An eco-friendly approach. *Journal of Cleaner Production*, 9:25–33.

Skerlos, S.J., et al. 2001. Challenges to achieving sustainable aqueous systems: A case study in metalworking fluids. In *Proceedings of the Second International Symposium on Inverse Manufacturing*, Tokyo, Japan, December 13–16, pp. 146–53.

Sun, Y. and Cheng, J. 2002. Hydrolysis of lignocellulosic material from ethanol production: A review. *Bioresource Technology*, 83:1–11.

Suslick, K.S. and Price, G.J. 1999. Applications of ultrasound to materials chemistry. *Annual Reviews of Materials Science*, 29:295–326.

Szmant, H.H. 1989. *Organic Building Blocks of the Chemical Industry*. Wiley: New York.

Tandon, V.K. and Kumar, M., 2004. BF_3Et_2O promoted one-pot expeditious and convenient synthesis of 2-substituted benzimidazoles and 3,1,5-benzoxadiazepines. *Tetrahedron Letters*, 45:4185–87.

Taranta, C. and Buckpesch, R. 2005. Bayer Crop Science, plant protection composition and use thereof. World Patent WO 2005/074685.

Thomas, V. 2003. Research issues in sustainable consumption: Toward an analytical framework for materials and the environment. *Environmental Science and Technology*, 37:5383–88.

Thurston, C.F. 1994. The structure and function of fungal laccases. *Microbiology*, 140:19–26.

Tundo, P., Anastas, P., StC. Black, D., Breen, J., Collins, T., Memoli, S., Miyamoto, J., Polyakoff, M., and Tumas, W. 2000. Synthetic pathways and processes in green chemistry. Introductory overview. *Pure and Applied Chemistry*, 72:1207–28.

Tzschucke, C.C., Markert, C., Glatz, H., and Bannwarth, W. 2002. Fluorous biphasic catalysis without perfluorinated solvents: Application to Pd-mediated Suzuki and Sonogashira couplings. *Angewandte Chemie International Edition*, 41:4500–03.

Van Wyk, J.P.H. 2001. Biowaste as a resource for bioproduct development. *TRENDS in Biotechnology*, 19:172–77.

Verhoef, S., Ruijssenaars, H.J., de Bont, J.A.M., and Wery, J. 2007. Bioproduction of *p*-hydroxybenzoate from renewable feedstock by solvent-tolerant *Pseudomonas putida* S12. *Journal of Biotechnology*, 132:49–56.

Villarroya, S., Thurecht, K.J., and Howdle, S.M. 2008. HRP-mediated inverse emulsion polymerisation of acrylamide in supercritical carbon dioxide. *Green Chemistry*, 10: 863–67.

Wang, J.-Q., Cai, F., Wang, E., and He, L.-N. 2007. Supercritical carbon dioxide and poly(ethylene glycol): An environmentally benign biphasic solvent system for aerobic oxidation of styrene. *Green Chemistry*, 9:882–87.

Wicks, G.G., Schulz, R.L., and Clark, D.E. 2000. Microwave remediation of hazardous and radioactive wastes. Document prepared in conjunction with work accomplished under Contract No. DE-AC09-96SR18500 with the U.S. Department of Energy.

Yang, Z. and Pan, W. 2005. Ionic liquids: Green solvents for nonaqueous biocatalysis. *Enzyme and Microbial Technology*, 37:19–28.

Yu., L. and Chen, L. 2009. *Renewable Resources. Biodegradable Polymer Blends and Composites from Renewable Resources*. Long Yu, ed., pp. 1–15. John Wiley & Sons, Inc., accessible at http://www3.interscience.wiley.com/cgi-bin/booktext/122186851/BOOKPDFSTART.

Zhang, W. 2003. Fluorous technologies for solution-phase high-throughput organic synthesis. *Tetrahedron*, 59:4475–89.

Zhang, W. 2004. Fluorous protecting groups and tags. *Handbook of Fluorous Chemistry*. In J.A. Gladysz, D.P. Curran, and I.T. Horvath, eds., pp. 222–36. Wiley-VCH: Germany.

Zhang, W. 2007. Fluorous-enhanced multicomponent reactions for making drug-like library scaffolds. *Combinatorial Chemistry & High Throughput Screening*, 10:219–29.

Zhang, W. 2009a. Green chemistry aspects of fluorous techniques—Opportunities and challenges for small-scale organic synthesis. *Green Chemistry*, 11:911–20.

Zhang, W. 2009b. Fluorous linker-facilitated chemical synthesis. *Chemical Reviews*, 109:749–795.

Zhang, W., Williams, J.P., Lu, Y., Nagashima, T., and Chu, Q. 2007. Fluorous synthesis of sclerotigenin-type benzodiazepine–quinazolinone library scaffold. *Tetrahedron Letters*, 48:563–65.

Zhang, W. and Chen, C.H.-T. 2005. Fluorous synthesis of biaryl-substituted prolines by 1,3-dipolar cycloaddition and Suzuki coupling reactions. *Tetrahedron Letters*, 46:1807–10.

Zhang, W. and Curran, D.P. 2006. Synthetic applications of fluorous solid-phase extraction (F–SPE). *Tetrahedron*, 62:11837–65.

Zhang, W. and Tempest, P., 2004. Highly efficient microwave-assisted fluorous Ugi and post-condensation reactions for benzimidazoles and quinoxalinones. *Tetrahedron Letters*, 45:6757–60.

Zheng, L. and Sun, D.-W. 2006. Innovative applications of power ultrasound during food freezing processes—A review. *Trends in Food Science & Technology*, 17:16–23.

Zhihui, A., Peng, Y., and Xiaohua, L. 2005. Degradation of 4-chlorophenol by microwave irradiation enhanced advanced oxidation processes. *Chemosphere*, 60:824–27.

Ziyaei Halimehjani, A., Pourshojaei, Y., and Saidi, M.R. 2009. Highly efficient and catalyst-free synthesis of unsymmetrical thioureas under solvent-free conditions. *Tetrahedron Letters*, 50:32–34.

2 Aliphatic Nitrocompounds as Versatile Building Blocks for the One-Pot Processes

Roberto Ballini, Serena Gabrielli, and Alessandro Palmieri

CONTENTS

2.1 INTRODUCTION

Chemistry has made an impact on almost every aspect of daily life from toothpaste to life-saving medicines. The essential feature of this central science is synthesis. Despite the great success and the importance of chemistry to life quality, its public image has deteriorated and in the twenty-first century, we can expect the drive toward cleaner technologies.

Effective organic synthesis is predicated on site isolation, the physical separation of reagents or catalysts from each other. Synthetic organic chemists typically achieve site isolation by using separate flasks or reactors. Separate vessels prevent incompatible catalysts or reagents from fouling or yielding intractable mixtures. This reliance on "multiple pots" is both a triumph and a curse. Iterative transformation and purification has been an enormously successful model, but it is plagued by wastes, mostly manifested in the form of solvents. Solvents are often incinerated and if the precursors to solvents are a finite resource, current chemical synthesis will only be possible for a limited amount of time. Beyond solvent, high-yielding solvent, high-yielding reactions often produce salts and other impurities that must be removed to avoid deleterious effects on downstream transformations. Serial reactions and purifications require massive amounts of solvents and materials (Sheldon, 1994, 1997).

This is why the preparation of fine chemicals and pharmaceuticals is frequently accompanied by the production of large amounts of waste reaching values of 25–100 times higher than those of the target compounds. Today it is not only a question of what we can synthesize, but how we do it. Major problems in chemical production are the handling of waste and the search for environmental tolerable procedures.

Environmental and economic pressures are now forcing the chemical community to search for more efficient ways of performing chemical transformations (Hall, 1994; Broadwater et al., 2005). Thus, it would be much more efficient if one could form two or more bonds or realize two or more transformations in one synthetic sequence without isolating the intermediates (Broadwater et al., 2005). It is obvious that this type of reaction would allow the minimization of waste and thus making its management unnecessary as compared with stepwise reactions the amount of solvents, reagents, adsorbents, and energy would be dramatically decreased and the amount of work would be reduced. Moreover, the one-pot process can drive the equilibria to the desired direction. Thus, these reactions would allow an ecologically and economically favorable production.

By carefully designing the reaction sequence so that the first step creates the conditions to trigger the next stage, and that in turn sets up the third reaction, and so on, chemists can plan this type of multistage reaction. In this context, the choice of starting materials with high chemical versatility is crucial.

In recent years, nitroalkanes have demonstrated to be highly versatile building blocks for the one-pot synthesis of a variety of important targets (Colvin et al., 1979; Ballini and Rosini, 1988; Ballini et al., 1990; Ono, 2001). In fact, nitroalkanes are a valuable source of stabilized carbanions since the high electron-withdrawing power of the nitro group provides an outstanding enhancement of the hydrogen acidity (Patai, 1996; Adams, 2002). Nitronate anions that can be generated from nitroalkanes using a wide range of bases act as carbon nucleophiles with common electrophiles including aldehydes (nitroaldol-Henry reaction) (Rosini, 1991; Luzzio, 2001), and electron-poor alkenes (Michael addition) (Ballini et al., 2005a), leading to carbon–carbon (C–C) bond formation.

On the other hand, conjugated nitroalkenes are very useful electron-poor alkenes, prone to act as nucleophilic acceptor, mainly in the Michael reaction (Berestovitaskaya et al., 1994) or in the Diels–Alder cycloaddition (Denmark and Thorarensen, 1996). Moreover, the nitro group can be easily turned into a respectable array of functional groups such as its reduction to a primary amine, replacement with hydrogen (Ballini et al., 1983; Ono, 2001), conversion into a carbonyl (Nef reaction) (Ballini and Petrini, 2004), and transformation into other important functionalities such as nitrile, nitrile oxide, oximes, hydroxylamines, and thiols (Colvin et al., 1979).

2.2   NITROALKENES AS BUILDING BLOCKS FOR ONE-POT PROCESSES

Nitro olefins are useful intermediates in organic synthesis. Conjugated nitroalkenes are excellent acceptors (Michael reaction and Diels–Alder cycloaddition) and the nitro group can be further transformed, *in situ*, into several reactive intermediates.

2.2.1   ONE-POT SYNTHESIS OF *TRANS*-β-ALKYLSTYRENES

One-pot synthesis of (E)-alkenes has been directly performed from nitromethane by a radical process (Liu and Yao, 2001). In fact, the strategy is to react β-nitrostirene (3) as shown in Figure 2.1, *in situ* prepared from aromatic aldehydes (1) with nitromethane (2) in the acetic acid solution with a catalytic amount of ammonium acetate at 70–100°C, with triethylborane in the biphase of diethyl ether and aqueous solution in the presence of oxygen in air as radical initiator, to generate (4).

The process gives overall satisfactory good yields (51–85%), and a variety of aromatics and heteroaromatic aldehydes can be employed. The role of nitro group is fundamental since, first, it helps the *in situ* generation of a C–C bond and the elimination of water in the conversion of (1) + (2) into (3), then acts as good leaving group in the radical substitution of nitro with the ethyl group.

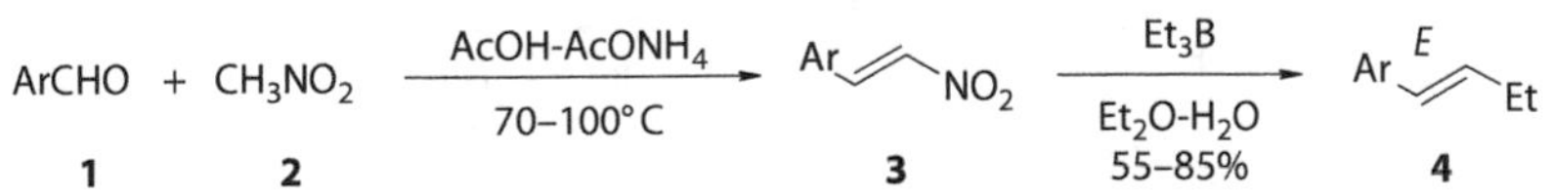

FIGURE 2.1

2.2.2   One-Pot Synthesis of *Trans*-Fused Bicyclic γ-Lactones

Medium ring lactones are increasingly important since they are contained in an ever-growing number of natural products (Rousseau, 1995; Collins, 1999), and their preparation becomes of great interest.

Trans-fused bicyclic γ-lactones (**7**) have been conveniently prepared, in a one-pot process, from conjugated nitroalkenes, by way of the expected transesterification–intramolecular hetero Diels–Alder (HAD) reaction, followed by the transformation of a functional group using (*E*)-1-ethoxy-2-nitroethylene (**5**) and *prim-*, *sec-*, and *tert-*γ,ε-unsaturated alcohols (**6**) having two methyl substituents at the terminal position as illustrated in Figure 2.2 (Wada and Yoshinaga, 2003).

This reaction proceeds in one flask as follows: (i) conjugate addition of unsaturated alcohols (**6**) to β-alkoxy-substituted nitroalkene (**5**), (ii) reversible elimination of ethanol, with the formation of the intermediate (**8**), (iii) intramolecular HAD reaction of the resulting transetherified compound leading the bicyclic nitronate (**9**), and (iv) further transformation of (**9**) to bicyclic lactones (**7**) as outlined in Figure 2.2.

Several triflates and metal salt hydrates were tested as Lewis acid catalysts (each 10 mol%) and the best results were obtained using Yb(OTf)$_3$ or Ni(ClO$_4$)$_2$·6H$_2$O. The stereoselective formation of bicyclic γ-lactones (**7**) could be obtained from bicyclic nitronate (**9**) via the hydrolytic process by the action of a strong acid generated from Lewis acid with a small amount of water in both nitroalkene (**5**) and the Lewis acid. This new methodology of one-pot reaction also involves a new type of intramolecular HAD reaction of nitroalkenes as heterodienes, which provides stereochemically defined bicyclic nitronates.

2.2.3   One-Pot Synthesis of Five- or Six-Membered Carbocycles

Five- or six-membered carbocycles are useful building blocks of relevant practical interest that provide an efficient entry to an important framework. These structures have been conveniently obtained, in a one-pot process, by the reaction of β-nitrostirenes with Grignard reagents (Hung et al., 1999). Thus, as reported in

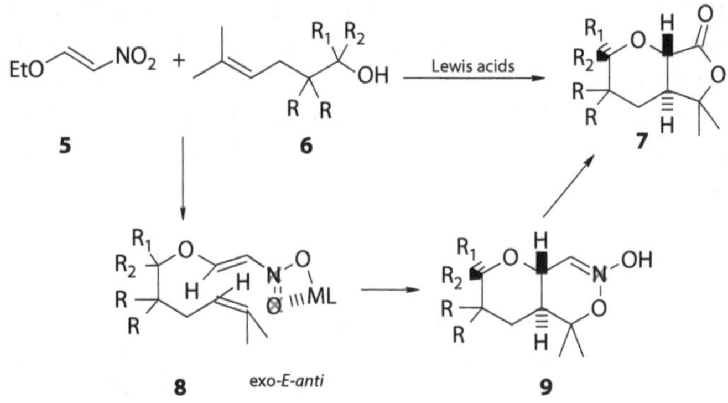

FIGURE 2.2

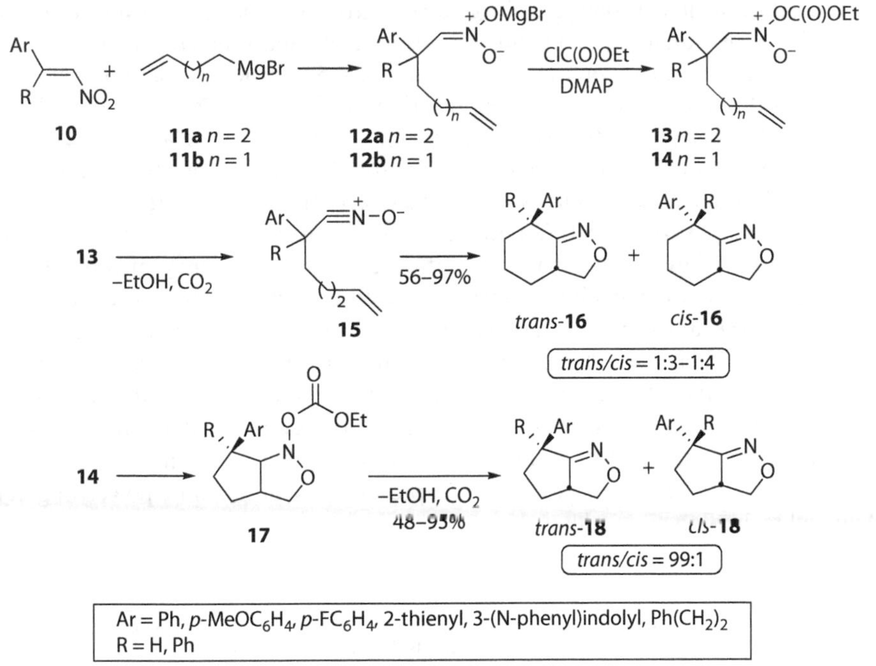

FIGURE 2.3

Figure 2.3, the Michael addition reactions of β-nitrostirenes (**10**) with 4-pentene-1-magnesium bromide (**11a**) or 3-butene-1-magnesium bromide (**11b**) generated the nitronates (**12**) or (**13**). Satisfactory to high yields of isoxazolidine derivatives (**16**) and (**18**) were obtained when nitronates (**12**) or (**13**) were treated, *in situ*, with ethyl chloroformate and catalytic amount of 4-dimethylaminopyridine (DMAP).

The formation of compound (**16**) is proposed to proceed through the nitrile oxide (**15**) as an intermediate to undergo intramolecular nitrile oxide-olefin cycloaddition (INOC). On the contrary, the generation of (**18**) is proposed to proceed through intramolecular alkoxycarbonyl nitronate-olefin cycloaddition (IAOC) step because only the *trans* isomers are formed.

2.2.4 ONE-POT SYNTHESIS OF 3-NITRO-1,2-DIHYDROQUINOLINES AND 3-NITROCHROMENES

Quinolines are widely used heterocyclic compounds in organic chemistry (Rees and Smithen, 1964; Popp, 1968), while 3-nitrochromenes are another important class of heterocyclic compounds due to their biological activity (Bianco et al., 1977) and their importance as precursors of flavonoids (Rao et al., 1984), amines (Booth et al., 1973), and other important targets (Adrieu et al., 1976; Aramaki et al., 1993).

Considering that α,β-unsaturated nitroalkenes are easily available starting material, it has been reported as a unique methodology for the efficient synthesis of both 3-nitro-1,2-dihydroquinolines and 3-nitrochromenes through nitroalkenes.

Thus, the reaction of conjugate nitroalkenes with 2-aminobenzaldehyde or salicyl-aldehyde using γ-alumina as a common, cheap, and neutral heterogeneous catalyst under solventless conditions (Ballini et al., 2005b) proceeds as illustrated in Figure 2.4 through the hetero Michael reaction of the amino (X = NH) or hydroxy (X = O) aldehydes (19) to the conjugated alkenes (20), giving the adducts (21). The latter are prone to give an intramolecular nitroaldol (Henry) reaction leading to the nitro cyclohexenols (22) which, by *in situ* water elimination, allow the one-pot synthesis of the target compounds (23) in satisfactory to good overall yields (55–85%).

The nature of nitroolefins has little influence on the efficiency of this one-pot process, and other functionalities such as cyano, ether, and chlorine can be preserved. The role of the nitro group is crucial in the reported process since its electron-withdrawing effect firstly helps the nucleophilic attack of the amino or hydroxyl functionality to the alkene, then it allows a stabilized carbanion with the consequent formation of an intramolecular C–C bond (nitroaldol reaction) and, finally, favors the elimination of water (E1cB). Moreover, the use of solvent-free conditions (SFC) in combination with heterogeneous catalyst represents one of the more powerful green chemical technology procedures.

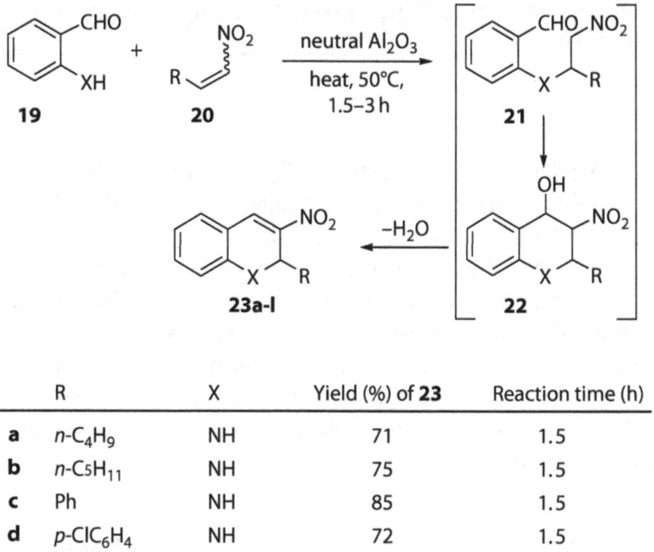

	R	X	Yield (%) of 23	Reaction time (h)
a	n-C_4H_9	NH	71	1.5
b	n-C_5H_{11}	NH	75	1.5
c	Ph	NH	85	1.5
d	p-ClC_6H_4	NH	72	1.5
e	c-C_6H_{11}	NH	55	1.5
f	$Ph(CH_2)_2$	NH	60	1.5
g	Ph	O	83	3
h	n-C_4H_9	O	81	3
i	n-C_5H_{11}	O	82	3
j	$Ph(CH_2)_2$	O	75	3
k	p-CNC_6H_4	O	72	3
l	p-$MeOC_6H_4$	O	72	3

FIGURE 2.4

2.2.5 One-Pot Synthesis of 1,7-Dioxaspiro[5.6]alkane Derivatives

Two important spiroketals, such as 1,7-dioxaspiro[5.6]undecane (**29a**), the major component of the olive fruit fly (*Dacus oleae*) sex pheromone (Baker et al., 1980; Fanelli et al., 1983), and (*E*)-2-methyl-1,7-dioxaspiro[5.6]dodecane (**29b**), a component of pheromone of *Andrena haemorrhoa* (Bergstroem et al., 1981; Katsurada and Mori, 1984), have been prepared in a cascade process from polyfunctionalized nitroalkanes (**26**) (Ballini and Petrini, 1992).

As reported in Figure 2.5, nitroolefins (**26**), easily obtained by nitroaldol condensation between δ-nitro ketones (**24**) and aldehydes (**25**), are converted directly into the spiroketals (**29**) by reduction with sodium boronhydride in methanol. The one-pot reduction–spiroketalization of nitroalkenes (**26**) probably proceeds via the nitronate (**27**) that by acidification is converted into carbonyl derivatives, which spontaneously cyclize to emiketals (**28**). Removal of the tetrahydropyranyl group, by heating the acidic mixture during the workup, affords, in a one-pot reaction from (**26**), the desired spiroketals in 64–66% overall yields. The spiroketalization of (**26**) (**29b**) proceeds in high (*E*) diastereoselectivity.

It is important to note that six different chemical transformations occur in a one-pot process: (i) reduction of C–C double bond of (**26**), (ii) reduction of carbonyl to alcohol [(**26**)–(**27**)], (iii) conversion of the nitro into carbonyl (Nef reaction), (iv) emiketalization [(**27**)–(**28**)], (v) removal of tetrahydropyranyl, and (vi) ketalization with the formation of (**29**).

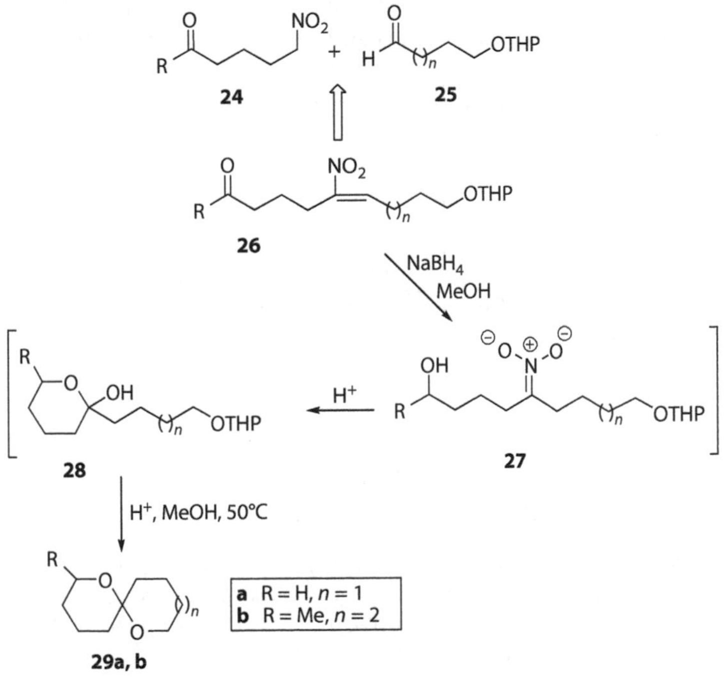

FIGURE 2.5

2.3 NITROALKANES AS BUILDING BLOCKS FOR ONE-POT PROCESSES, VIA NITROALDOL (HENRY) REACTION

Nitroalkanes are useful precursors in the formation of stabilized carbanions (nitronate), which could be reacted with a variety of carbonyl compound, mainly aldehydes, allowing the formation of β-nitroalkanols, with the consequent generation of a new C–C bond (Rosini, 1991). This reaction could represent the key starting step for one-pot processes.

2.3.1 ONE-POT SYNTHESIS OF 1,3-DINITROALKANES

1,3-Dinitroalkanes are powerful molecules due to their use as precursors of different targets such as 1,3-difunctionalized molecules, heterocycles (Alcantara et al., 1988), carbohydrate derivatives (Pham-Huu et al., 1999), and energetic materials (Carrol et al., 1987; Axenrod et al., 1995; Archibald et al., 1995; Dong, 1998).

1,3-Dinitro compounds are usually obtained through the conjugate addition of nitroalkanes to pre-prepared nitroalkenes (Alcantara et al., 1988, 1996; Pham-Huu et al., 1999), but the nitroolefins syntheses are often intricate, working in poor yields, due to their high reactivity and easy conversion into dimeric or polymeric derivatives; moreover, the synthesis of nitroalkenes needs two more steps: (i) nitroaldol reaction followed by (ii) dehydration of the obtained nitroalkanol (Barrett and Graboski, 1986). On the basis of previous experience on the use of basic alumina for the generation of new C–C bond starting from nitroalkanes (Ballini et al., 1985, 1986), it has been planned that the one-pot synthesis of 1,3-dinitroalkanes could be by using basic alumina as solid catalyst. In fact, the reaction of an aldehyde (30) with excess nitromethane as shown in Figure 2.6 (Ballini et al., 2004), in the presence of basic alumina and under reflux, provides the direct formation of 1,3-dinitroalkanes (33).

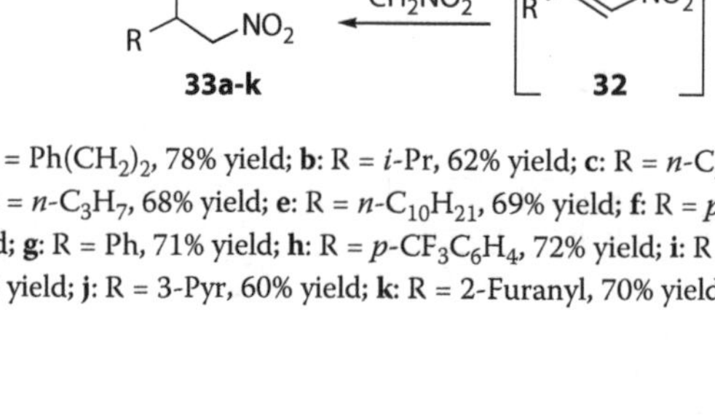

a: R = Ph(CH$_2$)$_2$, 78% yield; **b:** R = i-Pr, 62% yield; **c:** R = n-C$_5$H$_{11}$, yield 65%, **d:** R = n-C$_3$H$_7$, 68% yield; **e:** R = n-C$_{10}$H$_{21}$, 69% yield; **f:** R = p-NO$_2$C$_6$H$_4$, 60% yield; **g:** R = Ph, 71% yield; **h:** R = p-CF$_3$C$_6$H$_4$, 72% yield; **i:** R = p-MeOC$_6$H$_4$, 74% yield; **j:** R = 3-Pyr, 60% yield; **k:** R = 2-Furanyl, 70% yield.

FIGURE 2.6

The reaction originates from the nitroaldol reaction of nitromethane, which acts both as nucleophile and as solvent, with aldehydes (30), with the formation of β-nitroalkanol (31) as intermediate that converts into nitroalkene (32). Then, the carbanion derived from nitromethane gives the conjugate addition to the electron-poor alkene (32) giving the one-pot synthesis of (33). It is important to point out that the equilibrium step of conversion of (31) into (32) and the successive one for the generation of the nitroalkene (32) are both strongly helped by the *in situ* trapping of (32) with nitromethane, avoiding any possible decomposition of the formed conjugate nitroolefin.

The target compounds (33) were obtained in good yields from a variety of aliphatic, aromatic, and heteroaromatic aldehydes. Owing to mild reaction conditions, other functionalities such as ethers, trifluoromethyl, and heterocycles can be preserved.

2.3.2 ONE-POT SYNTHESIS OF 1,3-DIOLS

1,3-Diols are of potential utility for organic synthesis (Harada et al., 1987; Feldman, 1995; Patai, 1980). Moreover, polyfunctionalized 1,3-diols are of high interest, especially those functionalized in 2-position (Rieger, 1992; Majewski et al., 1995; Darabantu et al., 1995), and their preparation usually needs a long sequence of steps. Nitrocycloalkanones (34) can be conveniently used as precursors of polyfunctionalized 1,3-diol (35), a new interesting class of diols, through their one-pot double nitroaldol reaction-ring cleavage depicted in Figure 2.7 (Ballini et al., 1997).

Thus, reaction of (34) with a 30% aqueous solution of formaldehyde proceeds at room temperature, in the presence of potassium carbonate, allowing (35), in respectable yields, regardless of the ring size of cycloalkanones (34). Probably, the synthesis involves, firstly, the nitroaldol reaction of (34) with formaldehydes, followed by ring cleavage of the formed nitroalkanol promoted by water, then further nitroaldol reaction with a further molecule of formaldehyde.

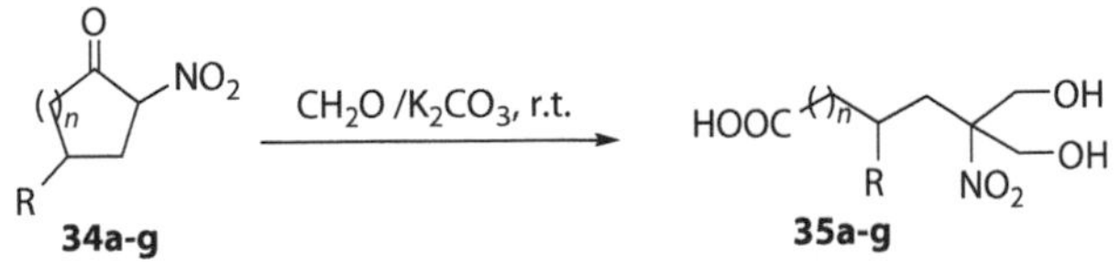

	n	R	Yield (%) of 35
a	1	H	60
b	2	H	75
c	3	H	60
d	7	H	50
e	11	H	48
f	2	Me	65
g	2	t-Bu	63

FIGURE 2.7

2.4   NITROALKANES AS BUILDING BLOCKS FOR ONE-POT PROCESSES, VIA MICHAEL REACTION

Nitroalkanes are a good source of carbanions for the conjugate addition to electron-poor alkenes (Ballini et al., 2005a), providing polyfunctionalized adducts that still retain the nitro group, and therefore, a suitable transformation of the nitro group very often follows the main addition process.

2.4.1   ONE-POT SYNTHESIS OF 1,4-DIFUNCTIONALIZED DERIVATIVES

The synergy between nitroalkanes and aqueous medium has been the key idea for the one-pot synthesis of δ-nitroalkanols (**38**) as shown in Figure 2.8, hydroxytetrahydrofurans (lactols) (**39**), 1,4-diketones (**40**), and 1,4-diols (**41**).

These compounds were obtained starting from common materials and using aqueous medium (Ballini et al., 2003a). Thus, the preparation of δ-nitroalkanols (**38**) was carried out in good yields (78–94%, as 1:1 diastereomeric mixture) by the conjugate addition of nitroalkanes (**36**) to unsaturated ketones (**37**), under aqueous K_2CO_3, followed by *in situ* reduction with $NaBH_4$ shown in Figure 2.9. The latter was chosen because it can be used in water and does not reduce the nitro group.

Lactols (**39**) have been synthesized in a one-pot process by *in situ* Nef reaction of the corresponding δ-nitroalkanols (**38**), obtained as reported above, by oxidation with 30% water solution of H_2O_2 as depicted in Figure 2.10.

The conversion of the nitro group into a carbonyl can be accomplished by several alternative methods although the use of H_2O_2/K_2CO_3 appeared to be the most compatible to develop environment-friendly processes. As reported in Figure 2.10, the γ-hydroxyketones could not be isolated, since they spontaneously cyclized to the corresponding lactols (**39**).

Owing to good results for the synthesis of lactols (**39**), the use of H_2O_2 for the one-pot preparation of 1,4-diketones (**40**) has been further investigated. Thus, after the addition of nitroalkanes (**36**) to enones (**37**), as shown in Figure 2.11, the mixture was treated with 30% water solution of H_2O_2, giving the corresponding diketones (**40**) in good yields.

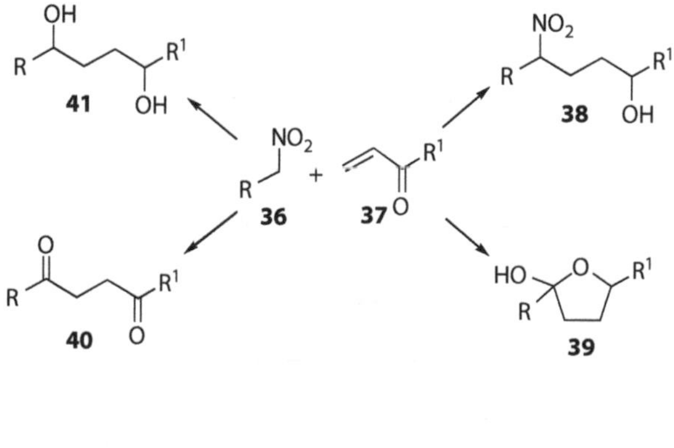

FIGURE 2.8

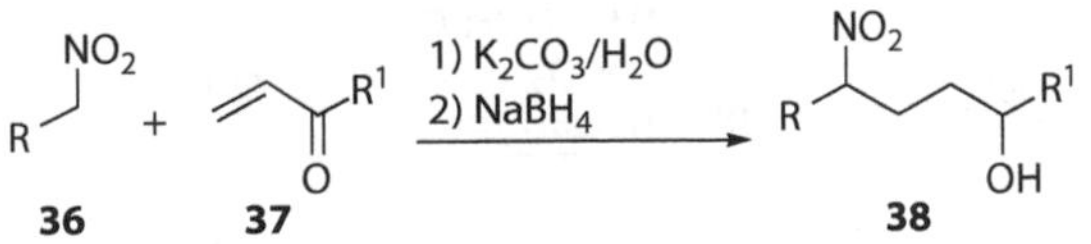

	R	R^1	Yield (%) of **38**
a	Me	Me	82
b	Et	Me	86
c	n-Pr	Me	83
d	Me	Et	94
e	Et	Et	82
f	n-Pr	Et	78
g	Ph	Et	79

FIGURE 2.9

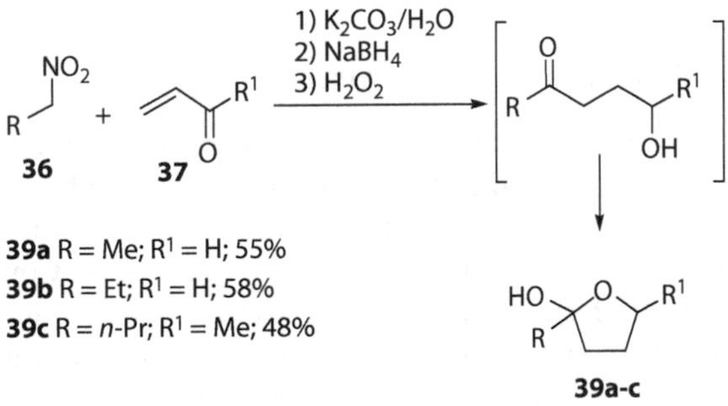

39a R = Me; R^1 = H; 55%
39b R = Et; R^1 = H; 58%
39c R = n-Pr; R^1 = Me; 48%

FIGURE 2.10

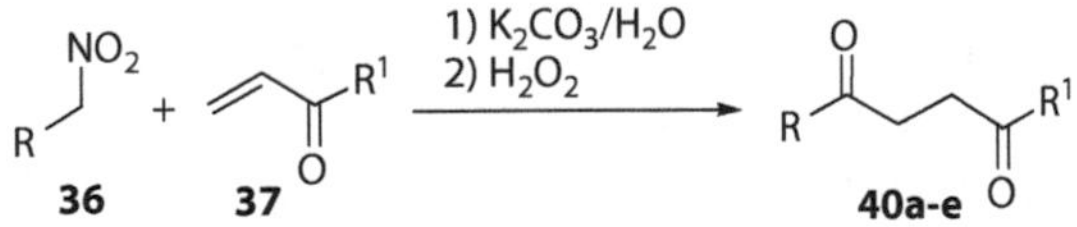

	R	R^1	Yield (%) of **40**
a	Me	Me	73
b	Me	Et	68
c	Et	Et	72
d	n-Pr	Et	70
e	Ph	Me	69

FIGURE 2.11

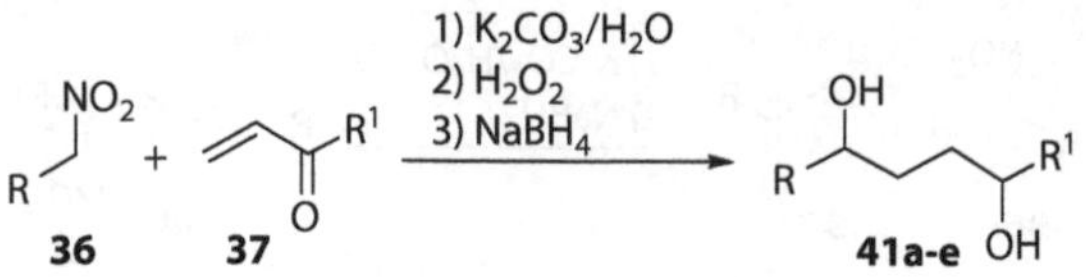

	R	R^1	Yield (%)^a of **41**
a	Me	Me	60
b	Et	Et	64
c	n-Pr	Et	61
d	Ph	Me	72
e	n-Bu	Me	60

^{a}The reported yields refer to the 1:1 diastereomeric inseparable mixture of diols

FIGURE 2.12

Finally, 1,4-diols (**41**) were prepared by *in situ* reduction of the 1,4-diketones (**40**) obtained as reported above, using an excess of NaBH$_4$ as reducing agent as shown in Figure 2.12.

2.4.2 ONE-POT SYNTHESIS OF ALLYLRETHRONE

Cyclopentenones are important moieties present in many natural products such as ally-lrethrone (**45**), which is an important component of insecticidal pyrethroid and an important intermediate for the synthesis of allethrolone and pyrethrins (Crombie et al., 1950; Berger et al., 1970; Cheng et al., 1982; Assereg et al., 1987). Commercial 5-nitro-1-pentene (**42**) has been chosen as the starting material for a new, improved synthetic approach for the preparation of allylrethrone, following two different ways: (i) multi-step sequence and (ii) one-pot process as shown in Figure 2.13 (Ballini, 1993).

The starting point of the multistep approach is the conjugate addition of (**42**) to methyl vinyl ketone (MVK), under basic alumina as solid catalyst, giving the

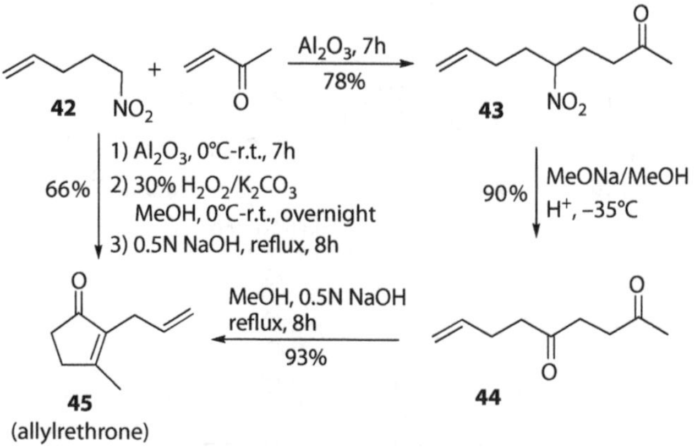

FIGURE 2.13

γ-nitro ketone (**43**), in 78% yield. The latter can be converted into the 1,4-diketone (**44**), in 90% yield, by the Nef reaction performed by the addition of the corresponding nitronate to a mixture of methanol and concentrated sulfuric acid at −35°C. Subsequent basic intramolecular aldolization–dehydration of (**44**) affords (**45**) in 65% overall yield. Alternatively, allylrethore can be obtained in a one-pot process by the conjugate addition of 5-nitro-1-pentene to MVK on alumina, followed by *in situ* oxidation with hydrogen peroxide in methanol (Nef reaction), and then *in situ* basic cyclization with 0.5N sodium hydroxide giving (**45**) in 66% overall yield. The one-pot process is strongly favored both by the choice of neat alumina as catalyst for the first step and by the versatility of the nitro group that serves both to generate a carbanion in the Michael reaction and as precursor of the carbonyl.

2.4.3 One-Pot Synthesis of Polyfunctionalized Nitrocyclopropanes

Nitrocyclopropanes are strained-ring nitro compounds of special attraction in the field of energetic materials because of the increased performance expected from the additional energy release upon opening the strained-ring system during decomposition. Moreover, nitrocyclopropane moiety is the key structure of important natural products or the key building block for other important targets (Ballini et al., 2007a).

The preparation of polyfunctionalized nitrocyclopropanes is of special interest and, recently, these molecules have been prepared, in a one-pot process, starting from bromonitromethane (**46**) and electrophilic alkenes (**47**) bearing two electron-withdrawing groups in α- and β-positions as shown in Figure 2.14 (Ballini et al., 2003b).

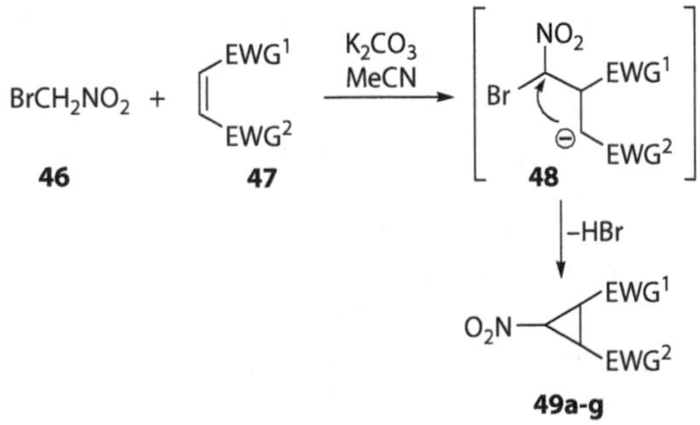

	EWG^1	EWG^2	Yield (%) of **49**	Diastereomeric ratio of **49**
a	MeCO	MeCO	82	23:77
b	PhCO	MeCO	86	88:12
c	p-MeC$_6$H$_4$CO	MeCO	83	84:16
d	PhCO	PhCO	94	15:85
e	COOMe	MeCO	82	69:31
f	COOMe	COOMe	78	40:60
g	CN	CN	79	70:30

FIGURE 2.14

The reaction proceeds via a "tandem Michael addition-HBr elimination" in which the carbanion (**48**) is the key intermediate to produce cyclopropanation. The success of the method depends from the correct addition of (**46**) to the basic solution of the alkenes (**47**). In fact, it is important to add bromonitromethane in several fractions in order to avoid its decomposition (that can happen under basic conditions). The cyclopropanation proceeds in good to excellent yields (75–96%) with a variety of substrates, but with moderate diastereoselectivity. The procedure works well even with cyclic alkenes, such as *N*-alkylmaleimides, with the exclusive formation of exo-cyclopropane derivatives.

2.4.4 One-Pot Synthesis of Acetophenone and Methyl Benzoate Derivatives

Aromatic compounds are important reagents in organic synthesis, in particular 3,5-alkylated acetophenones and methyl benzoate derivatives, which represent the key building blocks in the synthesis of a variety of molecules having important biological activities (Steinbaugh et al., 1996; Connolly et al., 1997; Zhang et al., 1996, 2003). Aromatization of acyclic precursors is undoubtedly a useful reaction in the synthesis of highly substituted aromatic rings (Bamfied and Gordon, 1984; Bradsher, 1987), and, in this context, 1,3-dinitroalkanes (**33**) have demonstrated to be the key starting materials for the synthesis of both acetophenone and methyl benzoate derivatives in a one-pot process (Ballini et al., 2005c).

In fact, the reaction of (**33**) with ene-dione (**50**), in the presence of diaza(1,3) bicyclo[5.4.0]undecane (DBU) as base, proceeds as depicted in Figure 2.15 as a tandem process in which a regioselective Michael addition (yielding **51**) is presumably followed by the elimination of nitrous acid, to give the corresponding nitro-enone (**52**).

The latter compounds are prone to an intramolecular nitroaldol (Henry) reaction, yielding nitrocyclohexanola (**53**) in less than 1 h. The formation of (**53**) can be easily observed by thin-layer chromatography (TLC), thus treatment of (**53**) with 4N hydrochloric acid favors the elimination of both water and a further molecule of nitrous acid, allowing the one-pot synthesis of target molecules (**54**) in 42–77% overall yields. It is important to note that this one-pot process formally includes five different transformations: (i) Michael addition, (ii) nitrous acid elimination, (iii) intramolecular nitroaldol reaction, (iv) water elimination, and (v) elimination of a further molecule of nitrous acid.

2.4.5 One-Pot, Diastereoselective Synthesis of Polyfunctionalized Bicyclo[3.3.1]nonanes

Polycarbocyclic compounds occupy a unique *niche* in the annals of synthetic organic chemistry and, among these, the chemistry of bicyclo[3.3.1]nonanes has received much attention from both synthetic and theoretical points of view (Peters, 1979; Butkus, 2001). The syntheses of bicyclononanes presently described in literature, mainly start from preformed cyclic precursors, while little attention has been devoted

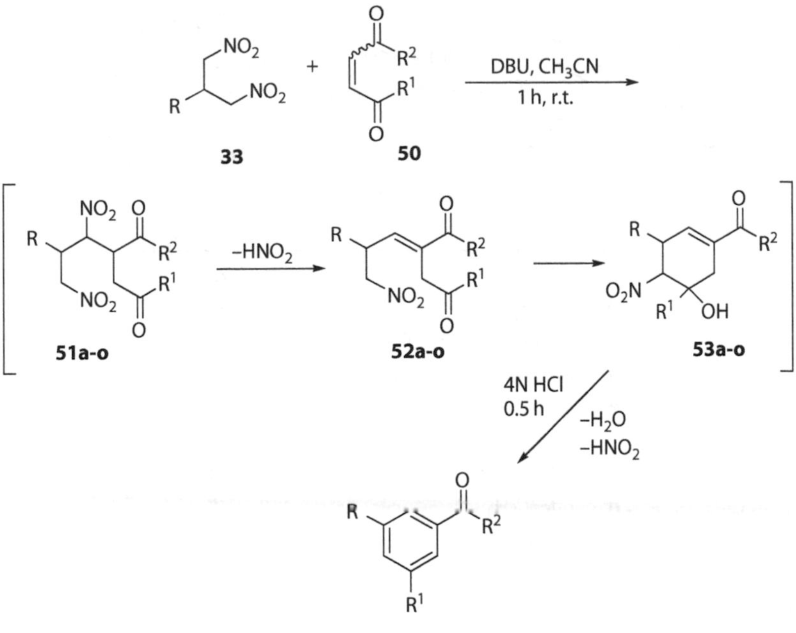

FIGURE 2.15

	R	R^1	R^2	Yield (%) of **54**
a	$CH_3(CH_2)_4$	Me	Me	55
b	$CH_3(CH_2)_4$	Ph	Me	62
c	$CH_3(CH_2)_7$	Me	Me	58
d	$Ph(CH_2)_2$	Me	Me	57
e	$CH_3(CH2)_7$	Ph	Me	61
f	$CH_3(CH_2)_4$	Me	MeO	60
g	$Ph(CH_2)_2$	Me	MeO	59
h	m-$CF_3C_6H_4$	Me	Me	58
i	p-$MeOC_6H_4$	Me	Me	61
j	m-$CF_3C_6H_4$	Ph	MeO	60
k	p-$MeOC_6H_4$	Me	MeO	77
l	p-$MeOC_6H_4$	Me	Me	66
m	m-$NO_2C_6H_4$	Me	MeO	42
n	Ph	Me	MeO	76
o	2-Py	Me	MeO	43

to the cyclization of acyclic precursors. 1,3-Dinitroalkanes are an emerging class of "bidentate" nitro derivatives, easily prepared from nitromethane and aldehydes as illustrated in Figure 2.16, able to afford double nucleophilic additions to electrophilic derivatives. This peculiarity prompted to explore an innovatory approach for the preparation of bicyclo[3.3.1]nonanes from 1,3-dinitroalkanes, with the aim to develop

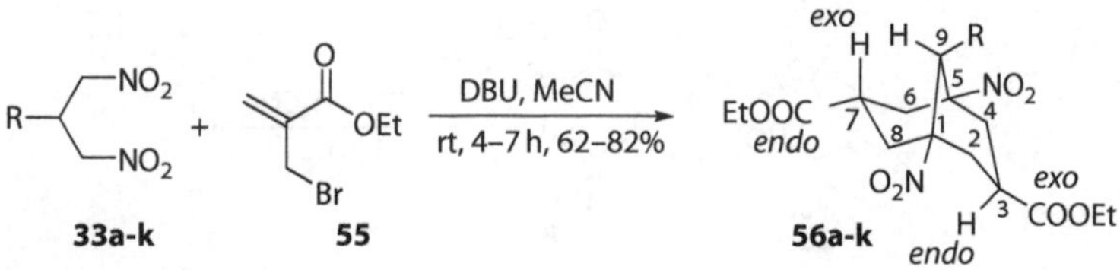

FIGURE 2.16

	R	Yield (%) of **56**
a	n-Pr	78
b	Ph(CH$_2$)$_2$	62
c	(Z)-Me(CH$_2$)4CH = CH(CH$_2$)$_2$	66
d	c-C$_6$H$_{11}$	73
e	n-C$_5$H$_{11}$	76
f	Ph	82
g	p-MeOC$_6$H$_4$	80
h	p-MeC6H$_4$	79
i	p-PhC$_6$H$_4$	76
j	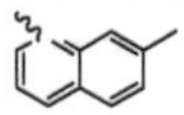	64
k	Pyr	70

a highly efficient and ecosustainable synthetic process. Thus, the synthesis has been performed (Barboni et al., 2009) simply by adding the acrylate (**55**, 2 equiv.) to a solution of dinitroalkane (**33**, 1 equiv.), under basic conditions (DBU), in acetonitrile and at room temperature and the reaction is shown in Figure 2.16.

The multistep process occurs with good overall yields (62–82%) of bicyclo[3.3.1] nonanes (**56**), in a one-pot way and under short reaction times (4–7 h), via an anionic domino process. The method works well with a variety of dinitroalkanes, including those possessing other functionalities such as (Z)–C,C double bond, ether and heterocyclic. Most of the bicyclononane derivatives were obtained exclusively as *exo*-3, *endo*-7 diastereoisomers. A plausible mechanism for the formation of (**56**) is reported in Figure 2.17 in which a double Michael reaction of the dinitrocompound (**33**) with two equivalents of (**55**), followed by the elimination of two molecules of bromidric acid, allows the intermediate **A**, which is prone to give a further double internal conjugate addition to the formed electron-poor alkenes, allowing the one-pot synthesis of target compounds (**56**) via an anionic domino process.

In addition, a structurally unique *bis*-bicyclic compound (**58**) can be obtained, one-pot, starting from tetranitro derivative (**57**) and four equivalent of ethyl(2-bromomethyl)acrylate (**55**), in 34% overall yield. The latter reaction is shown in Figure 2.18.

It is important to note that, since the 1,3-dinitroalkanes used as the key starting materials for the present synthesis were in turn prepared in a one-pot way from nitromethane and the appropriate aldehyde (Ballini et al., 2004), the polyfunctionalized

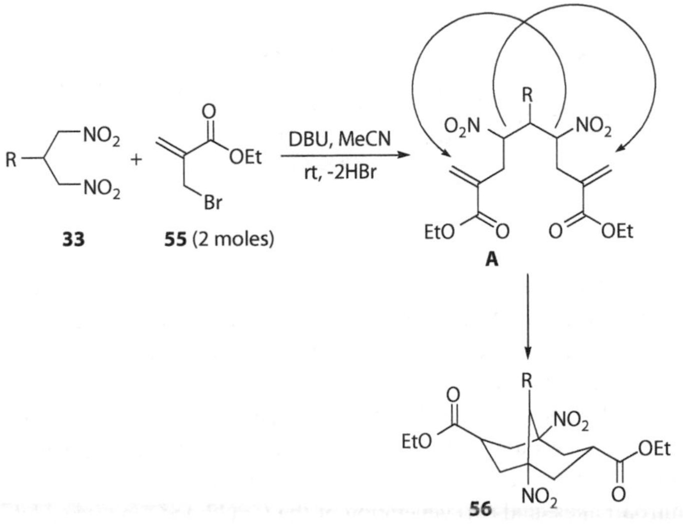

33 **55** (2 moles) **56**

FIGURE 2.17

bicyclo[3.3.1]nonanes (**56**) can actually be obtained from a very simple molecule such as nitromethane, just by two "one-pot" sequences, which are illustrated in Figure 2.19.

Thus, the synthetic process for the preparation of (**56**) represents a highly eco-friendly synthesis of bicyclononanes, in which at least four new C–C bonds are formed in a one-pot way (**33–56**), in very good overall yields and minimum consumption of energy (room temperature).

2.4.6 One-Pot Process for Both Nitroaldol (Henry) and Michael Reactions from Primary Nitroalkanes

Nitroalkanes are the key starting precursors for the formation of C–C bond by reactions with aldehydes (nitroaldol-Henry reaction) and electron-poor alkenes (Michael

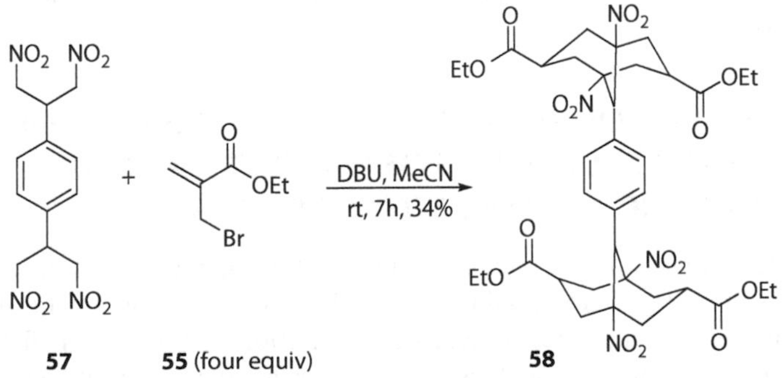

57 **55** (four equiv) **58**

FIGURE 2.18

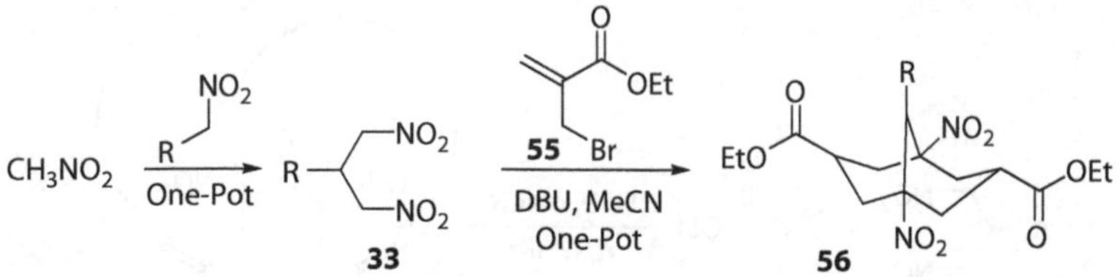

FIGURE 2.19

reaction). Usually, nitroalkanes must be prepared in a separate process and, in this context, direct nitrations of alkyl halides with metal nitrite remain the most used method (Kornblum, 1962; Feuer and Leston, 1963). Generally, long reaction times and tedious workup are required, and low yields are obtained. Moreover, a further and serious drawback is that the products obtained are usually a mixture, difficult to purify, of the desired nitroalkanes together with the undesired alkyl nitrite. Therefore, the synthetic sequence, (i) halo derivative conversion into nitroalkanes, (ii) nitronate generation from nitroalkanes, and (iii) generation of the C–C bond via nitroaldol of Michael reaction, suffers dramatically from the serious drawbacks derived from the first step.

Recently, a new one-pot process for the sequence (i)–(iii) has been realized under heterogeneous catalysis (Ballini et al., 2007b). In fact, treating at room temperature according to the reaction in Figure 2.20 one equivalent of primary haloalkanes (**57**) and one equivalent of aldehyde, or conjugate enone, in the presence of IRA-402 nitrite and Amberlyst A-21, the nitroalkanols (**60**) or γ-nitro ketones (**61**) were obtained, respectively, in a one-pot S_N2-nitroaldol (Henry) or S_N2-Michael reactions.

The conversion of (**57**) into (**60**) or (**61**) starts with the halogen/nitro S_N2 substitution from (**57**) to (**58**) promoted by IRA-402 nitrite, followed by in situ generation of the nitronate (**59**) under Amberlyst A-21 catalysis, then nucleophilic addition of (**59**) to the electrophilic substrates.

The overall yields are favored by the one-pot procedure, which drives the process to the formation of the target compounds (**60**) and (**61**). The nitroalkanols (**60**) are obtained as a diastereomeric mixture. Nevertheless, this seems not to be a problem since the main uses of nitroalkanols are the conversion into α-nitro ketones (Ballini et al., 2005d) or conjugated nitroalkenes (Berestovitaskaya et al., 1994; Barrett and Graboski, 1986; Kabalka and Varma, 1987). In addition, the process is mild and avoids complicated workup and complex purification steps with evident waste and energy consumption.

2.5 NITROCOMPOUNDS AS KEY BUILDING BLOCKS FOR THE ONE-POT SYNTHESIS OF COMPLEX MOLECULES CONTAINING MULTIPLE STEREOCENTERS

Asymmetric synthesis of complex molecules containing multiple stereocenters is a very important research area used for the synthesis of a variety of natural products and other molecules of interest (Nicolaou et al., 2003, 2006). In this field, a strategy that reduces the number of chemical steps for the final assembly on the desired products is highly demanded, thereby decreasing the formation of waste products

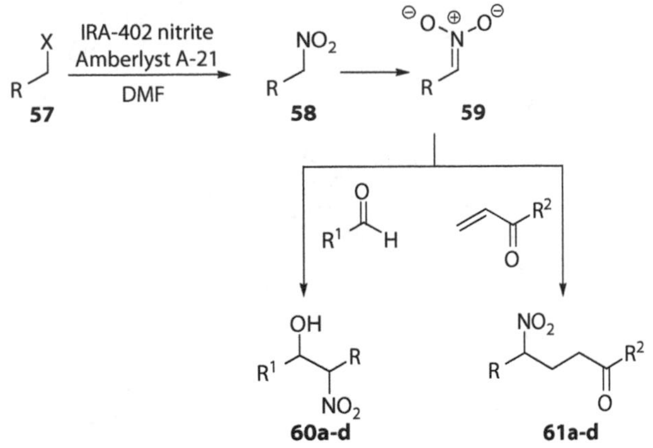

FIGURE 2.20

	R	R¹	R²	Yield (%)
60a	*n* Bu	*n*-C₅H₁₁		63
60a	*n*-Bu	*n*-C₅H₁₁		65
60b	*n*-Bu	Ph(CH₂)₂		62
60b	*n*-Bu	Ph(CH₂)₂		67
60c	PhthN(CH₂)₃	Et		61
60d	Ph(CH₂)₂	*c*-C₆H₁₁		60
61a	*n*-Bu		Me	56
61a	*n*-Bu		Me	55
61b	AcO(CH₂)₃		Me	53
61c	*n*-C₆H₁₃		Et	55
61d	Ph(CH₂)₂		Me	54

and increasing the economy of the process. One of the most important goals in this area is the generation of more than one bond in a multistep reaction concomitant with the creation of multiple stereocenters in a one-pot fashion. In this context, aliphatic nitrocompounds have been successfully employed, by the help of organocatalysis, in the formation of complex organic molecules with multiple stereogenic centers.

2.5.1 ASYMMETRIC ORGANOCATALYTIC SYNTHESIS OF PENTASUBSTITUTED CYCLOHEXANES

1,3-Dinitroalkanes (**33**) were recently used in combination with unsaturated aldehydes (**62**) for the organocatalytic synthesis of pentasubstituted cyclohexanes (**63**) by an asymmetric domino Michael/Henry process, which is shown in Figure 2.21 (Reyes et al., 2007).

The reaction mechanism has been explained by Reyes et al. (2007), as initial iminium activation of the unsaturated aldehyde (**62**) by pyrrolidine (**64**), followed by the

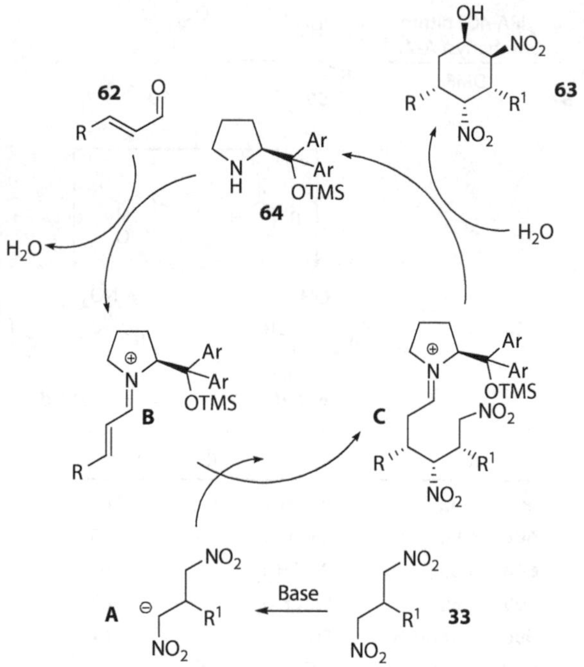

FIGURE 2.21

attack of the nitronate anion **A** from the less hindered face to give the Michael adduct **C**. After hydrolysis of the iminium ion, the nitro aldehyde intermediate undergoes a subsequent intramolecular Henry reaction, which affords the final product (**63**) with five contiguous stereocenters. The best reaction conditions were obtained at room temperature, using 20 mol% of (S)-2-[bis(3,5-bistrifluoromethylphenyl)(trimethyl-silyloxymethyl)]pyrrolidine (**64**), in the presence of 10 mol% of 1,4-diazabicyclo[2.2.2] octane (DABCO), and using dichloromethane as solvent. Although the process affords three diastereoisomers in most of the examples, the major isomer was isolated in good overall yields (38–61%) and with very considerable enantioselectivities (80–94% ee). The absolute configuration for the main stereoisomer was assigned by single-crystal x-ray analysis, and leads to a 1R,2S,3R,4R,5R assignment of the formed stereocenters.

2.5.2 CONTROL OF FOUR STEREOCENTERS IN A TRIPLE CASCADE ORGANOCATALYTIC REACTION

Another important one-pot process for the synthesis of chiral polysubstituted cyclo-hexenes, with the control of four stereocenters, involves the use of nitro compounds and it has been quite recently described by Enders et al. (2006, 2007). In particular, nitroalkenes (**65**) were used in combination with linear aldehydes (**66**) and α,β-unsaturated aldehydes (**62**) to give, by a three-component reaction, the target products (**68**). The reaction is shown in Figure 2.22.

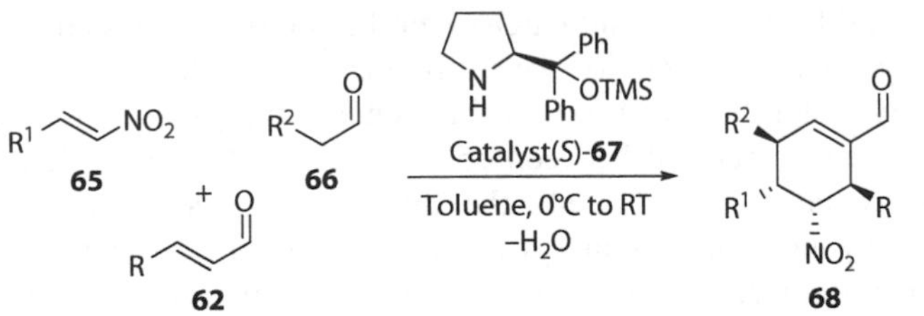

FIGURE 2.22

The process mechanism as shown in Figure 2.23 consists of an initial activation of the aldehyde (**66**) by the catalyst [(*S*)-**67**] with the formation of the corresponding chiral enamine, which then, selectively, adds to nitroalkene (**65**) in a Michael-type reaction. The following hydrolysis liberates the catalyst, which forms the iminium ion of the α,β-unsaturated aldehyde (**62**) to accomplish the conjugate addition with the nitroalkane **A**. In the third step, another enamine activation of the intermediate **B** leads to an intramolecular aldol condensation via **C**. Finally, the hydrolysis of it returns the catalyst and releases the desired chiral *tetra*-substituted cyclohexene carbaldehyde (**68**).

The optimized conditions require the use of toluene as solvent between 0°C and room temperature, nearly stoichiometric amounts of all three components

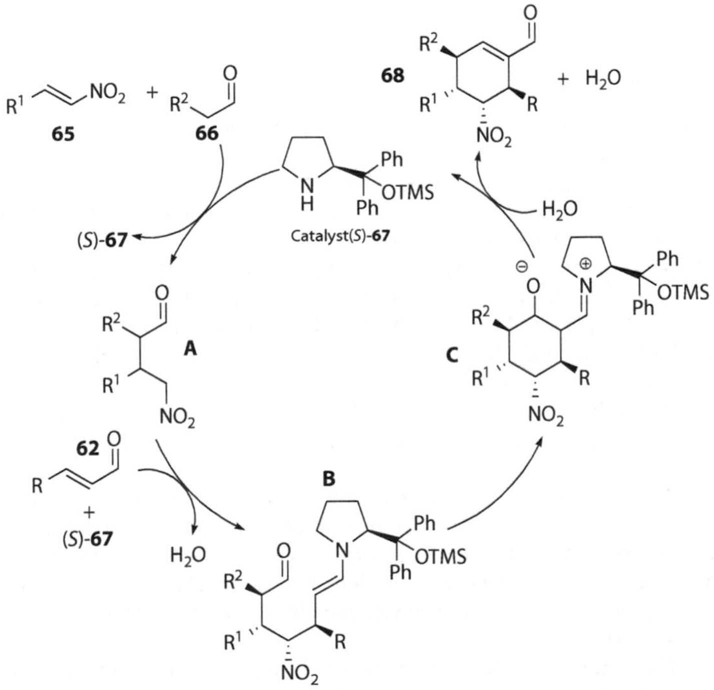

FIGURE 2.23

(**65**:**66**:**62** = 1.00:1.20:1.05) and although several catalysts were tested, the best results concerning yield and selectivity were obtained using 20 mol% of the OTMS-protected diphenylprolinol [(S)-**67**]. In this context, the yields of the isolated main diastereoisomer are included between 25% and 58%, the minor stereoisomer was determined as the 5-epimer of **68**, the diasteromeric ratio (determined by gas chromatography–mass spectrometry) are from good to excellent (from 6.8:3.2 to 9.9:0.1), and the diastereo- and enantiomeric excess (determined by high-performance liquid chromatography on a chiral stationary phase) are higher than 99%. Thus, this new protocol grants the generation of four new stereogenic centers in three consecutive C–C bond formations with high diastereo- and complete enantiocontrol, and opens up a simple entry to polyfunctional cyclohexene building blocks.

2.6 CONCLUSIONS

The aim of this chapter is to show some representative examples in which aliphatic nitrocompounds are the key starting materials for the synthesis of different targets in a one-pot process. Thus, the availability of a variety of nitroalkenes and nitroalkanes, their chemical versatility and their reactivity make these compounds highly powerful precursors and intermediates in environment-friendly organic synthesis.

ACKNOWLEDGMENTS

The authors acknowledge the University of Camerino and MUR-Italy for financial support.

REFERENCES

Adams, P. 2002. Nitro and related groups. *Journal of the Chemical Society, Perkin Transactions*, 1(23):2586–97.
Adrieu, L., Fatome, M., Laval, J.D., Rene, L., and Royer, R. 1976. Radioprotecting effects of Δ 3-chromenes substituted by an electron-attracting group in the 3-position. *European Journal of Medicinal Chemistry*, 11(1):81–2.
Alcantara, M.-P.D., Escribano, F.C., and Gomez-Sanchez, A. 1988. Heterocycle formation from 1,3-dinitroalkanes: A novel pyrazole synthesis. *Tetrahedron Letters*, 29(46):6001–4.
Alcantara, M.-P.D., Dianez, M.J., Escribano, F.C., Estrada, M.D., Gomez-Sanchez, A., Lopez-Castro, A., and Perez-Garrido, S. 1996. Synthesis of aliphatic 1,3-dinitro compounds. *Synthesis*, 1:64–70.
Aramaki, S., Ogawa, T., Ono, N., and Sugi, K. 1993. 3-Nitrochromenes for second order nonlinear optical applications. *Journal of the Chemical Society, Chemical Communications*, 23:1781–2.
Archibald, T.G.A., Bott, S.G., Marchand, A.P., and Rajagopal, D. 1995. Novel approach to the synthesis of 1,3,3-trinitroazetidine. *Journal of Organic Chemistry*, 60(15):4943–6.
Assereg, J.H., Glase, S.A., and Welch, S.C. 1987. 3-Chloro-2-[(diethoxyphosphoryl)oxy]-1-propene: a new reagent for a one-pot cyclopentenone annelation. Synthesis of desoxyallethrolone, *cis*-jasmone, and methylenomycin B. *Journal of Organic Chemistry*, 52(8):1440–50.
Axenrod, T., Watnick, C., and Yazdekhasti, H. 1995. Synthesis of 1,3,3-trinitroazetidine via the oxidative nitrolysis of *N-p*-tosyl-3-azetidinone oxime. *Journal of Organic Chemistry*, 60(7):1959–64.

Baker, R., Francke, W., Herbert, R., Howse, P.E., Jones, O.T., and Reith, W. 1980. Identification and synthesis of the major sex pheromone of the olive fly (*Dacus oleae*). *Journal of the Chemical Society, Chemical Communications*, 2:52–3.

Ballini, R., Marotta, E., Petrini, M., and Rosini, G. 1986. Conjugate addition of nitro derivatives to αβ-unsaturated carbonyl compounds on basic alumina. *Synthesis*, 3:237–8.

Ballini, R., Marotta, E., Petrini, M., Righi, P., and Rosini, G. 1990. Recent progress in the synthesis and reactivity of nitro ketones. A review. *Organic Preparations and Procedures International*, 22(6):707–46.

Ballini, R., Petrini, M., Rosini, G., and Sorrenti, P. 1985. A convenient synthesis of 1-(2-furyl)-2-nitroalk-1-enes on alumina surface without solvent. *Synthesis*, 5:515–7.

Ballini, R. and Rosini, G. 1988. Functionalized nitroalkanes as useful reagents for alkyl anion synthons. *Synthesis*, 11:833–47.

Ballini, R., Rosini, G., and Zanotti, V. 1983. Denitration of α-nitroketones by treatment of their tosylhydrazones with lithium aluminum hydride; new applications of the Henry reaction. *Synthesis*, 2:137–9.

Ballini, R. and Petrini, M. 1992. Hydroxy-functionalized conjugated nitro olefins as immediate precursors of spiroketals. A new synthesis of 1,7-dioxaspiro[5.5]undecane and (E)-2-methyl-1,7-dioxaspiro[5.6]dodecane. *Journal of the Chemical Society, Perkin Transactions*, 1(23):3159–60.

Ballini, R. 1993. 5-Nitro-1-pentene as a precursor for the synthesis of allylrethrone. *Synthesis*, 7:687–8.

Ballini, R., Barboni, L., and Pintucci, L. 1997. Synthesis of polyfunctionalized 1,3-diols through a one-pot double nitro aldol reaction ring cleavage of 2-nitro cycloalkanones. *Synlett*, 12:1389–90.

Ballini, R., Barboni, L., and Giarlo, G. 2003a. Nitroalkanes in aqueous medium as an efficient and eco-friendly source for the one-pot synthesis of 1,4-diketones, 1,4-diols, δ-nitroalkanols, and hydroxytetrahydrofurans. *Journal of Organic Chemistry*, 68(23): 9173–6.

Ballini, R., Fiorini, D., and Palmieri, A. 2003b. A general procedure for the one-pot preparation of polyfunctionalized nitrocyclopropanes. *Synlett*, 11:1704–6.

Ballini, R. and Petrini, M. 2004. Recent synthetic developments in the nitro to carbonyl conversion (Nef reaction). *Tetrahedron*, 60(5):1017–47.

Ballini, R., Bosica, G., Fiorini, D., and Palmieri, A. 2004. One-pot synthesis of 1,3-dinitroalkanes under heterogeneous catalysis. *Synthesis*, 12:1938–40.

Ballini, R., Bosica, G., Fiorini, D., Palmieri, A., and Petrini M. 2005a. Conjugate additions of nitroalkanes to electron-poor alkenes: Recent results. *Chemical Reviews*, 105(3): 933–71.

Ballini, R., Bosica, G., Fiorini, D., and Palmieri, A. 2005b. Neutral alumina catalyzed synthesis of 3-nitro-1,2-dihydroquinolines and 3-nitrochromenes, under solvent-free conditions, via a tandem process. *Green Chemistry*, 7(12):825–7.

Ballini, R., Barboni, L., Fiorni, D., Giarlo, G., and Palmieri, A. 2005c. One pot synthesis of 3,5-alkylated acetophenone and methyl benzoate derivatives via an anionic domino process. *Chemical Communications*, 20:2633–4.

Ballini, R., Bosica, G., Fiorini, D., and Palmieri, A. 2005d. Acyclic α-nitro ketones: A versatile class of α-functionalized ketones in organic synthesis. *Tetrahedron*, 61(38):8971–93.

Ballini, R., Fiorini, D., and Palmieri, A. 2007a. Synthesis and use of nitrocyclopropane derivatives, *Arkivoc*, 7:172–94.

Ballini, R., Barboni, L., and Palmieri, A. 2007b. A new heterogeneous one-pot process for both nitroaldol (Henry) and Michael reactions from primary haloalkanes via nitroalkanes. *Synlett*, 19:3019–21.

Bamfied, P. and Gordon, P.F. 1984. Aromatic benzene compounds from acyclic precursors. *Chemical Society Reviews*, 13(4):441–88.

Barboni, L., Gabrielli, S., Palmieri, A., Femoni, C., and Ballini, R. 2009. Diastereoselective, one-pot synthesis of polyfunctionalized bicyclo[3.3.1]nonanes via an anionic domino process. *Chemistry – A European Journal*, 32:7867–70.

Barrett, A.G.W. and Graboski, G.G. 1986. Conjugated nitroalkenes: Versatile intermediates in organic synthesis. *Chemical Reviews*, 86(5):751–62.

Berestovitaskaya, V.M., Efremov D.A., Lipina, E.S., and Perekalin, V.V. 1994. *Nitroalkenes Conjugated Nitro Compounds*. John Wiley & Sons, Chichester.

Berger, J., Le Mahieu, R.A., Kierstead, R.W., and Tabenkin, B. 1970, Microbiological hydroxylation of allethrone. *Journal of Organic Chemistry*, 35(5):1687–88.

Bergstroem, G., Francke, W., Reith, W., and Tengoe, J. 1981. Pheromone bouquet of the mandibular glands in *Andrena haemorrhoa* F. (Hym., Apoidea). *Zeitschrift fuer Naturforschung, C: Journal of Biosciences*, 36C(11–12):928–32.

Bianco, L., Rene, L., Royer, R., Cavier, R., and Lemoine, J. 1977. Studies of nitro-derivatives of biological interest. XII. Activities of 2-alkyl-3-nitro 2H-chromenes against microorganisms. *European Journal of Medicinal Chemistry*, 12(4):385–6.

Booth, H., Huckle, D., and Lockhart, I.M. 1973. Conformational studies on 2-methyl- and 2,*N*,*N*-trimethyl-3-chromanamine and derivatives. *Journal of the Chemical Society, Perkin Transactions*, 2:227–32.

Bradsher, C.K. 1987. Formation of six-membered aromatic rings by cyclialkylation of some aldehydes and ketones. *Chemical Reviews*, 87(6):1277–97.

Broadwater, S.J., Roth, S.L., Price, K.E., Kobašlija, M., and McQuade, D.T. 2005. One-pot multi-step synthesis: A challenge spawning innovation. *Organic & Biomolecular Chemistry*, 3(16):2899–906.

Butkus, E. 2001. Stereocontrolled synthesis and reactions of bicyclo[3.3.1]nonanes. *Synlett*, 12:1827–35.

Carrol, P.J., Dailey, W.P., and Wade, P.A. 1987. Polynitro-substituted strained-ring compounds. Synthesis, mechanism of formation, and structure of *trans*-dinitrocyclopropanes. *Journal of the American Chemical Society*, 109(18):5452–6.

Cheng, K.F., Nagakura, I., and Piers, E. 1982. Reaction of β-halo αβ-unsaturated ketones with cuprate reagents. Efficient syntheses of ββ-dialkyl ketones and β-alkyl αβ-unsaturated ketones. A synthesis of (Z)-jasmone. *The Canadian Journal of Chemistry*, 60(10):1256–63.

Collins, I. 1999. Saturated and unsaturated lactones. *Journal of the Chemical Society, Perkin Transactions*, 1(11):1377–96.

Colvin, E.W., Lehr, F., Seebach, D., and Weller, T. 1979. Nitroaliphatic compounds-ideal intermediates in organic synthesis. *Chimia*, 33(1):1–18.

Connolly, C.J.C., Hamby, J.M., Schroeder, M.C., Barvian, M., Lu, G.H., Panek, R.L., Amar, A., Shen, C., Kraker, A.J., Fry, D.W., Klohs, W.D., and Doherty, A.M. 1997. Discovery and structure–activity studies of a novel series of pyrido[2,3-d]pyrimidine tyrosine kinase inhibitors. *Bioorganic and Medicinal Chemistry Letters*, 7(18):2415–20.

Crombie, L., Edgar, A.J.B., Harper, S.H., Lowe, M.W., and Thompson, D. 1950. The synthesis of the pyrethrins. V. Synthesis of side-chain isomers and analogs of cinerone, cinerolone, and cinerin-I. *Journal of Chemical Society*, 3552–63.

Darabantu, M., Mager, S., Plé, G., and Puscas, C. 1995. Heterocyclic saturated compounds as derivatives or precursors of chloromycetine and of some related structures. *Heterocycles*, 41(10):2327–56.

Denmark, S.E. and Thorarensen, A. 1996. Tandem [4 + 2]/[3 + 2] cycloadditions of nitroalkenes. *Chemical Reviews*, 96(1):137–65.

Dong, H.-Z., Fan, J.-F., Gu, Z.-M., and Xiao, H.-M. 1998. Theoretical study on pyrolysis and sensitivity of energetic compounds. (3) Nitro derivatives of aminobenzenes. *Chemical Physics*, 226(1,2):15–24.

Fanelli, R., Gariboldi, P., and Verotta, L. 1983. Studies on the sex pheromone of *Dacus oleae*. Analysis of the substances contained in the rectal glands. *Experientia*, 39(5):502–5.

Feldman, K.S. 1995. The oxygenation of vinylcyclopropanes as an entry into stereoselective 1,3-diol synthesis. *Synlett*, 3:217–25.

Feuer, H. and Leston, G. 1963. *1,4-Dinitrobutane*. *Organic Synthesis Collection*, Vol IV. Wiley, New York, pp. 368–71.

Enders, D., Hüttl, M.R.M., and Grondal, C. 2007. Asymmetric organocatalytic Domino reactions. *Angewandte Chemie International Edition*, 46(10):1570–81.

Enders, D., Hüttl, M.R.M., Grondal, C., and Raabe, G. 2006. Control of four stereocenters in a triple cascade organocatalytic reaction. *Nature*, 441:861–3.

Hall, N. 1994. Chemists clean up synthesis with one-pot reactions. *Science*, 266, 32–4.

Harada, T., Oku, A., and Wada, I. 1987. Enantiodifferentiating functionalization of 2-substituted 1,3-propanediols via chiral spiro ketal: two methods for the preparation of (*S*)-2,3-dimethylbutyl phenyl sulfides. *Tetrahedron Letters*, 28(36):4181–4.

Hung, C., Jang, J.-J., Kao, K.-H., Lin, W.-W., Liu, J.-T., Liu, J.-Y., Wang, Y., Yan, M.-C., and Yao, C.-F. 1999. One-pot synthesis of five- or six-membered carbocycles through intramolecular cycloadditions by the use of ethyl chloroformate. *Tetrahedron*, 55(23): 7115–28.

Kabalka, G.W. and Varma, R.S. 1987. Syntheses and selected reductions of conjugated nitroalkene. *Organic Preparations and Procedures International*, 19(4–5):283–328.

Katsurada, M. and Mori, K. 1984. Pheromone synthesis. LXIV. Synthesis of the enantiomers of 2-methyl-1,7-dioxaspiro[5.6]dodecane, a component of the volatile secretion from the mandibular glands of *Andrena haemorrhoa* F. *Liebig Annals of Chemistry*, 1:157–61.

Kornblum, N. 1962. The synthesis of aliphatic and alicyclic nitro compounds. *Organic Reactions*, 12:101–56.

Liu, J.-T. and Yao, C.-F. 2001. One-pot synthesis of *trans*-β-alylstyrenes. *Tetrahedron Letters*, 42(35):6147–50.

Luzzio, F.A. 2001. The Henry reaction: Recent examples. *Tetrahedron*, 57(6):915–45.

Majewski, M., Gleave, D.M., and Nowak, P. 1995. 1,3-Dioxan-5-ones: Synthesis, deprotonation, and reactions of their lithium enolates. *The Canadian Journal of Chemistry*, 73(10): 1616–26.

Nicolaou, K.C., Montagnon, T., and Snyder, S.A. 2003. Tandem reactions, cascade sequences, and biomimetic strategies in total synthesis. *Chemical Communications*, 5:551–64.

Nicolaou, K.C., Edmonds, D.J., and Bulger, P.G. 2006. Cascade reactions in total synthesis. *Angewandte Chemie International Edition*, 45(43):7134–86.

Ono, N. 2001. *The Nitro Group in Organic Synthesis*. Wiley–VCH, New York.

Patai, S. 1996. *The Chemistry of Amino, Nitroso, Nitro and Related Groups*, Part 1, S. Patai, ed. John Wiley & Sons, Chichester.

Patai, S. 1980. *The Chemistry of Ethers, Crown Ethers, Hydroxyl Groups and their Sulphur Analogues*, Part 2, S. Patai, ed. Interscience Publisher, New York, p. 721.

Peters, J.A. 1979. Synthesis of bicyclo[3.3.1]nonanes. *Synthesis*, 5:321–36.

Pham-Huu, D.-P., Petruošvá, M., BeMiller, J.N., and Petruš, L. 1999. The first synthesis of a nitromethylene-linked C-(1,2)-disaccharide. *Tetrahedron Letters*, 40(15):3053–6.

Popp, F.D. 1968. Reissert compounds. *Advances in Heterocyclic Chemistry*, 9:1–25.

Rao, T.S., Singh, A.K., and Trivedi, G.K. 1984. A novel photochemical method for the synthesis of flavonols. *Heterocycles*, 22(6):1377–82.

Rees, C.W. and Smithen, C.E. 1964. The reactions of heterocyclic compounds with carbenes. *Advances in Heterocyclic Chemistry*, 3:57–78.

Reyes, E.R., Jiang, H., Milelli, A., Elsner, P., Hazell, R.G., and Jøegensen, K.A. 2007. How to make five continuous stereocenters in one reaction: Asymmetric organocatalytic synthesis of pentasubstituted cyclohexanes. *Angew. Chem. Int. Ed.*, 46:9202–05.

Rieger, M. 1992. Skin constituents as cosmetic ingredients. *Cosmetics & Toiletries*, 107(11): 85–92.

Rosini, G. 1991. The Henry (nitroaldol) reaction. In: *Comprehensive Organic Synthesis*. Trost, B.M., ed. Pergamon Press, Oxford, vol. 2, pp. 321–40.

Rousseau, G. 1995. Medium ring lactones. *Tetrahedron*, 51(10):2777–849.

Sheldon, R.A. 1994. Consider the environmental quotient. *Chemtech*, 24(3):38–47.

Sheldon, R.A. 1997. Catalysis: The key to waste minimization. *Journal of Chemical Technology and Biotechnology*, 68(4):381–8.

Steinbaugh, B.A., Hamilton, H.W., Vara Prasad, J.V.N., Para, K.S., Tummino, P.J., Fergusson, D., Lunney, E.A., and Blankley, C.J. 1996. A topliss tree analysis of the HIV-protease inhibitory activity of 6-phenyl-4-hydroxy-pyran-2-ones. *Bioorganic and Medicinal Chemistry Letters*, 6(10):1099–104.

Wada, E. and Yoshinaga, M. 2003. A new methodology of intramolecular hetero-Diels–Alder reaction with β-alkoxy-substituted conjugated nitroalkenes as heterodienes: Stereoselective one-pot synthesis of *trans*-fused bicyclic γ-lactones. *Tetrahedron Letters*, 44(43):7953–6.

Zhang, L., Nadzan, A.M., Heyman, R.A., Love, D.L., Mais, D.E., Croston, G., Lamph, W.W., and Boehm, M.F. 1996. Discovery of novel retinoic acid receptor agonists having potent antiproliferative activity in cervical cancer cells. *Journal of Medicinal Chemistry*, 39(14):2659–63.

Zhang, X., Pais, G.C.G., Svarovskaia, E.S., Marchand, C., Johnson, A.A., Karki, R.G., Nickklaus, M.C., Pathak, V.K., Pommier, Y., and Burke Jr, T.R. 2003. Azido-containing aryl β-diketo acid HIV-1 integrase inhibitors. *Bioorganic and Medicinal Chemistry Letters*, 13(6):1215–19.

3 Smart Biomaterials

Decontamination of Toxic Metals from Wastewater—A Green Approach

Pritee Goyal and Shalini Srivastava

CONTENTS

3.1 HEAVY METALS: BIO TERRORIST

Heavy metals are important members of Dirty Dozen Club of toxic pollutants encountered in various ecosystems of the environment. The dissolved metals (particularly heavy metals) escaping into the environment pose a serious health hazard. These metals have been classified as priority pollutants by the U.S. Environmental Protection Agency. Heavy metal pollution in the aquatic system has become a serious threat today and of great environmental concern as they are nonbiodegradable

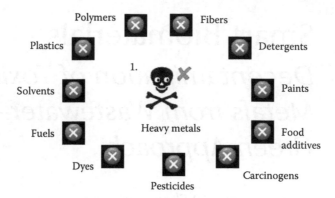

FIGURE 3.1 Pollution: a prize tag of modern society.

and thus persistent (Figure 3.1). They accumulate in living tissues throughout the food chain, which has humans at its top, multiplying the danger. Thus, it is necessary to control the presence and persistence of heavy metals in the environment.

The removal of heavy metals from water at point of use, entry, and for cowater system is generally accomplished by conventional methods such as electrochemical processes, membrane processes, ion exchange, and reverse osmosis. These currently practiced technologies for removal of pollutants from industrial effluents appear to be inadequate, often creating secondary problems with metal-bearing sludges, which are extremely difficult to dispose. The process cannot precipitate metals to low levels of solubility unless additional treatment reagents are employed, the use of which may significantly add to the volume of sludge. Most authorities consider such treatment to be best performed in specialized environment by trained personnel. Further, research findings have clearly raised strong doubts about the advisability of the use of synthetic coagulants used for metal removal. Their usage leads to several serious demerits concerning the effects on the central nervous system of human beings. For industries such as electroplating and finishers, the cost involved in the treatment of effluent produced is sometimes prohibitively expensive, especially for the smaller installations, and far outweighs the advantages of recycling and regenerating materials. Present treatment strategies require costly chemical and physical operations and involve a high degree of maintenance and supervision. While wastewater treatment by ion exchange resins is both effective and convenient, they are still too expensive to be used by the developing countries. Hence, in some way, their availability is limited to developed nations of the world.

3.2 BIOSORPTION: A PROMISING GREEN APPROACH

Unfortunately, despite the numerous contributions of chemistry to the well being and progress of humanity, it has been blamed for the present ills of the world. In fact, it is not chemistry or science or technology but our past mistakes of increasing the production only without considering the simultaneous generation of large amount of site products or wastes that have underlined us as culprit. Basically, the rapid, careless and unscientific urbanization, industrialization, and agricultural revolution are major threats to the environment as a whole. It is not the need of the poor but the greed of a rich person, which has been the main cause of environmental degradation of the world.

Chemists, since 1990 have started addressing complicated environmental issues in a safe and an economically profitable manner under various names such as Clean Chemistry, Environmentally Benign Chemistry, Sustainable Chemistry, Come back to Nature, Grey to Green Chemistry, Green Technologies, Eco-friendly Techniques, Green Processes, and more popularly as Green Chemistry. Green Chemistry is a special contribution of chemists to the conditions for sustainable development, incorporating an environmentally benign by design approach to all aspects of chemical industry. The words Green Chemistry, jointly coined by Professor Paul T. Anastas and Professor John C. Warner, may be understood as follows: "The invention, design and application of chemical products and processes to reduce or to eliminate the use and generation of hazardous substances."

To combine technology with environmental safety is one of the key challenges of the new millennium. There is a global trend of bringing technology into harmony with natural environment, thus aiming to achieve the goals of protection of ecosystems from the potentially deleterious effects of human activity, until finally improving its quality. The challenges of safe and various treating and diagnosing environmental problems require discovery of newer, more potent, specific, safe, and cost-effective (and synthetic) biomolecules. The "magic" plants are around us and are waiting to be discovered and commercialized. They are now recognized and accepted as storehouses of infinite and limitless benefits to human beings. These natural systems are often referred to as "Green Technologies," as they involve naturally occurring plant materials. Biosorption is one such important phenomenon, which is based on one of the 12 principles of Green Chemistry: "Use of renewable resources." It has gathered a great deal of attention in recent yesteryears due to a rise in environmental awareness and the consequent severity of legislation regarding the removal of toxic metal ions from wastewaters.

3.3   BIOSORPTION: MECHANISTIC ASPECTS

The complex structures of plant materials and microorganisms imply that there are many ways for the metal to be taken by the biosorbent. Numerous chemical groups have been suggested to contribute to the biosorptive binding of metal ions by either whole organisms or smaller molecules. These groups may comprise hydroxyl, carbonyl, carboxyl, sulfhydryl, thioether, sulfonate, amine, amino, imidazole, phosphonate, and phosphodiester. The importance of any given group for biosorption of certain metals by plant biomass depends on factors such as the number of sites in the biosorbent material, the accessibility of the sites, the chemical state of the sites (availability), and affinity between the site and the metal. Adsorption and desorption studies invariably yield important information on the mechanism of metal biosorption. This knowledge is essential for understanding the biosorption process and it serves as a basis for quantitative stoichiometric considerations, which constitute the foundation for the mathematical modeling of the process (Yang and Volesky, 2000).

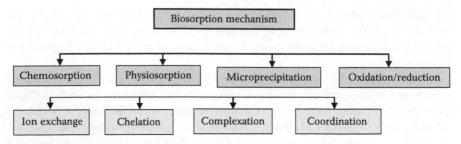

FIGURE 3.2 Mechanism of biosorption.

Various metal-binding mechanisms have been postulated to be active in biosorption process and presented in Figure 3.2.

Owing to the complexity of the biomaterials used, it is possible that at least some of these mechanisms are acting simultaneously to varying extents, depending on the biosorbent and the solution environment.

3.3.1 CHEMISORPTION

Chemisorption is of the following four types: ion exchange, complexation, coordination, and chelation. It is the adsorption in which the forces involved are valence forces of the same kind as those operating in the formation of chemical compounds. Some features that are useful in recognizing chemisorption include

- The phenomenon is characterized by chemical specificity.
- Since the adsorbed molecules are linked to the surface by valence bonds, they will usually occupy certain adsorption sites on the surface and only one layer of chemisorbed molecules is formed (monolayer adsorption).
- The energy of chemisorption is of the same order of magnitude as the energy change in a chemical reaction between a solid and a fluid.
- Chemisorption is irreversible.

3.3.1.1   Ion Exchange

Ion exchange is a reversible chemical reaction wherein an ion in a solution is exchanged for a similarly charged ion attached to an immobile solid particle. These solid ion-exchange particles are either naturally occurring inorganic zeolites or synthetically produced organic resins. Synthetic organic resins are the predominant type used today because their characteristics can be tailored to specific applications. Ion exchange reactions are stoichiometric, reversible and as such they are similar to other solution-phase reactions. For example, in the reaction

$$NiSO_4 + Ca(OH)_2 \rightarrow Ni(OH)_2 + CaSO_4,$$

the nickel ions of the nickel sulfate $(NiSO_4)$ are exchanged for the calcium ions of the calcium hydroxide $[Ca(OH)_2]$ molecule.

3.3.1.2   Chelation

The word *chelation* is derived from the Greek word *chele*, which means *claw*, and is defined as the firm binding of a metal ion with an organic molecule (ligand) to form a ring structure. The resulting ring structure protects the mineral from entering into unwanted chemical reactions. Examples include the carbonate (CO_3^{2-}) and oxalate ($C_2O_4^{2-}$) ions (Figure 3.3).

3.3.1.3   Coordination (Complex Formation)

A coordination complex is any combination of cations with molecules or anions containing free pairs of electrons. Bonding may be electrostatic, covalent, or a combination of both; the metal ion is coordinately bonded to organic molecules. An example of the formation of a coordination compound is

$$Cu^2 + 4H_2O \rightarrow [Cu(H_2O)]_4^{2+},$$

$$Cu^2 + 4Cl^- \rightarrow [CuCl_4]^{2-},$$

where coordinate covalent bonds are formed by the donation of a pair of electrons from H_2O and Cl^- (Lewis bases) to Cu^{2+} (Lewis acid).

In general, biosorption of toxic metals and radionuclide is based on nonenzymatic processes such as adsorption. Adsorption is due to the nonspecific binding of ionic species to polysaccharides and proteins on the cell surface or outside the cell. Bacterial cell walls and envelopes, and the walls of fungi, yeasts, and algae, are efficient metal biosorbents that bind charged groups. The cell walls of Gram-positive bacteria bind larger quantities of toxic metals and radionuclide than the envelopes of Gram-negative bacteria.

Bacterial sorption of some metals can be described by the linearized Freundlich adsorption equation: $\log S = \log K + n \log C$, where S is the amount of metal absorbed in $\mu mol\ g^{-1}$, C is the equilibrium solution concentration in $\mu mol\ L^{-1}$, and K and n are the Freundlich constants.

Biomass derived from several industrial fermentations may provide an economical source of biosorptive materials. Many species have cell walls with high concentrations of chitin, a polymer of *N*-acetyl-glucosamine that is a very effective biosorbent.

Biosorption uses biomass raw materials that are either abundant (e.g., seaweeds) or wastes from other industrial operations (e.g., fermentation wastes). The metal-sorbing performance of certain types of biomass can be more or less selective for

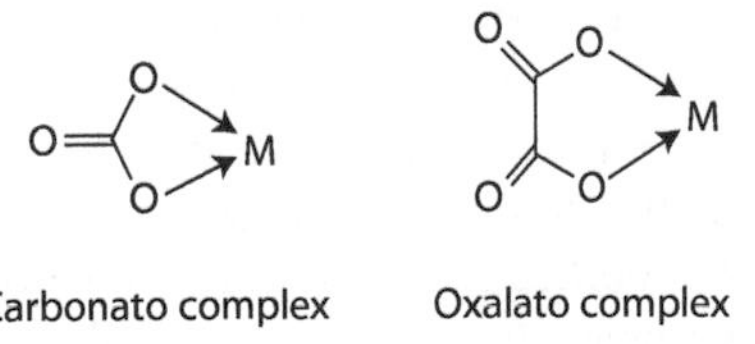

Carbonato complex   Oxalato complex

FIGURE 3.3   Chelation.

certain heavy metals, depending on the type of biomass, the mixture in the solution, the type of biomass preparation, and the chemical–physical environment.

It is important to note that the concentration of a specific metal in solution can be reduced either during the sorption uptake by manipulating the properties of the biosorbent or upon desorption during the regeneration cycle of the biosorbent.

3.3.2 PHYSIOSORPTION

In physiosorption, physiosorbed molecules are fairly free to move around the sample. As more molecules are introduced into the system, the adsorbate molecules tend to form a thin layer that covers the entire adsorbent surface. Some features, which are useful in recognizing physiosorption include

- The adsorbate molecules are held by comparatively weaker van der Waal's forces, thus resulting into lower activation energy.
- The process is, however, reversible as the substance adsorbed can be recovered from the adsorbent easily on lowering the pressure of the system at the same temperature.
- Physiosorption may extend beyond a monolayer also, since the physical forces can operate at any given distances.
- Physical adsorption is not specific in nature because it involves van der Waal's forces, which exist among the molecules of every two substances.

Physiosorption takes place with the help of van der Waal's forces. Kuyucak and Volesky (1988) hypothesized that uranium, cadmium, zinc, copper, and cobalt biosorption by certain plant materials takes place through electrostatic attraction between the metal ions in solution and functional groups present on the cell surface.

3.4 BIOSORBENTS USED SO FAR

Biosorption promises to fulfill the requirements, which are competitive, effective, and economically viable. Efforts have been made to use different forms of inexpensive plant materials for the removal of toxic metals from the aqueous media. Some typical biosorbents explored so far in removing toxic metal ions from water bodies have been listed in Table 3.1.

3.5 COMMERCIALIZATION OF BIOMATERIALS

Biomaterials have been found to be associated with drawbacks related to less sorption efficiency and stability, restricting their commercial use. Sincere efforts toward structural modifications onto the biomaterials leading to the enhancement of binding capacity or selectivity are, therefore, in greater demands. A special emphasis is to be paid on green chemical modifications resulting into tailored novel biomaterials improving their sorption efficiency and environmental stability making them liable for their commercial use as a simple, fast, economical, eco-friendly, and green

TABLE 3.1
List of Biosorbents Used for Metal Removal

Biosorbent	Metals Removed	References
Wood	Cu (II), Cr (III), As (III)	Clausen (2000)
Fruit peel of orange	Ni (II)	Ajmal et al. (2000)
Crab shell	Cd (II), Zn (II), Ni (II)	An et al. (2001)
Cone biomass	Cr (VI)	Ucun et al. (2002)
Orange juice residue	As (III), As (V)	Ghimire et al. (2002)
Portulaca oleracea	Cd (II), Cr (II), Pb (II), As (III), Ni (II)	Vankar and Tiwari (2002)
Olive pomace	Cd (II), Cu (II), Pb (II)	Pagnanelli et al. (2003)
Lemna minor	Pb (II), Ni (II)	Nicholas et al. (2003)
Chitosan	Cr (VI)	Boddu et al. (2003)
Cassava waste	Cu (II), Zn (II)	Horsfall et al. (2003)
Flyash	As (III)	Nagarnaik et al. (2003)
Waste crab shells	Au (II), Cr (VI), Se (II)	Niu and Volesky (2003)
Sphagnum peat moss	Cd (II), Pb (II), Ni (II)	Rosa et al. (2003)
Water lettuce	As (III), As (V)	Basu et al. (2003)
Cocoa shells	Cd (II), Cr (III), Ni (II)	Meunier et al. (2003)
Date pits	Cd (II)	Banat et al. (2003)
Sugarcane bagasse pith	Cd (II)	Krishnan and Anirudhan (2003)
Turbinaria ornata seaweed	Cr (VI)	Aravindhan et al. (2004)
Hazelnut shell	Cr (VI)	Kobya (2004)
Aquatic moss	Cd (II), Zn (II)	Martins et al. (2004)
Polysaccharides	Pb (II), Hg (II), Cu (II)	Son et al. (2004)
Grape stalk wastes	Ni (II), Cu (II)	Isabel et al. (2004)
Saltbush plant	Cr (III), Cr (VI)	Sawalha et al. (2005)
Jute fibers	Cu (II), Ni (II), Zn (II)	Shukla and Roshan (2005)
Chitosan	Cu (II), Cr (III), As (III)	Kartal and Imamura (2005)
Rice polish	Cd (II)	Singh et al. (2005)
Lechuguilla	Cr (III)	Gonzalez et al. (2006)
Rice husk ash	Cd (II), Hg (II)	Kumar and Bandyopadhyay (2006)
Datura innoxia	Pb (II), Ni (II)	Abia and Asuquo (2006)
Maize leaf	Pb(II)	Adesola Babarinde et al. (2006)
Okra waste	Pb (II)	Hashem (2007)
Tamarindus indica	Cr (VI)	Srinivasa et al. (2007)
Cassia fistula	Ni (II)	Hanif et al. (2007)
Juniper bark and wood	Cd (II)	Shin et al. (2007)
Sugarcane bagasse and maize cob	Cd (II)	Garg et al. (2008)
Calymperes erosum	Zn (II)	Adesola Babarinde et al. (2008)
Cassia fistula	Cr (III) and Cr (VI)	Abbas et al. (2008)

technology. Modifications onto biomaterials in order to increase their biosorption efficiency have been initiated by various research groups (Gardea-Torresdey et al., 1998; Tsezos, 1985; Saito et al., 1991).

3.5.1 Modifications onto Biomaterials to Increase Their Sorption Efficiency for Cationic Metals

Synthetic modifications onto biomaterials are to be carried out as follows.

It is inferred that carboxyl ligands are important in the binding of metal ions. Thus, increasing the number of such groups should increase the biomaterial-binding ability. This is achieved through succination, acetylation, and graft copolymerization of the biomaterial.

3.5.1.1 Succination

The succination of the amino groups on the biomaterials has been achieved by washing the biomaterial first in HCl to remove any debris, followed by washing in sodium acetate at pH 8.0. The biomaterials are to be resuspended in $NaC_2H_3O_2 \cdot H_2O$ at pH 8.0. Succinic anhydride should be then added to the suspended biomaterial. An additional succinic anhydride is to be added at 15 min interval for the next one and half hour (i.e., six additions of succinic anhydride to the biomaterial). The biomaterial should be washed with HCl, centrifuged, and washed again with deionized water. Although the amino group is neutralized, it now forms an additional carboxyl group. By the addition of a carboxylate group, there should be an enhancement of metal binding by those metals that bind to carboxyl ligands (Figure 3.4).

3.5.1.2 Acetylation

Acetylation of the amino group on the biomaterials has been achieved by washing the biomaterial first in HCl to remove any debris, followed by washing in sodium phosphate/sodium acetate buffer ($Na_3PO_4/NaC_2H_3O_2$) at pH 7.2. The biomaterial should be reacted with acetic anhydride and stirred while maintaining the pH of 7.2 for 1 h. The acetylated biomaterial is to be centrifuged for 5 min at 3000 rpm. After removing the supernatant, the biomaterial is to be resuspended in hydroxylamine to remove O-acetyl groups. The biomaterial should be washed with HCl to remove any more soluble materials and finally washed with deionized water. Acetylation of the biomaterial by acetic anhydride blocks the available amino ligands and decreases the number of positively charged sites on the biomaterial surface (Figure 3.5). This

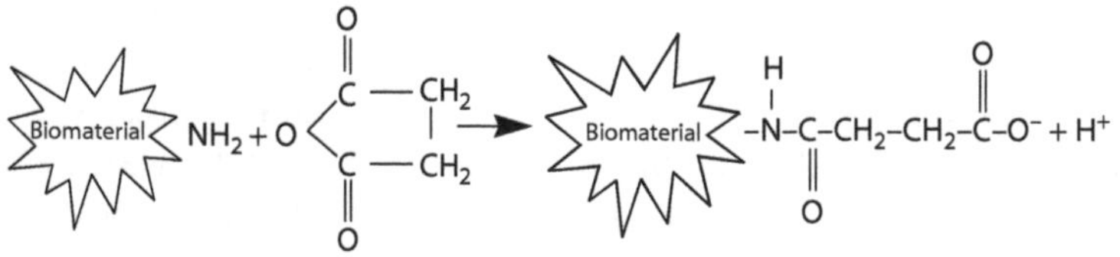

FIGURE 3.4 Succination of the amino groups.

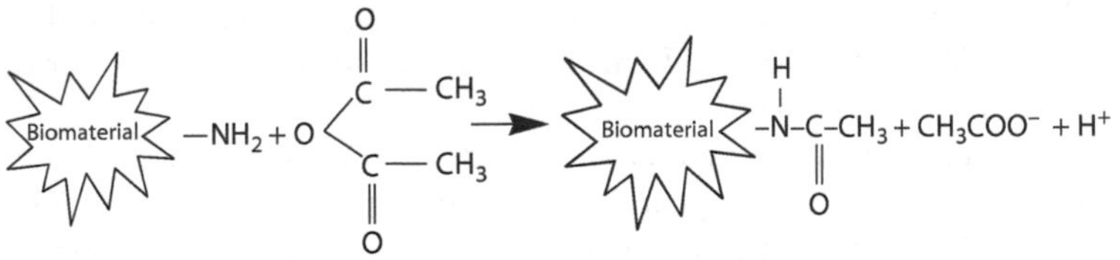

FIGURE 3.5 Acetylation of amino groups.

synthetic amendment reduces interference of the amino group, finally resulting in the increase of sorption of cationic metal species.

3.5.1.3 Graft Copolymerization

The biomaterial was dispersed in a definite amount of water. Ceric ammonium nitrate and nitric acid are then to be added slowly to the reaction mixture. Then, the monomer acrylic acid should be added drop wise to the reaction mixture. The reaction flask is to be placed in a water bath at 10–85°C for various time periods under stirring by a magnetic stirrer. After a definite time period, the reaction mixture is to be filtered and the homo-polymer should be removed with excess water. The grafted sample is to be dried to a constant weight and used for sorption studies. From the increase in weight of the biomaterial, the percentage of grafting should be calculated as follows:

$$\%Grafting = [(W_2/W_1) - 1] \times 100,$$

where W_1 and W_2 denote the weight of native and grafted biomaterial after complete removal of the homo-polymer, respectively (Figure 3.6).

3.5.2 FOURIER TRANSFORM INFRARED SPECTROSCOPY

A Fourier transform infrared spectroscopy (FTIR) analysis in solid phase in KBr has been performed using a Fourier transform infrared spectrometer to analyze and record the spectra before and after the modification of the biomaterial. Representative IR spectra of untreated and succinated biomaterial showed the presence of an

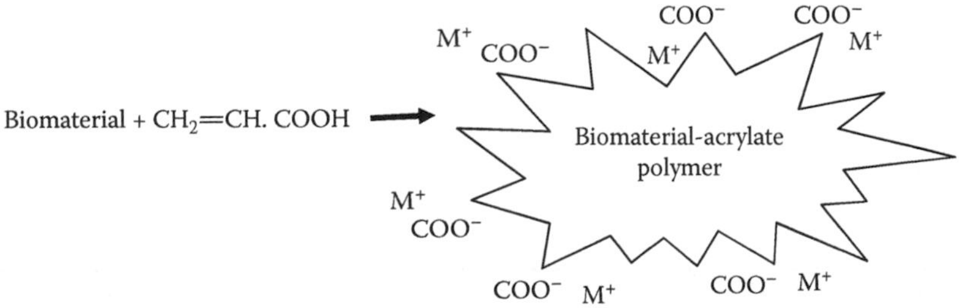

FIGURE 3.6 Evidence in support of chemical modifications occurred onto biomaterial leading to enhanced sorption.

additional peak of carboxylate ion (1744.70 cm^{-1}) and conversion of the amino group into amide (3289.77–3315.37 cm^{-1}) in succinated biosorbent confirming the succination of the biomaterial.

The conversion of the amino group into amide (3289.77–3365.14 cm^{-1}) confirmed the acetylation process. The appearance of a characteristic peak at 1725.2 cm^{-1} ($C = 0$) in the IR spectra of the graft copolymerized biomaterial compared with its unmodified biomaterial, and this confirmed the formation of the grafted biomaterial (Figure 3.7).

3.5.3 SORPTION EFFICIENCY OF MODIFIED BIOMATERIAL

All the modified biomaterials have been assessed for the sorption efficiency of cationic metals under previously standardized optimum conditions. Increased sorption efficiency in all the modified biomaterials is observed at biomaterial dosages of 2 and 4 g (Table 3.2).

The above experiment provides important information that structurally modified biomaterials are sufficient enough at the dosage of 2 g to show good sorption efficiency compared to the 4 g of unmodified biomaterials. This fact exhibits the cost effectiveness of a biomaterial when structurally modified.

3.5.4 IMPROVED ENVIRONMENTAL STABILITY OF THE BIOMATERIAL

3.5.4.1 Thermo Gravimetric Analysis

The thermo gravimetric analysis (TGA) was performed. The comparison of initial decomposition temperature (IDT) and final decomposition temperature (FDT) of

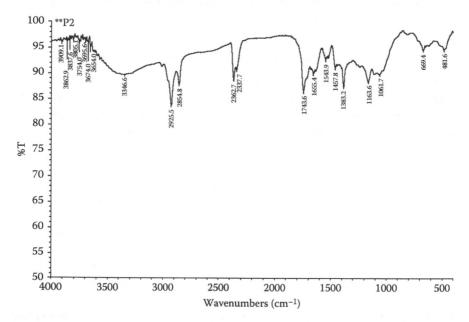

FIGURE 3.7 IR spectra of acrylic acid grafted *Zea mays* cob powder (ZMCP).

TABLE 3.2
Enhancement of Sorption Efficiency of Different Chemically Modified *Zea mays* Cob Powder (ZMCP) in Case of Single Metal Solution

Type of Biomaterials	Sorption Efficiency (%)		
	Cd (II)	Cr (III)	Ni (II)
Unmodified ZMCP	79.36[a]	76.43[a]	71.98[a]
	61.79[b]	59.24[b]	52.62[b]
Succinated	86.67[a]	84.78[a]	80.42[a]
	67.87[b]	64.23[b]	61.81[b]
Acetylated	83.64[a]	82.37[a]	78.56[a]
	63.58[b]	61.89[b]	58.56[b]
Graft copolymerized	92.52[a]	89.34[a]	85.87[a]
	88.69[b]	85.48[b]	81.19[b]

[a] At 4 g biomaterial dose.
[b] At 2 g biomaterial dose.

unmodified and graft copolymerized biomaterial was estimated based on the results of thermograms. TGA of the untreated (native) and graft copolymerized biomaterials showed significant difference in the IDT and FDT values. Upon grafting, the temperature of the biomaterial has been raised from IDT: 37.94–49.32°C and FDT: 600–609.23°C, indicating that grafting of acrylic acid improved the thermal stability of the biomaterial (Kardam et al., 2009).

The increased stability of the graft copolymerized biomaterial has been monitored on the basis of an increase in the number of regeneration cycles. Structurally modified biomaterial (polymerized) could be used up to six times as compared to only four times of unmodified biomaterial, thereby exhibiting the increased environmental stability of the polymerized biosorbent (Table 3.3).

TABLE 3.3
Sorption of Cationic Metal Ion on Regenerated Modified (Graft Copolymerized) Biomaterial

Cycles	% Sorption		
	Cd (II)	Cr (III)	Ni (II)
1	98.61	89.33	76.93
2	98.20	88.85	76.64
3	98.04	88.65	76.49
4	97.78	88.28	76.26
5	97.55	88.14	75.13
6	97.12	88.01	77.02

REFERENCES

Abia, A.A. and Asuquo, E.D. 2006. Lead and nickel adsorption kinetics from aqueous metal solutions using chemically modified and unmodified agricultural adsorbents. *African Journal of Biotechnology*, 5(16):1475–82.

Abbas, M., Nadeem, R., Zafar, M.N., and Arshad, M. 2008. Biosorption of chromium (III) and chromium (VI) by untreated and pretreated *Cassia fistula* biomass from aqueous solutions. *Water, Air and Soil Pollution*, 191:1–4.

Adesola Babarinde, A.A., Babalola, J.O., and Sanni, R.A. 2006. Biosorption of lead ions from aqueous solution by maize leaf. *International Journal of Physical Science*, 1(1):23–6.

Adesola Babarinde, A.A., Oyesiku, O.O., Oyebamiji Babalola, J., and Janet, O.O. 2008. Isothermal and thermodynamic studies of the biosorption of Zn (II) ions by *Calymperes erosum*. *Journal of Applied Science and Research*, 4(6):716–21.

Ajmal, M., Rao, A.K.R., Ahmad, R., and Ahmad, J. 2000. Adsorption studies on *Citrus reticulata* (fruit peel of orange): Removal and recovery of Ni (II) from electroplating water. *Journal of Hazardous Materials*, B79:117–31.

An, H.K., Park, B.Y., and Kim, D.S. 2001. Crab shell for the removal of heavy metals from aqueous solutions. *Water Research*, 35:3551–6.

Aravindhan, R., Madhan, B., Rao, J.R., and Nair, B.U. 2004. Recovery and reuse of chromium from tannery wastewaters using *Turbinaria ornata* seaweed. *Journal of Chemical Technology and Biotechnology*, 79:125–58.

Banat, F., Asheh, S.-Al., and Makhadmeh, L. 2003. Kinetics and equilibrium study of cadmium ion sorption onto date-pits, an agricultural waste. *Adsorption Science and Technology*, 21:245–60.

Basu, A., Kumar, S., and Mukhrjee, S. 2003. Arsenic reduction from aqueous environment by water lettuce (*Pistia stratiotes* L.). *Indian Journal of Environmental Health*, 45:143–50.

Boddu, V.M., Abburi, K., Talbott, J., and Smith, E.D. 2003. Removal of hexavalent chromium from wastewater using a new composite Chitosan biosorbent. *Environmental Science and Technology*, 37:4449–56.

Clausen, C.A. 2000. Isolating metal tolerant bacteria capable of removing copper, chromium, and arsenic from treated wood. *Waste Management Research*, 18:264–68.

Gardea-Torresdey, J.L., Tiemann, K.J., Dokken, K., and Gamez, G. 1998. Investigation of metal binding in Alfalfa biomass through chemical modification of amino and sulfhydryl ligands. *Proceedings of the 1998 Conference on Hazardous Waste Research*, pp. 111–21.

Garg, U., Kaur, M.P., Jawa, G.K., Sud, D., and Garg, V.K. 2008. Removal of cadmium (II) from aqueous solutions by adsorption on agricultural waste biomass. *Journal of Hazardous Materials*, 154(1–3):1149–57.

Ghimire, K.N., Inoue, K., Makino, K., and Miyajima, T. 2002. Adsorptive removal of arsenic using orange juice residue. *Separation Science and Technology*, 37:2785–99.

Gonzalez, R.L., Videa, J.R.P., Rodriguez, E., Delgado, M., and Torresday, G.J.L. 2006. Potential of *Agave lechuguilla* biomass for Cr (III) removal from aqueous solutions: Thermodynamic studies. *Bioresource Technology*, 97:178–82.

Hanif, M.A., Nadeem, R., Zafar, M.N., Akhtar, K., and Bhatti, H.N. 2007. Kinetic studies for Ni(II) biosorption from industrial wastewater by *Cassia fistula* (Golden Shower) biomass. *Journal of Hazardous Materials*, 145(3):501–5.

Hashem, M.A. 2007. Adsorption of lead ions from aqueous solution by okra wastes. *International Journal of Physical Science*, 2(7):178–84.

Horsfall Jnr M., Abia, A.A., and Spiff, A.I. 2003. Removal of Cu (II) and Zn (II) ions from waste water by cassava (*Manihot esculenta* Cranz) waste biomass. *African Journal of Biotechnology*, 2:360–4.

Isabel, V., Nuria, F., Maria, M., Nuria, M., Jordi, P., and Joan, S. 2004. Removal of copper and nickel ions from aqueous solutions by grape stalks wastes. *Water Research*, 38:992–1002.

Kardam, A., Goyal, P., Arora, J.K., Raj, K.R., and Srivastava, S. 2009. Novel biopolymeric material: Synthesis and characterization for decontamination of cadmium metal from waste water. *National Academic and Science Letters*, 32(5 and 6):179–81.

Kartal, S.N. and Imamura, Y. 2005. Removal of Cu (II), Cr (III) and As (III) from CCA-treated wood onto chitin and chitosan. *Bioresource Technology*, 96:389–92.

Kobya, M. 2004. Removal of chromium (VI) from aqueous solutions by adsorption onto hazelnut shell activated carbon: Kinetics and equilibrium studies. *Bioresource Technology*, 91:317–21.

Krishnan, K.A. and Anirudhan, T.S. 2003. Removal of cadmium from aqueous solution by steam activated sulphurised carbon prepared from sugar cane bagasse pith. *Water SA*, 29:1520–8.

Kumar, U. and Bandyopadhyay, M. 2006. Sorption of cadmium from aqueous solution using retreated rice husk. *Bioresource Technology*, 97:104–9.

Kuyucak, N. and Volesky, B. 1988. Biosorbents for recovery of metals from industrial solutions. *Biotechnology and Letters*, 10:137–42.

Martins, J.E., Pardo, R., and Boaventura, R.A.R. 2004. Cadmium (II) and zinc (II) adsorption by the aquatic moss *Fontinalis antipyretica*: effect of temperature, pH and water hardness. *Water Research*, 38:693–9.

Meunier, N., Laroulandie, J., Blais, J.F., and Tyagi, R.D. 2003. Cocoa shells for heavy metal removal from acidic solutions. *Bioresource Technology*, 90:255–63.

Nagarnaik, P.B., Bhole, A.G., and Natarajan, G.S. 2003. Adsorption of arsenic on flyash. *Indian Journal of Environmental Health*, 45:1–4.

Nicholas, R., Steven, P.K., and Claussen, K. 2003. Lead and nickel removal using *Microspora* and *Lemna minor*. *Bioresource Technology*, 89:41–8.

Niu, H. and Volesky, B. 2003. Characteristics of anionic metal species biosorption with waste crab shells. *Hydrometallurgy*, 71:209–15.

Pagnanelli, F., Sara, M., Veglio, F., and Luigi, T. 2003. Heavy metal removal by olive pomace: Biosorbent characterization and equilibrium modeling. *Chemical Engineering Science*, 58:4709–17.

Rosa, G.D., Torresday, G., Videa, J.R., Herrera, I., and Contreras, C. 2003. Use of silica-immobilized humin for heavy metal removal from aqueous solution under flow conditions. *Bioresource Technology*, 90:11–17.

Saito, K., Iwata, H., and Furusaki, S. 1991. Adsorption characteristics of an immobilized metal affinity membrane. *Biotechnology Progress*, 7:412–18.

Sawalha, M.F., Torresdey, G.J.L., Parsons, J.G., Saupe, G., and Peralta-Videa, J.R. 2005. Determination of adsorption and speciation of chromium species by saltbush (*Atriplex canescens*) biomass using combination of XAS and ICP-OES. *Microchemical Journal*, 81:122–32.

Shin, E.W., Karthikeyan, K.G., and Tshabalala, M.A. 2007. Adsorption mechanism of cadmium on juniper bark and wood. *Bioresource Technology*, 98:5885–94.

Shukla, S.R. and Roshan, S.P. 2005. Adsorption of Cu (II), Ni (II), and Zn (II) on modified jute fibres. *Bioresource Technology*, 96:1430–8.

Singh, K.K., Rastogi, R., and Hasan, S.H. 2005. Removal of cadmium from wastewater using agricultural waste rice polish. *Journal of Hazardous Materials*, 121(1–3):51–8.

Son, B.C., Park, K., Song, H.S., and Yoo, Y.J. 2004. Selective biosorption of mixed heavy metal ions using polysaccharides. *Korean Journal of Chemical Engineering*, 21:1168–72.

Srinivasa, R.P., Ajithapriya, J., Kachireddy, V.N., Reddy, S., and Krishnaiah, A. 2007. Biosorption of hexavalent chromium using tamarind (*Tamarindus indica*) fruit shell—A comparative study. *Environmental Biotechnology*, 10(3):358–67.

Tsezos, M. 1985. The selective extraction of metals from solution by microorganisms. A brief overview. *Canadian Metallurgy Quarterly*, 24:141–4.

Ucun, H., Bayhan, Y.K., Kaha, Y., Cakici, A., and Algur, O.F. 2002. Biosorption of chromium (VI) from aqueous solution by cone biomass of *Pinus sylvestris*. *Bioresource Technology*, 85:155–8.

Vankar, P.S. and Tiwari, V. 2002. Phytofiltration of toxic metals ions by *Portulaca oleracea*. *Journal of Analytical Chemistry*, 36:327–32.

Yang, J. and Volesky, B. 2000. Modeling of the uranium–proton ion exchange in biosorption. *Environmental Science and Technology*, 33:4072–9.

4 Opportunities and Challenges for Enzymatic Surface Modification of Synthetic Polymers

Vikash Babu, Shilpi, Meenu Gupta, and Bijan Choudhury

CONTENTS

4.1 INTRODUCTION

Synthetic polymers have wide application in various industries and in this chapter the application with respect to the textile industry and biomedical field are addressed. These fibers exhibit excellent physical properties such as strength, flexibility, toughness, stiffness, wear, and abrasion resistance (Ikada, 1994). However, synthetic polymers also exhibit some disadvantages that restrict their wider applications. A poor hydrophilicity is the main disadvantage with synthetic polymers and this makes it difficult for the application in textile industry and the biomedical field.

Synthetic polymers such as polyamides (PAs), polyethylene, poly(methyl methacrylate), polyacrylonitrile (PAN), polyethylene, and polyethyleneterepthalate are widely used polymers as biomaterials in the medical field for making sutures, joint replacement, dental implants, hemodialysis membrane, vascular prosthesis catheters, and surgical threads. For being an ideal biomaterial, its biocompatibility is the key requirement. When a foreign material is implanted in the body, it is attacked by microorganisms that are present in the body fluid and protein adsorption is also a main problem with the biomaterial. Similarly, proteins in the form of a "coagulation factor" are also adsorbed on the hydrophobic polymeric surface. Therefore, special treatments are required for making the ideal biomaterial by transforming the hydrophobic surface into a hydrophilic hemocompatible surface.

Synthetic polymers are also a major contributor in the textile industry. Synthetic fibers show excellent strength properties, chemical resistance, wrinkle resistance, and abrasion resistance. These fibers also show some undesired properties such as hydrophobicity due to which they show wearing discomfort as perspiration cannot penetrate the fabric and lower the reactivity with chemical agents, which normally act as barrier to other finishing agents. These problems are due to the presence of hydrophobic groups on the surface of synthetic polymers. Therefore, surface modification of synthetic polymers is very necessary.

The surface of the synthetic polymers can be modified by chemical, physical, and enzymatic methods (Figure 4.1). Chemical modification requires harsh reaction due to which strength properties of polymers get affected. Zeronian and Collins (1989) reported a 10–30% weight loss in polyester fibers after chemical treatment. Additionally, chemical treatments are difficult to control and have negative impacts on the environment.

Physical methods include plasma treatments, UV irradiation, corona discharge, and flame treatment. Among these, plasma treatment is widely used for the surface modification of synthetic polymers. Plasma can be obtained by exciting gases into an energetic state by radio frequency, microwave, or electrons from a hot filament discharge. Generation of plasma requires a vacuum, which normally poses several

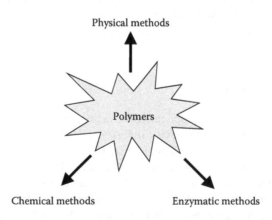

FIGURE 4.1 Methods of surface modification of synthetic polymers.

complications for continuous operation in a large-scale industrial process (Inagaki, 1996). Besides these disadvantages, plasma treatments are difficult to repeat, require complicated machinery, and are relatively very difficult to optimize. The main disadvantage of plasma treatments is the formation of heterogeneous and undesired functional groups on the surfaces. Therefore, to overcome these problems, enzyme treatments can be chosen as a third and green alternative for polymer surface modification as it offers many advantages over chemical and physical methods which include

1. They can act at very low concentration of catalyst (Menger, 1993).
2. They can act under moderate reaction conditions, which lead to less damage of strength properties of synthetic polymers.
3. They display selectivity such as (a) chemo-selectivity (act on a single type of functional group); (b) regioselectivity (enzymes can distinguish between functional groups); (c) enantioselectivity (specificity for asymmetric conversions) (Loughlin, 2000).
4. They can act on synthetic substrate.
5. They can work in nonaqueous environment, although some activity can be lost.
6. Enzymatic surface modification limits to the surfaces of polymers.

4.2   ENZYMES FOR SURFACE MODIFICATION OF SYNTHETIC POLYMERS

Very few reports are available for the enzymatic surface modification of synthetic fibers. Peroxidase, lipase, cutinase, nitrilase, nitrile hydratase, amidase, protease, and hydrolase have been reported for the surface modification of synthetic polymers (Table 4.1).

4.2.1   PEROXIDASES

Peroxidases belong to the class of oxidoreductases containing iron (III) and protoporphyrin IX as the prosthetic groups. Peroxidases catalyze the reduction of peroxides and the oxidation of many organic and inorganic compounds. These enzymes are widely used for the removal of phenolic compounds, decolorization of synthetic dyes, deodorization of swine manure, in enzyme immunoassays, for biofuel production and organic and polymer synthesis (Hamid and Rehman, 2009). Peroxidases have also been used for the surface modification of poly-p-phenylene-2,6-benzobisthiazole (PBO), polyethylene, and grafting of acrylamide onto kevlar fibers.

PBO fibers are widely used in aerospace industries and civil industries for their excellent properties such as good chemical and fire resistance, light weight, and good toughness (Wang et al., 2007). These fibers possess poor adhesive properties due to a certain degree of hydrophobicity. Therefore, surface modification of these fibers is required. Wang et al. (2007) modified these fibers using horseradish peroxidase so that they became hydrophilic. Acrylic acid was then grafted onto the modified PBO fibers, which were determined by FTIR and x-ray photoelectron spectroscopy (XPS) analysis (Wang et al., 2007).

TABLE 4.1
Enzymes Used for the Surface Modification of Synthetic Polymers

Enzyme	Source	Polymeric Substrate	Application of Polymer	Reference
Peroxidase	Soyabean peroxidase	High-density polyethylene	Hemofilteration membranes, sutures	Zhao et al. (2004)
	Horseradish	Poly-*p*-phenylene-2, 6-benzobisthiazole	In aerospace and civil industry	Wang et al. (2007)
Lipase	*T. lanuginosus*	PET	Textile fibers, vascular prosthesis	Brueckner et al. (2008)
	Rhizopus miehei	Poly(methyl acrylate)	Textile fibers	Inprakhon and Lalot (2007)
	Commercial lipase	PET	Textile fibers, vascular prosthesis	Kim and Song (2006)
Cutinase	*Fusarium solani pisi*	PET	Textile fibers, vascular prosthesis	Araujo et al. (2007)
Nitrilase	*Micrococcus luteus*	PAN (with 5% vinyl acetate)	Textile industry, hemodialysis membrane	Fischer-Colbrie et al. (2007)
	Commercial nitrilase	PAN (with 7% vinyl acetate)	Textile industry, hemodialysis membrane	Matama et al. (2007)
Nitrile hydratase	*Rhodococcus rhodochrous*	PAN 40, PAN 190 (with 7% vinyl actate groups)	Textile industry, hemodialysis membrane	Tauber et al. (2000)
	Brevibacterium imperiale	PAN (with 10% acetate groups)	Textile industry, hemodialysis membrane	Battistel et al. (2001)
	Corynebacterium nitrilophilus	PAN (with 10% acetate groups)	Textile industry, hemodialysis membrane	Battistel et al. (2001)
	Agrobacterium tumefaciens	PAN	Textile industry, hemodialysis membrane	Fischer-Colbrie et al. (2006)
Amidase	*Nocardia farcinica*	Polyamide 6	Textile industry, sutures	Heumann et al. (2009)
Protease	*Bacillus subtilis*	Nylon 6.6	Textile industry, hemofilteration membranes, sutures	Parvinzadeh et al. (2009)
Hydrolase	*Thermonospora fusca*	Polyethylene terephthalate	Textile industry, hemodialysis membrane	Alisch-Mark et al. (2006)
	Fusarium solani f. sp. pisi	Polyethylene terephthalate	Textile industry, hemodialysis membrane	Alisch-Mark et al. (2006)
	Fusarium oxysporum strain (LCH 1)	Polyethylene terephthalate	Textile industry, hemodialysis membrane	Nimchua et al. (2007)

Polyethylene is a widely used, inexpensive, and versatile polymer due to its abundant supply, good chemical resistance, good process ability, and low-energy demand for processing. Polyethylene exists in low-density polyethylene (LDPE), linear low-density polyethylene (LLDPE), and high-density polyethylene (HDPE) forms. HDPE has very poor adhesion properties to other materials because of PE's low surface energy, which limits its applications in gluing, painting, and printing. Peroxidase from soyabean was used to modify HDPE at the expense of hydrogen peroxide. Owing to the incorporation of —OH and —CO— functional group, the hydrophilicity and dyeing ability of water-soluble dyes was enhanced, which was determined by water contact angle measurement and XPS analysis (Zhao et al., 2004). Horseradish peroxidase was also used to graft acrylamide onto the Kevlar fibers for their surface modification. The modified fibers were analyzed by scanning electron microscopy and elemental analysis (Zhao et al., 2005).

4.2.2 LIPASES

Lipases are serine hydrolases, which catalyze the hydrolysis of long chain triglycerides. Lipases show chemo-, regio-, and enantio-selectivities (Saxena et al., 2003). A number of studies has been carried out on the surface modification of polyethylene terephthalate (PET) fibers using lipases (Figure 4.2). Kim and Song (2006) analyzed the effect of nine commercial lipases on PET fabrics. Moisture regain of PET fibers was improved by 2.4 times as compared to alkaline treatment and also the carboxyl and hydroxyl groups had been successfully introduced (Kim and Song, 2006). Similarly, lipase from *T. lanuginosus* was used to increase the hydrophilicity of PET fabrics by introducing hydroxyl and carboxyl functional groups (Brueckner et al., 2008) whereas lipase from *Rhizopus miehei* was used to introduce regioselective modification in a telomere of poly(methyl acrylate) by modifying the ester group (Inprakhon and Lalot, 2007). Hydrolase from *Thermomonospora (Thermobifida) fusca* and *Fusarium solani* f. sp. *pisi* was used to increase the hydrophilicity of PET fibers. Owing to an increase in the number of hydroxyl groups, reactive dye showed more intense color, which was confirmed by reflectance spectroscopy, and an enhancement in their water-absorption ability (Araujo et al., 2007).

4.2.3 CUTINASES

Cutinases are hydrolytic serine esterases that degrade cutin, a polyester of hydroxy and epoxy fatty acids (Purdy and Kolattukudy, 1975) and specific for primary alcohol esters (Murphy et al., 1996). The fatty acids of cutin are usually *n*-C16 and *n*-C18 oxygenated hydroxyacids (containing one to three hydroxyl groups). Cutins are lipid-based polymers of plants and ester bonds dominate in the cutins. Therefore, cutinases

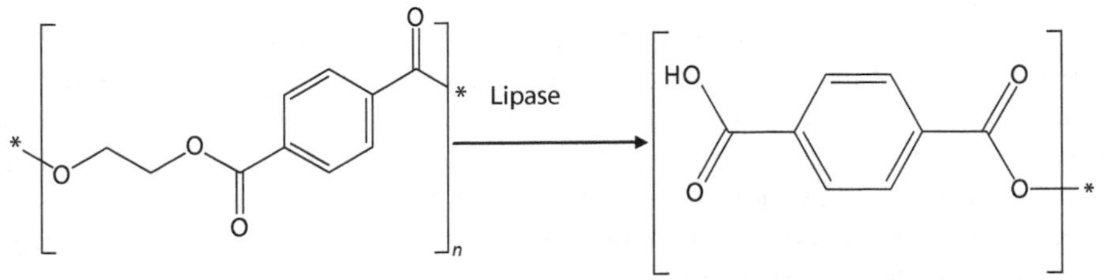

FIGURE 4.2 Hydrolysis of PET fibers.

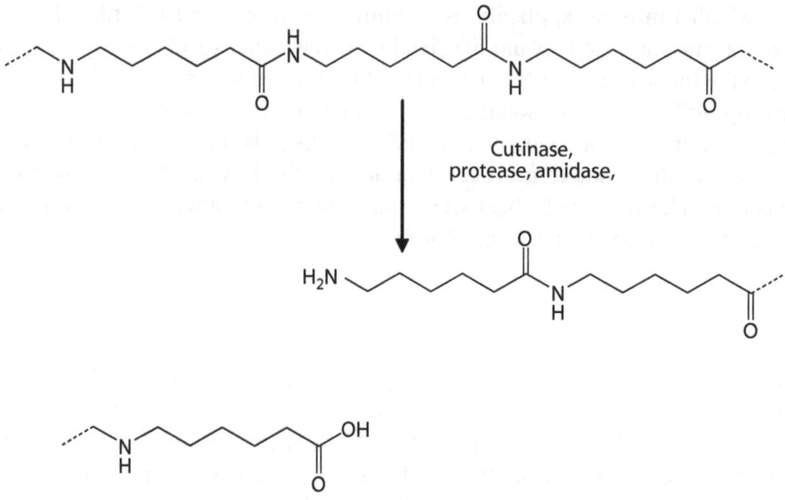

FIGURE 4.3 Mechanism of action of cutinase, protease, and amidase on PA.

are considered as esterases and they can be used for the hydrolysis of ester groups present in the synthetic fibers such as PET. There are many reports for the surface modification of polyester, polyamide 6,6 (PA 6,6), and acrylic fibers using cutinases. Cutinase from *Fusarium solani pisi* was able to hydrolyze polyester groups to modify the surface of polyester, PA, and acrylic fibers (introduction of hydroxyl groups in case of polyester and in the comonomer of acrylic, amino group in PA). This cutinase had higher activity on PA than on polyester (Silva et al., 2005a, 2005b) (Figure 4.3). Later on, this cutinase was genetically modified near the active site, by site-directed mutagenesis, to enhance its activity toward PET and PA 6,6 fibers by increasing the size of active site in order to fit a larger polymer chain (Araujo et al., 2007).

4.2.4 NITRILASES, NITRILE HYDRATASES, AND AMIDASES

Nitrilases and amidases belong to the class of hydrolases and nitrile hydratase belongs to the class of lyase. Nitrilases are an important class of nitrilase superfamily that convert nitrile to the corresponding carboxylic acids and ammonia, whereas nitrile hydratase first converts into the corresponding amide and then this amide is transformed by amidase. There are very few reports for the surface modification of PAN and PA for increasing its hydrophilicity using nitrilases, nitrile hydratases, and amidases.

PAN, a synthetic fiber, is a polymer of acrylonitrile monomers. Worldwide, 2.73 million tons of PAN are produced per year, of which over 98% are processed as filament yarn serving as material in the textile industry (Tauber et al., 2000). PAN usually has a molecular weight of 55,000–70,000 g mol^{-1} and is most commonly a copolymer produced by radical polymerization from acrylonitrile, 5–10 mol% vinyl acetate (or similar nonionic comonomers) to disrupt the regularity and crystallinity, and ionic comonomers, such as sulfuric or sulfonic acid salts. PAN is a hydrophobic polymer that affects the processability of the fibers. The surface is not easily wetted,

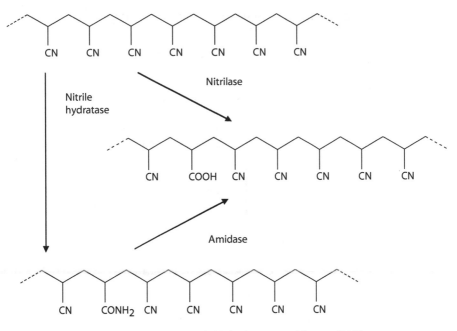

FIGURE 4.4   Mechanism of nitrilase, nitrile hydratase, amidase on PAN.

thus impeding the application of finishing compounds and coloring agent. Hydrophobicity also hinders water from penetrating into the pores of fabric. Therefore, for many applications surface modification is required to enhance its hydrophilicity. For the green chemistry route, nitrile metabolizing enzymes such as nitrilase, nitrile hydratase, and amidase can be used for the hydrolysis of PAN and PA (Figure 4.4). This bioconversion will lead to improved surface properties with minimal impact on the environment.

Rhodococcus rhodochrous was used for the hydrolysis of both granular PAN and acrylic fibers by nitrile hydratase and amidase (Tauber et al., 2000). Similarly, *Agrobacterium tumefaciens* (BST05) was found to convert polyacylonitrile into polyacrylic acid by nitrile hydratase and amidase (Fischer-Colbrie et al., 2006). Nitrilase was also used for the surface hydrolysis of PAN from *Micrococcus luteus* BST20 (Fischer-Colbrie et al., 2007). However, polyamidase from *Nocardia farcinica* leads to an increase of polar groups on the surface of PA, which was measured by tensiometry (Heumann et al., 2009).

4.2.5   PROTEASES AND TYROSINASES

Proteases are hydrolytic enzymes that catalyze the hydrolysis of peptide bond. Alkaline proteases are most commonly used in the textile industry as additives of detergents (Gupta et al., 2002) and can be used as for the surface modification of Nylon 6,6 fibers. Nylon 6,6 is a copolymer of hexamethylene diamine and adipic acid. Many proteases such as protex Gentle L, protex 40L, protex multiplus L, and protex 50FP were used to investigate changes in the nylon 6,6 polymer. Protease

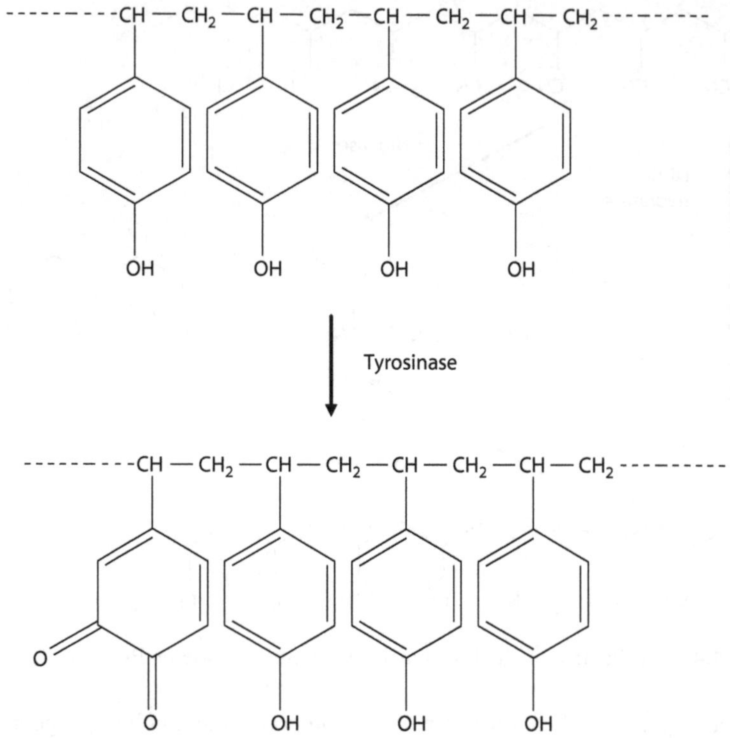

FIGURE 4.5　Oxidation of poly(4-hydroxystyrene) by tyrosinase.

treatment of Nylon 6,6 fiber showed significant decrease in thermal degradation temperature whereas reactive and acid dyes showed higher dyebath exhaustion on the protease-treated polymer (Parvinzadeh et al., 2009).

Tyrosinase catalyzes the oxidation of phenols. These enzymes are widespread in fungi, plants, and animals. Polyhydroxystyrene (PHS) is a phenol-containing polymer used as the excellent polymer matrix due to its good coating properties. Phenol moieties of PHS can be oxidized by tyrosinase. Mushroom tyrosinase was observed to catalyze the oxidation of 1–2% phenolic moieties of the synthetic polymer poly (4-hydroxystyrene) (PHS) (Shao et al., 1999) (Figure 4.5).

4.3　CHALLENGES

Surface modification of polymers exhibits many challenges during enzymatic treatment. Therefore, some parameters that should be kept in mind when modifying the surface of polymers are discussed below.

4.3.1　CRYSTALLINITY OF POLYMERS

Crystallinity of polymers is an important parameter for surface modification. Some researchers have reported that the high crystallinity of polymers has a negative influence on enzymatic hydrolysis. Fischer-Colbrie et al. (2007), however, reported that effect of low crystallinity of PAN supported a better efficiency during treatment of polymer with nitrile metabolizing enzyme catalysis of *M. luteus* (Fischer-Colbrie et al., 2007). A similar phenomenon was observed in the treatment of PET by commercial cutinase. This cutinase displayed higher activity toward less crystalline PET. Thus, it was suggested to use oligomer having low crystallinity.

4.3.2 PROTEIN ADSORPTION

When a polymer is treated with enzymes for surface modification, some of the undesired protein tends to adsorb on the polymer surface, which subsequently creates problems in the surface analysis and causes a slow down in the rate of catalysis. Adsorbed proteins can be removed from the surfaces by washing with large volumes of 1.5% Na_2CO_3 and water (Fischer-Colbrie et al., 2006) as part of a preparation for surface analysis. Protein-resistant molecules such as polyethylene glycol can be used to prevent the nonspecific protein adsorption. Surfaces can be precoated with an inert protein such as bovine serum albumin (Salisbury et al., 2002) for increasing the rate of catalysis.

4.3.3 ACCESSABILITY OF FUNCTIONAL GROUPS PRESENT ON POLYMER SURFACES

Synthetic polymers have a high number of functional groups. Some of the functional groups are embedded in the inner side of solid substrates. These functional groups are difficult to access by enzymes. Therefore, the rate of enzyme catalysis is slow on these solid substrates. Organic solvents can be used to increase the access of functional groups to enzymes. Matama et al. (2007) used di-methyl acetamide (4% v/v) and di-methyl formamide (4% v/v) for improving the access of nitrilase to –CN groups of PAN. This increased access to the –CN groups caused the color level of the nitrilase-treated acrylic fiber to increase to 156% (Matama et al., 2007). Similarly, di-methyl acetamide and benzyl alcohol were also used as swelling agents for PA fibers during cutinase treatment and these agents were reported to have increased the access of cutinase to amide. However, organic solvents do not fully meet the underlying characteristics of chemicals described under the 12 Principles of Green Chemistry (Silva et al., 2005a, 2005b). Therefore, replacement of organic solvents with green solvents is desirable.

4.3.4 POLYMER TREATMENT CONDITIONS

Enzymatic surface modification of polymers requires prolonged incubations due to which the desired enzyme gets denatured and inactivated due to their short half-life. Therefore, enzymes with better stability are required for surface catalysis. Even sometimes prolonged treatments lead to a weight loss of polymers, which is incidentally due to the degradation of polymers by the enzymes. Some of the additives can be used for increasing the half-life of polymer surface-modifying enzymes. Matama et al. (2007) used 1 M sorbitol and 4% DMA for increasing the half-life from 15 to 38 h at 30°C, pH 7.8. They noted a 249% increase in the half-life (Matama et al., 2007).

4.3.5 SURFACE ANALYSIS

Surface analysis is a big challenge for the modified polymers. It is well known that hydrophilicity and hydrophobicity can be measured in terms of water adhesion tension ($\tau°$). $\tau°$ greater than 30 dyne cm^{-1} are designated as hydrophilic and $\tau°$ less than 30 dyne cm^{-1} are designated as hydrophobic (Ma et al., 2007). Attenuated total reflectance-Fourier transformed infrared spectroscopy (ATR-FTIR) and XPS are two widely used techniques for the analysis of surfaces of modified polymers. ATR-FTIR is not very specific to the polymer surfaces as the signals are a combination of surfaces and underneath. Therefore, XPS can be used for the surface analysis because this technique has much smaller sampling depth (typically, <10 nm) (Ma et al., 2007).

Modified polymers can also be analyzed by the use of some reactive dyes. Some of the reactive dyes are specific to the functional groups. Rhodamine 6G can be used to determine the presence of carboxyl functional groups. Similarly, toluidine blue (TBO) and methylene blue can be used to determine the surface carboxyl group whereas methyl orange can be used to determine the presence of amino groups present on the surface of modified polymers.

4.4 CONCLUSIONS AND FUTURE PROSPECTS

Hydrophilicity is an important criterion for the use of synthetic polymers. Existing methods for surface modification of synthetic fibers are costly and complex. Therefore, the enzymatic surface modification of synthetic fibers is a new and green approach to synthesize polymers with improved surface properties. Use of enzymes for surface modification of polymers will not only minimize the use of hazardous chemicals but also minimize the environment pollution load. Besides these, the enzyme-modified polymers can also immobilize those enzymes which can only bind to the selective functional groups present on the polymeric surface such as $-COOH$ and $-NH_2$. Similarly, substrates can immobilize on the solid matrix (or polymer), which will be easily accessible to the enzymes. Genetic engineering can be employed for the modification of active sites of enzymes for better polymer catalysis.

Enzymatic surface modification of synthetic fibers can improve the present use of these fibers. A lot of experimental research work has been carried out on the surface modification of PA, PET, and PAN but their large scale, and maybe industrial, application is yet to be undertaken and optimized. Therefore, there is a need to commercialize surface modification of synthetic polymers using industrial enzymes.

REFERENCES

Alisch-Mark, M., Herrmann, A., and Zimmermann, W. 2006. Increase of the hydrophilicity of polyethylene terephthalate fibres by hydrolases from *Thermomonospora fusca* and *Fusarium solani f. sp. pisi. Biotechnology Letters*, 28:681–5.

Araujo, R., Silva, C., O'Neill, A., Micaelo, N., Guebitz, G., Soares, C.M., Casal, M., and Cavaco-Paulo, A. 2007. Tailoring cutinase activity towards polyethylene terephthalate and polyamide 6, 6 fibers. *Journal of Biotechnology*, 128:849–57.

Battistel, E., Morra, M., and Marinetti, M. 2001. Enzymatic surface modification of acrylonitrile fibers. *Applied Surface Science*, 17:32–41.

Brueckner, T., Eberl, A., Heumann, S., Rabe, M., and Guebitz, G.M. 2008. Enzymatic and chemical hydrolysis of poly(ethyleneterephthalate) fabrics. *Journal of Polymer Science: Part A: Polymer Chemistry*, 46:6435–43.

Fischer-Colbrie, G., Herrmann, M., Heumann, S., Puolakka, A., Wirth, A., Cavaco-Paulo, A., and Guebitz, G.M. 2006. Surface modification of polyacrylonitrile with nitrile hydratase and amidase from *Agrobacterium tumefaciens*. *Biocatalysis and Biotransformation*, 24:419–25.

Fischer-Colbrie, G., Matama, T., Heumann, S., Martinkova, L., Cavaco-Paulo, A., and Guebitz, G. 2007. Surface hydrolysis of polyacrylonitrile with nitrile hydrolysing enzymes from *Micrococcus luteus* BST20. *Journal of Biotechnology*, 129:62–8.

Gupta, R., Beg, Q., and Lorenz, P. 2002. Bacterial alkaline proteases: Molecular approaches and industrial applications. *Applied Microbiology and Biotechnology*, 59:15–32.

Hamid, M. and Rehman, K.U. 2009. Potential applications of peroxidases. *Food Chemistry*, 15:1177–86.

Heumann, S., Eberl, A., Fischer-Colbrie, G., Pobeheim, H., Kaufmann, F., Ribitsch, D., Cavaco-Paulo, A., and Guebitz, G. M. 2009. A novel aryl acylamidase from *Nocardia farcinica* hydrolyses polyamide. *Biotechnology and Bioengineering*, 102(4):1003–11.

Ikada, Y. 1994. Surface modification of polymers for medical applications. *Biomaterials*, 15:725–36.

Inagaki, N. 1996. *Plasma Surface Modification and Polymerization*. Lancaster, Technomic Pub. Co.

Inprakhon, P. and Lalot, T. 2007. Regioselectivity of enzymatic modification of poly(methyl acrylate). *Journal of Biotechnology*, 131(4):418–24.

Kim, H.R. and Song, W.S. 2006. Lipase treatment of polyester fabrics. *Fibers and Polymers*, 7(4):339–43.

Loughlin, W.A. 2000. Biotransformation on organic synthesis. *Bioresource Technology*, 74:49–62.

Ma, Z., Mao, Z., and Gao, C. 2007. Surface modification and property analysis of biomedical polymers used for tissue engineering. *Colloids Surface B*, 15:137–57.

Matama, T., Carneiro, F., Caparros, C., Gubitz, G.M., and Cavaco-Paulo, A. 2007. Using a nitrilase for the surface modification of acrylic fibres. *Biotechnology Journal*, 2:353–60.

Menger, F.M. 1993. Enzyme reactivity from an organic perspective. *Accounts in Chemical Research*, 26:206–12.

Murphy, C.A., Cameron, J.A., Huang S.J., and Vinopal R.T. 1996. Fusarium polycaprolactone depolymerase is cutinase. *Applied and Environmental Microbiology*, 62: 456–60.

Nimchua, T., Punnapayak, H., and Zimmermann, W. 2007. Comparison of the hydrolysis of polyethylene terephthalate fibers by a hydrolase from *Fusarium oxysporum* LCH I and *Fusarium solani* f. sp. *pisi*. *Biotechnology Journal*, 2:361–4.

Parvinzadeh, M., Assefipour, R., and Kiumarsi, A. 2009. Biohydrolysis of nylon 6,6 fiber with different proteolytic enzymes. *Polymer Degradation and Stability*, 94(8):1197–205.

Purdy, R.E. and Kolattukudy, P.E. 1975. Hydrolysis of plant cutin by plant pathogens purification, amino acids composition, and molecular weight of two isoenzymes of cutinase and a nonspecific esterase from *Fusarium solani* f. *pisi*. *Biochemistry*, 14:2824–31.

Salisbury, C.M., Maly, D.J., and Ellman, J.A. 2002. Peptide microarrays for the determination of protease substrate specificity. *Journal of the American Chemical Society*, 124:14868–70.

Saxena, R.K., Sheoran, A., Giri, B., and Davidson, W.S. 2003. Purification strategies for microbial lipases. *Journal of Microbiological Methods*, 52:1–18.

Shao, L., Kumar, G., Lenhart, J.L., Smith, P.J., and Payne, G.F. 1999. Enzymatic modification of the synthetic polymer polyhydroxystyrene. *Enzyme Microbiology and Technology*, 25:660–8.

Silva, C., Matama, T., Guebitz, G.M., and Cavaco-Paulo, A. 2005a. Influence of organic solvents on cutinase stability and accessibility to polyamide fibers. *Journal of Polymer Science: Part A: Polymer Chemistry*, 43:2749–53.

Silva, C.M., Carneiro, F., O'Neill, A., Fonesca, L.P., Cabral, J.S.M., Guebitz, G., and Cavaco-Paulo, A. 2005b. Cutinase—A new tool for biomodification of synthetic fibers. *Journal of Polymer Science: Part A: Polymer Chemistry*, 43:2448–50.

Tauber, M.M., Cavaco-Paulo, A., Robra, K.H., and Gubitz, G.M. 2000. Nitrile hydratase and amidase from *Rhodococcus rhodochrous* hydrolyze acrylic fibers and granular polyacrylonitrile. *Applied and Environmental Microbiology*, 66:1634–8.

Wang, J., Liang, G., Zhao, W., and Zhang, Z. 2007. Enzymatic surface modification of PBO fibres. *Surface & Coatings Technology*, 201:4800–4.

Zeronian, S.H. and Collins, M.J. 1989. Surface modification of polyester by alkaline treatments. *Textile Progress*, 20:1–34.

Zhao, J., Fan, G., Guo, Z., Zhang, Y., and Wang, D. 2005. Enzymatic surface modification of Kevlar fibers. *Science in China, Series B, Chemistry*, 48:37–40.

Zhao, J., Guo, Z., Ma, X., Liang, G., and Wang, J. 2004. Novel surface modification of high-density polyethylene films by using enzymatic catalysis. *Journal of Applied Polymer Science*, 91:3673–78.

5 Green Chemistry as an Expression of Environmental Ethics

George D. Bennett

CONTENTS

5.1 INTRODUCTION

In autumn 2006, the national power company of Iceland blocked two glacial rivers near the mountain of Kárahnjúkar in order to flood a wilderness area of 22 square miles and create a reservoir that would supply power for a new aluminum smelting facility built by the American company, Alcoa. The hydroelectric dam, the tallest of its kind in Europe, was expected to generate 4600 GWh of electricity each year, and the smelter had the capacity to process 344,000 metric tons of aluminum annually. In addition to the 400 jobs the project was expected to provide, Alcoa also claimed that theirs would be the safest, most environment-friendly smelter on Earth. Some Icelanders contested this assertion, however, by pointing out the 45 miles of tunnels, 31 miles of high-tension power lines, and the sulfur dioxide fumes associated with the smelter. To the protestors, the present-day costs to the environment were not worth the uncertain promise of future prosperity (del Guidice, 2007). In other words, this debate in Iceland was largely cast as an either–or conflict between the economy and the environment, which follows the same template that many debates about public policy on environmental issues follow.

Over the last three decades, most debates about public policy on environmental issues have been framed as variations of the question, "How do we save the earth without ruining the world?" Such a framing pits environmental interests against economic interests as contrary positions. This dichotomy, however, never quite rang true to this author. First, in the aftermath of the Cold War and the fall of the Iron Curtain, it became apparent that environmental problems resulting from pollution were worse in the formerly Communist countries of Eastern Europe and the Soviet Union that had centrally planned economies than in the countries of Western Europe, North America, and Japan with largely capitalist economies. Moreover, the latter countries had a greater capacity for remediating the environmental problems that did arise. An impartial observer would be hard-pressed to deny that environmental improvement and economic development correlated with each other, at least to some extent (some environmentalists would respond, with no small amount of justification that economic development created the environmental problems in need of remediation).

Second, the words "ecology" and "economy" have similar etymologies. Both are derived from the Greek *oikos*, for household or home. Thus, economics at its root refers to household management, and ecology refers to the science of the interactions within the Earth's biosphere, or the home of all life (Beers et al., 2000). Both terms deal with relationships of coexistence: cooperation, competition, collaboration, mutualism, symbiosis, advantage, disadvantage, and adaptation. By extension, if ethics is defined as competing visions of the good, then there must be economic ethics just as surely as there are ecological ethics. It stands to reason that a comprehensive, internally consistent ethical system would include both ecological and economic ethics.

Third, charity demands that everyone assume, until presented with overwhelming evidence to the contrary, that advocates of the "economic" side of the debate are not actively pursuing environmental destruction as a goal. In other words, even the greediest robber baron would desire clean air to breathe and clean water to drink. Nor are these advocates necessarily acting solely in their own interests. A business developer, for instance, could sincerely want to provide jobs for people in a community who are currently unemployed in addition to wanting to make a personal profit. Just because these advocates place a higher priority on economic growth, is not sufficient reason to label them as "anti-environment."

Fourth, one needs not go so far as to fear a green-industrial complex to recognize that even environmental activists can find activism to be a lucrative enterprise (O'Neill, 2009). That is, economic incentives can influence the behavior of environmentalists, too. No one can doubt that former U.S. Vice President Al Gore, Jr., to take one example, has experienced considerable financial gain as a result of his environmental initiatives. In the same way we allow that the economic advocates are not automatically motivated by environmental animus, we must also allow until proven otherwise that the environmental advocates are not automatically against financial benefit.

Fifth, technological advances can lead to more efficient use of resources and/or less pollution. Paul Ehrlich's formulation that environmental damage (defined as the combination of depletion of resources and emission of pollution) is the product of population, affluence, and technology, is empirically false (Beers et al., 2000). For example, scrubbers and getters in coal-burning power plants to reduce acid rain and substitutes for chlorofluorocarbons to preserve the ozone layer would generally be considered technological advances that lessened environmental damage. More recently, an industry-commissioned study concluded that innovations by the chemical industry, such as building insulation foam, low-temperature detergents, and lighter weight materials, save two units of greenhouse gas emissions for every single unit of greenhouse gases the chemical industry emits (Scott, 2009). To the extent growing affluence leads to technological innovation, both variables might potentially contribute to the reversal of environmental damage or the improvement of environmental health.

The combination of these observations of the real world and these logical inductions would seem to indicate that casting environmental policy discussions as "environment vs. economy" debates is inaccurate at best and dishonest at worst. Common ground for the basis of compromise might prove elusive in the end, but there is little justification for presupposing *a priori* that such common ground cannot and does not exist.

Trying to fit environmental issues into an inapt dichotomy might be convenient for the media that cover such issues and for the advocates at either end of the spectrum of opinion. It is not helpful, though, for promoting reconciliation among the sundry factions that lie along that spectrum. This sort of polarization fosters an adversarial relationship between those who claim to speak on behalf of the environment and those who claim to speak on behalf of the economy. Reconciliation of differing viewpoints is most facile when those viewpoints bear a conciliatory relationship with each other. The potential for reconciliation of different positions on environmental policy will be higher when economic interests and environmental interests are presumed to have compatibility.

5.2 DEVELOPMENT OF MODERN ENVIRONMENTALISM

The premise of enmity between environmental and economic considerations is an artifact of the development of modern environmentalism. The term "modern environmentalism" is in contrast to what could be referred to as "premodern" or "traditional" environmentalism. Concern regarding the environment is as old as the human race itself because people have always been part of and interacted with their natural surroundings. Until the early twentieth century, with a few notable exceptions, such as the teachings of St. Francis of Assisi regarding the brotherly treatment of animals, this concern was usually translated into action aimed at protecting humans from the environment. For example, establishing standard practices of hygiene among doctors and nurses during medical examinations, improving public sanitation, and controlling rat infestations to prevent the spread of flea-borne disease were some of the measures for cleaning up the environment that were motivated by a concern for human health.

Modern environmentalism, by contrast, has mostly aimed at protecting the environment from humans and has loaded this charge with moral freight. Walter

Lowdermilk sowed the seeds of modern environmentalism in 1940 with his essay on land usage, "The Eleventh Commandment" (Lowdermilk, 1940). Aldo Leopold's 1949 book, *A Sand County Almanac*, continued this theme of a land use ethic (Leopold, 2001). Lutheran seminary Professor Joseph Sittler, Jr., revived the arguments of St. Francis in a 1954 article in which he included all living things as bearers of God's image and worthy of treatment appropriate to that status (Sittler Jr., 1954). The turning point occurred in 1967 when Lynn White, Jr., laid the blame for environmental degradation squarely at the feet of monotheistic religion in general and Christianity in particular. His allegation was that this brand of theology subsumed a dualistic view of the world in which nature was not sacred and could thus be treated as a commodity (White Jr., 1967). Virtually all theologically derived environmental ethics during the modern era of environmentalism has been articulated in response to White, Jr. (Bennett, 2008).

Because of the strong historical and geographic association between capitalism and Christendom, White, Jr.'s bellicose critique of the Judeo-Christian environmental paradigm was tantamount to an attack on capitalist economics as well. Combined with Lowdermilk and Leopold's warnings about the unsustainability of some agricultural and industrial practices of modern civilization, this aggressive criticism set environmental goals and economic goals at odds with each other.

5.3 GREEN CHEMISTRY AS AN ETHIC

5.3.1 NORMATIVE NATURE OF GREEN CHEMISTRY

Green chemistry assumes that economic goals and environmental goals are compatible with one another and that both can be achieved simultaneously. It is a normative approach to chemistry built on the notion of source reduction, or the treatment of pollution through the prevention of its formation (Bennett, 2008). Indeed, the first of the 12 principles of green chemistry is "prevention" (Anastas and Warner, 1998). As a normative system, green chemistry is distinct within the field of chemistry because it requires a value judgment about the equal worthiness of the goal of decreasing hazards associated with chemical products and processes as compared to the goal of optimizing functionality. Because green chemistry requires a value judgment, it rests on several ethical presuppositions; however, it did not emanate from a particular environmental ethic (Bennett, 2008). It arose, rather, as an application of an ethic derived from political compromise. A period of divided government in the United States produced the Pollution Prevention Act in 1990 that declared source reduction to be the national policy of the country (U.S. Code, 2008). Prior to this legislation, environmental regulations in the United States focused on pollution management and control. In its capacity to implement the provisions of the Pollution Prevention Act, the U.S. Environmental Protection Agency (EPA) initiated research grant and industrial programs in 1991 aimed at rewarding efforts in the area of source reduction (U.S. Environmental Protection Agency; Ember, 1991).

Before we analyze the ethical presuppositions of green chemistry, we must be careful not to conflate the ethics of green chemistry with the ethics of other "green" initiatives and movements. For example, Green parties have been fixtures of

European politics for several decades and an addition to U.S. politics more recently. A superficial analysis could lead one to erroneously conclude on the basis of the similarity of the names that the ethical assumptions of green chemistry thoroughly match the ethical principles of the Green Party in their entirety. Similarly, green chemistry does not bear an organic (i.e., essential) relationship to green energy, green collar jobs, etc. A certain degree of overlap or similarity between the ethical assumptions of green chemistry and those of other green programs should come as no surprise, but green chemistry has a different origin. Because the origin of green chemistry is unique, the ethical assumptions of green chemistry deserve to be treated on their own terms.

5.3.2 Ethical Assumptions of Green Chemistry

As previously noted, green chemistry began as the implementation of a broadly supported political agreement. Therefore, its ethical assumptions also tend to command broad support. Among the ethical assumptions of green chemistry are the following:

- Environmental and economic goals are mutually compatible.
- The achievement of these goals will contribute to sustainable development.
- Preventing pollution is better than treating it after the fact.
- If pollution must be formed, that which does not persist in the environment is better than that which persists.
- Waste is bad; efficiency is good.
- Using the resources of the Earth is legitimate.
- Nondepleting resources are better than depleting resources.
- The welfare of the people who work with chemical products and processes is at least as important as the welfare of the environment.
- Reducing the inherent hazards of chemical products and processes is a superior way of reducing risk (to both people and the environment) than reducing the probability of exposure.
- Lastly, the profit motive is legitimate (Bennett, 2008).

Environmental and economic goals are mutually compatible. Green chemistry does not fit the template of most public debates about environmental issues because it does not insist that the economy and the environment are at odds with one another. A gain in one realm does not automatically mean the other realm will suffer loss. If a pollution-generating step of a process can be eliminated, for instance, the environmental and economic costs of the process will both decrease simultaneously. Not all "green" initiatives share this assumption of mutual compatibility between environmental and economic goals. Consider the current debate in the United States over the use of ethanol for fuel. Without government subsidies, corn-based ethanol would cost more than petroleum-based fuels, at least in the near term, and would either take corn away from food production, thus leading to higher food prices, or require more farmland, fertilization, and irrigation to meet the country's fuel needs, or both. As long as corn remains the primary source of ethanol, these tensions and incompatibilities will remain. The ethical principle of green chemistry, however, instructs us that the tensions and incompatibilities do not *necessarily* need to exist. By not assuming that tensions and incompatibilities are inevitable, we leave doors open to further research that might otherwise be closed prematurely. In the ethanol analogy, these avenues of investigation include efforts to make cellulosic ethanol cost-effective and to use perennial grasses and plants with minimal irrigation needs as sources of ethanol. Green chemistry does not promise to end every conflict between environmental and economic goals, but it predisposes us to look for resolutions to such conflicts.

The achievement of these goals will contribute to sustainable development. Sustainable development is usually defined as development that meets the needs of the present but preserves the ability of succeeding generations to continue meeting those needs in the future (United Nations, 1987). This concept implies that we in the present should not externalize environmental costs to those who are not yet born. At the same time, we should want future generations to inherit a standard of living from us that is at least as high as that which we currently enjoy. The role of green chemistry is not to redirect the imposition of environmental costs upon ourselves (or anyone else, for that matter) but rather to reduce or eliminate those costs in an economically feasible manner. When we apply green chemistry in this way, we maximize the value of both the environmental and economic legacies we are holding in trust for those who will come after us.

Preventing pollution is better than treating it after the fact. This notion might seem self-evident, but it did not govern environmental policy in the United States until comparatively recently. In economic terms, preventing pollution is a value-added approach. Management, disposal, remediation, and other forms of treatment of pollution cost something without increasing the value of the product or process that was responsible for the pollution. When chemists reduce or eliminate the need for treatment, they also reduce or eliminate the cost of treatment. Therefore, the total cost of the product or process more accurately reflects the value of the product or process. Although engineers and economists, among others, have known about value-added approaches for decades, the regulatory framework that American politicians erected mainly in the 1970s and 1980s emphasized crisis management, and businesses were apt to accept the same paradigm as the government. Once the costs of compliance (and the penalties for noncompliance) with "end-of-the-pipe" regulations became prohibitive for businesses, the failure of the regulations to coerce universal compliance became apparent to politicians, and the cross media impacts of the regulations, such as eliminating an air pollution problem by creating a water pollution problem, became apparent to environmentalists could the paradigm shift occur.

If pollution must be formed, that which does not persist in the environment is better than that which persists. This assumption is most overtly exhibited in the tenth principle of green chemistry, design for degradation (Anastas and Warner, 1998). One of the environmental costs that the present generation can externalize to future generations is in the form of persistent pollutants. If we can replace a persistent pollutant with something that will quickly degrade to innocuous by-products that can take part in normal biological or geological cycles, we can internalize the environmental cost, or keep the cost confined to our generation.

Waste is bad; efficiency is good. The former proposition can be construed from both an environmental and an economic standpoint. For example, the sixth principle of green chemistry is "Design for energy efficiency" (Anastas and Warner, 1998). From the standpoint of whoever is paying for the manufacturing process, energy must be purchased at a cost, so a lower demand for energy will keep that cost down. From an environmental standpoint, a more energy-efficient process better conserves resources so that they are available for others, whether in the present or in the future, to use. Other green chemistry principles that reflect this idea are two (atom economy), eight (fewer derivatives), and nine (catalysis) (Anastas and Warner, 1998). If we equate the term *waste* with *pollution*, then this concept ties back to the discussion above regarding pollution prevention.

The latter proposition is, perhaps, more controversial or more subject to disagreement. What is most efficient from an economic standpoint is not necessarily most efficient from an environmental standpoint. Furthermore, whether efficiency is always good is debatable. In the context of green chemistry, however, efficiency is best thought of as the absence of waste in terms of usage of raw materials and energy. In this sense, the latter proposition is simply the obverse of the former.

Using the resources of the earth is legitimate. Most people would likely agree that resource use beyond the needs for mere survival is necessary. The difference lies in whether they regard resource use as an evil to be tolerated for the sake of achieving a greater good (or a lesser evil), or as a good in its own right because of the creative and productive capacities of human beings. Green chemistry, by definition, focuses on products and productive processes, so resource use as a means to achieve the ends of green chemistry is not simply a lesser evil but instead an act of wisdom. It is considered primarily constructive rather than destructive.

Nondepleting resources are better than depleting resources. Green chemistry principle seven is, "Use of renewable feedstocks" (Anastas and Warner, 1998). As a practical matter, for a resource to be considered renewable, it must not only be capable of being replaced or restored, but also such replacement must occur within one human generation. Renewal on this timescale keeps the costs of depletion from being externalized to future generations.

The welfare of the people who work with chemical products and processes is at least as important as the welfare of the environment. Green chemistry is anthropocentric (as is sustainable development). Several green chemistry principles reflect this anthropocentrism. Principles 3 (less-hazardous chemical synthesis), 4 (design of safer chemicals), 5 (safer solvents and auxiliaries), and 12 (inherently safer chemistry for accident prevention) express concern for the health of the people who handle materials or attend to processes (Anastas and Warner, 1998). While many of these safety benefits also accrue to nonhuman organisms, the focus of the principles is on the people who are exposed to these materials and methods. Inasmuch as we cannot know all of the environmental needs of nonhuman things, it is hard to imagine how the focus could be on anything else.

Reducing the inherent hazards of chemical products and processes is a superior way of reducing risk (to both people and the environment) than reducing the probability of exposure. Risk is a function of both the potential harm or damage that can result (i.e., hazard) and the likelihood of encountering or experiencing such harm (i.e., probability of exposure). Mechanisms that reduce the probability of exposure are effective ways of reducing overall risk as long as they function properly. If a breakdown occurs, though, whether by accident or intent, the overall risk returns to its original level. Compliance with such mechanisms can sometimes be costly, as well. On the other hand, diminishing the potential harm makes overall risk lower without the uncertainty or costs associated with mechanisms to prevent exposure. This method of risk reduction is effective in general, but it takes on heightened significance in this age of security concerns. A facility where green chemistry is practiced would be less inviting as a target for saboteurs.

The profit motive is legitimate. Green chemistry is ultimately results-oriented because it makes a value judgment about the results. Whether the practitioners of green chemistry are motivated by a concern for the environment, a desire for profit, or something else does not really matter as long as the results are obtained. At the same time, the principles that lead to those results take into account the fact that people often respond to economic incentives. Getting people to respond to economic rewards that they can personally enjoy is a more straightforward task than getting them to respond to environmental costs that they will not have to personally pay. Human nature comprises more than just the materialistic, but the materialistic is an aspect of that nature. Green chemistry appeals to this materialistic aspect of human nature because it regards the profit motive as legitimate.

5.3.3   POTENTIAL OBJECTIONS TO THE ETHICS OF GREEN CHEMISTRY

Because bipartisan political solutions in the United States are typically compromises that result from an attempt to reconcile competing points of view in the interests of maintaining social harmony, and because the Pollution Prevention Act of 1990 was a bipartisan political solution, this legislation was an attempt to reconcile environmental and economic interests. Given that implementation of the Pollution Prevention Act initially gave rise to green chemistry, it should come as no surprise that the ethical assumptions of green chemistry also harmonize with the environmental and economic ethics of varied factions. Of the 10 ethical assumptions listed and described above, the most controversial would seem to be the last one, "The profit motive is legitimate." In the minds of some people, every profit is accompanied by someone else's loss, or profit is interpreted in terms of its victims rather than its beneficiaries. For others, the term *profit motive* is a euphemism for greed. From this viewpoint, although profit might be applied toward good ends, the desire for profit is not good. Still others, such as certain Muslims, believe that the appearance of wealth created by credit and lending schemes is illusory (Bennett, 2008). No matter what the origin of the objection, these individuals will encounter an inconsistency between their ethical presuppositions and the ethical assumptions of green chemistry.

Some people argue that the prospect of sustainable development is a delusion. According to them, if we want sustainability, we must abandon the conceit of development (Dardozzi, 2009). They maintain that population (or, more accurately, population growth) is the dominant variable in the aforementioned Ehrlich equation. They argue that the only meaningful way to engender a sustainable environment is to stem population growth (Bartlett, 2004). What they fail to acknowledge is that affluence and/or technology beyond a certain level correlates with stabilization, if not absolute decreases, of population in countries. The population trends in such countries as Japan, Italy, and others suggest that the most effective way to reduce worldwide population growth would be for affluent countries to transfer efficient technology to countries that want to undergo development so that they may develop as quickly and efficiently as possible. Commentators who reject the possibility of sustainable development or who assume that technological efficiency, affluence, and population increase together indefinitely will not be comfortable with the ethical assumption of green chemistry that "the achievement of (environmental and economic) goals will contribute to sustainable development."

Although it is hard to imagine anyone's disagreeing with the premise that waste is bad, believers in Jevons' paradox would vigorously object to the corollary that "efficiency is good." According to Jevons' paradox, or the rebound effect, more efficient use of resources results in greater consumption of those resources and, therefore, greater total environmental damage (Dardozzi, 2009). An increase in efficient use of a resource effectively increases the supply of that resource, which drives down the cost. According to the laws of supply and demand, the lower cost leads to an increased demand for the resource, so net consumption of the resource increases. In the case of renewable resources, this effect would seem to be a useful driver of their incorporation into the energy portfolio of a nation. The paradox assumes that the supply and demand curves are elastic only in the direction of increase. If demand were to decrease, such as when a population declines through attrition or migration, or if supply were to decrease, such as when a producer leaves the market, the cost of the resource would rise until it reaches a new equilibrium point. To be sure, the individuals who tout Jevons' paradox do not support wasteful use of resources as a viable alternative. The aspect of green chemistry that they would find most disconcerting is that green chemistry includes no mechanism for blocking the savings caused by greater efficiency from being directed toward more resource consumption.

Another ethical assumption of green chemistry that could meet with resistance or objection is, "Using the resources of the earth is legitimate." The objection would be over the aspect of legitimacy. Assorting those advocates who accept the legitimacy of resource use and those who do not into different classifications requires a somewhat unconventional demarcation. When trying to categorize various commentators or advocates on the basis of their environmental ethics, it is tempting to try to divide them into liberal and conservative camps, or perhaps statists and localists (because of the types of remedies they prefer), or, as Lynn White, Jr., tried to do, Judeo–Christian dualists and everyone else (White Jr., 1967). None of these dichotomies quite captures the appropriate distinction, however. The labels "liberal" and "conservative" do not mean the same things in different countries or at different times, so they could be misinterpreted. Furthermore, even within the same polity, not all liberals and not all conservatives are equally sanguine about resource usage. Categorizing advocates according to the remedies they favor sheds little light on what premises give rise to their preferences. Lynn White, Jr.'s blaming Judeo–Christian dualists (and Islamic dualists by extrapolation) is an example of the informal logical fallacy of the sweeping generalization. Christian, Jewish, and Muslim commentators on environmental issues do not by any means monolithically subscribe to a dualist interpretation through which nature is not considered sacred. Those who could be characterized as espousing a dualist interpretation are seldom as reckless in applying that interpretation as White, Jr. makes them out to be (Gottlieb, 2006). In addition, people who consider nature sacred, whether monotheists or not, do not have uniformly benign interactions with nature (Royal, 2006). Relying on such conventional forms of categorization takes the focus off the relevant point of difference.

The single most important question that serves to demarcate an individual's environmental views is whether that individual thinks of humans as primarily producers or primarily consumers. The question is one of priority. Virtually all individuals surely recognize that humans play both roles. The issue is whether we as humans create more than we use or use more than we create. Those who think we consume more than we produce are more likely to regard the Earth and its other inhabitants as victims of human consumption and to propose that we ratchet back our consumption in order to restore or achieve some sort of equilibrium. Those who think we produce more than we consume are more likely to regard the Earth and its other inhabitants as dynamically changing and to propose that we try to harness or direct those changes toward the betterment of the human condition. Jordan Ballor has coined the useful labels *preservationist stewardship* and *productivity stewardship* to describe these models of environmentalism (Ballor, 2004).

The assumption of green chemistry that resource use is legitimate could prove difficult for someone who advocates preservationist stewardship to accept. In the preservationist stewardship model, resource use is only legitimate insofar as it does not permanently shift the ecological equilibrium. Once our resource use exceeds that amount, it ceases to be legitimate. Because advocates of the preservationist stewardship model tend to think we exceeded the limit of our legitimate resource use a long time ago, they would be uncomfortable supporting something that encouraged additional use of resources. On the other hand, most of them would welcome green chemistry innovations that replace nonrenewable feedstocks with renewable feedstocks or that replace stoichiometric methods with catalytic methods because such innovations would apparently suppress the overall human appetite for resources.

The assumption is much more consistent with the productivity stewardship model. Advocates of the productivity stewardship model tend to think that we should use the resources that are available to us to build a better world. They reason that sustainable development is still development, and development necessitates resource use. If those resources can be put to a safer, cleaner, or less expensive use, as green chemistry promises, so much the better.

In summary, the ethical assumptions that underlie green chemistry show it to be a methodology for optimizing the stewardship of economic, ecological, and social capital simultaneously. With the exception of those individuals who reject either the possibility of achieving such collective optimization simultaneously or the desirability of exercising stewardship over one or more of the particular forms of capital, most people will find the ethical assumptions of green chemistry to be consistent with their own ethical assumptions. Chemists from a broad range of ethical traditions, therefore, should be able to practice green chemistry without compromising their ethical principles. Chemists from a productivity stewardship tradition should especially find green chemistry appealing from the standpoint of ethics. That is not to say that chemists cannot, and have not, found other grounds for objecting to green chemistry. There is disagreement about what role green chemistry ought to play in the education and training of chemists, as well as differing attitudes about what criteria must be met before a product or process can merit the "green" label. These disagreements are beyond the scope of this chapter. The point is that the normative nature of green chemistry has not been a significant source of controversy within the chemistry community thus far.

5.4 THRIFTY CHEMISTRY

If green chemistry did not grow directly from any one environmentalist school of thought, from where did its ethical assumptions come? As outlined previously, the instrumental cause was a political process of negotiation among parties with competing interests and priorities. That only explains, however, how the ethical assumptions that were already present in large segments of society came to be reflected in the law. It does not identify the formal cause of those assumptions: thrift.

5.4.1 GREEN CHEMISTRY AS THE APPLICATION OF THRIFT

Our contemporary usage of the term *thrift* has stripped it of much of its historical meaning. Etymologically, it is derived from the verb *to thrive*. More than mere saving or stinginess, the term refers to a normative concept. Thrift is both an end and a means. While it describes a state of growth and prosperity that enables generosity, it presupposes certain tenets that habituate us toward making the wisest use of what is available to us. It assumes that we are trustees, not owners, of what we have. It assumes that fulfillment of our duties as trustees requires productive work and that such work is inherently good. It operates at the level of the individual or household, in the realm of private commercial enterprise, and in the public sector. In its fullest sense, thrift is a moral virtue, available to anyone and everyone, that opposes waste (Blankenhorn, 2008).

Green chemistry is the application of the virtue of thrift to the practice of chemistry, or thrifty chemistry. If this thesis about the correspondence between green chemistry and thrift is correct, the ethical assumptions of thrift should bear similarity to the ethical assumptions of green chemistry. Among the ethical assumptions of thrift are the following:

- Waste is bad
- Saving (money and/or goods) is good as long as doing so does not otherwise compromise health or welfare
- Use of resources is legitimate
- Prosperity is legitimate as both a motive and an outcome

Three of the four assumptions listed are nearly verbatim restatements of ethical assumptions of green chemistry. The second assumption listed is close to being a synthesis of several ethical assumptions of green chemistry.

5.4.2 CHARACTERISTICS GREEN CHEMISTRY AND THRIFT HAVE IN COMMON

Green chemistry and thrift share a number of common traits. Both are normative. They regard certain goals as worthy of attainment, and they furnish a framework of guidelines for how those goals ought to be attained. Within each framework, individuals retain the latitude to decide how best to implement the guidelines. In this manner, both green chemistry and thrift adhere to the principle of subsidiarity because decisions about implementation are left to the least centralized authorities.

Both green chemistry and thrift are idealistic in the sense that they express visions of ideal scenarios in which waste ceases to occur. Our attempts to reach such a standard will fall short, but through the prisms of green chemistry and thrift these attempts will still have value if they represent movement closer to the standard. The trajectory toward the standard is asymptotic (Goodwin, 2004). Green is not an all-or-nothing status for a chemical product or process. There is a continuum or spectrum of greenness. The more relevant measure of an innovation that results from the application of the principles of green chemistry is against what that innovation supersedes because that measure shows the extent of the shift along the green spectrum. In a similar fashion, movement along the continuum of thrift is best measured against the prodigality of the habits that the practice of thrift replaces.

Both green chemistry and thrift are democratic. Any chemical product or process is, in theory, subject to influence by the application of the principles of green chemistry. In reality, green chemistry has a higher profile in such areas as synthetic organic chemistry, but that is probably more reflective of the prevalence of hazards, such as volatile, flammable organic solvents, that offer a high potential for replacement or elimination than anything else. A chemist (or chemistry student) does not need any extra credentials to begin practicing green chemistry. Green chemistry is not the exclusive province of one subdiscipline of chemistry, nor is it a subdiscipline in its own right. It is more of a methodology that can infiltrate any subdiscipline. A product or process does not need to meet or surpass any particular threshold before the principles of green chemistry can be applied to it. Practitioners of green chemistry begin right where they are with what they have. Likewise, thrift is something just about anyone can practice. It does not require a degree in accounting or finance. Although its most obvious or overt application is in the area of finances, it can be applied at the personal level to diet, health, husbandry, and other areas and at the commercial level to different forms of capital. This egalitarian character means that green chemistry has the potential to be practiced very widely among the community of chemists, and thrift has the potential to be practiced broadly among the population as a whole.

Both green chemistry and thrift are productive in the sense of encouraging growth or improvement through industrious, creative labor. The affinity between the ethical assumptions of green chemistry and the ethical assumptions of the productivity stewardship model of environmentalism should make this productive character obvious. There is also a streak of meliorism in both green chemistry and thrift that creates the expectation that the effort expended in such labor will produce a positive outcome for the practitioner and, ideally, for the practitioner's surroundings. At the same time, there is a utilitarian aspect to both green chemistry and thrift in that bodies of empirical evidence continue to grow and show that they function more or less irrespectively of the value sets the practitioners hold.

Both green chemistry and thrift are voluntarily practiced. One of the more striking aspects of green chemistry at the industrial level compared to other activities for the sake of environmental improvement is that the implementation of green chemistry in the United States has not happened by coercion through the threat of legally sanctioned punishment for noncompliance. Companies that have adopted green chemistry techniques have done so because they have seen it as more fitting of their corporate culture, whether because it improves the bottom line, enhances their reputation for social responsibility, or some combination of the two. Thrift is a similar act of self-interest, self-expression, or both.

Both green chemistry and thrift value the cumulative impact of multiple smaller actions. For instance, green chemistry principles 4–11 describe a menu of discrete tasks from which one may choose in order to achieve measurable gains. To be sure, sometimes progress in one principle can be accompanied by regress in another, but at other times progress in one principle begets progress in others. Perusal of the abstracts of projects that have won Presidential Green Chemistry Challenge Awards reveals multiple examples of innovations in which more than one green chemistry principle were applied in a manner that achieved a dramatic outcome (American Chemical Society, 2009). Similarly, thrift can produce seemingly out-sized results. Many of us have firsthand experience of saving money on a regular basis over a period of time in order to afford a major purchase that our budgets would not normally accommodate.

Both green chemistry and thrift are anthropocentric. This anthropocentrism is the case both in the sense that they aim to benefit the practitioner(s) and in the sense that the benefits of their practice can be extended to other people. Anthropocentrism is also the case in the sense that green chemistry and thrift both accept the entirety of human nature rather than one facet. The owner of a company that implements the principles of green chemistry will likely benefit from higher profit margins and/or a higher volume of business. The employees of the company will benefit from increased safety. The owner might choose to pass the cost savings on to the workers in the form of higher pay or benefits, to the stockholders in the form of dividends, to the customers in the form of lower prices, to the company itself in the form of capital improvements, to new employees in the form of business expansion, or to the community in the form of philanthropy. The beneficence associated with thrift is evident in the institutions of thrift. For example, a vegetable garden can provide the gardener with the aesthetic benefits of pleasing colors, shapes, and fragrances; the health benefits of exercise while tending the garden and nutrition from eating the fruits of the garden; the intellectual benefits of planning, cultivating, inspecting, and protecting the garden; the social benefits of conversing with other gardeners, sharing surpluses of produce with friends and neighbors, and teaching prospective gardeners; and the economic benefits of less expensive food.

Both green chemistry and thrift are conservative. That is, they hold that valuable things, such as money and resources, ought to be conserved as much as possible (as opposed to being preserved unchanged from their current state or recklessly squandered). This sort of conservatism is neither an ideology nor inertia, but rather it is a predilection or predisposition toward a particular way of handling things that matter. Green chemistry and thrift are also conservative in the sense of being relatively focused and not cutting wide swaths with sweeping agendas. Whereas, for instance, the United Nations Educational, Scientific and Cultural Organization has an "Educating for a Sustainable Future" program as part of the United Nations Decade of Education for Sustainable Development that attempts to encompass all levels of education and promote specific values and lifestyles with respect to religion, citizenship, social justice, health, and tourism, among other areas, the agenda of green chemistry is limited to the practice of chemistry and does not require what could be interpreted as indoctrination (United Nations Educational, Scientific and Cultural Organization, 2005).

Whether green chemistry and thrift are conservative in the political sense is harder to determine. The political meaning of the word "conservative" differs among countries and among systems of government. What all political conservatives have in common is a desire that public policy maintains and supports what they view as the best traditions and legacies from the past. Such conservation of the best of the past is inherently conservative. One of the ironies of politics in the United States is that the term "conservation" has become the province of environmentalists who tend to favor policies that are politically liberal in the American context. Had conservatives rejected the premise of antipathy or antagonism between environmental and economic goals and instead cast the debate over environmental issues as one of "conservation vs. preservation," perhaps they would own the rhetorical high ground now. We will, of course, never know. Because green chemistry and thrift are anthropocentric, assume the value of human and economic capital as well as natural capital, are productive, and are voluntarily practiced, political conservatives in the United States, Canada, and Western Europe should raise few objections against either.

If green chemistry and thrift are not clearly conservative in the political sense, are they liberal? If all political conservatives want public policy to conserve the best traditions and legacies of the past, then all political liberals want public policy to liberate us from what they view as the worst traditions and legacies of the past. To the extent that waste and hazard qualify as traditions in need of breaking, green chemistry and thrift are liberal, or liberating. Green chemistry and thrift both oppose waste and aim to reduce risk by reducing hazard. In the case of green chemistry, an example would be the substitution of a nonflammable solvent for a flammable one or the substitution of a nontoxic alternative for a toxic reagent rather than mandating a new layer of personal protective equipment. In the case of thrift, an example might be an individual's avoiding debt rather than avoiding debt collectors or a credit union's lending money to an account holder who would not likely be able to repay a loan from another source.

These common traits show that the ethics of green chemistry are the ethics of thrift applied to the chemical enterprise. If the practice of thrift produces a condition of thriving, then the application of thrift to chemistry through the practice of green chemistry should contribute to a thriving chemical enterprise and a thriving environment.

5.4.3 AN ILLUSTRATIVE CASE STUDY

A case study will show how the ethical principles of green chemistry and the common traits between green chemistry and thrift can be put into action. Donaghys Industries of New Zealand has developed a nitrogen response enhancer called LessN, a derivative of several microbes and organic compounds that aid in the uptake and utilization of nitrogen. When mixed with urea, LessN almost doubles the amount of nitrogen uptake by pasture plants per unit of nitrogen input. Because the uptake is more rapid and efficient, loss of nitrogen to the air in the form of ammonia is largely eliminated. Because the pasture plants use the nitrogen before livestock graze, the concentration of nitrogen in the urine of livestock is reduced, so pollution of groundwater through nitrate leaching and pollution of the air through nitrous oxide emissions are also reduced (Underhill and Smellie, 2009).

LessN accomplishes environmental and economic goals simultaneously. The ostensible environmental benefit is that it decreases atmospheric nitrogen emissions and nitrate contamination of groundwater and/or streams. The economic benefit is that dairy farmers can cut their urea usage almost in half without any diminishment of plant growth (i.e., feed). The accomplishment of these goals contributes to sustainable development. Donaghys' testing indicates that the use of LessN can cut dairy farmers' carbon footprint by 10–20%. Preventing or greatly reducing such pollution as nitrate leaching eliminates or greatly reduces the need to extract the nitrates from water supplies. The LessN technology reduces waste by making nitrogen uptake almost twice as efficient. Manufacturing LessN requires using resources of the earth, but the microbial antecedents are nondepleting resources, and the urea with which the LessN is combined is theoretically available from renewable sources. The actual production of urea is energy intensive, so the greater the extent to which the use of LessN decreases the overall demand for urea, the greater the reduction in demand for carbon-rich depleting resources such as natural gas. This additive promotes the welfare of the dairy farmers who use it by reducing their input costs and by sustaining the growth, or even increasing the abundance, of pasture plants available for their livestock to graze. LessN reduces the health risk to infants associated with nitrate-contaminated water by reducing the amount of nitrates that leach or run off into water supplies. The fact that Donaghys Industries has commercialized LessN shows that the people who run the company expect to turn a profit.

LessN bears the marks of thrift. It implicitly acknowledges that reducing the emissions of carbon and carbon equivalents is a worthy goal to pursue. Because LessN shortens the timescale of uptake and utilization and thereby makes these processes more competitive than emission, it tacitly confirms that reducing emissions at the source is more effective than capture or treatment at the end. LessN affirms the principle of subsidiarity because the dairy farmers who use it are free to decide how to fit its use into their overall efforts to reduce carbon emissions. The use of LessN is conservative because it decreases the loss or waste of urea. The use of LessN is productive because it allows the dairy farmers to shrink their carbon footprints without culling their herds or decreasing milk output. Inasmuch as LessN prevents hazardous materials, such as ammonia, nitrous oxide, and nitrates, from entering the air and water, it reduces risk associated with air or water pollution by reducing the inherent hazard. Donaghys Industries voluntarily undertook the development and commercialization of LessN. The development of LessN was democratic in that Donaghys Industries did not need special expertise beyond the biotechnology that is integral to their work. The use of LessN is democratic in that it could appeal to constituencies besides dairy farmers who have an interest in increasing plant uptake and utilization of nitrogen. LessN affirms the cumulative impact of smaller actions because individual dairy farmers can incorporate it into a more extensive program of reducing urea inputs and nitrogen emissions while the dairy farming industry of New Zealand could potentially meet most, if not all, of the collective carbon savings the government would like to see. By offering economic benefits to the owners and employees of Donaghys Industries as well as to the dairy farmers who use it, and health benefits to those who live downstream of dairy farms, the development of LessN was anthropocentric. Finally, the development and use of LessN represents an improved, less wasteful way to make and utilize urea. Therefore, it also represents movement toward the goal of an emission-free process of fertilization.

5.5   GREEN CHEMISTRY AND OUR MORAL OBLIGATIONS

Understanding the ethical principles that undergird the normative character of green chemistry helps show how green chemistry can play a role in contributing to the honoring of our various moral obligations and the resolution of conflicts between or among those obligations. Here the term "moral obligations" refers to the tenets of right or just conduct with respect to other people. The Hippocratic tenet, do no harm, is about as close to a universally accepted principle governing conduct among people as one could hope to find. Those of us who are affluent according to global standards have moral obligations to do no harm to our successors in future generations, to our less affluent contemporaries, to ourselves, and to our predecessors.

5.5.1   Moral Obligations to Our Successors in Future Generations

Our moral obligations to our successors in future generations are arguably those of which we are most keenly aware. Our comparative affluence has enabled us to accumulate wealth in order to create a legacy. We can easily envision passing that legacy to our heirs. Just as the stereotypical parents want their children to enjoy more opportunities and greater success in life than they had, many of us sincerely want future generations to enjoy a standard of living at least as high as that to which we are accustomed. The corollary to this desire is that most of us do not want to impair the ability of our heirs to experience such a standard of living by saddling them with undue burdens. The reason we do not want to burden them is that we recognize they are not here to object to our doing so. Those generations of people who have not yet been born are defenseless and cannot help themselves. The only people who can advocate on their behalf are the people who live now: us. Our advocacy takes the form of protecting their inheritance and not externalizing our costs onto them.

Green chemistry helps to both protect the inheritance and prevent the externalization of costs. Green chemistry helps to protect the economic, natural, and social aspects of the inheritance. It contributes to the protection of the economic inheritance by placing economic goals on an equal footing with environmental and other goals and by accepting the profit motive. It allows economics as a social science to take part in a recursive feedback web with the natural sciences of chemistry and environmental science, ethics, and our praxis. Green chemistry contributes to the protection of the natural inheritance by upholding specific environmental goals, such as preventing pollution, designing for biodegradability, and replacing depleting resources with nondepleting resources. Green chemistry's contributions to the protection of the social aspect of the inheritance come through its anthropocentric emphasis on the welfare of the people who work with or are otherwise exposed to chemical products and processes. This emphasis makes social objectives, such as safety (especially in terms of lower toxicity) and accident prevention, prominent in the principles of green chemistry.

Green chemistry helps to prevent the externalization of both economic and environmental costs to future generations. If economic costs were to be externalized, they would manifest themselves in a lower standard of living. By honoring economic goals, green chemistry takes economic value and viability into consideration. Because green chemistry often adds economic value to products and processes, it promises to make maintenance and enhancement of our standard of living easier. Consequently, maintenance and enhancement of the standard of living of our successors should also be easier. Such an outcome would indicate that we did not impose our costs upon them.

Environmental costs would manifest themselves in the forms of persistent pollution and depletion of resources. Green chemistry directly addresses both of these potential costs by stipulating that no pollution is preferable to pollution, nonpersistent pollution is preferable to persistent pollution, efficient use of resources is preferable to wasteful use of resources, and renewable resources are preferable to nonrenewable resources. The combination of preventing pollution and designing for degradation can potentially abbreviate the lifetime of pollution to such an extent that it fits within the lifetime of our generation.

5.5.2 MORAL OBLIGATIONS TO OUR LESS AFFLUENT CONTEMPORARIES

Our moral obligations to our less affluent contemporaries also include environmental and economic components. Just as we need to guard against externalizing economic and environmental costs to future generations, we need to guard against externalizing those same kinds of costs to people in other parts of the world. Although those people are here and capable of speaking on their own behalf, they do not necessarily have an equal opportunity to do so. Furthermore, their priorities can easily differ from ours. The fact of the matter is that attention to environmental concerns is a luxury. Only after basic needs, such as a reliable food supply, adequate nutrition, the eradication of preventable disease, and so forth, are met can the people of a country shift their focus to more abstract matters. Most commentators on environmental issues are Western-educated individuals addressing predominantly Western audiences. The history of the West suggests that development until a threshold level of affluence is reached is a prerequisite for environmental improvement. On the other hand, this history also suggests that the process of economic development leads to a certain amount of environmental degradation.

This history points to a paradox: If we want other countries to adhere to our environmental standards, we need them to become affluent enough to ensure that their basic needs are met, but if we insist that they comply with our environmental standards, we apparently cannot let them engage in the economic development that will provide them with that affluence. A pernicious consequence of this paradox is that some rich Westerners use various mechanisms to force some poor non-Westerners to stay in poverty under the guise of keeping their carbon footprints small. Expecting others to live a certain way while exempting oneself from the limitations of that way of life is, of course, nothing new. Such behavior was even reflected in sumptuary laws as early as the fourteenth century Italy (Stark, 2005). The staying power of this behavior, however, does not make it any more justified. Blocking economic development in these countries could also be construed as a form of environmental imperialism because it would amount to the West's imposing its environmental values on the people who live in these countries regardless of whether they share the values. Blocking economic development would also open the West to charges of hypocrisy for denying the opportunity to follow the same historical path that the West took.

Green chemistry contributes to the resolution to the paradox. Resolving the paradox requires that economic development take place without the concomitant environmental degradation that has historically accompanied it. Technological innovations stemming from the principles of green chemistry will accomplish environmental and economic goals simultaneously. Through a combination of transfer of this kind of technology from more affluent countries to less affluent countries (or, more accurately, from industries in more affluent countries to industries in less affluent countries) and the generation of such affordable technology within less affluent countries, a more environmentally benign form of economic development can occur to create the affluence that will ensure the basic needs of the people are met. In essence, our less affluent contemporaries will be able to learn from our experience and at least partially bypass the more environmentally harmful phases of our pattern of economic development. They will benefit from the application of thrift without suffering all of the burdens of austerity. The people in these countries who currently live in poverty but want to escape poverty will find that green chemistry, because of its anthropocentric concern for the welfare of humans, affirms their desire. They will not be made to feel guilty for wanting more comfortable lives.

5.5.3 MORAL OBLIGATIONS TO OURSELVES

Our moral obligations to ourselves operate at both the communal and individual levels. At the communal level, our moral obligations are to our affluent contemporaries. Just as we can externalize environmental and economic costs to our affluent successors in future generations and to our less affluent contemporaries in other parts of the world, we can externalize costs to our neighbors and fellow residents. Because the distance of separation is shorter, we perhaps see our costs more clearly and avoid externalizing them more easily. Our fellow residents certainly have an easier task of seeking restitution when we do externalize our costs to them. Their ability to speak for and defend themselves against such costs does not really mitigate against the immorality of externalizing our costs. Our acceptance of our neighbors' right to defend themselves and seek restitution amounts to tacit recognition that externalizing our costs is harmful.

Green chemistry helps to prevent against the externalization of environmental, economic, and social costs to our fellow residents. Environmental costs would most likely manifest themselves in the form of pollution and depletion of resources. Green chemistry directly addresses both costs through its focus on pollution prevention, waste minimization, and renewability. Economic costs would most likely manifest themselves through regulatory mandates and the costs associated with site remediation. The green chemistry goal of pollution prevention should lead to fewer sites in need of remediation and, therefore, should lead to fewer costs associated with such remediation. Green chemistry will not necessarily lower the number of regulatory mandates, but it should make compliance with them less financially burdensome. Most regulations deal with end-of-the-pipe concerns, such as capture and storage or treatment of pollutants. Green chemistry attempts to clean the end of the pipe by cleaning the front and middle of the pipe. When the end of the pipe is cleaned in this fashion, compliance with the regulatory mandates becomes a nonissue. Social costs would most likely manifest themselves in the form of accidental releases of substances with acute toxicity. Green chemistry directly addresses these costs through its focus on safer materials, safer methods, and real-time analysis for accident prevention.

At the individual level, we are morally obligated to not harm ourselves apart from acts of altruism and sacrifice. Green chemistry helps us avoid the physical harm from toxic substances and accidents through its emphasis on inherently safer methods and materials. Green chemistry helps us to stave off economic harm because its application and practice adds value to products and processes and makes them more profitable.

5.5.4   MORAL OBLIGATIONS TO OUR PREDECESSORS

Some people might disagree that we have moral obligations to our predecessors because our predecessors are no longer here and will not be returning to struggle under the load of environmental and economic circumstances we create. Furthermore, we cannot externalize our costs to the past. This ratiocination would be valid if our only moral obligations had to do with avoiding the externalization of environmental and economic costs. We have a different sort of moral obligation to our predecessors. We owe them gratitude for bestowing to us a legacy consisting of economic, social, and environmental capital. Squandering any or all components of that legacy would amount to ingratitude. Because our predecessors took care of the legacy they left for us, we honor them by taking care of that legacy for our successors, for our contemporaries, and for ourselves.

As noted above, not all readers will concur that our moral obligations fall into these four particular categories or will define our obligations within each category exactly the same way. The point of this discussion is not to demonstrate that green chemistry requires complete agreement among its practitioners about these moral obligations. The point, rather, is to demonstrate that someone who acknowledges any or all of these moral obligations can view green chemistry as a useful strategic element in the fulfillment of the obligations. Those who reject the notion that any of us has a moral obligation to anyone else are no more denied the right to practice green chemistry than are people who would add that we have moral obligations to other living things or to a deity.

5.6 CONCLUSIONS

Green chemistry stands in contrast to other areas of Chemistry in that it is normative. Normative systems have a foundation in ethics, and green chemistry is no different. The principles of green chemistry represent the application of the ethics of thrift to the practice of chemistry. This application of the ethics of thrift serves environmental, economic, and social goals. Whereas much of modern environmentalism holds that environmental and economic goals conflict with each other, green chemistry holds that they can and do mesh. This concordism positions green chemistry to serve as an agent of reconciliation between political factions engaged in debate over environmental policy and as a contributor to the resolution of the moral obligations we owe to those who came before us, those who live with us, and those who will come after us.

The episode from Iceland that began this chapter illustrated a dilemma that has reared its head countless times in countless locations. The changes in the infrastructure of society that appear to be necessary to sustain environmental health seem to require a length of time that does not match the urgency of either the need for change or the demands of the lives and livelihoods of the people. Green chemistry offers an avenue by which conservation can be profitable. When conservation is profitable, people and nature can both thrive. Green chemistry is by no means a panacea, but it is well suited to be an integral part of the solution.

REFERENCES

American Chemical Society. 2009. Presidential green chemistry challenge awards. Green chemistry Institute. http://portal.acs.org/portal/acs/corg/content?_nfpb=true&_page Label=PP_SUPERARTICLE&node_id=1341&use_sec=false&sec_url_var=region1&_ uuid=0e632791-8d7f-4fb8-ba6e-66af313a9662 (accessed August 15, 2009).

Anastas, P.T. and Warner, J.C. 1998. *Green chemistry: Theory and Practice*. New York: Oxford University Press, Inc.

Ballor, J. 2004. Preserved garden or productive city? Two competing views of stewardship. http://www.acton.org/commentary/commentary_201.php (accessed August 15, 2009).

Bartlett, A.A. 2004. Thoughts on long-term energy supplies: Scientists and the silent lie: The world's population continues to grow—shouldn't physicists care? *Physics Today*, 57(7):53–55. http://fire.pppl.gov/energy_population_pt_0704.pdf (accessed August 29, 2009).

Beers, M.J., Hittinger, R., Lamb, M., Neuhaus, R.J., Royal, R., and Sirico, R.A. 2000. The Catholic church and stewardship of creation. In *Environmental Stewardship in the Judeo–Christian Tradition: Jewish, Catholic, and Protestant Wisdom on the Environment*, M.B. Barkey (ed.), pp. 25–62. Grand Rapids: Acton Institute for the Study of Religion and Liberty.

Bennett, G.D. 2008. A comparison of green chemistry to the environmental ethics of the Abrahamic religions. *Perspectives on Science and Christian Faith*, 60:16–25.

Blankenhorn, D. 2008. *Thrift: A Cyclopedia*. West Conshohocken, PA: Templeton Foundation Press.

Dardozzi, J. 2009. The spector of Jevons' paradox. Population Press. http://www.population-press.org/publication/2009-2-dardozzi.html (accessed August 29, 2009).

del Guidice, M. 2007. Power struggle: The people of Iceland awaken to a stark choice: Exploit a wealth of clean energy or keep their landscape pristine. *National Geographic*, 213(3):62–89.

Ember, L.R. 1991. Strategies for reducing pollution at the source are gaining ground. *Chemical & Engineering News*, 8:7–16.

Goodwin, T.E. 2004. An asymptotic approach to the development of a green organic chemistry laboratory. *Journal of Chemical Education*, 81:1187–90.

Gottlieb, R.S. (ed.) 2006. *The Oxford Handbook of Religion and Ecology*. New York: Oxford University Press, Inc.

Leopold, A. 2001. *A Sand County Almanac with Essays on Conservation*, illustrated ed. New York: Oxford University Press, Inc.

Lowdermilk, W.C. 1940. The eleventh commandment. *American Forests*, 46:12–15.

O'Neill, B. 2009. Green–industrial complex: Al Gore and his allies know the color of money. *The American Conservative*, 8(11):8–10.

Royal, R. 2006. *The God that Did Not Fail: How Religion Built and Sustains the West*. New York: Encounter Books.

Scott, J. 2009. New study confirms key role for chemical industry in reducing greenhouse gas emissions. http://www.americanchemistry.com/s_acc/sec_news_article.asp?CID=206&DID=9860 (accessed August 17, 2009).

Sittler, J. Jr. 1954. A theology for the earth. *The Christian Scholar*, 37:367–74.

Stark, R. 2005. *The Victory of Reason: How Christianity Led to Freedom, Capitalism, and Western Success*. New York: Random House.

Underhill, J. and Smellie, P. eds. 2009. Homegrown solution helps farmers reduce emissions. Scoop Independent News. http://www.scoop.co.nz/stories/BU0907/S00608.htm (accessed August 17, 2009).

United Nations. 1987. Report of the world commission on environment and development. http://www.un.org/documents/ga/res/42/ares42-187.htm (accessed August 10, 2009).

United Nations Educational, Scientific and Cultural Organization. 2005. Teaching and learning for a sustainable future: A multimedia teacher education programme. http://www.unesco.org/education/tlsf/ (accessed August 21, 2009).

U.S. Code. 2008. http://frwebgate.access.gpo.gov/cgi-bin/usc.cgi?ACTION=RETRIEVE&FILE=$$xa$$busc42.wais&start=47775858&SIZE=3235&TYPE=TEXT (accessed August 15, 2009).

U.S. Environmental Protection Agency. Green chemistry. http://www.epa.gov/opptintr/greenchemistry/pubs/whats_gc.html (accessed March 7, 2007).

White, L. Jr. 1967. The historical roots of our ecologic crisis. *Science*, 155:1203–7.

6 Phytoremediation of Arsenic-Contaminated Environment

An Overview

Ackmez Mudhoo, Sanjay K. Sharma,
Zhi-Qing Lin, and Om Parkash Dhankher

CONTENTS

6.1 INTRODUCTION

Human evolution has led to immense scientific and technological progress. Global development, however, poses new challenges, especially in the field of environmental protection, pollution prevention and control, and ecosystem conservation. Technological ingenuity has enhanced the potential for improving industrial development and rapid progress has been made not only in the field of electronics but also in engineering, biological, medical, and pharmaceutical applications. In recent decades, increasingly precise knowledge of basic biological structures and functions has brought about biotechnological advances (Barceló and Poschenrieder, 2003). The possibility to produce transgenic organisms has opened up new fields of experimentation and perspectives for scientific and technological development, which go beyond the limits of natural evolution. Metal and metalloids [such as lead, arsenic (As), cadmium, copper, zinc, nickel, and mercury] are continuously being discharged in the soil through agricultural activities (such as agrochemicals usage and long-term application of biosolids in agricultural soils) and from industrial sources (such as waste disposal, waste incineration, and vehicle exhausts). Many of the metal pollutants are also known carcinogens (Ensley, 2000). All these sources cause accumulation of metals and metalloids in our soils and water bodies, and pose a serious threat to food safety and public health due to As accumulation and trans-formation in food chains (Khan, 2005). Pollutants in both organic and inorganic forms severely impact human health, productivity of agricultural lands, and sustain-ability of natural ecosystems (Bridge, 2004). Widespread contamination of agricul-tural lands has significantly decreased the extent of arable land available for cultivation worldwide (Grêman et al., 2003). Unlike some organic pollutants, most inorganic pollutants, such as heavy metals and radionuclides, cannot be eliminated by chemical or biological transformation. Although it is possible to reduce their tox-icity by influencing their chemical speciation, heavy metals do not degrade and are generally persistent in the environment. The costs associated with the cleanup of organic and inorganic pollutants can be overwhelming, even for developed countries. Given the nature and extent of contamination worldwide and the costs involved in remediation, there has been a strong drive in recent years in developing alternative yet effective remediation technologies for the cleanup of polluted sites, including microbe-based bioremediation and plant-based phytoremediation (Mrak et al., 2008). Phytoremediation technologies have generated much interest as cost-effective and environmental-friendly technologies for the cleanup of a broad spectrum of hazard-ous organic and inorganic pollutants (Pilon-Smits, 2005). Plant-based environmen-tal remediation has been widely pursued by academic and industrial scientists as a sustainable cleanup technology applicable in both developed and developing nations (Raskin and Ensley, 2000; Robinson et al., 2003).

6.2 PRINCIPLES OF PHYTOREMEDIATION

Phytoremediation is a newly evolving field of science and technology (Salt et al., 1998) that uses plants to cleanup polluted soil, groundwater, and wastewater. This solar energy-driven green technology is often favored over more conventional meth-ods of cleanup due to its low cost, low impact, and wider public acceptance (Gleba et al., 1999; Suresh and Ravishankar, 2004; Rajakaruna et al., 2006). Substantial research efforts are currently underway to realize the economic potential of these green technologies (Ghosh and Singh, 2005) with several plant species now recog-nized as suited for the phytoremediation of nickel, cobalt, thallium, lead, copper, zinc (Anderson et al., 1999; Boominathan et al., 2004), radionuclides, and even As and gold (Mohan, 2005; Anderson et al., 2005; Visoottiviseth et al., 2002; Ma et al., 2001). Phytoremediation is defined as the engineered use of green plants, including grasses and woody species, to remove, contain, or render harmless such environmen-tal contaminants as heavy metals, metalloids, trace elements, organic compounds, and radioactive compounds in soil or water. This definition includes all plant-influenced biological (Zouboulis and Katsoyiannis, 2005), chemical, and physical processes that aid in the uptake, sequestration, degradation, and metabolism of contaminants, either by plants, soil microbes, or plant and microbial interactions. Phytoremediation takes advantage of the unique and selective uptake capabilities of plant root systems, together with the translocation, bioaccumulation, and contaminant storage/degradation abilities of the entire plant body. Plant-based soil remediation systems can be viewed as biological treatment systems with an extensive, self-extending uptake network (i.e., the root system) that enhances the below-ground ecosystem for subsequent productive use. Phytoremediation avoids excavation and transport of polluted media thus reducing the risk of spreading the contamination and has the potential to treat sites polluted with more than one type of pollutant. Some drawbacks associated with phytoremediation are dependency on the growing conditions required by the plant (i.e., climate, geology, altitude, temperature); large-scale operations require access to agricultural equipment and knowledge; tolerance of the plant to the pollutant affect the success for remediation; contaminants collected in senescing tissues may be released back into the environment in certain seasons; contaminants may be collected in woody tissues used as fuel; time taken to remediate sites far exceeds that of other technologies and contaminant solubility may be increased leading to greater environmental damage and the possibility of leaching.

Phytoremediation is essentially comprised of two interactive processes, one associated with soil microbes and the other by plants, which degrades the toxic organic compounds or transforms chemical properties of metals or metalloids (such as As) (Lin, 2005; Suresh and Ravishankar, 2004). Biological degradation of organic pollutants attenuates the level of contamination and generally reduces the toxicity of the compounds via dehalogenation and denitrification leading to breakdown of a complex compound to simple and less or nontoxic products. Phytoremediation of contaminated environments primarily involves the following processes: (1) phytostabilization, (2) phytoextraction, (3) rhizofiltration, (4) phytodegradation, and (5) phytovolatilization for some pollutants.

Phytostabilization can occur through the sorption, precipitation, complexation, or metal valence reduction (Ghosh and Singh, 2005). The plants' primary purpose is to decrease the amount of water percolating through the soil matrix, which may result in the formation of hazardous leachate and prevent soil erosion and transport of the toxicants to the surrounding environment via dense canopies of root systems (Berti and Cunningham, 2000). It is very effective when rapid immobilization is needed to preserve ground and surface water and disposal of biomass is not required. However, the major disadvantage is that, the contaminant remains in soil as it is, and therefore requires regular monitoring.

Phytoextraction is the best approach to remove the contamination primarily from soil and isolate it, without substantially alternating the soil structure and fertility. It is also referred as phytoaccumulation. As the plant absorbs, concentrates, and accumulates toxic metals and radionuclides from contaminated soils and waters into plant tissues, it is best suited for the remediation of diffusely polluted areas, where pollutants occur only at relatively low concentrations and superficial distribution in soil (Rulkens et al., 1998). Several approaches have been studied to enhance the effectiveness of phytoextraction, including the use of chelators to increase the bioavailability and plant uptake of metal contaminants. In order to make this technology feasible, the plants must extract large concentrations of pollutants into aboveground biomass, and produce a large quantity of plant biomass. The removed heavy metals can be recycled from the contaminated plant biomass. Factors such as growth rate, element selectivity, resistance to disease, method of harvesting are also important. However, slow growth, shallow root system, small biomass production, final disposal may limit the use of phytoextraction. Some plants and algae species have the ability to hyperaccumulate various metals in plant cells to very high concentrations (e.g., over thousands of pert per billion). Plant hyperaccumulation of metals involves different detoxification or tolerance mechanisms, such as the action of phytochelatins (PCs) (Cobbett and Goldsbrough, 2002) and metallothioneins forming complexes with metals stored in vacuoles (Suresh and Ravishankar, 2004).

Phytodegradation is the breakdown of organics, taken up by the plant to simpler molecules that are incorporated into the plant tissues (Chaudhry et al., 1998). Plants contain enzymes that can breakdown and convert ammunition wastes, chlorinated solvents (such as trichloroethylene), and other herbicides. The enzymes include usually dehalogenases, oxygenases, and reductases (Black, 1995). Rhizodegradation is the breakdown of organics in the soil through microbial activity of the root zone (rhizosphere). Soil microorganisms can utilize organic pollutants as their carbon and energy sources. Indeed, all phytoremediation processes or technologies are not exclusive and may be used simultaneously. For example, a constructed treatment wetland may involve all the phytoremediation processes for the cleanup of wastewaters contaminated with both metals and organic compounds.

Phytovolatilization involves the use of plants and plant-associated soil microbes to take up contaminants from the soil, transform them into volatile forms, and release them into the atmosphere (Lin, 2008). Phytovolatilization occurs as growing trees and other plants take up water and the organic and inorganic contaminants. Metalloids, such as selenium, As, and tin, can be methylated to volatile compounds or mercury that can be biologically transformed to elemental Hg. Phytovolatilization has been primarily used for the removal of mercury and selenium.

Indeed, all phytoremediation processes or technologies are not exclusive and may be used simultaneously. For example, a constructed treatment wetland may involve all the phytoremediation processes for the cleanup of wastewaters contaminated with both metals and organic compounds.

6.2.1 Plant and Microbial Interactions in Phytoremediation

The phytoremediation process may be viewed as a symbiotic process between plants and soil microbes that involved in phytoremediation (Lasat, 2002). Plant and bacterial interaction can enhance the effectiveness of phytoremediation technology because plants provide carbon and energy sources or root exudates in the rhizosphere that will support microbial community in the degradation and transformation of soil pollutants (Siciliano and Germida, 1998). In addition, the presence of soil microbes can increase the water solubility or bioavailability of pollutants in soils, which facilitates the uptake of pollutants by plants (Lasat, 2002; Siciliano and Germida, 1998). However, the specificity of the plant–bacteria interactions besides being much intricate is dependent upon soil and the aqueous conditions, which can alter contaminant bioavailability, root exudates composition, and nutrient levels (Siciliano and Germida, 1998). In addition, the metabolic requirements for contaminant degradation may also compose the form of the plant–bacteria interaction, that is, plant-associated bacteria versus other general soil bacteria. No systematic framework of reactions and metabolic mechanisms that can predict the plant–bacteria interactions in a contaminated stratum has emerged, but it appears that the development of plant–bacteria associations that degrade contaminants in soil may be related to the presence of allelopathic chemicals in the rhizosphere (Siciliano and Germida, 1998; Lasat, 2002). However, as research progressed in yesteryears, a modest but uncertain understanding of the physiological mechanisms of metal uptake and transport in these plants (Yang et al., 2005) has been developed. Clemens et al. (2002) propose the following mechanism: metal ions (M^{n+}) are mobilized by the secretion of chelators and by the acidification of the rhizosphere; uptake of hydrated metal ions or metal–chelate complexes is mediated by various uptake systems residing in the plasma membrane. Inside the cell, metals are chelated and excess metal is sequestered by transport into the vacuole; from the roots, transition metals are transported to the shoot via the xylem. Presumably, the larger portion reaches the xylem via the root symplast. Apoplastic passage might occur at the root tip. Inside the xylem, metals are present as hydrated ions or as metal–chelate complexes; after reaching the apoplast of the leaf, metals are differentially captured by different leaf cell types and move cell-to-cell through plasmodesmata (Figure 6.1). Storage appears to occur preferentially in trichomes; and uptake into the leaf cells again is catalyzed by various transporters.

Intracellular distribution of essential transition metals is mediated by specific metallochaperones and transporters localized in endomembranes. In other words, the major processes involved in hyperaccumulation of trace metals from the contaminated medium to the shoots by hyperaccumulators as proposed by Yang et al. (2005) include bioactivation of metals in the rhizosphere through root–microbial interaction; enhanced uptake by metal transporters in the plasma membranes; detoxification of metals by distributing metals to the apoplasts such as binding to cell walls and chelation of metals in the cytoplasm with various ligands (such as PCs, metallothioneins, metal-binding proteins); and sequestration of metals into the vacuole by tonoplast-located transporters.

Plants affect the water balance of a site; they change the redox potential and pH, and stimulate microbial activity of the soil (Trapp and Karlson, 2001). These indirect influences may accelerate degradation in the root zone or reduce leaching of compounds to groundwater. Compounds taken up into plants may be metabolized, accumulated, or volatilized into air.

6.3 REMEDIATION OF As POLLUTION—A WORLDWIDE CHALLENGE

In recent years, As pollution has dramatically increased worldwide predominantly from anthropogenic activities. Since As is a potent toxin and highly carcinogenic, the exceptionally high concentrations in the environment have drawn much public attention (Matschullat, 2000). Consequently, remediation of As-contaminated sites through a concerted effort by researchers has become one of the top priorities in developing

Transport of metal ion, M^{n+}, in xylem vessel

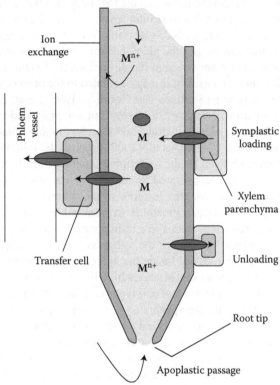

FIGURE 6.1 Molecular mechanisms involved in metal accumulation by plants in the xylem.

environmental remediation technology. Remediation strategies adopted to cleanse the As-contaminated environments consist of physico-chemical processes, which are normally expensive and cumbersome. Historically, the most common technologies for As removal have been coagulation with metal salts, lime softening, and iron/manganese removal. Since the World Health Organization (WHO) Guideline value for As in drinking water was lowered from 50 to 10 μg L^{-1} in 1993, many countries, such as the United States and India (Chen et al., 2007), have also revised their drinking water standards accordingly. As a result, various alternate technologies have been developed or adapted that are capable of removing As from the contaminated environment. While those physical and chemical remediation technologies have all been shown to be effective in lab or pilot studies, there is still relatively limited experience with large-scale treatment in the field. In addition, a number of novel cleanup technologies are under development, and show great promise.

An alternative remediation strategy is to exploit higher plants that are capable of taking up exceptional amounts of metals and metalloids from the soil solution and confine it in the aboveground plant tissues. These plants have been termed as hyperaccumulators and their unique property is exploited in the *in situ* remediation technology of phytoremediation, which is ecologically safe, cost effective, and easy to adopt. These hyperaccumulators have been employed for the phytoextraction of a variety of metals, including As, cadmium, chromium, copper, mercury, nickel, selenium, and zinc (Antosiewicz et al., 2008), and some of these plants have been used in field applications.

6.4 As REMOVAL PROCESSES

6.4.1 General Properties of As

Pure As is a silver-gray, brittle, crystalline, metallic-looking substance that exists in three allotropic forms (yellow, black, and gray) and 35 isotopes. Arsenic has synonyms such as the arsenic-75, metallic arsenic, arsenic black, arsenicals, and colloidal arsenic; however, As is odorless and nearly tasteless. Chemically, As compounds are of two types—inorganic and organic As. Inorganic As is divided into two types—trivalent (arsenite) and pentavalent As (arsenate). As is found mostly in compounds with oxygen, chlorine, and sulfur, which form inorganic As compounds (Lindberg et al., 2008). As in plants and animals combines with carbon and hydrogen to form organic As compounds (Leermakers et al., 2006), and organic As is usually less toxic than more prevalent inorganic arsenite.

6.4.2 As Speciation and Toxicity

In natural waters, As concentrations up to a few milligrams per liter have been reported by Bissen and Frimmel (2003). The natural content of As found in soils varies between 0.01 mg kg^{-1} and a few hundred milligrams per kilogram. As concentrations in soil vary considerably with geographical regions. In the Canadian environment background, values seldom exceed 15 mg kg^{-1}. Rosas et al. (1999) determined the As levels in 50 soil samples collected in the Comarca Lagunera located in Coahuila and Durango, Mexico. The agricultural soil texture of the sampled area was sandy clay loam type with total As levels up to 30 mg kg^{-1}. Following the release of an estimated 5–6 million m^3 of metal-rich sludge and acidic waters into the Rio Guadiamar, SW Spain, in April 1998, Taggart et al. (2004) conducted an extensive set of 0–5 cm soil analyses to determine the residual As levels in these soil samples. It was found that the pseudo-total levels of As in the sludge ranged from 1521 to 3510 mg kg^{-1}. Soils in the Guadiamar Valley and Entremuros areas were found to contain 85.4–782 mg kg^{-1} and 7.1–196 mg kg^{-1} pseudo-total As, respectively. Palumbo-Roe et al. (2005) determined the total As concentrations for 73 ironstone-derived soils (Jurassic ironstones, eastern England) and bioaccessible As determined using an *in vitro* physiologically based extraction test. Palumbo-Roe et al. (2005) found that the bioaccessible As concentration for these soils was found to be well below the soil guideline value (20 mg kg^{-1}) with a mean concentration of 4 mg kg^{-1} and a range of 2–17 mg kg^{-1}.

The toxicity of As in its inorganic form has been known for centuries under the following forms: acute toxicity (Abernathy et al., 1999), subchronic toxicity (Saha et al., 1999), genetic toxicity, developmental and reproductive toxicity (Chakraborti et al., 2003), immunotoxicity, biochemical and cellular toxicity, and chronic toxicity

(Khan et al., 2003). Today, the chronic exposure to arsenic is also known to have an adverse effect on human health. Chronic As ingestion from drinking water has been found to cause carcinogenic and noncarcinogenic health effects in humans (Mandal and Suzuki, 2002). An example of vascular effects of As exposure is the so-called *blackfoot disease*, which was found to be endemic in the southwestern coast of Taiwan associated with As exposure from drinking water (Tseng, 2005). This relationship has been well established in epidemiological studies, most commonly case–control studies, in which the health effects of populations exposed to inorganic As are compared with those who are not exposed. The primary exposure route of inorganic As is ingestion from drinking water due to natural contamination in groundwater from dissolution of natural mineral deposits, industrial effluent, and drainage problems (Mandal and Suzuki, 2002), and of geogenic origin and caused by natural anoxic conditions in the aquifers (Buschmann et al., 2008). Long-term chronic exposure to As increases health risks ranging from conjunctivitis, skin (Ghosh et al., 2007) and internal cancers to diabetes, vascular, reproductive, and neurological effects (Buschmann et al., 2008). As has been determined to be a class A human carcinogen by the Environmental Protection Agency, the Department of Health and Human Services (DHHS), and the WHO from evidence substantiated from global studies. As concentrations in soils, natural waters, and wastewaters are now a worldwide agreed problem and often referred to as a twentieth to twenty-first century calamity (Mohan and Pittman Jr., 2007; Hansen, 2008). The pressing need for environmental protection from excessive levels of As has led to the development of As removal processes all over the world (Chen et al., 2008). Existing As removal technologies reported in the literature of over 2500 may be lumped in the following main categories: oxidation (Yoon and Lee, 2007), precipitation (Mercer and Tobiason, 2008), coagulation, membrane separation (Iqbal et al., 2007), ion exchange (Anirudhan and Unnithan, 2007), biological treatment and removal systems (Pokhrel and Viraraghavan, 2006), chemisorption filtration (Solozhenkin et al., 2007), and adsorption (Pokhrel and Viraraghavan, 2006). Most of the established technologies for As removal make use of several of these processes, either at the same time or in sequence, but remain costly and produce colateral waste streams (e.g., As-laden sludges or residues). Oremland and Stolz (2003) pointed out that As pollution is a problem of critical importance currently affecting the health of millions of people worldwide and hence requires prompt action with regards to the development of green and benign As removal methods. Bearing this line of thought in mind, phytoremediation has been proposed as an effective tool in As cleanup (Alkorta et al., 2004; Dhankher, 2005; Kertulis-Tartar et al., 2006). The subsequent sections of this chapter shall now substantiate on the removal of As from polluted media using phytoremediation and also probe into the related prospects of genetic engineering in this field (Rugh, 2004; Eapen and D'Souza, 2005).

6.4.3 As Hyperaccumulators

As we mentioned earlier, certain plant species are capable of taking up very high amounts of metals from the soil and water. Although As hyperaccumulators bioconcentrate As over 2000 mg kg^{-1} in plant tissues, the biomass production rates of most of hyperaccumulator species are low. Therefore, a lack of rapid growth, large biomass production, and high accumulation capacity render these hyperaccumulator species unsuitable for phytoextraction under field conditions (Bondada and Ma, 2003). In addition, a good As phytoextraction species should accumulate more As in shoots than in roots because of easy harvest or removal of As-laden aboveground biomass. The concentration of the contaminant is generally very high in these plants when grown in contaminated media. To compare the levels of bioconcentration and distribution of As in plants a bioconcentration factor (BF) and transfer factor (TF) can be used. The BF of As is the ratio of the As concentration in plants to the concentration in soil, while the TF is the ratio of the As concentration in roots to the concentration in shoots. Ma et al. (2001) discovered an As-hyperaccumulator species—Chinese brake fern (*Pteris vittata*) that accumulates As in the shoots to a concentration as high as 22,000 mg kg^{-1} (Huang et al., 2004a). Research has demonstrated that other species in the *Pteris* genus also hyperaccumulate As in their shoots. Greenhouse studies (Salido et al., 2003) indicated that *P. vittata* accumulated an As concentration in the aboveground plant tissue more than 200-fold higher than most other plant species tested using As-contaminated soil. In addition, this perennial fern species grows rapidly and generates substantial amounts of biomass, thus making *P. vittata* an excellent candidate to rapidly remove As from As-contaminated environments. The Chinese brake fern is a primitive plant that thrives on As, doubling its biomass in 1 week when subjected to 100 mg L^{-1} As. The striking difference between *P. vittata* and As nonaccumulators is the remarkable transport of As from roots to shoots in *P. vittata*, accumulating up to 95% of the As in the aboveground tissue (Doucleff and Terry, 2002; Zhang et al., 2002). Despite this remarkable As hyperaccumulation ability of *P. vittata*, the molecular and biochemical mechanisms of hyperaccumulation has not been fully explored yet. Once the molecular mechanisms involved in As uptake and transport by *P. vittata* are known, genes responsible for the hyperaccumulation abilities of this plant could be used to transform fast-growing, high-biomass phytoremediators (Doucleff and Terry, 2002). Many other ferns in the *Pteris* genus, *P. longifolia*, *P. cretica*, and *P. umbrosa* (Zhao et al., 2002), as well as a non-*Pteris* fern, *Pityrogramma calomelanos* (Francesconi et al., 2002; Visoottiviseth et al., 2002) also have recently been determined to hyperaccumulate As.

However, not all members of the *Pteris* genus are able to hyperaccumulate As. For example, Meharg (2003) found that *Pteris tremula* and *Pteris stramina* do not hyperaccumulate As. To date the only non-*Pteris* fern to exhibit this ability is *P. calomelanos* (Francesconi et al., 2002). Srivastava et al. (2006) designed experiments to search for new As hyperaccumulators under greenhouse conditions. Their research showed that *Pteris biaurita* L., *P. quadriaurita* Retz, and *P. ryukyuensis Tagawa* could be used as new hyperaccumulators of As. The average As concentration ranged from 1770 to 3650 mg kg^{-1} dry weight (DW) in the fronds and 182–507 mg kg^{-1} DW in the roots of *P. cretica*, *P. biaurita*, *P. quadriaurita*, and *P. ryukyuensis* after having been grown in 100 mg As kg^{-1} soil.

6.4.4 AS HYPERACCUMULATION MECHANISMS

The mechanism of As hyperaccumulation is of great interest because, to most plants, inorganic arsenite species are more toxic. As hyperaccumulation is a trait inherent to

P. vittata because of the similarities in hyperaccumulative behavior noted when comparing ferns from As-contaminated soils or from uncontaminated soils (Wang et al., 2002). Being a nonessential element for plants, As can prove to be fatal at high concentrations due to its interference with metabolic processes for most plants. At low concentrations, As appears to be taken up along with other nutrients, because greater concentrations of As are observed in young fronds. This pattern of localization is correlated with nutrient uptake because nutrients are preferentially directed to actively growing parts of plants. The level of As concentration that *P. vittata* is exposed to have an effect on its growth. When soil As concentrations exceed 500 mg kg^{-1}, the plant biomass was then found to significantly decrease. This indicates that a soil As concentration above 500 mg kg^{-1} has a toxic effect versus the enhancing effect noted at lower concentrations. No significant difference has been noted between adding As in the form of arsenite or arsenate. This may be supposedly due to the oxidation of arsenite to arsenate in aerobic soil (Smith et al., 1998).

Generally, the hyperaccumulation mechanisms of different metals varied. Metal hyperaccumulating species and plant-associated soil microbes may have the ability to solubilize metals from the soil, efficiently take up metal via specific ion transporters, and detoxify specific metal effects on cellular processes by chelation and compartmentation in vacuoles. Physiological, biochemical, and molecular research approaches are continually being applied to identify the underlying mechanisms of metal tolerance by hyperaccumulators. Though research has been carried out to understand the mechanisms of As detoxification and tolerance in *P. vittata* (Luongo and Ma, 2005), it is still relatively uncertain why *P. vittata* is so efficient in As accumulation (Wang et al., 2002). In an 18-d hydroponics experiment with varying concentrations of arsenate and phosphate, *P. vittata* accumulated As in fronds up to 27,000 mg As kg^{-1} DW, and the frond As to root As concentration ratio ranged from 1.3 to 6.7. It was observed that an increase in the phosphate supply decreased the As uptake markedly, with the effect being greater on root As concentration than on shoot As concentration. All the more, the presence of phosphate in the uptake solution decreased arsenate influx largely. An in-depth speciation analysis using high-performance liquid chromatography–inductively coupled plasma–mass spectroscopy (HPLC–ICP–MS) showed that about 85% of the extracted As was in the form of arsenite, and the remaining mostly as arsenate. Wang et al. (2002) concluded that arsenate was taken up by *P. vittata* via the phosphate transporters, and finally sequestered in the fronds primarily as reduced to arsenite (As^{3+}). A similar mechanism was later proposed by Huang et al. (2004b) where synchrotron radiation-based extended x-ray absorption fine structure (EXAFS) was employed to monitor the speciation and transformation of As in the *P. cretica* L. var. *nervosa* Thunb. hyperaccumulator. Huang et al. (2004b) demonstrated that the As in the plant mainly coordinated with oxygen, except that some As coordinated with sulfur as As–glutathione (GSH, γ-glutamylcysteinylglycine) in the roots. Nevertheless, Huang et al. (2004b) brought forward that the complexation of As with GSH might not be the major detoxification mechanism in *P. cretica* L. var. *nervosa* Thunb. It was observed that arsenate tended to be reduced to arsenite in roots after it was taken up, and As was kept as As(III) when it was transported to the aboveground tissues. In their study, on As influx kinetics, Poynton et al. (2004) have conducted short-term unidirectional As influx and translocation experiments with [73]As-radiolabeled arsenate, and found that the concentration-dependent influx of arsenate into roots was significantly larger in *P. vittata* (L.) and *P. cretica* cv. Mayii (L.) than in *Nephrolepis exaltata* (L.). The arsenate influx could be described by the Michaelis–Menten kinetics and the kinetic parameter K_m being lower in the *Pteris* species supported a higher affinity of the transport protein for arsenate. Further quantitative analysis of the kinetic parameters demonstrated that phosphate inhibited the arsenate influx in a directly competitive manner, consistent and in agreement with the inference of Wang et al. (2002) that arsenate enters plant roots on a phosphate-transport protein. Luongo and Ma (2005) have conducted research to understand the mechanisms of As hyperaccumulation in *P. vittata* by comparing the characteristics of As accumulation in *Pteris* and non-*Pteris* ferns. Seven *Pteris* (*P. vittata*, *P. cretica rowerii*, *P. cretica parkerii*, *P. cretica albo-lineata*, *P. quadriavrita*, *P. ensiformis*, and *P. dentata*) and six non-*Pteris* (*Arachnoides simplicor*, *Didymochlaena truncatula*, *Dryopteris atrata*, *Dryopteris erythrosora*, *Cyrtomium falcatum*, and *Adiantum hispidulum*) ferns were exposed to 0, 1, and 10 mg L^{-1} As as sodium arsenate for 14 days in hydroponics. As a group, the *Pteris* ferns were more efficient in As accumulation than the non-*Pteris* ferns, with *P. vittata* being the most efficient followed by *P. cretica*. When exposed to 10 mg L^{-1} As, As concentrations in the fronds and roots of *P. vittata* were 1748 and 503 mg kg^{-1}, respectively. The fact that frond As concentrations in the controls were highly correlated with those exposed to As may then suggest that they may be used as a preliminary tool to screen potential As hyperaccumulators. Therefore, Luongo and Ma (2005) confirmed that the ability of *P. vittata* to translocate As from the roots to the fronds (at an average of 75% As in the fronds) reduce arsenate to arsenite in the fronds, and maintain high concentrations of phosphate in the roots, all contributing synergistically (Glick, 2003) to its very high As tolerance and hyperaccumulation.

6.4.5   PHYTOEXTRACTION OF As BY OTHER PLANT SPECIES

Visoottiviseth et al. (2002) assessed the potential of the native plant species (collected from two areas in Thailand that had histories of As pollution from mine tailings) for phytoremediation. The areas under analysis were the Ron Phibun District and Bannang Sata District. They found that the total As concentrations in soil ranged from 21 to 14,000 µg g^{-1} in Ron Phibun, and from 540 to 16,000 µg g^{-1} in Bannang Sata. Of the 36 plant species, only two species of ferns (*P. calomelanos* and *P. vittata*), a herb (*Mimosa pudica*), and a shrub (*Melastoma malabrathricum*) seemed suitable for potential phytoremediation of As-laden soils. In concert with the findings of Francesconi et al. (2002), Visoottiviseth et al. (2002) equally concluded that the ferns were by far the most proficient plants at hyperaccumulating As from soil, attaining concentrations of up to 8350 µg g^{-1} DW in the frond.

Srisatit et al. (2003) assessed *Vetiveria zizanioides* (Linn.) *Nash* and *Vetiveria nemoralis* (Balansa) for As removal. Both plants were grown for one month, then transplanted in pots containing the soil treated with sodium arsenate (Na$_2$HAsO$_4$·7H$_2$O) at different concentrations (control, 50, 75, 100, 125, and 150 mg As kg^{-1} soil). It was observed that all of the plants grew well in the highest concentration of As.

Accumulation of As in the root of both species was higher than in the leaf. The amount of As accumulation in *V. zizanioides* (Linn.) *Nash* was more than in *V. nemoralis* (Balansa) *A. Camus*. In a different study, Azizur Rahman et al. (2007) have explored the potential of an aquatic macrophyte *Spirodela polyrhiza* L. for phytofiltration of As, and investigated the mechanism of the As uptake, showing that the As uptake into *S. polyrhiza* L. was negatively correlated with phosphate uptake when arsenate was applied to the culture solutions owing to similar sorption mechanism between AsO_4^{3-} and PO_4^{3-}. Thus, the *S. polyrhiza* L. macrophyte was deduced to bioaccumulate As by physio-chemical adsorption and via the phosphate uptake pathway when arsenate was added to the solutions. In contrast, the dimethylarsenic acid (DMAA) accumulation in *S. polyrhiza* L. was neither affected by the phosphate concentration in the culture nor correlated with an iron accumulation in plant tissues, which led to the hypothesis that *S. polyrhiza* L. used different mechanisms for the DMAA uptake. Baldwin and Butcher (2007) found that As preferentially accumulated in the fronds and stems of *P. cretica* cv. *Mayii* compared to roots. Fereshteh and Yassaman (2007) investigated the efficiency of As removal from surface water using macroalga—*Chara vulgaris*—at four initial As concentrations (50, 100, 200, and 300 µg L^{-1}). It was found that the As concentration in algae was high due to a high affinity of the algae for As uptake. The As content of alga increased approximately to about 62.7 mg kg^{-1} DW in 19 days of exposure with initial concentration of 300 µg L^{-1}. The high As accumulation ability of *C. vulgaris* could actually reduce As concentrations by an average of 66.25% in the contaminated water of the bioreactor used in the experiments. In order to shed some light on the hyperaccumulation mechanism of As by *P. vittata*, Fayiga et al. (2007) examined the effects of heavy metals and plant As uptake on soil As distribution. Chemical fractionation of an As-contaminated soil spiked with 50 or 200 mg kg^{-1} nickel, zinc, cadmium, or lead was performed before and after growing the As hyperaccumulator *P. vittata* for 8 weeks using NH_4Cl (water-soluble plus exchangeable, WE–As), NH_4F (Al–As), NaOH (Fe–As), and H_2SO_4 (Ca–As). After 8 weeks of plant growth, the Al–As and Fe–As fractions were significantly greater in the metal-spiked soils than the control, with changes in the WE–As fraction being significantly correlated with plant As removal. Fayiga et al. (2007) deduced that the plant's ability to solubilize soil As from recalcitrant fractions may have enhanced its ability to hyperaccumulate As. Further, Sekhar et al. (2007) tested the hyperaccumulating plant *Talinum cuneifolium*, belonging to the family Portulacaceae, for removal of As from soil in the roots and shoots under laboratory experimental conditions. After 1 month, Sekhar et al. (2007) observed that the leaf As concentration increased, compared to the roots and stems, and that the plant could actually tolerate As concentrations up to 2000 mg kg^{-1} DW, though phytotoxic symptoms appeared later on. Kachenko et al. (2007) broadened the scope of their research by comparing the As hyperaccumulation of *P. calomelanos* var. *austroamericana* with *P. vittata* L. *P. calomelanos* var. *austroamericana* accumulated up to 16,415 mg As kg^{-1} DW. The spatial distribution of As generated using micro-PIXE analysis were further quantitatively analyzed by region selection analysis (RSA). Kachenko et al. (2007) observed that the As concentrations in the pinnule cross-sections were 2.3 times more than in the stipe ones. It was concluded that *P. calomelanos* var. *austroamericana* has the potential for use in phytoremediation of soils with contamination levels up to 50 mg As kg^{-1}. Recently, Alvarado et al. (2008) monitored the removal of As by water hyacinth (*Eichhornia crassipes*) and lesser duckweed (*Lemna minor*) under a concentration of 0.15 mg L^{-1} As and with plant densities of 1 kg m^{-2} for lesser duckweed and 4 kg m^{-2} for water hyacinth (wet basis). No significant differences in the bioaccumulation capability of both species were observed. The removal rate by *L. minor* was 140 mg As ha^{-1} day^{-1} with a removal recovery of 5% while the water hyacinth had a removal rate of 600 mg As ha^{-1} day^{-1} and a removal recovery of 18%, under the conditions of the assay. The removal efficiency of water hyacinth was higher due to the biomass production and the more favorable climatic conditions and could represent a reliable alternative for As phytoremediation in waters (Alvarado et al., 2008).

6.5 CHELATING AGENTS IN As PHYTOREMEDIATION

One of the challenges in phytoremediation is that some metals such as lead are largely immobile in soil and their extraction rate is limited by their solubility or bioavailability (Lombi et al., 2001). To overcome such difficulties, chemically enhanced phytoextraction has been developed (Blaylock, 2000). This approach makes use of high-biomass crops that are induced to take up large amounts of metals when their mobility in soil is enhanced by chemical treatments. PCs, citric acid, ethylenediaminetetraacetic (EDTA), cyclohexanediaminetetraacetic acid (CDTA), diethylene triamine pentaacetic acid (DTPA), ethyleneglycol-bis(β-aminoethyl ether)-*N,N,N',N'*-tetraacetic (EGTA), ethylenediaminedi(*o*-hydroxyphenylacetic) acid (EDDHA), and nitrilotriacetic acid (NTA) have been tested for their ability to mobilize metals in soil and increase metal accumulation in plants (Lombi et al., 2001). Metals chemically bond to organic compounds at multiple active sites to make a ring structure involving the metal and the agent. Metal chelators have been used for plant nutrition and were used originally for correction of iron deficiencies in citrus. EDTA is the most widely used metal chelator, which can be blended into mixed dry fertilizers, slurry, and liquid fertilizers. In the soil, water molecules are coordinated with a metal ion, and are replaced with more stable bi-, tri-, or polydentate groups resulting in a ring formation. Metals are therefore prevented from inactivation in the soil and remain available to plants (Lombi et al., 2001). Synthetic chelators have to be resistant to decomposition by soil microorganisms. Binding is stronger for some metals than for others, and may form hydroxy complexes that are difficult to be absorbed by plants. Several studies have been conducted to analyze the role of chelating agents in the phytoaccumulation of As in plants. Bagga and Peterson (2001) conducted a greenhouse study to identify plants capable of tolerating and accumulating high concentrations of As. The application of 5 mmol kg^{-1} of CDTA to As-contaminated loam soil enhanced the accumulation of As in the test plants. The plants accumulated up to 1400 mg L^{-1} of As as compared to 950 mg L^{-1} when the plants were grown in soil containing 1200 mg L^{-1} of As. Wongkongkatep et al. (2004) studied the As hyperaccumulation in *P. calomelanos* and reported that the accumulation of As in *P. calomelanos* shoot was doubled with the EDTA chelating agent addition. However, dimercaptosuccinic acid (DMSA) resulted in the reverse effect to As accumulation in *P. calomelanos*. According to Wongkongkatep et al. (2004), the reverse effect between EDTA and DMSA chelating agent treatment might be due to strong binding of the thiol group to As ion. *P. calomelanos* plants preferred to take up and translocate As in the form of arsenate and arsenite rather than As–DMSA complex. Azizur Rahman et al. (2008) have also investigated the influence of EDTA and chemical species on As accumulation in aquatic floating macrophyte *S. polyrhiza* L. (giant duckweed). The uptake of inorganic As species in plant tissues and their adsorption on iron plaque of plant surfaces were significantly higher than those of organic species such as monomethylarsenic acid (MMAA) and DMAA. Azizur Rahman et al. (2008) further observed that the addition of EDTA to the culture media increased the uptake of As(V) and As(III) into the plant tissue although the MMAA and DMAA uptake were unaffected. An average of 5% of the inorganic As species were desorbed or mobilized from iron plaque by EDTA.

6.6 As PHYTOREMEDIATION WITH TRANSGENICS

Increasing the quantity of plant degradative enzymes (e.g., peroxidases, laccases, oxygenases), both within the root tissue and excreted into the soil, should increase their utility. Understanding the physiology and biochemistry of metal accumulation in plants is important for several reasons: (1) to allow the identification of agronomic practices capable of optimizing the potential for phytoextraction (Lasat, 2002) and (2) to identify and isolate genes responsible for the expression of the hyperaccumulating phenotype. Metal-hyperaccumulating plants and microbes with unique abilities to tolerate, accumulate, and detoxify metals and metalloids will form an important and unique gene pool (Danika and LeDuc, 2005). These genes could be transferred to fast-growing plant species to enhance the phytoremediation efficiency (Dhankher, 2005; Fulekar et al., 2009). Through genetic engineering modification of physiological and molecular mechanisms of plants, As uptake and tolerance can be successfully achieved (Fulekar et al., 2009), which will open up new avenues for enhancing efficiency of phytoremediation (Lasat, 2002; Eapen and D'Souza, 2005).

A number of transgenic plants have been generated in an attempt to modify the tolerance, uptake, or homeostasis of As. The phenotypes of these plants provide important insights for the improvement of engineering strategies. A better understanding, both of micronutrient acquisition and homeostasis, and of the genetic, biochemical, and physiological basis of metal hyperaccumulation in plants, will be of importance for the success of phytoremediation. It was previously observed by Nie et al. (2002) that transgenic tomato plants that express the *Enterobacter cloacae* UW4 1-aminocyclopropane-1-carboxylate (ACC) deaminase (EC 4.1.99.4) gene, and thereby produce lower levels of ethylene, were partially protected from the toxic effects of six different metals. However, since tomato plants are unlikely to be utilized in the phytoremediation of contaminated terrestrial sites, transgenic canola (*Brassica napus*) plants that constitutively express the same gene were generated and tested by Nie et al. (2002) for their ability to proliferate in the presence of high levels of arsenate in the soil and to accumulate it in plant tissues. The ability of the plant growth-promoting bacteria *E. cloacae* CAL2 to facilitate the growth of both nontransformed and ACC deaminase-expressing canola plants was also tested. In the presence of arsenate, in both the presence and absence of the added plant growth-promoting bacteria, transgenic canola plants grew to a significantly greater extent than nontransformed canola plants. Further, transgenic plants with strong tolerance to As were developed by coexpressing bacterial arsenate reductase, ArsC, and γ-glutamylecysteine synthetase, γ-ECS (Dhankher et al., 2002). The ArsC was expressed under a light-regulated leaf-specific expression vector, SRS1pt, in order to reduce As(V) to As(III) only in aboveground tissues, whereas γ-ECS was expressed constitutively under a strong constitutive expression vector, ACT2pt, to provide As(V) tolerance in the whole plant body. The double transgenic plants were super-resistant to arsenate as compared to the plants expressing ECS alone (Dhankher et al., 2002). These double transgenic plants obtained almost 17-fold higher biomass and hyperaccumulated threefold more As in the aboveground biomass than wild-type plants when grown on 125 mM As(V). This work was the first proof-of-concept for phytoremediation of As-contaminated soil or water by transgenic plants. However, in contrast to *P. vittata* where most of the As is hyperaccumulated in the aboveground frond, most angiosperms retained major fraction (~95%) of As in the roots and only a small amount of total As extracted is translocated to shoot. The speciation of As by x-ray absorption fine structure of shoots and roots of wild-type samples revealed that most of the As was in the form of arsenite–glutathione complexes, As(GS)$_3$ (Dhankher et al., 2002). Therefore, these plants appeared to have high levels of endogenous arsenate reductase activity that convert As(V) into As(III) in roots, immobilizing this toxic species below ground. Thus, the enzymatic reduction of As(V) to As(III) in roots is a major barrier for efficient As translocation from roots to shoots (Pickering et al., 2000; Dhankher et al., 2002). To engineer plants with enhanced phytoremediation potential, Dhankher et al. (2006) focused their efforts on identifying and blocking arsenate reductase activity to further enhance translocation and hyperaccumulation of As in the aboveground tissues. First, plant endogenous arsenate reductase designated as AtACR2 from *Arabidopsis thaliana* was cloned and characterized (Dhankher et al., 2006). The knockdown of AtACR2 by RNAi in Arabidopsis caused translocation of significantly higher As levels from roots to shoots. The AtACR2 RNAi knockdown lines translocated 10- to 16-fold more As in shoots and retained slightly less As in their roots than wild type. Bleeker et al. (2006) also characterized the AtACR2 (named as Arath CDC25) using T-DNA insertion lines and showed the T-DNA knockout lines were sensitive to arsenate. These results suggested that blocking AtACR2 function enhances arsenate transport from roots to shoots.

Recently, endogenous reductase enzymes have been identified and cloned from *P. vittata* (Ellis et al., 2006), and rice (Duan et al., 2007). Rathinasabapathi et al. (2006) cloned another gene from *P. vittata* that controls arsenate reductase activity. The isolated gene had homology to a plant cytosolic triosphosphate isomerase. *Escherichia coli* expressing the fern triose phosphate isomerase gene (TP1) rather than bacterial TP1 had more of the arsenate than the arsenite form, indicating the arsenate reductase activity (Rathinasabapathi et al., 2006). Further, to understand the genetic and biochemical mechanisms underlying As resistance in *P. vittata*, Rathinasabapathi et al. (2006) developed an efficient functional cloning method to rapidly identify *P. vittata* cDNAs based on their ability to increase arsenic resistance when expressed in *E. coli* (Rathinasabapathi et al., 2006). A cDNA for an unusual arsenate-activated glutaredoxin (PvGRX5) was implicated in As metabolism because it improved As tolerance when expressed in *E. coli* (Sundaram et al., 2008). More recently, the overexpression of PvGRX5 in transgenic *A. thaliana* increases plant As tolerance and decreases As accumulation in leaves (Sundaram et al., 2009).

Sizova et al. (2002) inoculated sorghum with *Pseudomonas* bacteria, including strains harboring an As-resistance plasmid, pBS3031, to enhance As extraction by the plant. *Pseudomonas* strains (*P. fluorescens* 38a, *P. putida* 53a, and *P. aureofaciens* BS1393) were chosen because they are antagonistic to many phytopathogenic fungi and bacteria, and they can stimulate plant growth. Thereafter, the resistance of natural rhizospheric pseudomonads to sodium arsenite needs to be assessed. The genetically modified *Pseudomonas* strains resistant to As (III)/As(V) were obtained via a conjugation process. On the basis of their results, Sizova et al. (2002) concluded that their investigated experimental procedures constitute a mechanism making it possible to develop a genetically modified bacterium-assisted phytoremediation technology for the cleanup of As-contaminated soils. Catarecha et al. (2007) have identified an As(V)-tolerant mutant of *A. thaliana* lacking PHT1;1–3, which harbors a semidominant allele coding for the high-affinity phosphate transporter PHT1;1. The pht1;1–3 knockout plants displayed a slow rate of As(V) uptake that ultimately enables the mutant to accumulate double the amount of As found in wild-type plants. Overexpression of the mutant protein in wild-type plants was found to provoke phenotypic effects similar to pht1;1–3 with regard to As(V) uptake and accumulation. In addition, gene-expression analysis of wild-type and mutant plants by Catarecha et al. (2007) has revealed that, in *Arabidopsis*, As(V) represses the activation of genes specifically involved in phosphate uptake, while inducing others transcriptionally regulated by As(V), suggesting that converse signaling pathways are involved in plant responses to As(V) and low phosphate availability. Furthermore, the repression effect of As(V) on phosphate starvation responses may reflect a regulatory mechanism to protect plants from the extreme toxicity of As (Catarecha et al., 2007). For efficient phytoremediation of As, future goals should be focused on combining the expression of those various genes in a tissue-specific manner in a high biomass plant species.

6.7   CONCLUDING REMARKS

The long-term goal remains in developing, testing, and validating vigorous, field-adapted plant species that can restrain As from entering the food chain by extracting As to aboveground tissues, where they can be harvested and disposed of. In concert with the suggestions of Meagher and Heaton (2005), to be able in reaching these goals and apply phytoremediation to As, research and development on native hyperaccumulators and genetic-engineered model plants need to proceed in the following focus areas: (1) increasing the efficiency of plant roots in penetrating and extracting pollutants from contaminated soils; (2) studying modifications in rhizosphere secretion of various enzymes and small molecules, and adjustments in pH to improve the extraction of both essential nutrients and toxic elements; (3) developing short distance and long-distance transport systems for nutrients (Li et al., 2006) in the root system; (4) speciation of the electrochemical state of elemental pollutants to improve their transport from roots up to shoots; (5) developing chemical sinks that may well increase the storage capacity for essential nutrients such as iron, zinc, sulfate, and phosphate; and (6) optimizing the efficiency of physical sinks to store large quantities of a few toxic pollutants aboveground in various native hyperaccumulators. Moreover, Fulekar et al. (2009) identify recombinant DNA technology as a promising approach for manipulating plant characteristics. The future of phytoremediation is promising in research, development, and filed scale applications, but there are certain technical barriers and limitations of phytoextraction (Thangavel and Subburaam, 2004) that need to be addressed. Both agronomic management practices and plant genetic abilities (Lasat, 2002) need to be further explored and optimized for the development of large-scale and commercially feasible phytoremediation practices.

REFERENCES

Abernathy, C.O., Liu, Y.P., Longfellow, D., Aposhian, H.V., Beck, B., Fowler, B., Goyer, R., et al. 1999. Arsenic: Health effects, mechanisms of actions, and research issues. *Environmental Health Perspectives*, 107:593–7.

Alkorta, I., Hernandez-Allica, J., Becerril, J.M., Amezaga, I., Albizu, I., and Garbisu, C. 2004. Recent findings on the phytoremediation of soils contaminated with environmentally toxic heavy metals and metalloids such as zinc, cadmium, lead, and arsenic. *Reviews in Environmental Science and Bio/Technology*, 3:71–90.

Alvarado, S., Guédez, M., Lué-Merú, M.P., Nelson, G., Alvaro, A., Jesús, A.C., and Gyula, Z. 2008. Arsenic removal from waters by bioremediation with the aquatic plants water hyacinth (*Eichhornia crassipes*) and lesser duckweed (*Lemna minor*). *Bioresource Technology*, 99(17):8436–40.

Anderson, C.W.N., Brooks, R.R., Chiarucci, A., LaCoste, C.J., Leblanc, M., Robinson, B.H., Simcock, R., and Stewart, R.B. 1999. Phytomining for nickel, thallium, and gold. *Journal of Geochemical Exploration*, 67:407–15.

Anderson, C.W.N., Moreno, F., and Meech, J. 2005. A field demonstration of gold phytoextraction technology. *Minerals Engineering*, 18:385–92.

Anirudhan, T.S. and Unnithan, M.R. 2007. Arsenic (V) removal from aqueous solutions using an anion exchanger derived from coconut coir pith and its recovery. *Chemosphere*, 66:60–6.

Antosiewicz, D.M., Escudě-Duran, C., Wierzbowska, E., and Skłodowska, A. 2008. Indigenous plant species with the potential for the phytoremediation of arsenic and metals contaminated soil. *Water, Air & Soil Pollution*, 193(1–4):197–210.

Azizur Rahman, M., Hasegawa, H., Ueda, K., Maki, T., Okumura, C., and Rahman, M.M. 2007. Arsenic accumulation in duckweed (*Spirodela polyrhiza* L.): A good option for phytoremediation. *Chemosphere*, 69(3):493–9.

Azizur Rahman, M., Hasegawa, H., Ueda, K., Maki, T., and Rahman, M.M. 2008. Influence of EDTA and chemical species on arsenic accumulation in *Spirodela polyrhiza* L. (duckweed). *Ecotoxicology and Environmental Safety*, 70(2):311–18.

Bagga, D.K. and Peterson, S. 2001. Phytoremediation of arsenic-contaminated soil as affected by the chelating agent CDTA and different levels of soil pH. *Remediation Journal*, 12(1):77–85.

Baldwin, P.R. and Butcher, D.J. 2007. Phytoremediation of arsenic by two hyperaccumulators in a hydroponic environment. *Microchemical Journal*, 85(2):297–300.

Barceló, J. and Poschenrieder, C. 2003. Phytoremediation: Principles and perspectives. *Contributions to Science*, 2(3):333–44.

Berti, W.R. and Cunningham, S.D. 2000. In Raskin, I. (ed.), *Phytoremediation of Toxic Metals: Using Plants to Clean up the Environment*. Wiley-Interscience, John Wiley and Sons, Inc., New York, NY, pp. 71–88.

Bissen, M. and Frimmel, F.H. 2003. Arsenic—A review. Part I: Occurrence, toxicity, speciation, mobility. *Acta Hydrochimica and Hydrobiologica*, 31:9–48.

Black, H. 1995. Absorbing possibilities: Phytoremediation. *Environmental Health Perspectives*, 103(12):1106–8.

Blaylock, M.J. 2000. Field demonstration of phytoremediation of lead contaminated soils. In Terry, N. and Banuelos, G. (eds), *Phytoremediation of Contaminated Soil and Water*. Lewis Publishers, Boca Raton, FL, pp. 1–12.

Bleeker, P.M., Hakvoort, H.W., Bliek, M., Souer, E., and Schat, H. 2006. Enhanced arsenate reduction by a CDC25-like tyrosine phosphatase explains increased phytochelatin accumulation in arsenate-tolerant *Holcus lanatus*. *Plant Journal*, 45:917–29.

Bondada, B. and Ma, L.Q. 2003. Tolerance of heavy metals in vascular plants: Arsenic hyperaccumulation by Chinese brake fern (*Pteris vittata* L.). In Chandra S. and Srivastava M. (eds), *Pteridology in the New Millennium*. Kluwer Academy Publishers, the Netherlands, pp. 397–420.

Boominathan, R., Saha-Chaudhury, N.M., Sahajwalla, V., and Doran, P.M. 2004. Production of nickel bio-ore from hyperaccumulator plant biomass: Applications in phytomining. *Biotechnology and Bioengineering*, 86:243–50.

Bridge, G. 2004. Contested terrain: Mining and the environment. *Annual Review of Environment and Resources*, 29:205–59.

Buschmann, J., Berg, M., Caroline Stengel, C., Winkel, L., Sampson, M.K., Trang, P.T.K., and Viet, P.H. 2008. Contamination of drinking water resources in the Mekong delta floodplains: Arsenic and other trace metals pose serious health risks to population. *Environment International*, 34:756–64.

Catarecha, P., Segura, M.A., Franco-Zorrilla, J.M., Garcia-Ponce, B., Lanza, M., Solano, R., Paz-Ares, J., and Leyva, A. 2007. A mutant of the Arabidopsis phosphate transporter PHT1;1 displays enhanced arsenic accumulation. *The Plant Cell*, 19:1123–33.

Chakraborti, D., Mukherjee, S.C., Pati, S., Sengupta, M.K., Rahman, M.M., Chowdhury, U.K., Lodh, D., Chanda, C.R., Chakraborti, A.K., Basu, G.K., and Chakraborti, D. 2003. Arsenic groundwater contamination in Middle Ganga Plain, Bihar, India: A future danger? *Environmental Health Perspectives*, 111:1194–1201.

Chaudhry, T.M., Hayes, W.J., Khan, A.G., and Khoo, C.S. 1998. Phytoremediation—focusing on accumulator plants that remediate metal-contaminated soils. *Australasian Journal of Ecotoxicology*, 4:37–51.

Chen, Y.-N., Chai, L.-Y., and Shu, Y.-D. 2008. Study of arsenic (V) adsorption on bone char from aqueous solution. *Journal of Hazardous Materials*, 160:168–72.

Chen, Y., van Geen, A., Graziano, J.H., Pfaff, A., Madajewicz, M., Parvez, F., Hussain, A.Z., Slavkovich, V., Islam, T., and Ahsan, H. 2007. Reduction in urinary arsenic levels in response to arsenic mitigation efforts in Araihazar, Bangladesh. *Environmental Health Perspectives*, 115(6):917–23.

Clemens, S., Palmgren, M.G., and Krämer, U. 2002. A long way ahead: Understanding and engineering plant metal accumulation. *TRENDS in Plant Science*, 7(7):309–15.

Cobbett, C. and Goldsbrough, P. 2002. Phytochelatins and metallothioneins: Roles in heavy metal detoxification and homeostasis. *Annual Reviews in Plant Biology*, 53:159–82.

Danika, L. and LeDuc, N.T. 2005. Phytoremediation of toxic trace elements in soil and water. *Journal of Industrial Microbiology and Biotechnology*, 32:514–20.

Dhankher, O.P. 2005. Arsenic metabolism in plants: An inside story. *New Phytologist*, 168:503–5.

Dhankher, O.P., Li, Y., Rosen, B.P., Shi, J., Salt, D., Seneco, J.F., Sashti, N.A., and Meagher, R.B. 2002. Engineering tolerance and hyperaccumulation of arsenic in plants by combining arsenate reductase and c-glutamylcysteine synthetase expression. *Nature Biotechnology*, 20:1140–5.

Dhankher, O.P., McKinney, E.C., Barry P. Rosen, and Meagher, R.B. 2006. Hyperaccumulation of arsenic in the shoots of *Arabidopsis* silenced for arsenate reductase, ACR2. *Proceedings of the National Academy of Sciences USA*, 103:5413–18.

Doucleff, M. and Terry, N. 2002. Pumping out the arsenic. *Nature Biotechnology*, 20: 1094–5.

Duan, G.L., Zhou, Y., Tong, Y.P., Mukhopadhyay, R., Rosen, B.P., and Zhu, Y.G. 2007. A CDC25 homologue from rice functions as an arsenate reductase. *New Phytologist*, 174:311–21.

Eapen, S. and D'Souza, S.F. 2005. Prospects of genetic engineering of plants for phytoremediation of toxic metals. *Biotechnology Advances*, 23(2):97–114.

Ellis, D.R., Gumaelius, L., Indriolo, E., Pickering, I.J., Banks, J.A., and Salt, D.E. 2006. A novel arsenate reductase from the arsenic hyperaccumulating fern *Pteris vittata*. *Plant Physiology*, 141:1544–54.

Ensley, B.D. 2000. Rationale for use of phytoremediation. In Raskin, I. and Ensley, B.D. (eds), *Phytoremediation of Toxic Metals: Using Plants to Clean up the Environment*. John Wiley and Sons, New York, pp. 3–11.

Fayiga, A.O., Ma, L.Q., and Zhou, Q. 2007. Effects of plant arsenic uptake and heavy metals on arsenic distribution in an arsenic-contaminated soil. *Environmental Pollution*, 147:737–42.

Fereshteh, G. and Yassaman, B. 2007. Phytoremediation of arsenic by macroalga: Implication in natural contaminated water, Northeast Iran. *Journal of Applied Sciences*, 7(12):1614–19.

Francesconi, K., Visoottiviseth, P., Sridokchan, W., and Goessler, W. 2002. Arsenic species in an arsenic hyperaccumulating fern, *Pityrogramma calomelanos*: A potential phytoremediator of arsenic-contaminated soils. *The Science of the Total Environment*, 284(1–3):27–35.

Fulekar, M.H., Singh, A., and Bhaduri, A.M. 2009. Genetic engineering strategies for enhancing phytoremediation of heavy metals. *African Journal of Biotechnology*, 8(4):529–35.

Ghosh, M. and Singh, S.P. 2005. A review on phytoremediation of heavy metals and utilization of its byproducts. *Applied Ecology and Environmental Research*, 3:1–18.

Ghosh, P., Banerjee, M., De Chaudhuri, S., Chowdhury, R., Das, J.K., Mukherjee, A., Sarkar, A.K., Mondal, L., Baidya, K., Sau, T.J., Banerjee, A., Basu, A., Chaudhuri, K., Ray, K., and Giri, A.K. 2007. Comparison of health effects between individuals with and without skin lesions in the population exposed to arsenic through drinking water in West Bengal, India. *Journal of Exposure Science and Environmental Epidemiology*, 7:215–23.

Gleba, D., Borisjuk, N.V., Borisjuk, L.G., Kneer, R., Poulev, A., Skarzhinskaya, M., Dushenkov, S., Logendra, S., Gleba, Y.Y., and Raskin, I. 1999. Use of plant roots for phytoremediation and molecular farming. *Proccedings of the National Academy of Sciences USA*, 96(11):5973–77.

Glick, B.R. 2003. Phytoremediation: Synergistic use of plants and bacteria to clean up the environment. *Biotechnology Advances*, 21(5):383–93.

Grêman, H., Vodnik, D., Velikonja-Bolta, Š., and Leštan, D. 2003. Heavy metals in the environment. *Journal of Environmental Quality*, 32:500–6.

Hansen, H.K., Nunez, P., and Jil, C. 2008. Removal of arsenic from wastewaters by airlift electrocoagulation. Part 1: Batch reactor experiments. *Separation Science and Technology*, 43:212–24.

Huang, J.W., Poynton, C.Y., Kochian, L.V., and Elless, M.P. 2004a. Phytofiltration of arsenic from drinking water using arsenic-hyperaccumulating ferns. *Environmental Science and Technology*, 38:3412–17.

Huang, Z., Chen, T., Lei, M., Hu, T., and Huang, Q. 2004b. EXAFS study on arsenic species and transformation in arsenic hyperaccumulator. *Science in China Series C Life Sciences*, 47(2):124–9.

Iqbal, J., Kim, H.-J., Yang, J.-S., Baek, S., and Yang, J.-W. 2007. Removal of arsenic from groundwater by micellar-enhanced ultrafiltration (MEUF). *Chemosphere*, 66:970–6.

Kachenko, A.G., Bhatia, N.P., Singh, B., and Siegele, R. 2007. Arsenic hyperaccumulation and localization in the pinnule and stipe tissues of the gold-dust fern (*Pityrogramma calomelanos*) (L.) Link var. *austroamericana* (Domin) Farw. using quantitative micro-PIXE spectroscopy. *Plant and Soil*, 300:207–19.

Kertulis-Tartar, G.M., Ma, L.Q., Tu, C., and Chirenje, T. 2006. Phytoremediation of an arsenic-contaminated site using *Pteris vittata* L.: A two-year study. *International Journal of Phytoremediation*, 8:311–22.

Khan, A.G. 2005. Role of soil microbes in the rhizospheres of plants growing on trace metal contaminated soils in phytoremediation. *Journal of Trace Elements in Medicine and Biology*, 18(4):355–64.

Khan, M.M., Sakauchi, F., Sonoda, T., Washio, M., and Mori, M. 2003. Magnitude of arsenic toxicity in tube-well drinking water in Bangladesh and its adverse effects on human health including cancer: Evidence from a review of the literature. *Asian Pacific Journal of Cancer Prevention*, 4:7–14.

Lasat, M.M. 2002. Phytoextraction of toxic metals: A review of biological mechanisms. *Journal of Environmental Quality*, 31:109–20.

Leermakers, M., Baeyens, W., De Gieter, M., Smedts, B., Meert, C., De Bisschop, H.C., Morabito, R., and Quevauviller, Ph. 2006. Toxic arsenic compounds in environmental samples: Speciation and validation. *Trends in Analytical Chemistry*, 25:1–10.

Li, Y., Dankher, O.P., Carreira, L., Smith, A.P., and Meagher, R.B. 2006. The shoot-specific expression of g-glutamylcysteine synthetase directs the long-distance transport of thiol-peptides to roots conferring tolerance to mercury and arsenic. *Plant Physiology*, 141:288–98.

Lin, Z.-Q. 2008. Ecological process: Volatilization. In Jorgensen, S.E. and Fath, B. (eds), *Encyclopedia of Ecology*. Oxford, Elsevier, pp. 3700–5.

Lin, Z.-Q. 2005. Bioaccumulation. In Lehr, J.H. and Keeley, J. (eds), *Water Encyclopaedia: Surface and Agricultural Water*. John Wiley & Sons, Inc., Hoboken, NJ, pp. 34–36.

Lindberg, A.-L., Ekström, E.-C., Nermell, B., Rahman, M., Lönnerdal, B., Persson, L.-A., and Vahter, M. 2008. Gender and age differences in the metabolism of inorganic arsenic in a highly exposed population in Bangladesh. *Environmental Research*, 106:110–20.

Lombi, E., Zhao, F.J., Dunham, S.J., and McGrath, S.P. 2001. Phytoremediation of heavy metal-contaminated soils: Natural hyperaccumulation versus chemically enhanced phytoextraction. *Journal of Environmental Quality*, 30:1919–26.

Luongo, T. and Ma, L.Q. 2005. Characteristics of arsenic accumulation by *Pteris* and non-*Pteris* ferns. *Plant and Soil*, 277:117–26.

Ma, L.Q., Komar, K.M., Tu, C., Zhang, W., Cai, Y., and Kennelley, E.D. 2001. A fern that hyperaccumulates arsenic. *Nature* (London), 409:579.

Mandal, B.K. and Suzuki, K.T. 2002. Arsenic round the world: A review. *Talanta*, 58:201–35.

Meagher, R.B. and Heaton, A.C.P. 2005. Strategies for the engineered phytoremediation of toxic element pollution: Mercury and arsenic. *Journal of Industrial Microbiology and Biotechnology*, 32(11–12):502–13.

Meharg, A.A. 2003. Variation in arsenic accumulation-hyperaccumulation in ferns and their allies. *New Phytology*, 157:25–31.

Mercer, K.L. and Tobiason, J.E. 2008. Removal of arsenic from high ionic strength solutions: Effects of ionic strength, pH, and preformed versus *in situ* formed HFO. *Environmental Science and Technology*, 42:3797–802.

Mohan, B.S. 2005. Phytomining of gold. *Current Science*, 88:1021–2.

Mohan, D. and Pittman Jr, C.U. 2007. Arsenic removal from water/wastewater using adsorbents—A critical review. *Journal of Hazardous Materials*, 142:1–53.

Matschullat, J. 2000. Arsenic in the geosphere—A review. *Science of the Total Environment*, 249:297–312.

Mrak, T., Šlejkovec, Z., Jeran, Z., Jaćimović, R., and Kastelec, D. 2008. Uptake and biotransformation of arsenate in the lichen *Hypogymnia physodes* (L.) Nyl. *Environmental Pollution*, 151(2):300–7.

Nie, L., Shah, S., Rashid, A., Burd, G.I., Dixon, D.G., and Glick, B.R. 2002. Phytoremediation of arsenate contaminated soil by transgenic canola and the plant growth-promoting bacterium *Enterobacter cloacae* CAL2. *Plant Physiology and Biochemistry*, 40(4): 355–61.

Oremland, R.S. and Stolz, J.F. 2003. The ecology of arsenic. *Science*, 300:939–44.

Palumbo-Roe, B., Cave, M.R., Klinck, B.A., Wragg, J., Taylor, H., O'Donnell, K.E., and Shaw, R.A. 2005. Bioaccessibility of arsenic in soils developed over Jurassic ironstones in eastern England. *Environmental Geochemistry and Health*, 27:121–30.

Pickering, I.J., Prince, R.C., George, M.J., Smith, R.D., George, G.N., and Salt, D.E. 2000. Reduction and coordination of arsenic in Indian mustard. *Plant Physiology*, 122: 1171–7.

Pilon-Smits, E.A.H. 2005. Phytoremediation. *Annual Review of Plant Biology*, 56:15–39.

Pokhrel, D. and Viraraghavan, T. 2006. Arsenic removal from an aqueous solution by a modified fungal biomass. *Water Research*, 40:549–52.

Poynton, C.Y., Huang, J.W., Blaylock, M.J., Kochian, L.V., and Elless, M.P. 2004. Mechanisms of arsenic hyperaccumulation in *Pteris* species: Root As influx and translocation. *Planta*, 219(6):1080–8.

Rajakaruna, N., Tompkins, K.M., and Pavicevic, P.G. 2006. Phytoremediation: An affordable green technology for the clean-up of metal-contaminated sites in Sri Lanka. *Ceylian Journal of Science*, 35(1):25–39.

Raskin, I. and Ensley, B.D. 2000. *Phytoremediation of Toxic Metals—Using Plants to Clean up the Environment*. J. Wiley & Sons, New York, p. 304.

Rathinasabapathi, B., Wu, S., Sundaram, S., Rivoal, J., Srivastava, M., and Ma, L.Q. 2006. Arsenic resistance in *Pteris vittata* L.: Identification of a cytosolic triosephosphate isomerase based on cDNA expression cloning in *Escherichia coli*. *Plant Molecular Biology*, 62:845–57.

Robinson, B.H., Green, S., Mills T., Clothier, B., Velde, M, Laplane, R., Fung, L., Deurer, M., Hurst, S., Thayalakumaran, T., and Dijssel, C. 2003. Phytoremediation: Using plants as blowups to improve degraded environments. *Australian Journal of Soil Research*, 41:599–611.

Rosas, I., Belmont, R., Armienta, A., and Baez, A. 1999. Arsenic concentrations in water, soil, milk and Forage in Comarca Lagunera, Mexico. *Water, Air & Soil Pollution*, 112:133–49.

Rugh, C.L. 2004. Genetically engineered phytoremediation: One man's trash is another man's transgene. *Trends in Biotechnology*, 22(10):496–8.

Rulkens, W.H., Tichy, R., and Grotenhuis, J.T.C. 1998. Remediation of polluted soil and sediment: Perspectives and failures. *Water Science and Technology*, 37:27–35.

Saha, J.C., Dikshit, A.K., Bandyopadhyay, M., and Saha, K.C. 1999. A review of arsenic poisoning and its effects on human health. *Critical Reviews in Environmental Science and Technology*, 29:281–313.

Salt, D.E., Smith, R.D., and Raskin, I. 1998. Phytoremediation. *Annual Review of Plant Physiology and Plant Molecular Biology*, 49:643–68.

Salido, A.L., Hasty, K.L., Lim, J.-M., Butcher, D.J. 2003. Phytoremediation of arsenic and lead in contaminated soil using Chinese brake ferns (*Pteris vittata*) and Indian mustard (*Brassica juncea*). *International Journal of Phytoremediation*, 5(2):89–103.

Sekhar, K.C., Kamala, C.T., Chary, N.S., and Mukherjee, A.B. 2007. Arsenic accumulation by *Talinum cuneifolium*—Application for phytoremediation of arsenic-contaminated soils of Patancheru, Hyderabad, India. *Trace Metals and Other Contaminants in the Environment*, 9:315–37.

Siciliano, S.D. and Germida, J.J. 1998. Mechanisms of phytoremediation: Biochemical and ecological interactions between plants and bacteria. *Environmental Reviews*, 6(1):65–79.

Sizova, O.I., Kochetkov, V.V., Validov, S.Z., Boronin, A.M., Kosterin, P.V., and Lyubun, Y.V. 2002. Arsenic-contaminated soils genetically modified *Pseudomonas* spp. and their arsenic-phytoremediation potential. *Journal of Soils and Sediments*, 2(1):19–23.

Smith, E., Naidu, R., and Alston, A.M. 1998. Arsenic in the soil environment: A review. *Advancements in Agronomy*, 64:149–95.

Solozhenkin, P.M., Zouboulis, A.I., and Katsoyiannis, I.A. 2007. Removal of arsenic compounds from waste water by chemisorption filtration. *Theoretical Foundations of Chemical Engineering*, 41:772–9.

Srisatit, T., Kosakul, T., and Dhitivara, D. 2003. Efficiency of arsenic removal from soil by *Vetiveria zizanioides* (Linn.) *Nash* and *Vetiveria nemoralis* (Balansa) *A. Camus*. *Science Asia*, 29:291–6.

Srivastava, M., Ma, L.Q., and Santos, J.A.G. 2006. Three new arsenic hyperaccumulating ferns. *Science of the Total Environment*, 364:24–31.

Sundaram, S., Rathinasabapathi, B., Ma, L.Q., and Rosen, B.P. 2008. An arsenate-activated glutaredoxin from the arsenic hyperaccumulator fern *Pteris vittata* L. regulates intracellular arsenite. *Journal of Biological Chemistry*, 283:6095–101.

Sundaram, S., Wu, S., Ma, L.Q., and Rathinasabapathi, B. 2009. Expression of a *Pteris vittata* glutaredoxin PvGRX5 in transgenic *Arabidopsis thaliana* increases plant arsenic tolerance and decreases arsenic accumulation in the leaves. *Plant, Cell and Environment* (doi: 10.1111/j.1365-3040.2009.01963.x).

Suresh, B. and Ravishankar, G.A. 2004. Phytoremediation—A novel and promising approach for environmental clean-up. *Critical Reviews in Biotechnology*, 24(2–3):97–124.

Taggart, M.A., Carlisle, M., Pain, D.J., Williams, R., Osborn, D., Joyson, A., and Meharg, A.A. 2004. The distribution of arsenic in soils affected by the Aznalcóllar mine spill, SW Spain. *Science of the Total Environment*, 323:137–52.

Thangavel, P and Subburaam, C.V. 2004. Phytoextraction: Role of hyperaccumulators in metal contaminated soils. *Proceedings of the Indian National Science Academy*, B70(1):109–30.

Trapp, S. and Karlson, U. 2001. Aspects of phytoremediation of organic pollutants. *Journal of Soils & Sediments*, 1:1–7.

Tseng, C.-H. 2005. Blackfoot disease and arsenic: A never-ending story. *Journal of Environmental Science and Health, Part C*, 23:55–74.

Visoottiviseth, P., Francesconi, K., and Sridokchan, W. 2002. The potential of Thai indigenous plant species for the phytoremediation of arsenic contaminated land. *Environmental Pollution*, 118:453–61.

Wang, J., Zhao, F.J., Meharg, A.A., Raab, A., Feldmann, J., and McGrath, S.P. 2002. Mechanisms of arsenic hyperaccumulation in *Pteris vittata*. Uptake kinetics, interactions with phosphate, and arsenic speciation. *Plant Physiology*, 130:1552–61.

Wongkongkatep, J., Parkpian, P., Polprasert, C., Supaibulwatana, K., Iida, T., and Fukushi, K. 2004. Phytoremediation of arsenic by *Pityrogramma calomelanos*: Do synthetic chelating agents increase or decrease arsenic phytoextraction efficiency? *Annual Report of Interdisciplinary Research Institute of Environmental Sciences*, 22:61–72.

Yang, X., Feng, Y., He, Z., and Stoffella, P.J. 2005. Molecular mechanisms of heavy metal hyperaccumulation and phytoremediation. *Journal of Trace Elements in Medicine and Biology*, 18(4):339–53.

Yoon, S.-H. and Lee, J.-H. 2007. Combined use of photochemical reaction and activated alumina for the oxidation and removal of arsenic(III). *Journal of Industrial and Engineering Chemistry*, 3:97–104.

Zhang, W., Cai, Y., Tu, C., and Ma, L.Q. 2002. Arsenic speciation and distribution in an arsenic hyperaccumulating plant. *Science of the Total Environment*, 300:167–77.

Zhao, F.J., Dunham, S.J., and McGrath, S.P. 2002. Arsenic hyperaccumulation by different ferns species. *New Phytology*, 156:27–31.

Zouboulis, A.I. and Katsoyiannis, I.A. 2005. Recent advances in the bioremediation of arsenic-contaminated groundwaters. *Environment International*, 31(2):213–19.

7 Influence of Nitrates from Agricultural Mineral Fertilizers on Erythrocyte Antioxidant Systems in Infants

Catalina Pisoschi, Ileana Prejbeanu, Camelia Stanciulescu, Oana Purcaru, and Daniela Tache

CONTENTS

7.1 INTRODUCTION

Demographic rise over the last decades has triggered the necessity for an adequate increase of food supply; the problem of alimentary production being generally solved through technical options: new lands given to agriculture and a rise in production on already exploited areas. The latter option was the most important factor that increased agricultural production.

The food supply could be increased due to the use of different hybrid varieties, improved techniques of farming, better seeds, insecticides and herbicides for the eradication of parasites and weeds, but mostly due to the intensive use of natural and synthetic fertilizers. Therefore, the dependence on chemical fertilizers, especially those containing nitrogen, is a hallmark of high-intensity agriculture and since 1960 the worldwide rate of application of nitrogen fertilizers has been increasing (Tilman, 1998).

Soon after this development, several adverse effects became pronounced through the pollution of land, water, and atmosphere. The pollution reached such levels that different governments made laws to reduce it. Land fertilization using nitrogen fertilizers had as a result the pollution of nonagricultural ecosystems through the accumulation of nitrate in vegetables and fodder as well as in the contamination of underground water and well water used for drinking purposes (Bouwer, 2000; Testud, 2004; Ward et al., 2005). In an earlier research, we noted that in 65.3% of the water samples collected from the rural communities of the Dolj County, the nitrates concentration exceeded the maximum contaminant level of 50 mg L^{-1} stipulated by the environmental limits under the laws Law 311/2004 and Law 458/2002 (Prejbeanu et al., 2008). Therefore, groundwater pollution by nitrates is still a real public health hazard due to mineral fertilization of the agricultural lands using nitrogen fertilizers in Romania.

7.2　NITRATE AND NITRITE TOXICITY

The toxicity of nitrate and nitrite is known for a long time. Their biological action became a real problem when many babies consuming vegetables or drinking liquids with an increased content of nitrate and nitrite presented signs of acute and/or chronic intoxication. Nitrate is a naturally occurring ion formed in the soil from nitrite during the nitrogen cycle, being also an important nutrient for plant growth and the main nitrogen compound found in groundwater and surface waters. Nitrate itself is not toxic to humans and animals. In humans, ingested nitrates are absorbed from the small bowel into the body and over 80% are released in urine (Bensoltane et al., 2006). Nitrite is formed by bacterial reduction of nitrate, which escaped elimination at different levels of the digestive system, the reduction reaction being catalyzed by nitrate reductase. Microorganisms contain another enzyme involved in the regulation of nitrate/nitrite balance, the nitrite reductase. If nitrite reductase action correlates with that of nitrate reductase, nitrite will be transformed to other compounds such as nitric oxide and ammonia. Generally, the activity of nitrate reductase overwhelms that of nitrite reductase and nitrite, a more toxic ion, could then be generated in bigger amounts.

Nitrite is reabsorbed into the blood where it could react with the ferrous ion (Fe^{2+}) of hemoglobin to form the ferric ion (Fe^{3+}) of methemoglobin, a form that is unable to transport oxygen. In normal conditions, methemoglobin reductase, a NADH-dependent enzyme is responsible for the reduction of methemoglobin to hemoglobin. Unless favorable conditions exist to reduce nitrate in the digestive system, ingested nitrate is metabolized and excreted without producing adverse effects. However when ingested nitrates are increased or the intestinal flora ascends at sites where the nitrate levels and pH are higher, then the signs of nitrate/nitrite intoxication start to appear. Methemoglobinemia is the most important adverse effect caused by excessive nitrate/nitrite exposure. Depending on the percentage of methemoglobin, clinical signs vary from cyanosis to coma and confusion.

Several *in vitro* studies proved that treatment of intact erythrocytes with nitrites causes the oxidation of hemoglobin to methemoglobin by radical generation along with a decrease in reduced glutathione (GSH) level associated with erythrocyte membrane dysfunctions and namely altered cell ionic flux, lipid peroxidation, and perturbation of membrane transport (Batina et al., 1990; May et al., 2000). Nitrate/nitrite-induced oxidation of biological molecules potentiates reactions, which interfere in the oxidative chain and which can affect some antioxidant systems.

Superoxide dismutases (SODs) are enzymes, which catalyze the dismutation of superoxide into hydrogen peroxide and oxygen. There are multiple SODs that depend on their activity for the presence of copper, zinc, manganese, iron, and even nickel at the active site (Fridovich, 1998). SODs could be cytosolic, localized in different cell organelles and secreted from the cell. Peroxidases are enzymes that use a variety of electron donors to reduce hydrogen peroxide to water. The most important peroxidase in mammals is glutathione peroxidase (GPx). Reduced glutathione is oxidized to the disulfide form during its action and glutathione reductase (GR) reconverts it into reduced glutathione using NADPH, H^+ as the hydrogen donor. GPx is necessary not only for the elimination of hydrogen peroxide but also for acting on hydroperoxides. As SODs, there are several GPx with special localizations, all of them containing selenium (Fridovich, 1998). In this chapter, we evaluate the effect of acute nitrate intoxication on several parameters linked to the oxidative stress: SOD, GPx, GR, malondialdehyde (MDA), which is a marker of lipid peroxidation. This work has been completed as part of a research supported by the Romanian Academy during 2007–2008.

7.3 STUDY PROTOCOL

7.3.1 PATIENTS AND SAMPLE COLLECTION

Twelve babies aged between 3 and 8 weeks had been selected from those hospitalized for infant methemoglobinemia in the Pediatrics Clinic of the Emergency Clinical Hospital Craiova from October 2007 to June 2008, according to their clinical signs (cyanosis, dyspnea, lethargy) correlated with rapid evaluation of methemoglobin and hemoglobin. They were from several rural communities of Dolj County, Romania, areas known for the intense use of nitrogen mineral fertilizers. Information about the type of water used to prepare tea and baby milk was obtained from mothers and when it was possible, samples of well water were collected for chemical analysis on nitrates. Other five babies appropriate for age living in areas where manure was used instead of mineral fertilizers served as the control group for this evaluation. Informed consent for all diagnostic procedures was obtained from mothers of all infants. Venous peripheral blood samples was drawn in heparinized tubes and processed for plasma separation, hemolysis, and evaluation of some oxidative stress markers.

7.3.2 ERYTHROCYTE HEMOLYSIS

Aliquots of 0.5 mL whole blood were centrifuged for 10 min at 3000 rpm. Plasma was aspirated and the cells were washed four times with 3 mL physiological solution, and centrifuged for 10 min at 3000 rpm after each wash. The washed and centrifuged erythrocytes pellet was made up to 2 mL with cold redistilled water, mixed and kept at 4°C for 15 min in order to induce cell lysis. Erythrocyte lysate was then diluted to the appropriate dilution order and subsequent analysis of the antioxidant enzymes was performed.

7.3.3 MEASUREMENT OF GPx ACTIVITY

GPx activity was measured using the Ransel kit (Randox Laboratories Ltd.). According to this method, GPx from the sample catalyzes transformation of GSH to the oxidate form (GSSG) in the presence of cumene hydroperoxide and then GR reduces GSSG using NADPH, H+ as coenzyme. The decrease in absorbance at 340 nm is proportional with enzyme activity. GPx activity was expressed in U L^{-1} of whole blood and reported to the normal range for this method.

7.3.4 MEASUREMENT OF GR ACTIVITY

GR activity was measured using a kit from Randox Laboratories. According to this method, GR catalyzes transformation of oxidized glutathione (GSSG) in reduced glutathione (GSH) in the presence of NADPH, H+. The decrease in absorbance at 340 nm after the oxidation of the coenzyme is proportional with enzyme activity. GR activity was expressed in U gHb^{-1} and reported to the normal range.

7.3.5 MEASUREMENT OF SOD ACTIVITY

SOD activity was assessed using a Ransod kit (Randox Laboratories Ltd.) that quenches the rate of inhibition of 2-(4-iodophenyl)-3-(4-nitrophenol)-5-phenyltetrazolium (INT) reduction by the superoxide anion released after xanthine oxidation with xanthine oxidase. One unit of SOD causes a 50% inhibition rate and the activity was expressed as U mL^{-1} of whole blood and reported to the normal range. In each case, appropriate control serum for quality control had been used.

7.3.6 MEASUREMENT OF MDA LEVEL

For measurement of MDA, venous blood samples were centrifuged for 10 min at 3000 rpm at room temperature. MDA was estimated using a solution of thiobarbituric acid in trichloroacetic acid 20% according to the reaction mixture absorbance at 532 nm. All the chemicals used for this procedure were supplied from Sigma (St. Louis, USA). The results were expressed as nmol MDA mL^{-1} of plasma.

7.3.7 STATISTICAL ANALYSIS

All measurements are expressed as mean ± SD (standard deviation). A Pearson correlation was performed for testing relationships between certain parameters. The data were analyzed by analysis of variance (ANOVA). Differences between groups were considered significant at $P < 0.05$.

7.4   RESULTS

Table 7.1 presents the most important biochemical features obtained for the infants included in the study. All water samples collected from the areas known for an intense use of nitrogen mineral fertilizers exceeded the maximum contaminant level for nitrates of 50 mg L^{-1}, thereby indicating that the acute nitrate/nitrite intoxication caused the clinically evaluated methemoglobinémia.

We noted a decreased activity of GPx in infants which ingested nitrates via drinking liquids (N group) compared to the control group (Figure 7.1), six of them having values under the minimum limit of the normal range. Regarding GR, its activity was in the normal range for infants from the N group but lower when compared to the controls (Figure 7.2).

TABLE 7.1
Activities of Blood Antioxidant Enzymes and Markers of Lipid Peroxidation

	Control ($n = 5$)	N group ($n = 12$)
SOD activity (U mL^{-1})	176.8 ± 37.03[a]	111.33 ± 28.6
GPx activity (mU mL^{-1})	12,144 ± 1957	4071 ± 749
GR (U gHb^{-1})	8.74 ± 1.86	5.06 ± 1.55
MetHb (%)	0.95 ± 0.47	47.5 ± 4.16
Plasma MDA (nmol mL^{-1})	0.98 ± 0.13	2.4 ± 0.35

[a] Data represent means ± SD.

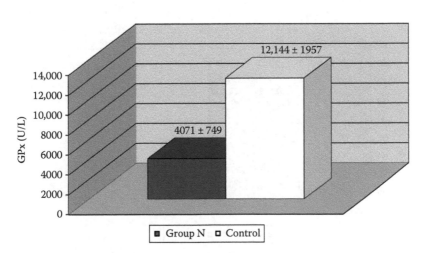

FIGURE 7.1   Erythrocyte GPx activities in N group (nitrate intoxicated infants) and controls ($P < 0.0001$). Data represent means ± SD.

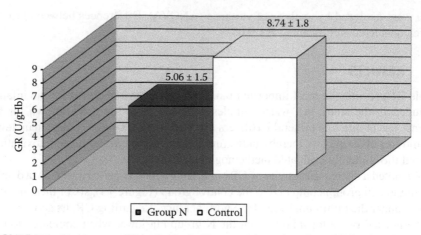

FIGURE 7.2 Erythrocyte GR activities in N group (nitrate intoxicated infants) and controls ($P < 0.001$). Data represent means ± SD.

Nitrates intake had a bigger influence on SOD activity, all infants from the N group displaying lower activity of this enzyme comparing with the control group (Figure 7.3) and with the normal range.

The intensity of lipid peroxidation estimated as thiobarbituric acid reactive substances, mainly MDA, was higher in infants subjected to nitrate intake (N group) comparing to the controls (2.4 ± 0.35 nmol mL^{-1} vs. 0.98 ± 0.13 nmol mL^{-1}) (Figure 7.4).

From Pearson's correlation coefficients, we noted a negative weak correlation between methemoglobin and antioxidant enzymes for the N group ($r = -0.35$ for

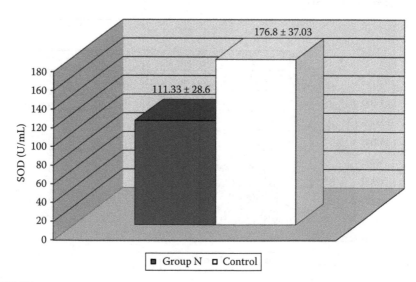

FIGURE 7.3 Erythrocyte SOD activities in N group (nitrate intoxicated infants) and controls ($P < 0.001$). Data represent means ± SD.

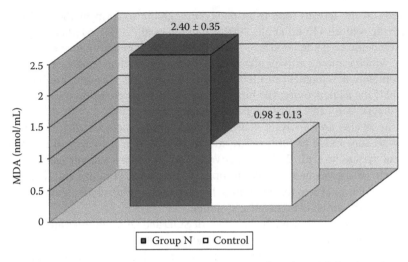

FIGURE 7.4 Plasma MDA levels in N group (nitrate intoxicated infants) and controls ($P < 0.0001$). Data represent means ± SD.

GPx, $r = -0.38$ for SOD, respectively, $r = -0.21$ for GR), although the number of subjects is not big enough. GPx and GR had a significant positive correlation ($r = -0.78$). A weak positive correlation was obtained between methemoglobin and MDA level. SOD and GPx showed a significant negative correlation with MDA ($r = -0.75$ for GPx, respectively, $r = -0.83$ for SOD), meanwhile GR did not correlate well with MDA level. We did not obtain a significant correlation between the levels of nitrates in water (between 175 and 610 mg L^{-1} for the samples collected) and the activity of the antioxidant enzymes. We cannot conclude that their influence does not exist because we tested only a small number of water samples.

7.5 DISCUSSION AND CONCLUSIONS

Our investigation compares the levels of some oxidative stress markers in infants drinking milk and tea prepared with nitrates-polluted water with those obtained for babies living in areas where people do not use mineral fertilizers to improve agricultural performances. The use of polluted water to prepare babies food suggests the lack of information in those rural communities according to the danger involved by the use of nitrates-polluted water. The results displayed suggest that acute exposure to nitrates could alter some markers of the oxidative balance in a different manner in correlation with methemoglobinemia. There are other similar studies that have examined the relationship between nitrate levels in drinking water and methemoglobinemia in infants (Mensinga et al., 2003; Ward et al., 2005), between the rate of nitrite uptake and oxidative stress in human erythrocytes *in vitro* (May et al., 2000) but we have not found any study related to the effects presented in our work.

Nitrate/nitrite-induced methemoglobinemia is an important event that generates free radicals (Kohn et al., 2002). It is well known that nitrate/nitrite could enhance oxidation of the ferrous ions from deoxyhemoglobin to form methemoglobin and superoxide radicals (Kohn et al., 2002; Gladwin et al., 2004). Methemoglobinemia observed could be potentiated by several conditions: a predominance of fetal hemoglobin, which is more easily oxidized to methemoglobin than adult hemoglobin, and an incomplete development of methemoglobin reductase, this enzyme having only about half the adult activity; the bigger quantity of water containing nitrates that an infant ingests daily kg^{-1} body weight comparing to an adult.

Nitrite is also an important source of nitric oxide, molecule that could rapidly react with superoxide to form peroxynitrite (ONOO$^-$), a potent cytokine which is very reactive (Kohn et al., 2002). Reactive species of oxygen and nitrogen could initiate a toxic oxidative chain, including lipid peroxidation, protein oxidation, directly inhibiting some enzymes from the mitochondrial respiratory chain, and causing dysfunctions of the antioxidant defense systems.

Our investigations showed the decrease of SOD and GPx activities and an increase of MDA levels, which consequently caused acute nitrate/nitrite intoxication in infants.

The decrease of GPx activity could be related to peroxynitrite generation. Peroxynitrite could react with selenium from the GPx active site and inactivate this enzyme (Gonchar et al., 2006). Decrease of this enzyme activity has been correlated with that of GR, another important enzyme involved in the regulation of the GSH/GSSG ratio, proving that glutathione system is altered. SOD decrease could be explained by the use of this enzyme for superoxide metabolism. The absence of a good correlation between antioxidants and drinking water nitrate level suggests that it is possible that other mechanisms could be involved in this relation.

In conclusion, nitrates pollution of rural domestic wells continues to represent an important concern for the human and environment health in the Dolj County. The occurrence of new cases of infant methemoglobinemia in these rural communities suggests serious deficiencies in the conditions of life in villages and in the sanitary educational status of their inhabitants.

Methemoglobinemia induced by the acute nitrates/nitrites exposure of infants through drinking water enhances the oxidative reactions and a significant perturbation of enzyme antioxidant systems. Further assessments taking into account the relation between drinking water nitrates and other systems involved in the antioxidant protection are hence much necessary.

7.6 PERSPECTIVES OF GREEN CHEMISTRY TO SUSTAINABLE AGRICULTURE

One way to minimize the effects of nitrate fertilizers toxicity is to pay increased attention to practices that can provide sustainable yields, comparable to those of high-intensity agriculture. Sustainable agriculture, a concept that is not new, is part of a larger movement toward environmental sustainability, which recognizes that natural resources are finite, acknowledges limits on economic growth and encourages equity in resource allocation (Horrigan et al., 2002). One of the goals of sustainable agriculture is to create farming systems that eliminate environmental harms associated with high-intensity agriculture. Soil management is one method which is able to increase agricultural sustainability and involves the management of soil

chemical, biological, and physical properties. Mineral fertilizers alter all soil chemical and physical properties, and at the end, plant growth and water/soil quality. A
more novel concept is that green chemistry provides substance to scientific research
and contributes to the increase in agricultural productivity and sustainability
(Hjeresen and Gonzales, 2002). In sustainable agriculture, desired qualities and
areas of interest are well described and these offer green chemistry novel possibilities for fruitful research according to its basic principles.

In comparison with conventional agricultural methods, organic alternatives/
additives can improve soil fertility and have fewer detrimental effects on the environmental ecosystems. These alternatives can also produce equivalent crop yields to
conventional methods.

Several organic alternatives represent a return to traditional agriculture in agreement with the practices for environmental sustainability. One of these alternatives is
to use, as a crops fertilizer, manure which is rich in nitrogen because of the intense
feeding of cattle with grasses and legumes (Drinkwater et al., 1998). It has been
reported that the soil organic matter and nitrogen contents increased markedly in the
manure system than in the legume system (Drinkwater et al., 1998). The organic
methods could be superior to conventional agriculture because they supply nitrogen
in organic forms that gradually release mineral nitrogen, and probably better correlate
the nutrient availability with plant needs, and being harmless for the environment at
all times since the organic alternatives have the advantage of being biodegradable.

Biofertilizers or bioinoculants are formulates of beneficial microorganisms (bacteria and fungi), which when applied to soil, roots, or seeds enhance the availability
of different nutrients to the plant because they are able to fix atmospheric nitrogen,
solubilize phosphorus, and promote uptake of immobile ions and other micronutrients (Gupta et al., 2007).

The green chemistry principles can be precisely applied to a greener and far less
harmful utilization of fertilizers. In this perspective, green chemistry procedures are
involved in the synthesis of enhanced-efficiency nitrogen fertilizers, most of them
being based on the use of urea, in such forms that impede too fast a degradation and
elimination of nitrogen.

Urea undergoes microbial hydrolysis catalyzed by urease, leading to loss of as
much as 30% of its nitrogen from ammonia volatilization. The reduced nitrogen
availability in the soils appears particularly when urea is surface broadcast on soils.
The factors that influence ammonia volatilization include levels of urease activity,
moisture availability, nitrification rate, and soil texture (Bernard et al., 2009).

Several alternatives are used to reduce ammonia elimination. Applied in relatively small quantities, urease inhibitors such as N-(n-butyl) thiophosphoric acid tri-
amide reduce the rate of microbial hydrolysis of urea and increase its efficiency as a
fertilizer (Manahan, 2005). Ammonia volatilization could also be reduced using a
mixture of urea with tropical peat soil or free humic substances, such as humic and
fulvic acids, isolated from peat soils (Bernard et al., 2009). Another application of
green technologies is the use of thermal polyaspartate, a product formed by the condensation and base treatment of a natural compound, aspartic acid. This has been
found to be effective in stimulating plant uptake of fertilizer thus reducing the amount
of fertilizer required (Manahan, 2005).

In recent years, the conclusions of many scientific meetings proved that green chemistry provides solutions and strategies to improve agricultural sustainability. Green chemistry is also a platform for a collaborative effort among the chemical industry, research sector, and the authorities, which are all actually involved in environmental health management projects designed to promote the adoption of green methods and practices.

ACKNOWLEDGMENT

This work was supported by the Romanian Academy, Grants 137/2007 and 133/2008.

REFERENCES

Batina, P., Fritsch, P., de Saint, B., and Mitjavila, M.T. 1990. *In vitro* kinetics of the oxidative reactivity of nitrate and nitrite in the rat erythrocyte. *Food Additives & Contaminants Part A*, 1:145–9.

Bensoltane, S., Messerer, L., Berrebbah, H., Djekoun, M., and Djebar, M.R. 2006. Effects of acute and subchronic ammonium nitrate exposure on rat liver and blood tissues. *African Journal of Biotechnology*, 5(9):749–54.

Bernard, R., Ahmed, O.H., Majid, Ab., N.M., and Jalloh, M.B. 2009. Reduction of ammonia loss from urea through mixing with humic acids isolated from peat soil (Saprists). *American Journal of Environmental Sciences*, 5(3):393–7.

Bouwer, H. 2000. Integrated water management: Emerging issues and challengers. *Agricultural Water Management*, 45:217–28.

Drinkwater, L.E., Wagoner, P., and Sarrantonio, M. 1998. Legume-based cropping systems have reduced carbon and nitrogen losses. *Nature*, 396:262–5.

Fridovich, I. 1998. Oxygen toxicity: A radical explanation. *Journal of Experimental Biology*, 201:1203–9.

Gladwin, M.T., Crawford, J.H., and Patel, R.P. 2004. The biochemistry of nitric oxide, nitrite and hemoglobin: Role in blood flow regulation. *Free Radical Biology & Medicine*, 36:707–17.

Gonchar, O., Mankovskaya, I., and Klyuchko, E. 2006. Role of complex nucleosides in the reversal of oxidative stress and metabolic disorders induced by acute nitrite poisoning. *Indian Journal of Pharmacology*, 38(6):414–18.

Gupta, R.P., Kalia, A., and Kapoor, S. 2007. *Bioinoculants: A Step Towards Sustainable Agriculture.* New India Publishing Agency, New Delhi.

Hjeresen, D.L. and Gonzales, R. 2002. Can green chemistry promote sustainable agriculture. *Environmental Science and Technology*, 36:103–7.

Horrigan, L., Lawrence, R.S., and Walker, P. 2002. How sustainable agriculture can address the environmental and human health harms of industrial agriculture. *Environmental Health Perspectives*, 110:445–56.

Kohn, M.C., Melnick, R.L., Ye, F., and Portier, C.J. 2002. Pharmacokinetics of sodium nitrite-induced methemoglobinemia in the rat. *Drug Metabolism and Disposition*, 30:676–83.

Manahan, S.E. 2005. *Environmental Chemistry.* 8th Edition, CRC Press, Taylor & Francis Group, Boca Raton, Florida, USA, pp. 466–9.

May, J.M., Qu, Z.C., Xia, L., and Cobb, C.E. 2000. Nitrite uptake and metabolism and oxidant stress in human erythrocytes. *American Journal of Physiology—Cell Physiology*, 279:1946–54.

Mensinga, T.T., Speijers, G.J.A., and Meulenbelt, J. 2003. Health implications of exposure to environmental nitrogenous compounds. *Toxicological Reviews*, 22(1):41–51.

Prejbeanu, I., Pisoschi, C., Rada, C., Stanciulescu, C., Purcaru, O., Puiu, I. and Tache, D. 2008. Nitrate exposure via groundwater pollution and its influence on the antioxidant defense mechanisms. *Proceedings of the International Conference "Environmental Pollution and its Impact on Public Health,"* organized by the Balkan Environmental Association, the 16th–18th July, Brasov, Romania.

Testud, F. 2004. Inorganic fertilizers. *EMC—Toxicologie Pathologie*, 1:21–8.

Tilman, D. 1998. The greening of the green revolution. *Nature*, 396:211–12.

Ward, M.H., de Kok, T.M., Levallois, P., Brender, J., Gulis, G., Nolan, B.T., and Van Derslice, J. 2005. Workgroup report: Drinking-water nitrate and health—Recent findings and research needs. *Environmental Health Perspectives*, 113(11):1607–14.

ADDITIONAL SOURCES OF INFORMATION

Law 458/2002 regarding the drinkable water quality.
Law 311/2004 to modify and complete the Law 458/2002.

8 Microwave-Assisted Organic Reactions

Abdul Rauf and Shweta Sharma

CONTENTS

8.1 BACKGROUND TO MICROWAVE CHEMISTRY

The microwave region of the electromagnetic spectrum lies between 1 cm and 1 m and in order to avoid interfering with radar and telecommunication activities that operate within this region, most domestic and commercial microwave instruments operate at 2.45 GHz. The heating effect utilized in microwave-assisted organic transformations is due in the main, to dielectric polarization, although conduction losses can also be important particularly at higher temperatures. When a molecule is irradiated with microwaves, it rotates to align itself with the applied field. The frequency of molecular rotation is similar to the frequency of microwave radiation and consequently the molecule continually attempts to realign itself with the changing field and energy is thus absorbed.

Microwave-assisted organic synthesis (MAOS) can be divided into two main categories:

a. *Nonsolid-state reactions*: Reactions conducted in solvent (organic synthesis in homogenous media). The reactions utilizing organic solvent can be carried out efficiently under microwave irradiations. The experimental procedure can be extremely simple when reactions can be carried out in open Erlenmeyer flasks and subjected to short periods of irradiation. It is necessary that such reactions should be carried out if neither the solvent nor the reactants or products are flammable, otherwise there is a serious risk of fire or explosion due to sparking. The experiments utilizing this protocol cover a number of organic solvents among which, results with dimethylformamide (DMF) are highly appreciable. In addition to the significant practical simplicity of this type of procedure, it is sometimes found that volatile reaction by-products evaporate rapidly thus avoiding methods for their specific removal. Reactions may also be carried out in sealed systems.

b. *Solid-state reactions*: Solvent-free reactions.

Within this category of solvent-free reactions are three subdivisions:

1. Solvent-free reactions without support or catalyst (neat reagents)
2. Solvent-free reactions using supported reagents (adsorbed on the support, noncovalently linked to it), such as clays, silica-gel, zeolites and graphite
3. Solvent-free phase transfer catalysis

A number of interesting reports describing the utility of solid-state reactions are available in literature. There are commonly two types of microwave-assisted dry reactions; one in which the reagents are "supported" on a microwave inactive (or poorly active) material such as alumina/silica. In this type of reaction at least one of the reagents must be polar if the reaction is to be performed under microwave irradiation. The second type of dry reactions are those which utilize microwave active solid support. Thus, the reactants do not have to be microwave active. Solid-state reactions are of course very convenient from a practical view point. In general, the reagents and solid support are efficiently mixed in an appropriate solvent, which is then evaporated. The adsorbed reagents are then placed in a vessel and subjected to microwave irradiation after which the organic products are simply extracted from the support by washing and filtration. The absence of solvent coupled with the high yields and short reaction times often associated with reactions of this type make these procedures very attractive for synthesis.

Microwave-assisted synthetic methodologies can also be divided into two main categories: (a) methodologies in which the reactant(s) or reagent(s) are previously covalently linked to a soluble (e.g., polyethylene glycol) or nonsoluble (e.g., Wang resin, Merrifield resin, Tenta Gel, and cellulose) polymer support, and (b) parallel, simultaneous or sequential synthesis without use of a polymer support.

Parallel synthesis of combinatorial libraries is a synthetic sequence where the assembly of library is performed using an ordered array of spatially separated reaction vessels under the same reaction conditions. In sequential synthesis, the general reaction conditions can be modified for each building block combination according to the reactivity of the reagents in each reaction vessel. Parallel synthesis can be carried out with both conventional and MAOS and either in solid-phase or solution-phase synthesis.

8.1.1 SOLVENTS

As previously mentioned, if the solvents are to be used as the source of heat then they must couple effectively under microwave radiation. The various solvents can be used and a number of factors need to be considered. The type of vessel used may be incompatible with certain solvents and with open vessel systems flammability and volatility are both important issues. Solvents such as DMF, MeCN, and CH_2Cl_2 are all useful for particular kinds of reaction but obviously there are some limitations to each. Thus, solvents such as water, methanol, DMF, ethyl acetate, acetone, chloroform, acetic acid, and dichloromethane are all heated when irradiated with microwaves. Solvents such as hexane, toluene, diethyl ether, and CCl_4 do not couple and therefore do not heat with microwave irradiation although it is of course possible to use mixtures comprising of microwave active/microwave inactive solvents. The use of water is appealing and has been noted by several groups. However, a number of interesting opportunities are still to be developed and this may be a fruitful area of future activity. To increase heating rates for samples that do not contain polar reactants or polar solvents, it is common practice to add a small amount of a solvent with a large loss tangent. Salts can also be used to increase rates of heating, but heterogeneous mixtures usually do not provide a uniform heating pattern. Ionic liquids may be used instead of salts. Ionic liquids are fairly inert materials that are stable at temperatures up to 200°C. It has been shown that with microwave irradiation solvents can be heated above their boiling points and the temperature can be elevated up to 26°C beyond its conventional value. This is called as superheating effect.

8.1.2 DOMESTIC MICROWAVE OVENS

The relatively low cost of modern domestic microwave ovens makes them reasonably readily available to academic and industrial chemists. However, somewhat surprisingly only a relatively small number of organic synthesis research groups have reported their use. One disadvantage is that the variable power levels are produced by simply switching the magnetron on and off; this may be problematic if reaction mixtures cool down rapidly. Notwithstanding this limitation, there are a number of useful reactions that one can carry out in a domestic microwave oven.

8.1.3 MODIFIED DOMESTIC MICROWAVE OVENS

There are a number of published methods for the safe modification of domestic microwave ovens. The improved safety of this type of system is particularly attractive, the advantage of modifying an oven in this way is that the reaction vessel is neither sealed nor directly open to the microwave oven. Thus, reactions can be carried out using a flask attached to a reflux condenser. There are two advantages of using this type of system: (i) a wide variety of solvents can be used because the fire hazard is significantly reduced compared to an open vessel system, (ii) reactions can be carried out under inert conditions.

8.1.4 COMMERCIAL SYSTEMS

The commercial system is a more expensive way of carrying out microwave-assisted organic reactions. A number of systems are currently available and one such system is manufactured by CEM Microwave Technology Ltd. (www.cem.com). This type of system operates with a rotating carousel, so that a number of reaction vessels can be irradiated and agitated at the same time. The vessels are made of poly(etherimide) with Teflon lining and it is possible to monitor both the internal temperature and pressure of the reaction.

8.1.5 MULTIMODE REACTORS

The most widely used equipment for organic synthesis on a laboratory scale is the domestic oven, which is a multimode reactor. The distribution of electric field inside the cavity results from multiple reflections on the walls and on the products and hence the heating is totally nonuniform.

8.1.6 SINGLE-MODE REACTORS

Recently, commercial single-mode (or focused) microwave ovens have been developed. These ovens have cavities with optimized shape and volume to guarantee a uniform heating pattern. This feature improves reaction reproducibility, thereby extending the applicability of microwave technology beyond one-off reactions and providing an opportunity for microwave use in methodology studies for yield optimization. As an added safety measure, single-mode systems offer automated temperature, pressure and voltage controls. They are also equipped with a pressure releasing device designed to prevent explosions when used in combination with sealed vessels. Thus, single-mode ovens can safely carry out reactions under pressure, allowing for reaction temperatures well beyond the solvent boiling point. This, in turn, can increase the rate of a given synthetic transformation. The improved features of the single-mode microwave oven make it simple and safe for use in chemistry. For these reasons, microwave-assisted synthesis is gaining popularity in all fields of synthetic chemistry.

8.1.7 BATCH REACTORS

Most of the single-mode reactors commercially available have been designed for small to medium scale reactions (250 µL–120 mL). Single mode and multi-mode batch reactors that allow for microwave-assisted synthesis up to 500 mL have been recently introduced. Importantly, reactions optimized in smaller cavities can be directly reproduced in these larger reactors.

8.1.8 CONTINUOUS FLOW REACTORS

In addition, a decade ago, single-mode continuous flow systems were first introduced. In the continuous flow systems, reagents are pumped through the microwave cavity, allowing only a small portion of the sample to be irradiated at a time. In this way, it becomes possible to maintain a consistent heating profile for even large-scale synthesis.

8.1.9 VESSELS

Microwave-assisted reactions can, under the appropriate conditions, be carried out using conventional vessels Pyrex and PTFE. However, if reactions are to be carried out under pressure in sealed systems the major concern is the ability of the vessel to withstand the changes in pressure and temperature associated with the particular transformation. The rapid increases in temperature and pressure associated with microwave promoted superheating of organic materials makes it very difficult to ensure the safety of this type of procedure. It is essential that great precautions are taken when carrying out reactions in sealed vessels. The technology of vessel design is improving and a range of vessels are now available for carrying out reactions under pressure; many incorporate pressure releasing systems as an additional safety measure. An alternative simple procedure suitable for small-scale experiments is to seal the reagents in Pyrex vials, surround them in vermiculite and irradiate with microwaves. However, great care must be taken when using this type of protocol.

Thus, MAOS continues to affect synthetic chemistry significantly by enabling rapid, reproducible and scalable chemistry development. The use of microwave irradiation has been an established tool in organic synthesis for achieving better selectivity, rate enhancement and reduction of thermally degradative by products. Moreover, they also provide an opportunity to work with open vessels and enhance the possibility of upscaling the reactions on a preparative scale. Numerous reactions including condensations, cycloadditions, heterocycle formation and metal catalyzed cross coupling have been explored under microwave conditions, some of which have been applied to medicinal chemistry and total synthesis of natural product.

This chapter remains relatively confined to some recent and significant reports. There is no gainsaying the fact that there are a vast number of studies based on the abovementioned class of reactions. But, at the same time, one has to accept the fact that all those reports cannot be included in the present chapter due to restrictions of space and being reasonably comprehensive. Therefore, this study presents and discusses the findings of representative papers, publications and reports in these fields.

8.2 MICROWAVE IRRADIATED PERICYCLIC REACTIONS

A pericyclic reaction is a type of organic reaction wherein the transition state of the molecule has a cyclic geometry, and the reaction progresses in a concerted fashion (electrons move in a circular fashion and there is no positive and negative charges). Pericyclic reactions are usually rearrangement reactions. The major classes of pericyclic reactions are:

Cycloadditions
Sigmatropic reactions
Electrocyclic reactions
Group transfer reactions
Cheletropic reactions

This section consists of cycloaddition reactions and sigmatropic reactions conducted under microwave irradiation. Few selected examples of Diels–Alder and hetero-Diels–Alder reaction taken from literature are represented here.

8.2.1 1,3-DIPOLAR CYCLOADDITION OF NITRILE IMINE AND AZOMETHINE IMINES/OR YLIDES

Bougrin et al. (1995) reported the first practical utilization of microwave irradiation with nitrile imine as 1,3-dipole using solvent-free conditions. In this work, a comparative study of the reactivity of diphenylnitrilimine (DPNI) (1) with some dipolarophiles was made in dry media using microwave irradiation. The good yields were obtained on mineral support in a monomode reactor.

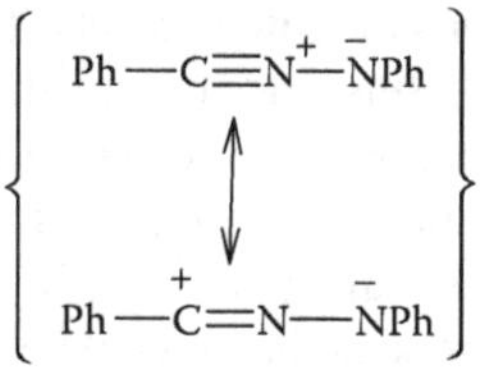

$$\left\{ \begin{array}{c} Ph-C \equiv N^{+}-\overset{-}{N}Ph \\ \uparrow \\ \downarrow \\ Ph-\overset{+}{C}=N-\overset{-}{N}Ph \end{array} \right\}$$

(1)

Alloum et al. (1998) described new spiro-rhodanine-pyrazolines (2) by 1,3-dipolar addition of DPNI to some 5-arylidenerhodanines on alumina or KF/alumina in "dry media" under microwave irradiation. In all the cases, the yields were quantitative and obtained in 10 min whereas no reaction occurred under classical thermal conditions. These observations showed some specific role of microwave irradiation in product synthesis.

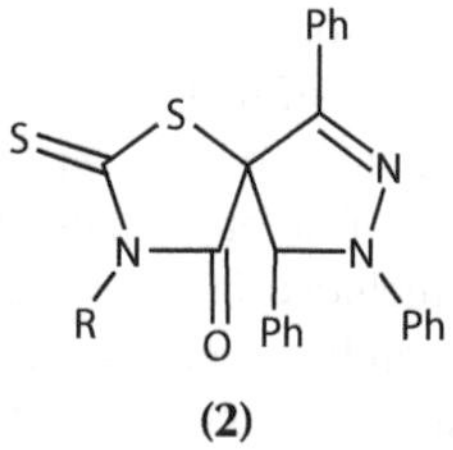

(2)

Arrieta et al. (1998) illustrated the synthesis of (4,3′)-bipyrazole (**3**) by utilizing the mixture of dimethylhydrazone and ethyl phenylpropiolate under microwave irradiation. They also described the microwave-assisted 1,3-dipolar cycloadditions with electron-deficient dipolarophiles to afford the corresponding cycloadducts. The use of pyrazolyl hydrazones led to valuable compounds, such as bipyrazoles (**4**), in good yields, providing a new approach to the preparation of these heterocyclic derivatives.

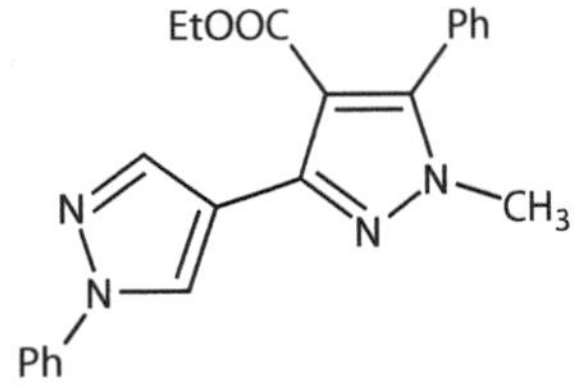

(**3**)

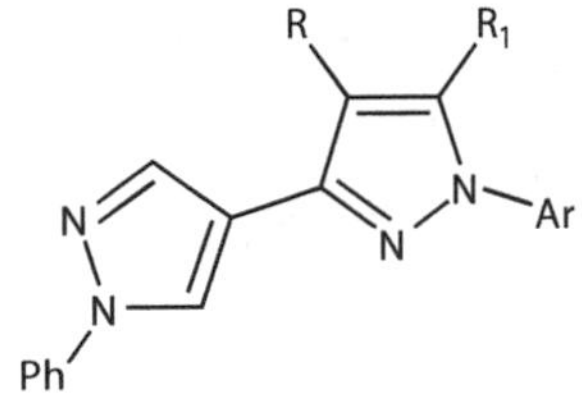

(**4**)

Microwave-mediated 1,3-dipolar cycloadditions under solvent-free conditions involving azomethine ylides have been widely reported in the literature (Lerestif et al., 1995; Kerneur et al., 1997; Fraga-Drubreuil and Cherouvrier, 2000). Dinica et al. (2000) showed that the reaction of 4,4′-bipyridinium ylides, generated *in situ* from 4,4′-bipyridinium diquaternary salts, with activated alkynes under microwaves, on KF/alumina in the absence of solvent, to give 7,7′-bis-indolizines (**5**) in good yields. Cheng et al. (2001) described the highly stereoselective intramolecular cycloadditions of unsaturated N-substituted azomethine ylides under microwave irradiation. The mixture of aldehyde and N-methyl- or N-benzylglycine ethyl ester impregnated on silica gel, exposed under microwave for 15 min, generated azomethine ylides that next underwent *in situ* intramolecular cycloadditions to afford the corresponding tricyclic compounds (**6**) in good yield. Syassi et al. (1998) reported an efficient and rapid one-pot synthesis of pyrazolines or pyrazoles using arylaldehydes, hydrochloride phenylhydrazine and dipolarophiles on montmorillonite K10 in dry media under microwave irradiation. The N-aminoiminium intermediates generated *in situ* were trapped by dipolarophiles, and in all cases, the yields in cycloadducts (**7**) or (**8**) are much better than under conventional heating. Azizian et al. (2001) described a one-pot diastereoselective synthesis of cycloadducts with good yields and diastereomeric excesses using microwave irradiation. Isatin derivatives and proline reacted to give stereospecific formation of an azomethine ylide intermediate via the decarboxylation route. The resulting 1,3-dipole undergoes a cycloaddition reaction with *N*-substituted maleimide, as a dipolarophile, to produce the adduct stereoselectively (9).

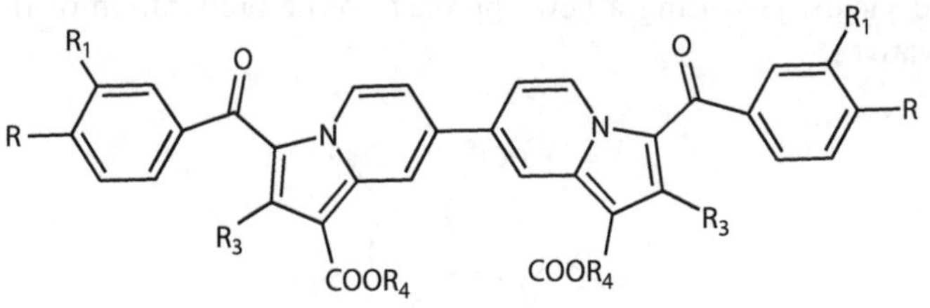

(5)

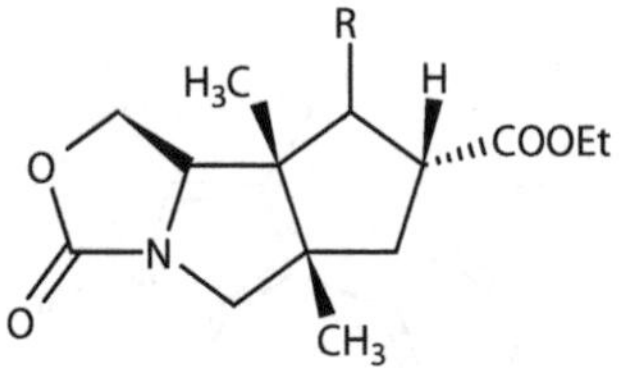

(6)

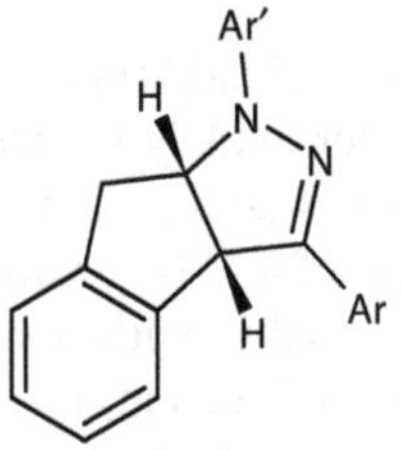

or

(7)

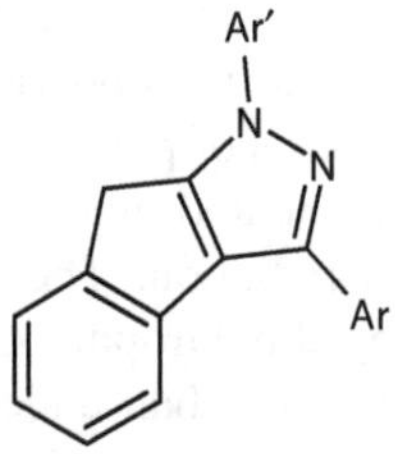

(8)

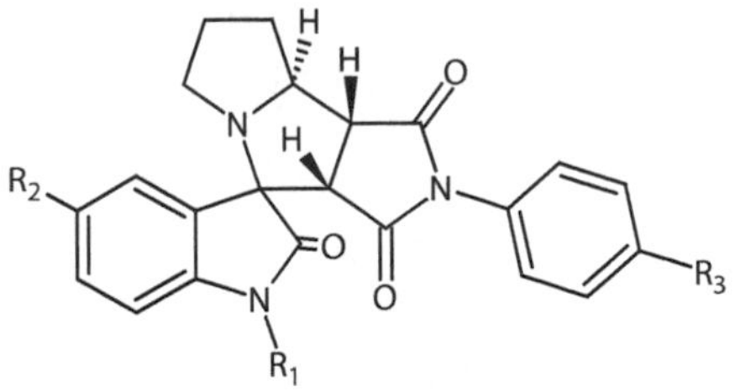

(9)

8.2.2   1,3-Dipolar Cycloaddition of Nitrile Oxides, Nitrones, or Azides

Microwave irradiation has been extensively employed to generate nitrile oxides or to promote 1,3 cycloaddition of the previously prepared dipole. Touaux et al. (1994) reported for the first time, the use of microwave irradiation in combination with dry media for the generation of nitrile intermediates. In this example, methyl nitroacetate was mixed with dimethylacetylenedicarboxylate (DMAD) in the presence of catalytic amounts of p-toluene sulfonic acid (p-TSA) (10% weight) and alkyne leading to the formation of the corresponding heterocyclic adducts (**10**) in good yield (91%). Bougrin et al. (1995) reported that arylnitrile oxides (**11**) can be generated *in situ* under microwaves by two methodologies—by the action of alumina and by the addition of chlorination agent, N-chlorosuccinimide (NCS). Touaux et al. (1998) synthesized some isoxazoles utilizing p-chlorobenzyloxime and alkynes as dipolarophiles. The yields of products (**12**) were 57–68% after 4–5 min. Khaddar et al. (1999) reported the synthesis of cycloadducts (**13**) over alumina with hydroxamoic acid chloride without solvent and under microwaves, showing that alkynes could give moderate yields (40–60%). The corresponding procedure with trisubstituted alkenes did not give any products. These reactions were also performed with N-methylmorpholine (NMM) under focused microwaves; the same products were obtained with better yields as compared to the same reactions in dry media. Inter- and intramolecular 1,3-dipolar cycloadditions of nitrones under microwave irradiation have been reported to afford five-membered heterocyclic compounds. The application of solvent-free conditions to this cycloaddition has been limited to only a few examples. Loupy et al. (1995) showed the utility of microwave irradiation as an energy source in the 1,3-dipolar cycloaddition of N-methyl-α-phenylnitrone with fluorinated dipolarophiles under solvent-free conditions leading to the formation of isoxazolidines (**14**) of biological interest. Yields and experimental conditions were improved in solvent-free conditions and even more under microwave.

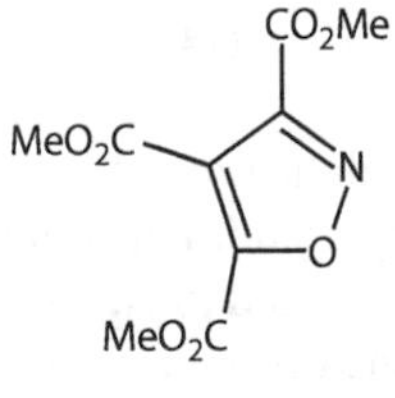

(10)

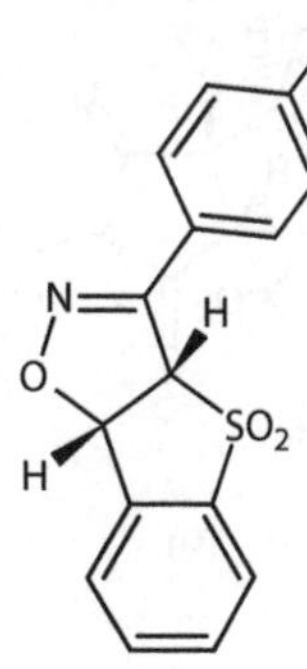

(11)

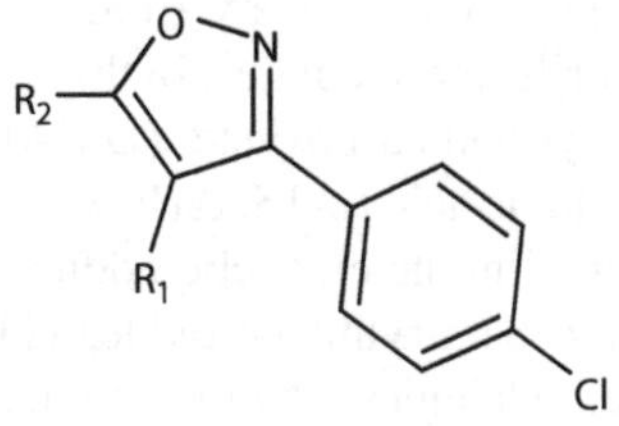

(12)

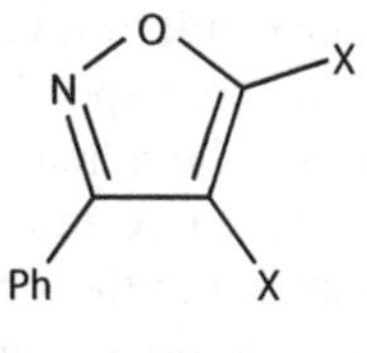

(13)

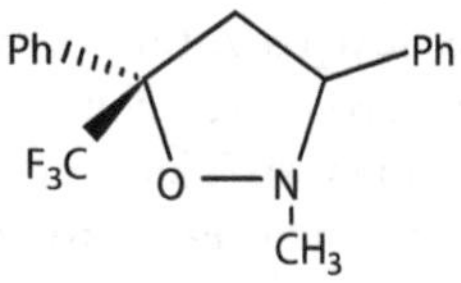

(14)

Diaz-Ortiz et al. (1996) extended the procedure for accessing to 2,3-dihydro-1,2,4-oxadiazoles (**15**) by 1,3-dipolar cycloaddition of nitriles with nitrone under solvent-free conditions. The use of microwave led to yields that were always higher than those obtained with classical heating, the differences being more significant with the less reactive nitriles. Baruah et al. (1997a,b) reported that the regiospecific 1,3-dipolar cycloaddition of unreactive nitrones with alkene without solvent under microwave activation was more rapid (6–15 min) than the thermolytic or sonochemical reactions. All reactions under microwave irradiation afforded various isoxazolidines (16) in high yields (76–90%) as a one-pot procedure. Cheng et al. (2001) illustrated that the microwave-assisted synthesis of cycloadduct of aldehyde and N-methylhydroxylamine on silica gel in solvent-free conditions afforded the corresponding adduct (17) in high yields for 15 min heating time. With traditional heating in a sealed tube at 120°C in alcohol as solvent, the mixture of aldehyde and N-methylhydroxylamine afforded the products (17) and (18) in a 8:1 ratio. The authors proposed that the selectivity observed under microwave-mediated reaction resulted from a faster transformation leading to the kinetically controlled product. The examples of 1,3-dipolar cycloaddition of azides employed in solvent-free conditions under microwave irradiation remain very rare. Two main results are discussed in this section.

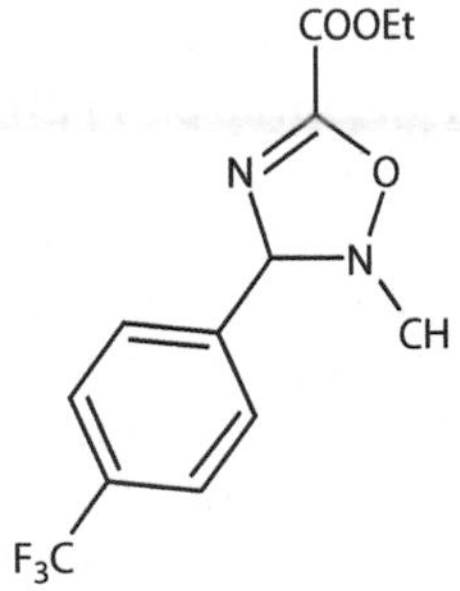

(15)

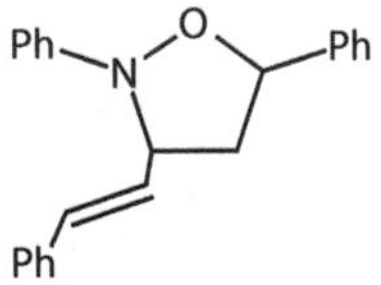

(16)

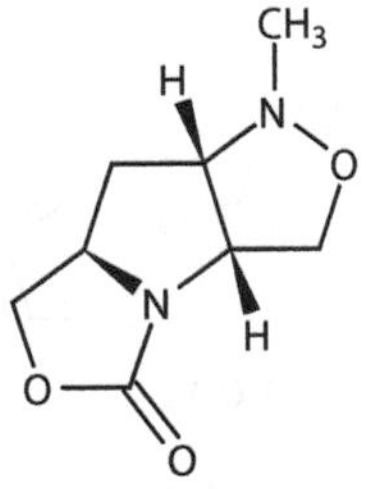

(17)

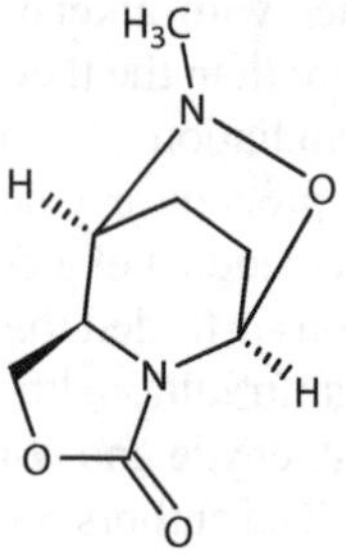

(18)

N-Substituted 1,2,3-triazoles are very important compounds that can be prepared by 1,3-dipolar cycloadditions between azides and alkynes. Louërat et al. (1998) reported that the acetylenic esters reacted with phosphonate azide to form the regioisomeric *N*-substituted 1,2,3-triazoles (**19**) and (**20**) under microwaves in solvent-free conditions. This procedure avoided the harsh reaction conditions associated with thermal cycloadditions (toluene at reflux) as well as the very long reaction time. Katritzky and Singh (2002) reported the synthesis of variety of 1,2,3-triazoles (**21**) by cycloaddition between organic azides and acetylenic amides under dry media microwave irradiation. The same Katritzky's group has been able to prepare bis-triazoles (**22**) [41–73% yields] from mono-azide and bis-amidopropiolates-2,4-tolylene and *n*-C_6H_{14} under solvent-free microwave conditions. This method could be utilized for the preparation of polymers utilizing bis-triazoles and bis-amidopropiolates.

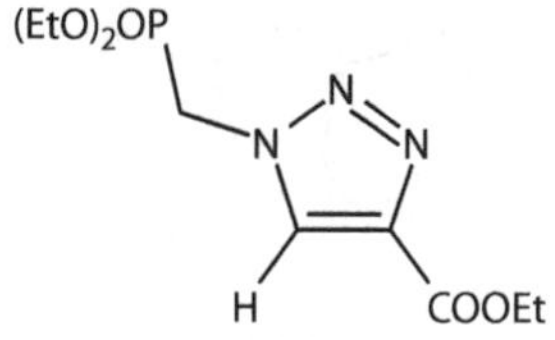

(19)

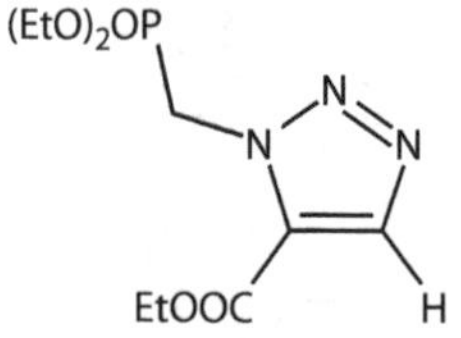

(20)

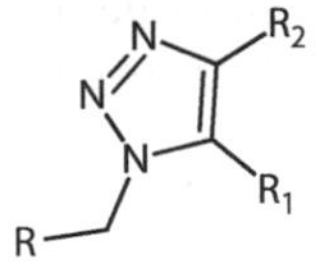

(21)

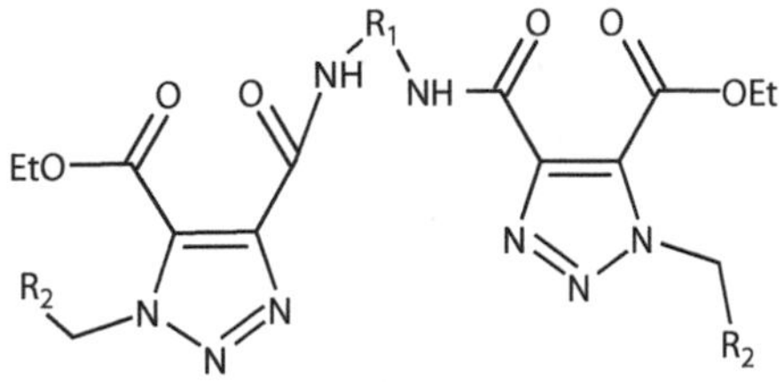

(22)

8.2.3 Hetero-Diels–Alder [4 + 2] Reactions

The hetero-Diels–Alder reaction is amongst the most efficient processes for the synthesis of six-membered heterocyclic ring systems. Solvent-free conditions have been used to improve reactions of heterodienophiles and heterodynes with low reactivities. Cado et al. (1997) have described the hetero-Diels–Alder reaction of ethyl 1*H*-perimidine-2-acetate as heterocyclic ketene aminal with ethyl propiolate under solvent-free conditions with focused microwave irradiation. The new fused perimidines (**23**) were obtained in good yields (67–98%).

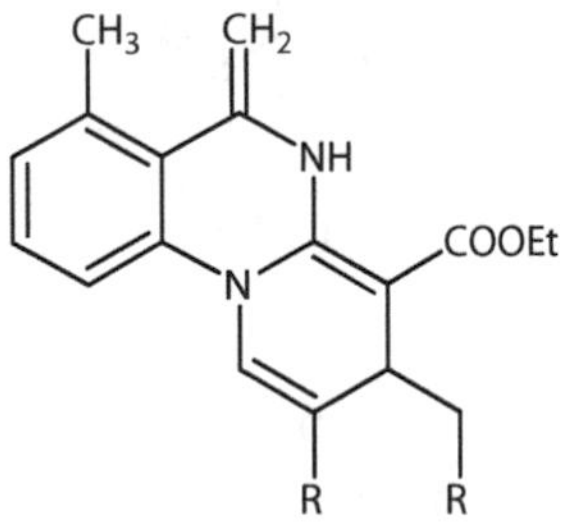

(23)

Diaz-Ortiz et al. (2000) reported the first example of a [4 + 2] cycloaddition involving a pyrazole ring and it represents a new, interesting, and versatile approach to the preparation of pyrazolo[3,4-b]pyridines. Microwave irradiation under solvent-free conditions induces pyrazolyl 2-azadienes to undergo Diels–Alder cycloadditions with nitroalkenes within 5–10 min, to give good yields of adducts (**24**), (**25**), and (**26**). Under classical heating conditions, only traces of adducts were detected. Das et al. (2000) described the Diels–Alder reaction of the naturally occurring alkaloids, camptothecin and mappicine ketone with maleic anhydride under solvent-free for 9 min of microwave irradiation in a commercial domestic oven to produce adducts (27) and (28).

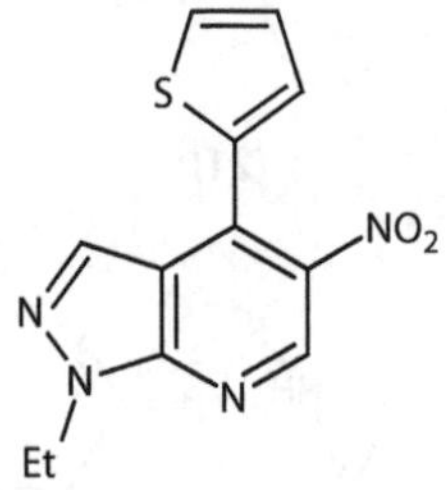

(24)

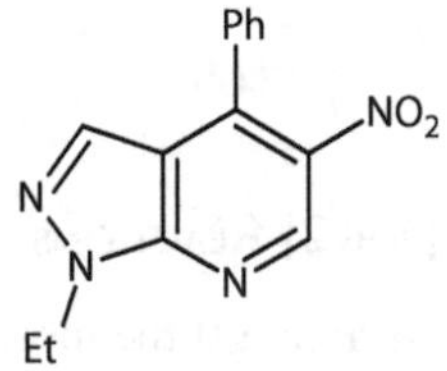

(25)

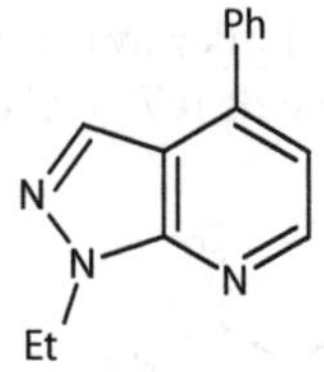

(26)

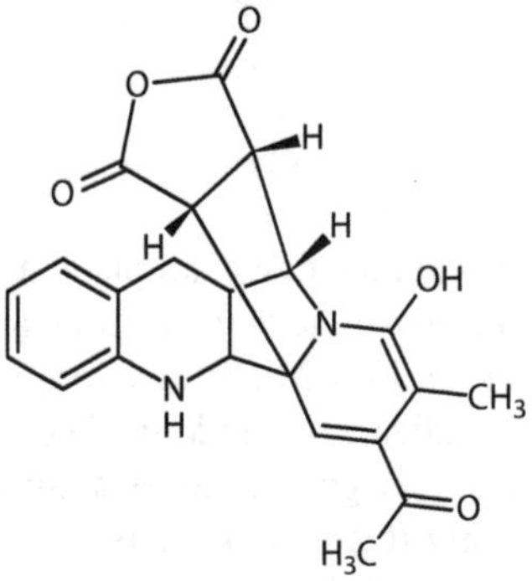

(27)

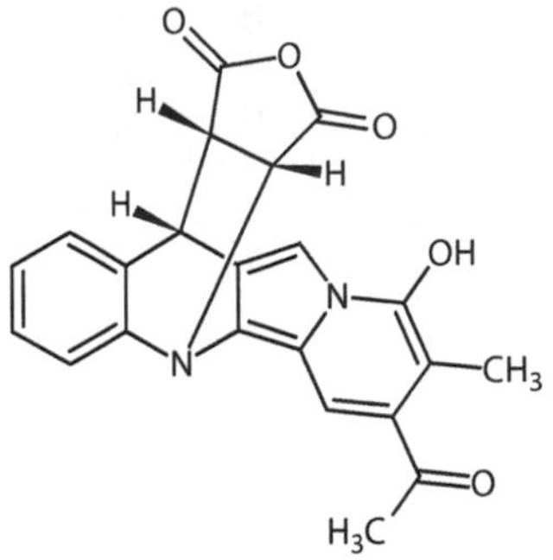

(28)

De la Cruz et al. (1997) illustrated an interesting example of the applications in C_{60} hetero-Diels–Alder reaction in solvent-free microwaves conditions. The reaction of C_{60} with o-quinone methide, generated *in situ* from o-hydroxybenzyl alcohol, was performed in a modified domestic microwave oven to give (29). Jacob et al. (2003) developed a green and efficient method for the synthesis of several octahydroacridines (OHAs) with the use of a simple one-pot intramolecular hetero-Diels–Alder reaction between (+)-citronella and N-arylamines in the presence of a solid-supported catalyst ($SiO_2/ZnCl_2$), under microwave irradiation without any solvent (30 and 31). Avalos et al. (1999) described the Diels–Alder reaction between 1-aryl-1,2-diaza-1,3-butadienes and azadicarboxylic ester as giving tetrazines (32) for a reaction rate speeded-up by a factor of 1000 by microwave enhancement.

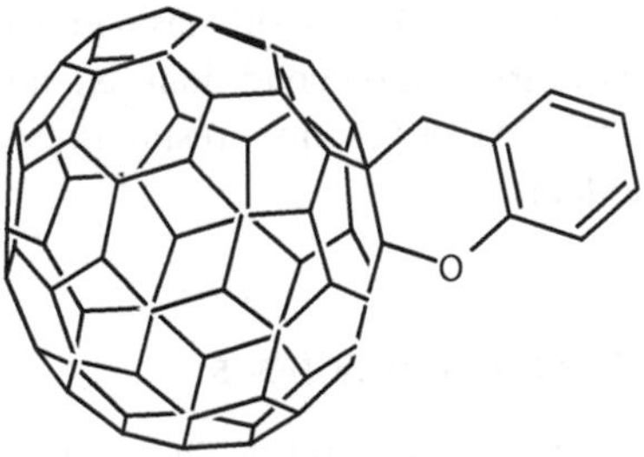

(29)

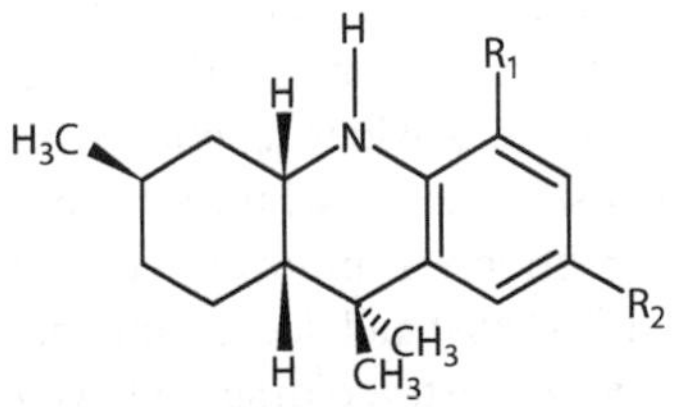

(30)

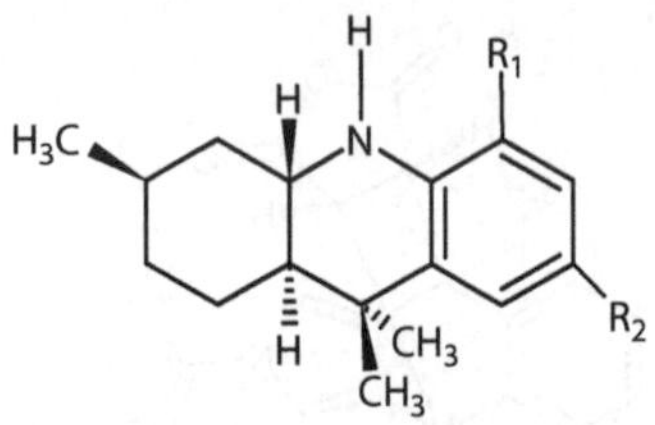

(31)

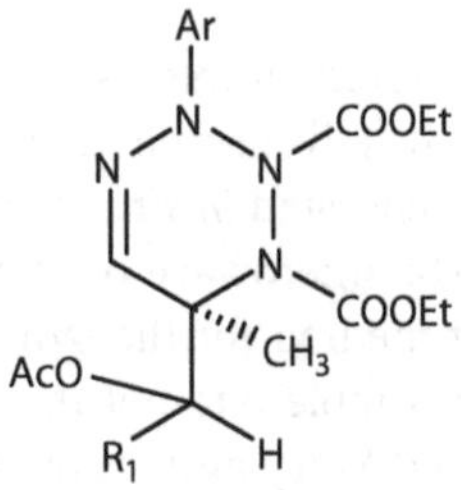

(32)

Narsaiah and Krishnaiah (2002) reported that the hetero-Diels–Alder reaction of *N*-acetyl perfluoroalkyl substituted 2(1*H*) pyridones with DMAD on neutral alumina under solvent-free microwave irradiation conditions extended to undergo an inverse electron demand hetero-Diels–Alder reaction, resulting exclusively in *E, Z* isomers (3:1) of Michael-type *N*-adducts.

8.2.4   Miscellaneous Pericyclic Reactions

Recently, much attention has been paid to a family of naturally occurring 2-azaanthraquinones having benz[*g*]-isoquinoline-5,10-dione (**33**). Choshi et al. (2008) reported the total synthesis of the 2-azaanthraquinone alkaloid, scorpinone, isolated from the mycelium of a Bispora-like tropical fungus. The synthesis of scorpinone was started with isoquinoline derivative. The 5,7-dibromoisoquinoline was immediately oxidized with cerium ammonium nitrate (CAN) to afford the isoquinoline-5,8-dione, which undergo Diels–Alder reaction with diene in toluene at 110°C yielded the known 6-deoxybostrycoidine with the elimination of a methoxy group. The microwave irradiated condition was more effective than the conventional condition for improving the yield (73–86%) and decreasing the reaction time (30–10 min). Finally, the alkylation of 6-deoxybostrycoidine with MeI in the presence of K_2CO_3 afforded scorpinone (**34**).

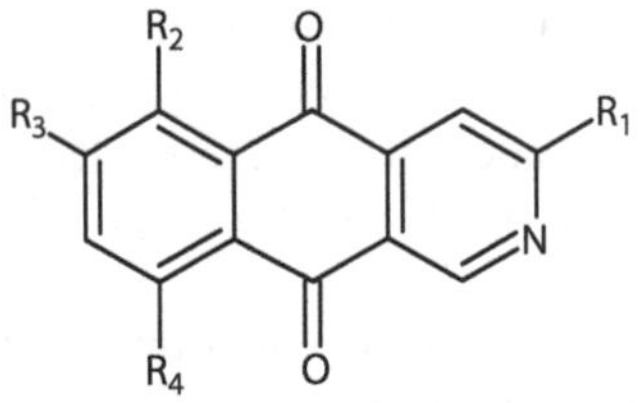

benz[g]isoquinoline-5,10-dione
$R_1 = R_2 = R_3 = R_4 = H$

Bostrycoidin
$R_1 = Me, R_2 = OH, R_3 = OMe, R_4 = OH$

6-deoxybostrycoidin
$R_1 = Me, R_2 = H, R_3 = OMe, R_4 = OH$

Scorpinone
$R_1 = Me, R_2 = H, R_3 = R_4 = OMe$

(33)

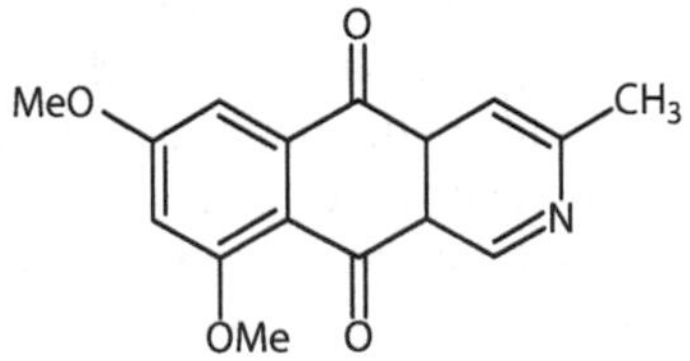

(34)

Azafulvenium methides and diazafulvenium methides have been generated under microwave irradiation from 2,2-dioxo-1H, 3H-pyrrolo[1,2-c]thiazoles and 2,2-di-oxo-1H, 3H-pyrazolo[1,5-c]thiazoles, respectively. Pericyclic reactions of these 1,7-dipole intermediates, namely, sigmatropic [1,8]-H shifts, 1,7-electrocyclization or [8π + 2π] cycloaddition led to the synthesis of a range of pyrrole and pyrazole derivatives. The first evidence for the azafulvenium methides by intermolecular trapping via [8π + 2π] cycloaddition has been reported by Soares et al. (2008). Starting from 3-methyl-2,2-dioxo-1H,3H-pyrrolo[1,2-c]thiazoles, the reaction carried out in 1,2,4-trichlorobenzene under microwave irradiation afforded the corresponding N-vinylpyrroles (**35**) (Soares et al., 2008). The synthesis of these heterocycles results from the SO$_2$ extrusion of sulfones giving azafulvenium methides followed by a sigmatropic [1,8]-H shift.

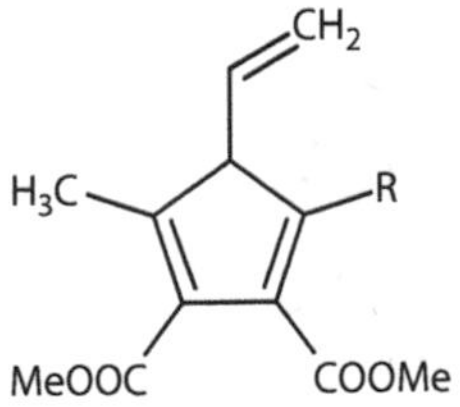

(35)

Azafulvenium methide was generated from 3-phenyl-2,2-dioxo-1*H*,3*H*-pyrrolo-[1,2-c]thiazole under microwave irradiation (Giguere et al., 1986). In this case, the 1,7-dipole undergoes electrocyclization to give bicycloderivative, which is converted into C-vinylpyrrole (**36**). Giguere et al. (1986) reported the Diels–Alder reaction of anthracene with dimethyl fumarate in *p*-xylene to give (**37**) under microwave conditions. The cycloaddition reaction of cyclohexadiene with dimethyl-but-2-ynedioate gives (**38**) via the Alder–Bong reaction was performed by Giguere et al. (1987). The product was synthesized in reduced reaction time as compared to conventional heating method. Stambouli et al. (1991) reported the hetero-Diels–Alder reaction under microwave condition to give (**39**) and (**40**). Then Abrarnovitch and Bulman (1992) demonstrated the application of microwave technology to [3,3]-sigmatropic rearrangements involved in Fischer cyclizations. They have found that irradiation of *N*-(cyclohexylidenemethyl) aniline in formic acid in a Parr bomb produced product (**41**) in excellent yields. Lei and Fallis (1990, 1993) have reported the synthesis of (+) Longifolene (**42**) under microwave-assisted intramolecular Diels–Alder reaction. Linders et al. (1988) have utilized the microwave-assisted synthesis of morphinan analogs. They simplified the synthesis of cycloadducts (**43**) and (**44**) from the reaction of thebaine derivative with methyl vinyl ketone under microwave condition. Wang and Rosekamp (1992) reported the synthesis of cycloadducts (**45**) and (**46**) by absorbing the furan onto silica gel, saturating with water and irradiating with microwaves. Jones et al. (1993) utilized the KSF clay as catalyst and accelerate ortho-ester Claisen rearrangements. Microwave irradiation of cyclohept-2-ene-1-ol with triethyl orthoacetate and KSF clay in DMF led to the isolation of rearranged product (**47**) in good yield.

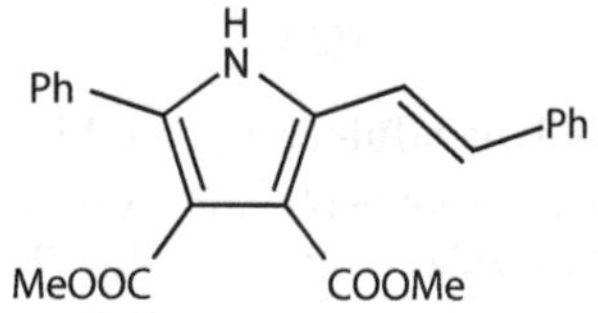

(**36**)

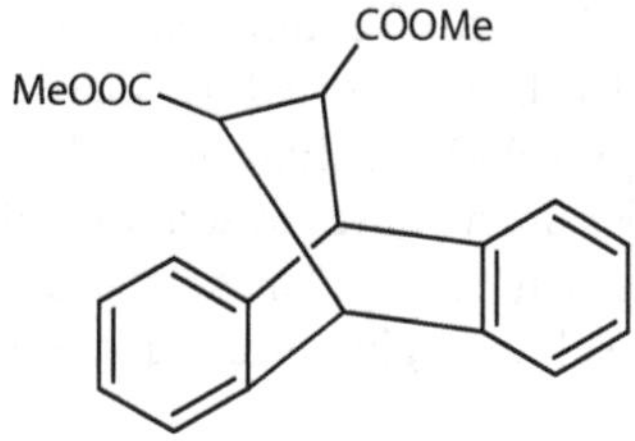

(**37**)

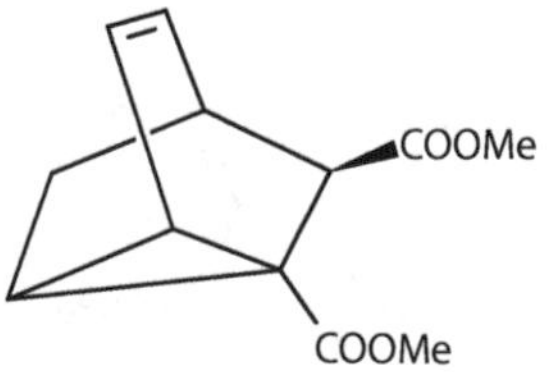

(**38**)

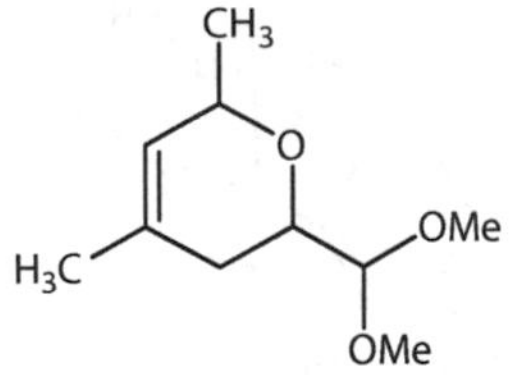

(39)

(40)

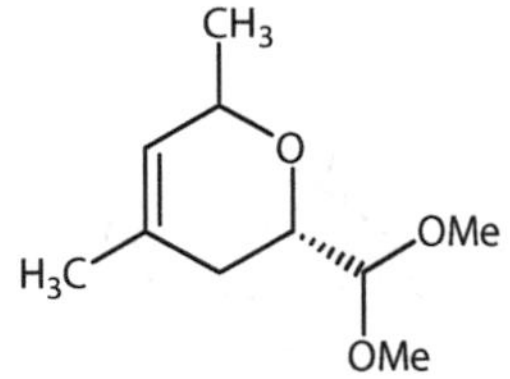

(41)

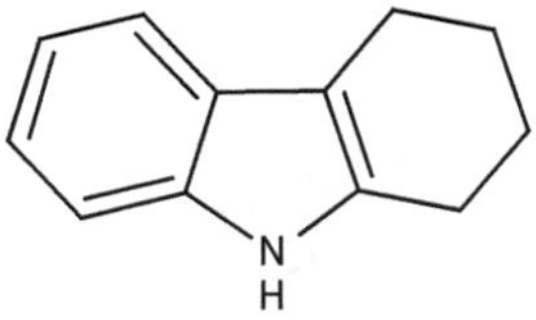

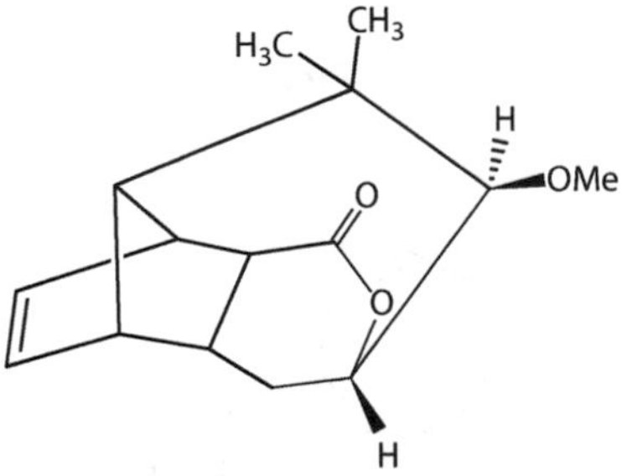

(42)

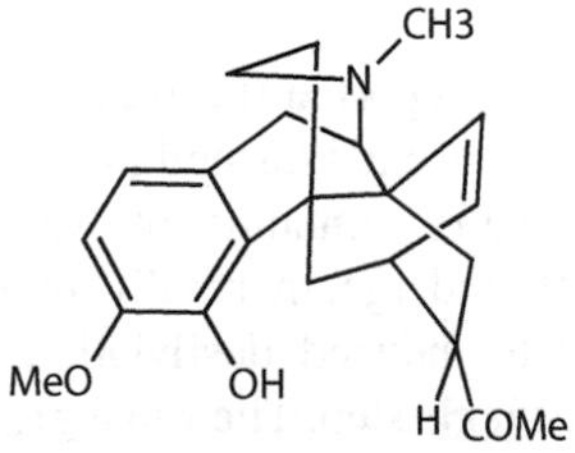

(43)

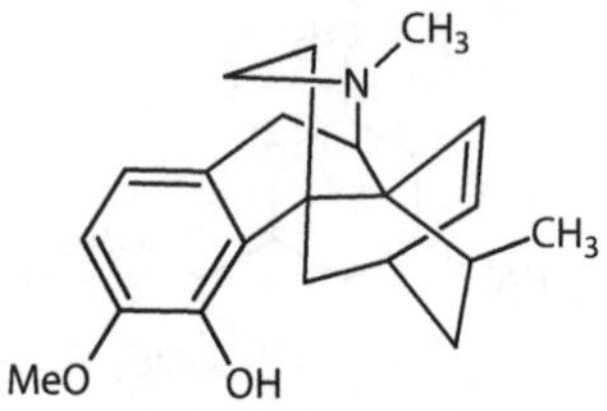

(44)

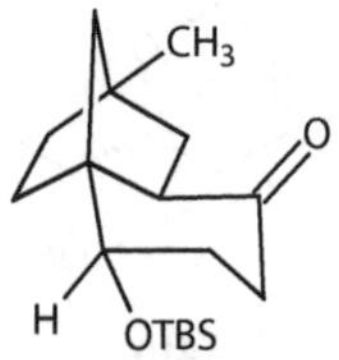

(45)

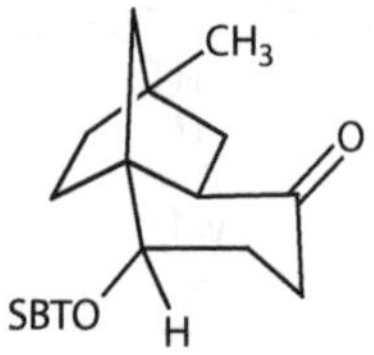

(46)

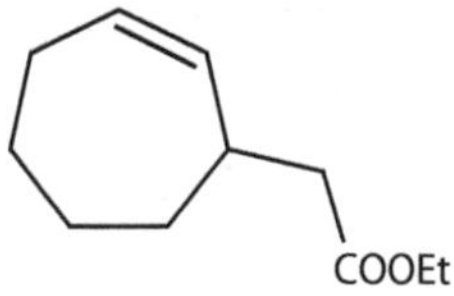

(47)

Bourgeois et al. (2005, 2006) reported the Ireland-Claisen rearrangement reaction in which α-tosyl silyl keteneacetals formed *in situ* from allylic tosylacetates in the presence of sub-stoichiometric amounts of N,O-bis(trimethylsilyl)acetamide (BSA) and potassium acetate undergo thermally induced [3,3]-sigmatropic rearrangement followed by acetate induced desilylation–decarboxylation to provide homoallylic sulfones (**48**) in a single step. The same group showed that the introduction of additional electron-withdrawing α-substituent enables malonyl substrates to

undergo decarboxylative Claisen rearrangement (dCr) to provide α-carboxyhomo-
allyl sulfones (**49**) at ambient temperature. The unsymmetrical 2-tolylsulfonylmalonyl
compounds (diallyl esters) are able to undergo two Claisen rearrangements within
the same molecule to give (**50**) and (**51**) (Craig and Grellepois, 2005). Srikrishna and
Nagaraju (1992) illustrated that Claisen rearrangement of the ketene-acetal to give
(**52**) in 83% yield in 10 min in microwave irradiation as compared with 48 h under
conventional heating. Alloum et al. (1989) carried out the [2,3] sigmatropic rearrangements where the adsorption of but-2-ene-ol onto KSF clay followed by microwave irradiation for 5 min led to the isolation of product (**53**) in 75% yield.

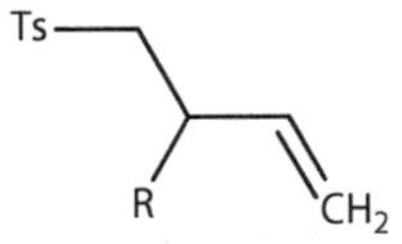

(48)

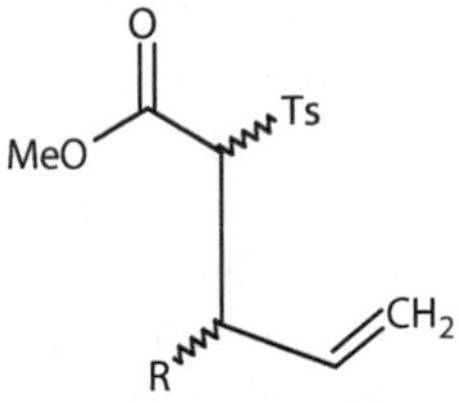

(49)

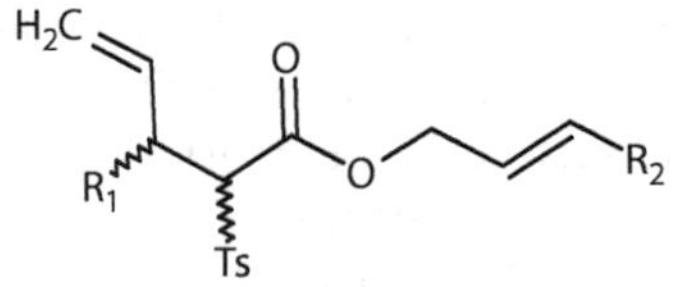

(50)

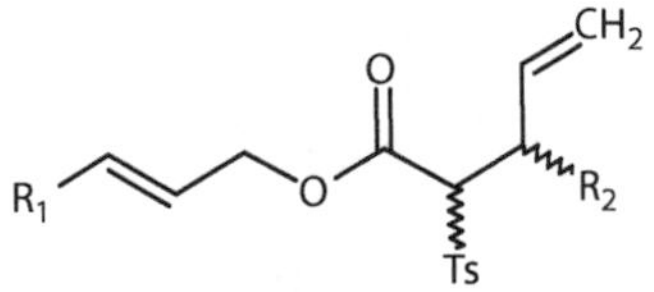

(51)

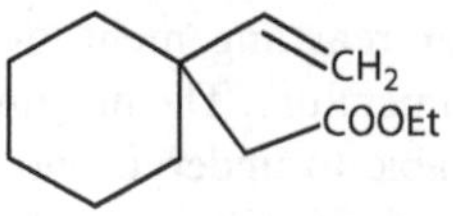

(52)

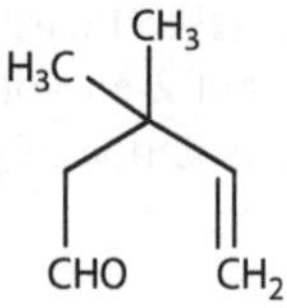

(53)

8.3 NUCLEOPHILIC SUBSTITUTIONS UNDER MICROWAVE IRRADIATION

Diethyl arylphosphonates (54) are efficiently and rapidly prepared by palladium catalyzed nucleophilic substitution of aryl halides performed in a Teflon autoclave under microwave irradiation (Villemin et al., 1997). Hu et al. (1999) described that the nitro group of α-nitrostyrene were readily substituted by organic moiety of organozinc halides under microwave irradiation to give derivatives of styrene (55) in excellent yields. Sagar et al. (2000) described the rapid synthesis of triaryl cyanurates (2,4,6-triaryloxy-1,3,5-triazine) (56) by reacting cyanuric chloride with the sodium salt of hydroxyaryl compounds in water using focused microwaves. Effective spatially addressed parallel assembly of trisamino- and amino-oxy-1,3,5-triazines was achieved by applying the SPOT-synthesis technique on cellulose and polypropylene membranes. In addition to developing a suitable linker strategy and employing amines and phenolate ions as building blocks, a highly effective microwave-assisted nucleophilic substitution procedure at membrane-bound monochlorotriazines was developed. The 1,3,5-triazines obtained could be cleaved in parallel from the solid support by trifluoroacetic acid (TFA) vapor to give compounds adsorbed on the membrane surface in a conserved spatially addressed format for analysis and screening. The reaction conditions developed were employed for the synthesis of 8000 cellulose-bound 1,3,5-triazines, which were probed in parallel for binding to the antitransforming growth factor—α monoclonal antibody in order to identify epitope mimics (Scharn et al., 2002).

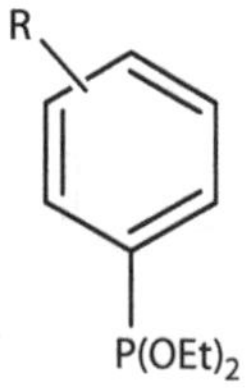

(54)

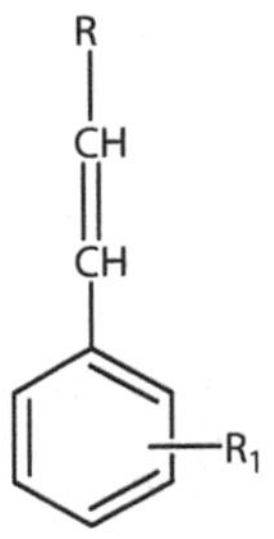

R
|
CH
||
CH
|

(benzene ring with R₁ substituent)

(55)

OAr
(triazine ring)
ArO   N   OAr

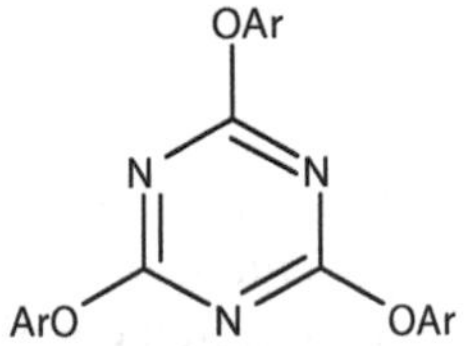

(56)

Biaryl ethers and thio ethers (**57**) were formed in high yields in solvent-free micro-wave irradiation by the substitution of electron-deficient aryl halides by phenols and thiophenols with using CsF supported on Al₂O₃ (Yadav et al., 2000). Heber and Stoyanov (1999) showed that on irradiating 4-hydroxy-6-methyl-2(1*H*)-pyridones with araliphatic amino compounds in an ordinary domestic microwave oven gave rise to the 4-alkylamino-6-methyl-2(1*H*)-pyridones (**58**). 1-Arylpiperazines (**59**) were synthesized under microwave irradiation from bis-(2-chloroethyl) amine hydro-chloride and substituted anilines without any solvent. 1-Arylpiperazines were syn-thesized in 53–73% yields (Jaisinghani and Khadilkar, 1997). Potent serotonin ligands such as trifluoromethylphenylpiperazine (TFMPP) and 3-chlorophenylpiperazine (mCPP) were also prepared in a short reaction time.

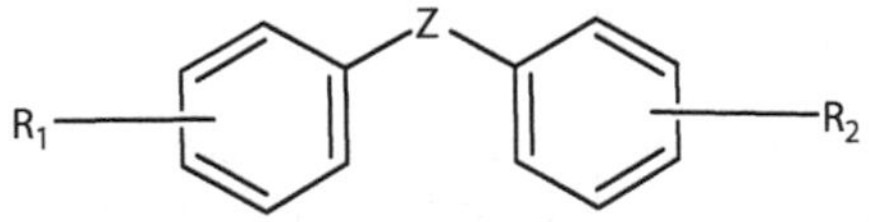

R_1 — (phenyl) — Z — (phenyl) — R_2

(57)

NH-R_2

(pyridone ring)
H_3C N   O
             |
             R_1

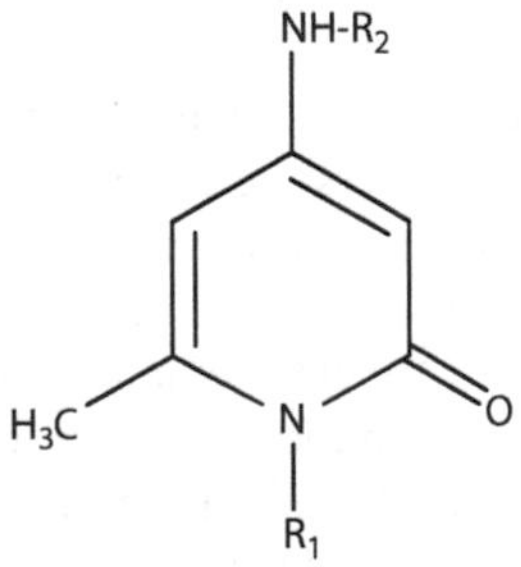

(58)

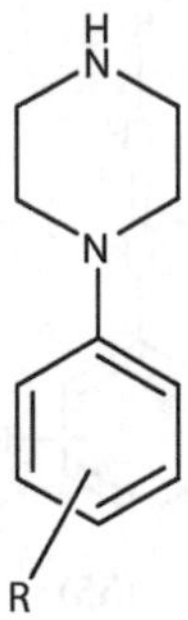

(59)

Quinoline derivatives (**60**) have been synthesized by the reaction of 5-alkyl-1,3,4-thia-diazol/oxadiazol-2-thiols with 7-chloro-6-uoro-2,4-dimethylquinoline and by the reaction of 2-hydroxy-1,4-naphthoquinone with 2-chloro-3-formyl-4-methyl/6-methyl/7-methyl/8-methylquinolines, respectively, on basic alumina using microwaves (Kidwai et al., 2000).

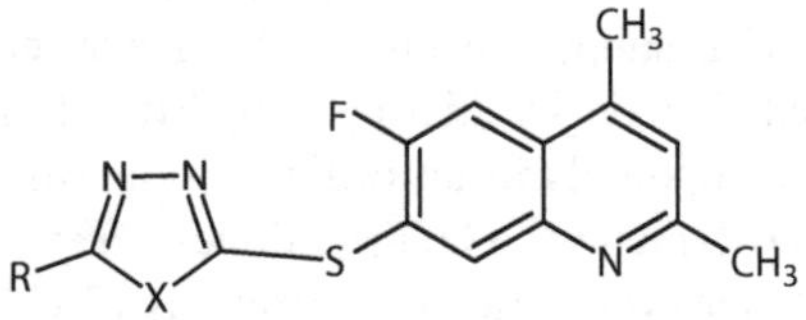

(60)

8.4 ALKYLATION REACTIONS UNDER MICROWAVE IRRADIATION

Alkylation reactions are reactions where an alkyl group is introduced into a molecule. Alkylation reactions are widely used in chemistry because the alkyl group is probably the most common group encountered in organic molecules. The alkyl group may be transferred as a carbocation, carbanion, free radical, or as a carbene. Alkylating agents are classified according to their nucleophilic and electrophilic character. Nucleophilic alkylating agents deliver the equivalent of alkyl anion and electrophilic alkylating reagent delivers the equivalent of alkyl cation. Carbenes are extremely reactive and are known to attack even unactivated C–H bonds.

One very important alkylation reaction is the Friedel–Crafts alkylation reaction, used for the alkylation of aromatic species, which was discovered in 1877. Alkylation is used industrially to produce basic building blocks for the synthesis of more elaborate materials. One commonly used application is in the production of antiknock gasoline. In medicine, alkylation of deoxyribonucleic acid (DNA) is used in chemotherapy to damage the DNA of cancer cells. In this section, the three principle types of alkylation reactions without which the study of organic chemistry is partial have been discussed. These are:

C-alkylation reactions
O-alkylation reactions
N-alkylation reactions

8.4.1 C-ALKYLATION

Carbon–carbon bond formation via the Michael addition of α,β-unsaturated ketone and 1,3-diketone is achieved in high yields and short times to give (**61**) by employing catalytic amounts of $EuCl_3$ in dry media under microwave irradiation (Soriente et al., 1997). Ranu et al. (1997) reported the Michael addition of ethyl acetoacetate, acetyl acetone, and ethyl cyanoacetate to cycloalkenones, β-substituted enones and enal. The reaction accomplished efficiently on the surface of alumina under microwave irradiation in dry media. Baruah et al. (1997a,b) also demonstrated the $BiCl_3$ and CdI_2 catalyzed solvent-free Michael addition of 1,3-dicarbonyl compounds under microwave irradiations with good yields.

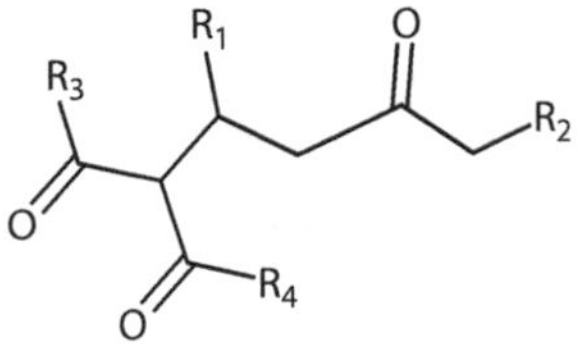

(**61**)

Nitromethane reacts readily and selectively with a variety of electrophilic alkenes activated by cyano, ester, or amide group to give the corresponding Michael mono-adduct within a few minutes, in good yields over alumina without solvent and under focused microwaves (Michaud et al., 1997a). Nitromethane reacts via a diastereo-selective double Michael addition with electrophilic alkenes activated by cyano and methoxycarbonyl groups $[XC_6H_4CH=C(CN)CO_2Me]$ in the presence of catalytic amounts of piperidine under solvent-free conditions coupled with focused micro-wave irradiation to afford new, highly functionalized cyclohexenes (**62**) (Michaud et al., 1997b). Olofsson et al. (1999) demonstrated the radical Michael addition reac-tion of highly fluorous compound $HSn(CH_2CH_2C_{10}F_{21})_3$ to yield (**63**) in presence of microwave irradiation.

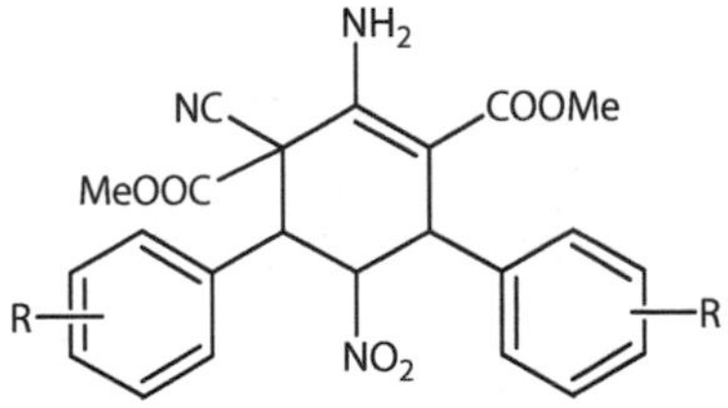

(**62**)

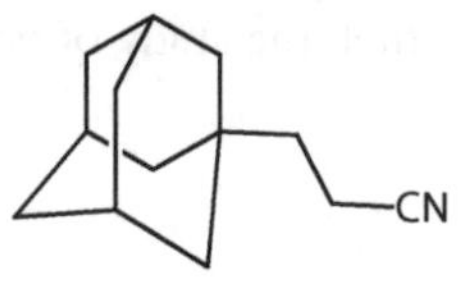

(63)

A rapid Michael addition of secondary amines to α,β-unsaturated carbonyl compounds has been achieved in good to excellent yields to provide (**64**) in the presence of water under microwave irradiation. In the absence of water and under conventional method, the reaction does not proceed, or rather, takes place with a very low yield after a long reaction time (Moghaddam et al., 2000). Deng et al. (1994) performed the rapid alkylation of ethyl acetoacetate with a series of halides in a domestic microwave oven and reported the isolated yield of monoalkylated product (**65**) varying from 59 to 82%.

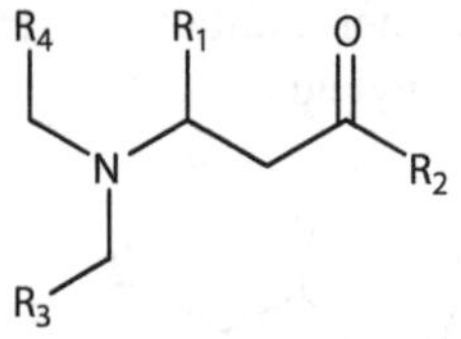

(64)

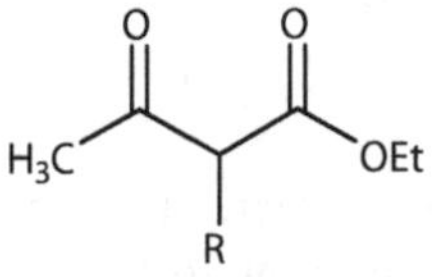

(65)

Alkylations of nitrophenylacetates, indole, and methyl benzylidene glycinate under dry conditions induced by microwaves have been studied, leading to the synthesis of arenes (**66**) linked by a long hydrocarbon chain (Abramovitch et al., 1995). Kumar et al. (1999) reported the environmental-friendly process for the synthesis of 2-alkylated hydroquinones (**67**) under microwave irradiation using 1,4-cyclohexanedione and aldehydes catalyzed by KF-Al$_2$O$_3$. An attractive methodology using microwave irradiation is described by Vanelle et al. (1999) for S$_{RN}$[1] reactions of different reductive alkylating agents with 2-nitropropane anion yield (**68**).

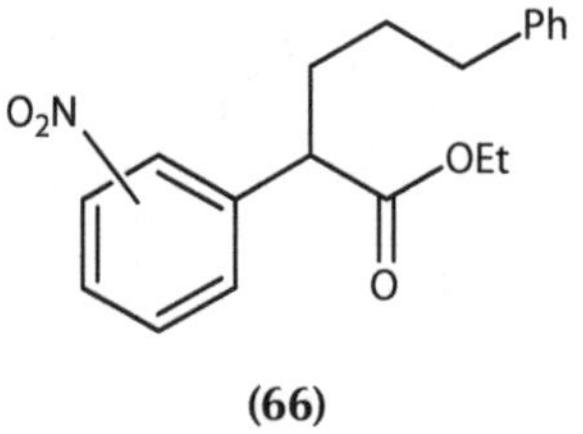

(66)

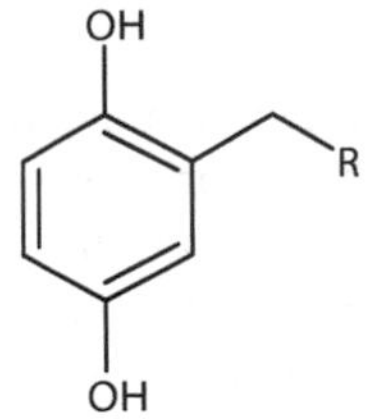

OH

R

OH

(67)

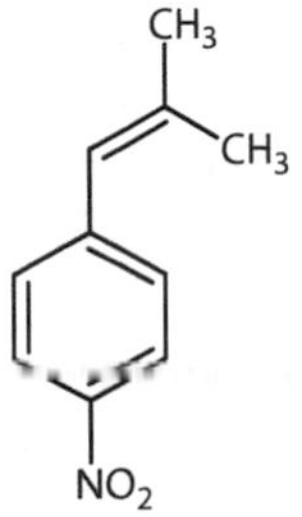

CH$_3$

CH$_3$

NO$_2$

(68)

Microwave irradiation of a mixture of cyclohexenones and ethyl acetoacetate adsorbed on the surface of solid lithium *S*-(2)-prolinate leads to the stereoselective construction of bicyclo[2.2.2]octanone (69) systems through Michael addition and subsequent intramolecular aldolization (Ranu et al., 2000). Electron-rich aromatic compounds react with formaldehyde and a secondary amine under solvent-free condition and microwave irradiation in a microwave oven to produce amino ethylated products (70) in good to excellent yields (Mojtahedi et al., 2000).

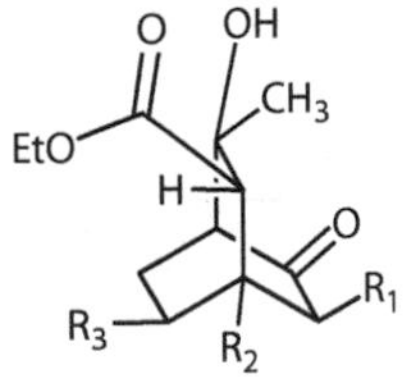

O   OH

EtO   CH$_3$

H   O

R$_1$

R$_3$ R$_2$

(69)

NEt$_2$

HO

N

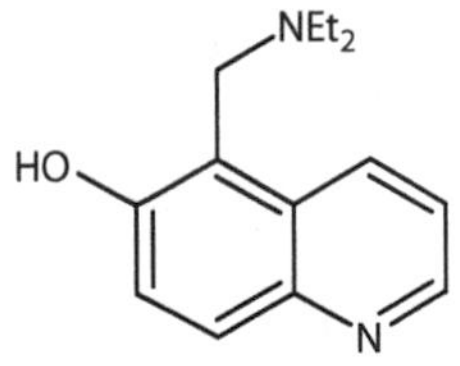

(70)

8.4.2   O-Alkylation

Selective O-alkylation of nonactivated primary alcohols in the presence of amide function is achieved to yield (**71**) by adsorption on the surface of chromatographic alumina. Microwave irradiation enhances considerably the rate of these reactions (Motorina et al., 1996). The synthesis of epoxypropoxyphenols (**72**) was carried out with phenols and epichlorohydrin under microwave irradiation (Khadilkar and Bendale, 1997). Bagnell et al. (1999) reported the synthesis of symmetrical ether (**73**) with excess of alcohol (R—OH) and a catalytic amount of R—X under microwave irradiation. Bogdal et al. (1998) reported that under microwave irradiation, phenols react remarkably fast with a number of primary alkyl halides to give aromatic ethers (**74**). A simple rapid and efficient procedure for the synthesis of 8-quinolinyl ethers (**75**) via microwave irradiation is reported by Wang et al. (1998). Naphthyl ethers (**76**) and naphthyl esters were synthesized in good yield by using a mixture of copper powder and cupric chloride in a conventional microwave oven (Zadmard et al., 1998).

$$R \diagdown\diagup O \diagdown\diagup \diagup{}_{CH_2}$$

(**71**)

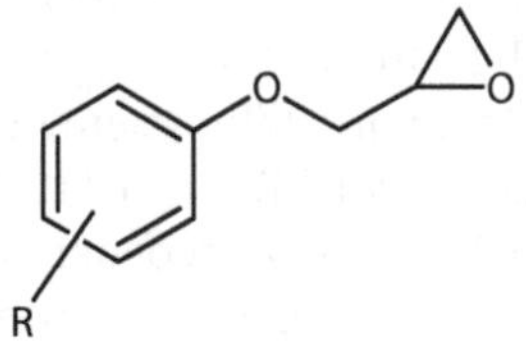

(**72**)

$$R \diagdown \diagup O \diagdown \diagup {}_{R}$$

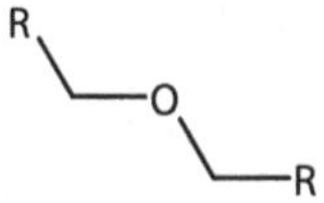

(**73**)

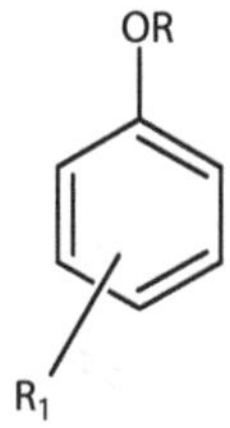

(**74**)

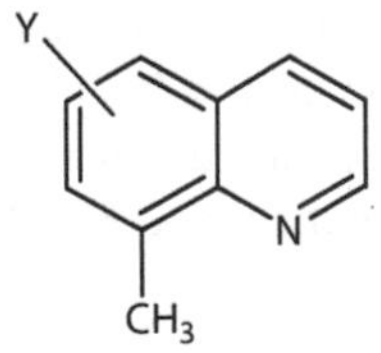

(75)

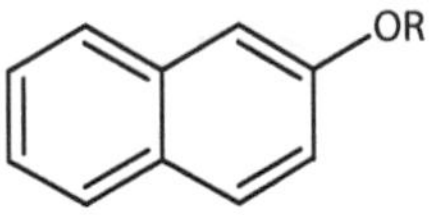

(76)

Solvent-free synthesis of substituted phenoxyacetic acids (77) with substituted phenols and chloroacetic acid was carried out under microwave irradiation by Nagy et al. (1997). 2-Chlorophenol reacted with ethanol in the presence of sodium hydroxide and a phase transfer catalyst under microwave irradiation to give 2-ethoxyphenol (78) conveniently within a few minutes (Pang et al., 1996).

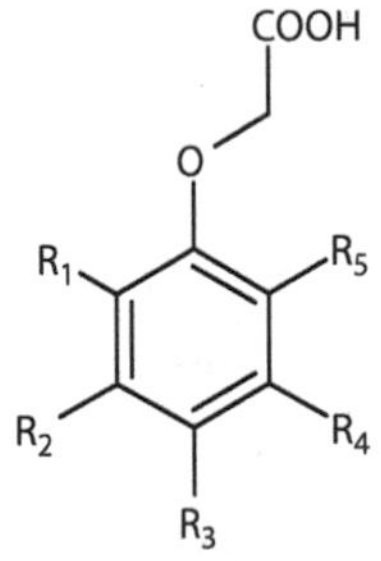

(77)

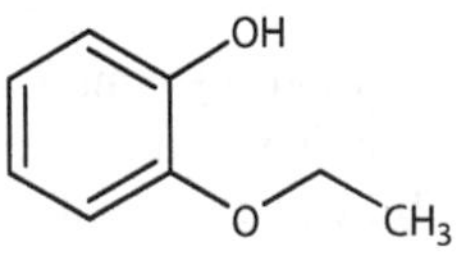

(78)

8.4.3 N-Alkylation

A new method of synthesis of N-alkylphthalimides (79) via alkylation of phthalimide in dry media under microwave irradiation was carried out by Bodgał et al. (1996). The reactions were performed by simply mixing phthalimide with an alkyl halide adsorbed on potassium carbonate to give a good yield of the product. Yadav and Subba Reddy (2000) reported a novel and efficient synthesis of N-arylamines (80) by the reaction of activated aryl halides with secondary amines in the presence of basic Al_2O_3 under microwave irradiation in solvent free conditions. Secondary amines and thiophenols were alkylated with alkyl and benzyl halides rapidly on alumina supported potassium carbonate under solvent-free conditions using microwaves yielding product (81) (Jaisinghani and Khadilkar, 1999).

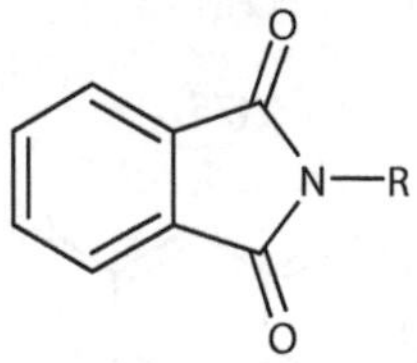

(79)

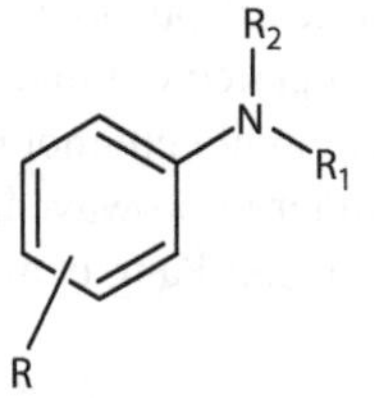

(80)

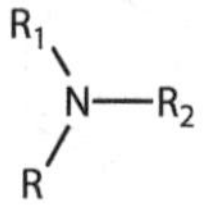

(81)

Under microwave irradiation, carbazole reacted remarkably fast with a number of alkyl halides to give N-alkyl derivatives of carbazole (82) (Bogdal et al., 1997). The reaction was carried out with high yields by simply mixing carbazole with an alkyl halide, which was adsorbed on potassium carbonate. A facile synthesis of a series of N-alkylpyrrolidino fullerenes (83) by phase transfer catalysis without solvent under microwave irradiation has been described by De la Cruz et al. (1998), while Adamczyk and Rege (1998) have illustrated the dramatic rate acceleration for N-sulfopropylation of heterocyclic compounds using 1,3-propane sultone under microwave irradiation affording the N-sulfopropylated compounds in 68–95% yield.

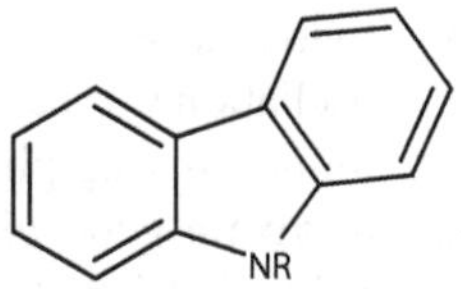

(82)

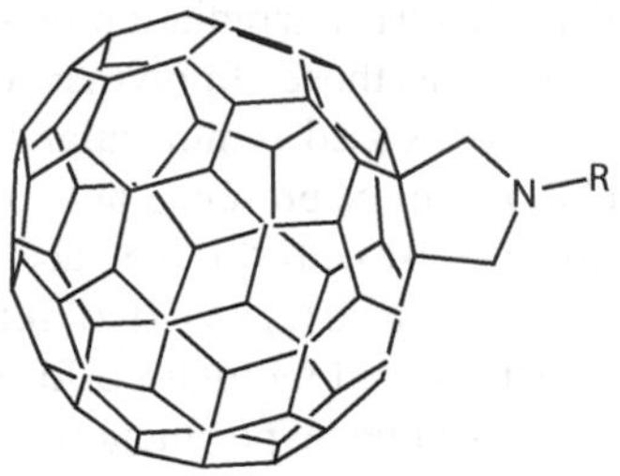

(83)

8.5 MICROWAVE IRRADIATED RADICAL REACTIONS

In chemistry, radicals or free radicals are atoms, molecules, or ions with unpaired electrons in the outermost shell. Such species with unpaired electrons are highly reactive, so radicals are important reaction intermediates. In chemistry, free radicals take part in radical addition and radical substitution as reactive intermediates. The species which is formed by the homolytic cleavage of covalent bond is the intermediate in vital chemical reactions such as combustion reaction and polymerization. As a result many plastics, enamels, and other polymers are formed through radical polymerization. Free radicals play an important role in a number of biological processes such as the intracellular killing of bacteria by neutrophil granulocytes and certain cell signaling processes. Free radicals are also involved in Parkinson's disease, senile and drug-induced deafness, schizophrenia, and Alzheimer's disease.

The rapid and homogeneous heating in the case of MAOS is advantageous over the conventional thermal heating techniques. The addition of azidotrimethylsilane to arylnitrileboronate esters was found to proceed rapidly in dimethoxyethane to give the aryltetrazoleboronates (84) in moderate to good yield, with dibutyltin oxide as catalyst (Schulz et al., 2004).

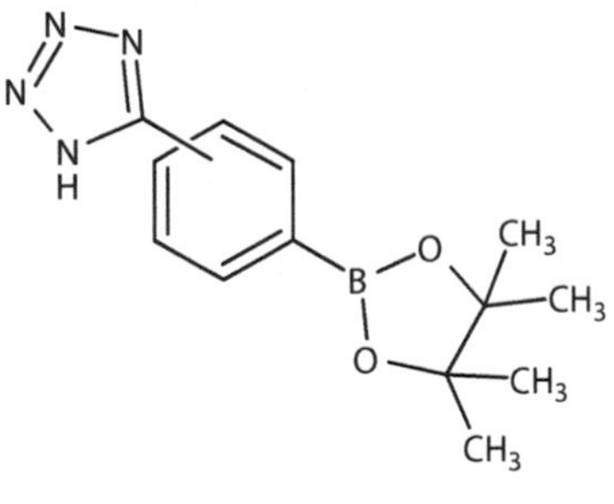

(84)

Microwave irradiation was applied to the atom transfer radical polymerization (ATRP) of azo-containing acrylates. The polymerization was greatly promoted and the reaction time was shortened from several days to about 1 h. The functional polymers (**85**) obtained under microwave irradiation process have good third-order nonlinear optical (NLO) properties as those of polymers obtained under CH process (Najun et al., 2007). Salicylic acid was used as molecular probe to determine hydroxyl radical (•OH) in aqueous solutions in the presence of activated carbon under microwave radiation. Rapid reaction of •OH with salicylic acid produced a stable fluorescence product, 2,3-dihydroxybenzoic acid (2,3-DHBA) (**86**) as well as its homologue, 2,5-DHBA, (**87**) which were determined by high-performance liquid chromatography (HPLC) (Quan et al., 2007). Another experiment was carried out to confirm that organic substances may degrade in aqueous solution by the microwave-assisted catalytic process. Pentachlorophenol (PCP) was used as target molecule. Results showed that nearly 72–100% degradation was observed in 60 min, which implied that continuous production of •OH during microwave irradiation supported free radical reactions that benefit PCP degradation.

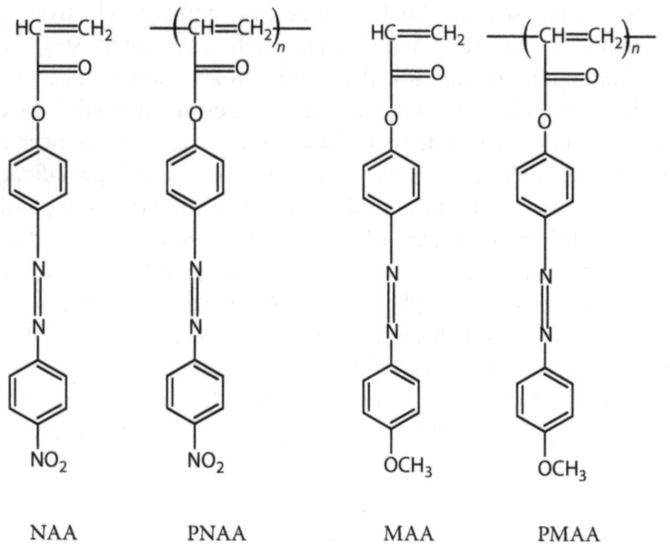

NAA   PNAA   MAA   PMAA

4-(4-nitrophenyl-diazenyl) phenyl acrylate (NAA)

4-(4-methoxyphenyl-diazenyl) phenyl acrylate (MAA)

(**85**)

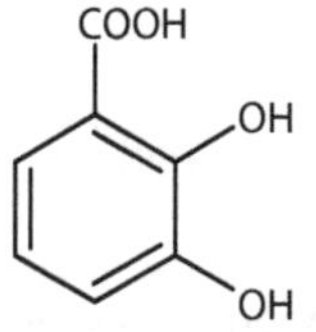

(**86**)

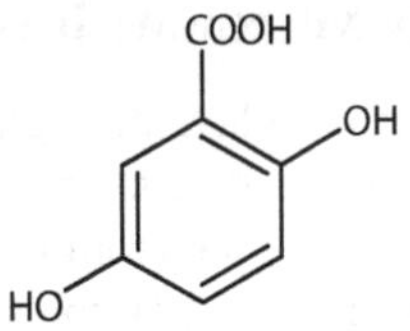

(87)

Solid-phase synthesis of various indol-2-ones (**88**) was carried out by means of aryl radical cyclization of resin-bound *N*-(2-bromophenyl)acrylamides using Bu$_3$SnH in DMF by Akamatsu et al. (2004). The reaction proceeded smoothly under microwave irradiation to give the desired indol-2-ones within a very short reaction time, in comparison to conventional thermal heating. Various alkoxyamines were isomerized using microwave irradiation at high temperatures (210°C) and short reaction times (2.5 min). Under these conditions, allyltosylamide isomerized efficiently to give (**89**) (70%, *trans:cis* = 3.3:1) (Wetter and Studer, 2004). Lamberto et al. (2003) have reported the microwave-assisted free radical cyclization of alkenyl and alkynyl isocyanides with thiols to give the five-membered nitrogen heterocycles. In a typical reaction, a thiyl radical (RS·) was found to add to an alkenyl isocyanide, generating a thioimidoyl radical, which underwent 5-exo cyclization and subsequent hydrogen atom abstraction to afford *cis*- and *trans*-pyrrolines. By using 2-mercaptoethanol *cis*- and *trans*-pyroglutamates were obtained, through the intermediate and also through the intermediate of a cyclic derivative, which underwent hydrolysis during the reaction.

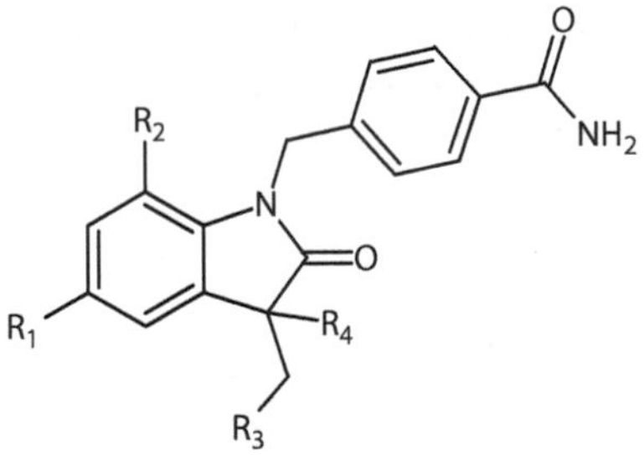

(88)

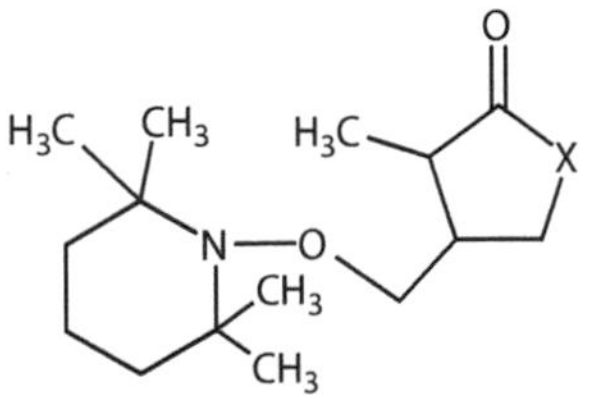

(89)

8.5.1 Rearrangements under Microwave Irradiation

A rearrangement reaction is a broad class of organic reactions where the carbon skeleton of a molecule is rearranged to give a structural isomer of the original molecule. Often, a substituent moves from one atom to another atom in the same molecule. In the example given below (Figure 8.1), the substituent R moves from carbon atom 1 to carbon atom 2.

The most common rearrangement reactions that are encountered in organic chemistry are:

1,2-Rearrangements: The rearrangement is intramolecular and the starting compound and reaction product are structural isomers. The 1,2-rearrangement belongs to a broad class of chemical reactions called rearrangement reactions. Examples are the Wagner–Meerwein rearrangement, Beckmann rearrangement, and Claisen rearrangement.

1,3-Rearrangements: 1,3-rearrangements take place over three carbon atoms. Examples are the Fries rearrangement and 1,3-alkyl shift of verbenone to chrysanthenone.

Intermolecular rearrangements also take place. The rearrangement reaction may involve the formation of any of the reaction intermediate such as carbocation, carbanion, free radical, and nitrene. The heterocycles containing sulfur moieties, benzo[*b*][1,4]thiazin-3(4H)-one derivatives are privileged structures due to the large number of biologically active molecules and natural products containing this moiety (Fujita et al., 1990; Kajino et al., 1991; Matsui et al., 1994; Fringuelli et al., 1998; Guarda et al., 2003). The synthesis of benzo[b][1,4]thiazin-3(4H)-one (**90**) derivatives in a simple and efficient method from the one-pot reaction of substituted 2-chlorobenzenthiols, chloroacetyl chloride and primary amines via Smiles rearrangement under microwave irradiation gave high yields of 65–92% of the products with short reaction time of 15–20 min (Zuo et al., 2008).

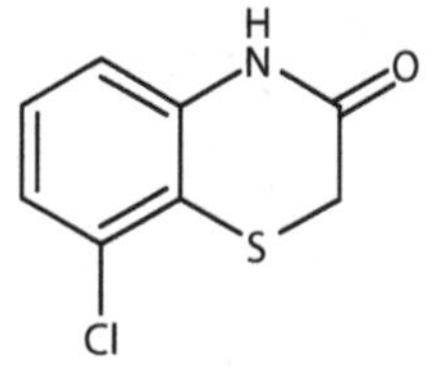

(90)

FIGURE 8.1 Rearrangement reaction where substituent R moves from carbon atom 1 to carbon atom 2.

An efficient synthesis of functionalized pyrimidones (**91**) via microwave-accelerated rearrangement reaction of amidoxime DMAD adducts has been described by Summa et al. (2004). In most cases, the pyrimidone formation was furnished in reasonable yield after 2 min of microwave irradiation. Pyrimidones of this type were recently found to inhibit a series of hepatitis C virus (HCV) NSSB polymerase. 3-Formyl-chromones readily react with primary amines in an alcoholic medium yielding an enamine adduct, which rarely reacts further to give the corresponding Schiff base (El-Shaaer et al., 1998; Gaplovsky et al., 2000). The important role of 3-formyl-chromones as versatile synthons in heterocyclic chemistry as well as their pharma-ceutical importance is well known (Gasparova et al., 1997; El-Shaaer et al., 1998; Margita et al., 2005). These compounds also have interesting photochemical properties (Gaplovsky et al., 2000). The rearrangement of 3-formylchromone enamines as a simple and facile route to novel pyrazolo[3,4-b]-pyridines (**92**) was carried out using a microwave (Sabitha, 1996). Three (1 → 6)-linked disaccharides, melibiose, isomaltose, and palatinose were converted into their isomers (**93**) via intramolecular rearrangement catalyzed by Mo(VI). The reaction proceeded with excellent stereose-lectivity under microwave irradiation. It is well known that many oligosaccharides and glycoconjugates play important roles in fundamental biological systems (Varki, 1993; Dwek, 1996). They are involved in metabolic processes, signal transduction and the immune response (Rudd et al., 2001). The rapidly growing development of glycobiology hence sets demands for the increased development and methodological advances for the selective construction of natural and unnatural carbohydrates (Hricovíniova, 2008).

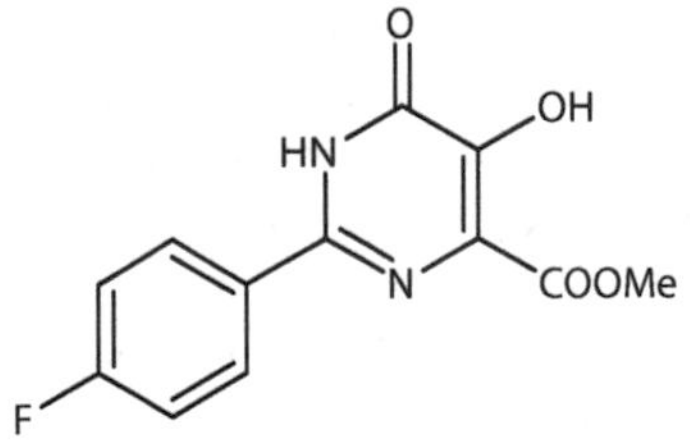

(**91**)

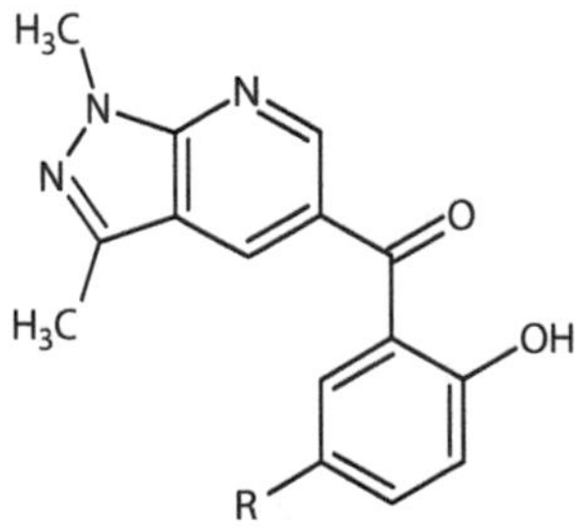

(**92**)

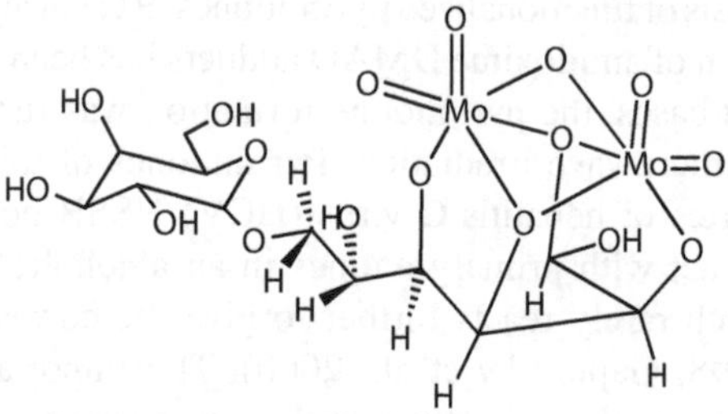

(93)

The one-pot conversion of 1,4-bis-allyloxybenzene derivatives to 2,3-bis-allyl-1,4-quinone (**94**) derivatives was carried out by taking advantage of the cumulative effect of the silica gel support and microwave irradiation under solvent-free conditions. A fast, efficient and environmentally benign solvent-free procedure has been developed for microwave-assisted Claisen rearrangement on a silica gel support (Castro, 2004). Various bis-allyl ketones (**95**) were prepared using this protocol. [3,3]-Sigmatropic rearrangements, when used as a tool for the formation of carbon–nitrogen bonds, have an enormous potential for the synthesis of molecules containing nitrogen-bearing stereocenters. The prototype of this reaction is the Overman rearrangement, which provides protected allylic amines with excellent stereocontrol and constitutes one of the few possibilities for introducing an amino group at a tertiary carbon atom (Mehmandoust et al., 1992; Calter et al., 1997; Hollis and Overman, 1997; Overman and Zipp, 1997; Nishikawa et al., 2001). The microwave-assisted thermal aza-Claisen rearrangement of allylic imidates and thiocyanates to the corresponding amides (**96**) and isothiocyanates (**97**) was investigated by Gonda et al. (2007). Significant accelerations of the rearrangement of allylic imidates to amides and of allylic thiocyanates to isothiocyanates in comparison with standard thermal reactions have been observed. The microwave irradiation of thiocyanate in *o*-xylene under sealed vessel conditions at 150°C led to the rearranged products with substantial shortening of the reaction time (24 times) compared to the thermal reaction performed at 90°C. Acceleration of the rearrangement 1 → 2 (24–80 times) in a variety of solvents was observed giving 59–72% of (**98**) and (**99**) (Williams and Wander, 1980).

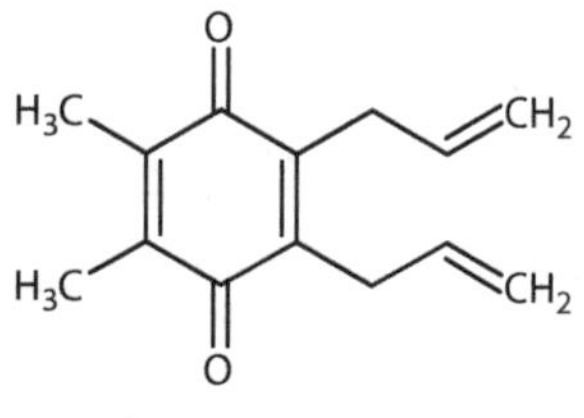

(94)

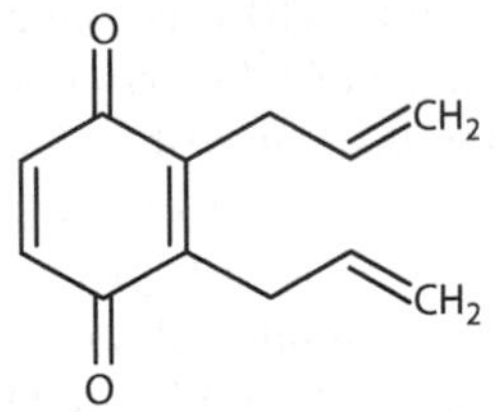

(95)

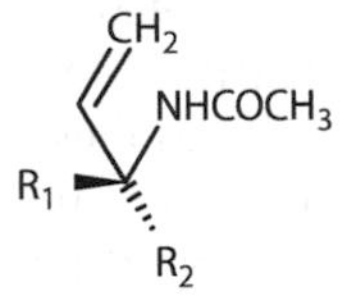

(96)

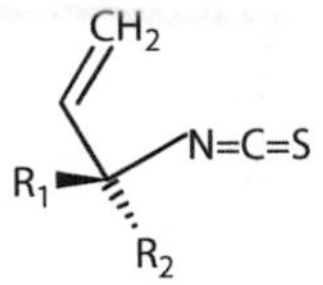

(97)

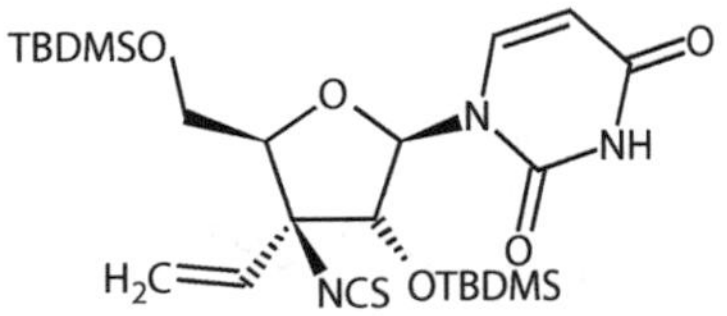

(98)

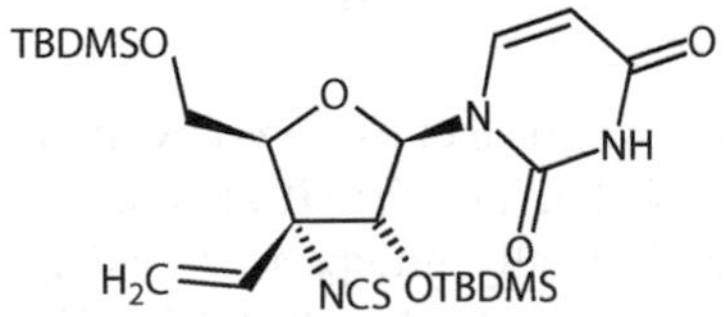

(99)

2,3-Unsaturated glycosides (pseudoglycals) are versatile synthetic intermediates and also constitute the structural units of several antibiotics (Williams and Wander, 1980). Allylic rearrangement of glycols is known as the Ferrier rearrangement (Ferrier and Prasad, 1969; Ferrier, 2001) in the presence of a nucleophile generally leads to the formation of 2,3-unsaturated glycosides. 2,3-Unsaturated O-aryl glyco-sides, in turn, can be transformed to C-aryl glycosides through (Ramesh and Balasubramanian, 1992) sigmatropic rearrangement. Shanmugasundaram et al.

(2002) reported that montmorillonite K10 was found to catalyze, under microwave irradiation, the rapid O-glycosidation of 3,4,6-tri-O-acetyl-D-galactal to afford exclusively the alkyl and aryl 2,3-dideoxy-D-*threo*-hex-2-enopyranosides (**100**) with very high selectivity and without the formation of the 2-deoxy-D-*lyxo*-hexopyranosides. Under these conditions, 3,4,6-tri-O-acetyl-D-glucal as usual also underwent the Ferrier rearrangement. A unique example of microwave-assisted aza-Cope rearrangement of allylanilines has been reported by Yadav et al. (2000) utilizing microwave irradiation for the preparation of indolines from allylanilines with the help of montmorillonite as a catalyst. These compounds were formed by means of a [3,3] sigmatropic rearrangement, followed by *in situ* cyclization. Gonzalez et al. (2008) reported the microwave-assisted reaction conditions for the transformation of N-allylanilines into o-allylanilines (**101**), as these are quite interesting starting materials for the preparation of different bicyclic nitrogen-containing heterocycles through the cyclization of the nitrogen onto the double bond.

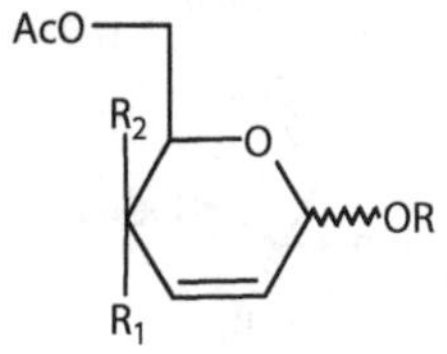

(**100**)

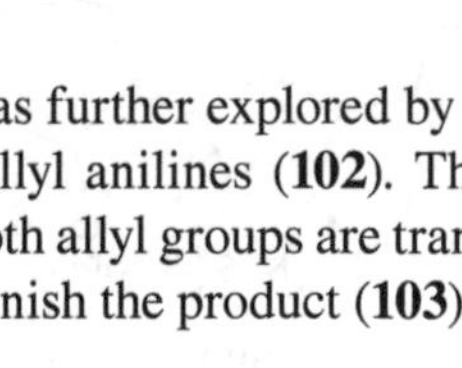

(**101**)

The scope of the reaction was further explored by the substitution on the aryl ring for different N-benzylated N-allyl anilines (**102**). The reaction was also performed with N-bisallylated anilines. Both allyl groups are transferred to the benzyl ring under usual reaction conditions to furnish the product (**103**), (**104**) in appreciable yields.

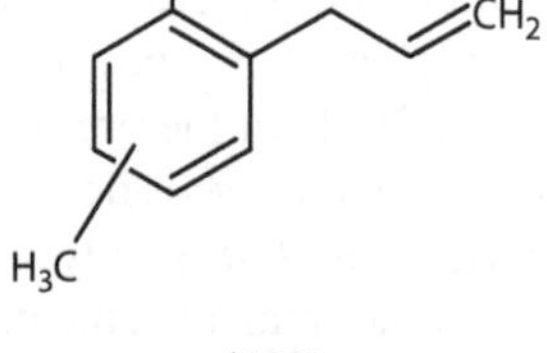

(**102**)

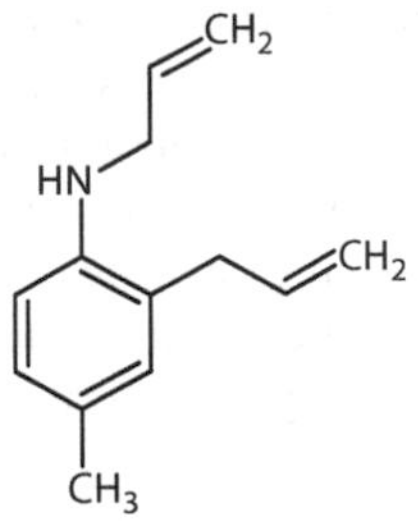

(103)

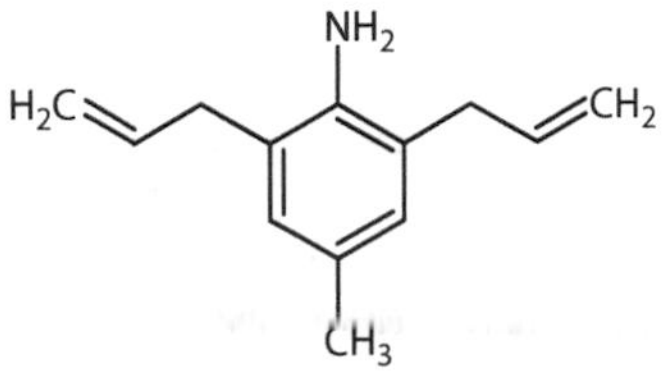

(104)

8.6 MICROWAVE IRRADIATED RING OPENING

Compounds containing epoxide (oxirane) ring are versatile synthetic intermediates for the preparation of various synthetic compounds of biological as well as industrial importance. The condition under which epoxide is attacked by the reagent determines the course of reaction. Generally, under basic conditions nucleophile attack at the less substituted carbon whereas in acidic conditions, a greater proportion of attack occurs at more substituted carbon of epoxide. Sometimes, solvent may play an important role in deciding the stereochemical outcome of the reaction. Epoxy compounds on treatment of amines yield aminoalcohols, which can be utilized in the manufacture of cosmetics, drugs, antifungal, and antibacterial compounds. Kureshy et al. (2006) reported for the first time asymmetric ring opening of meso-stilbene oxide and cyclohexene oxide with aniline and substituted anilines catalyzed by a Ti-(S)-BINOL complex to give enantio-rich syn-β-amino alcohols and trans-β-amino alcohols (105) in high yields (up to 95%) with up to 55% enantiomeric excess under microwave irradiation at 60°C. A microwave accelerated epoxide ring-opening process with N-biaryl sulfonamides has been described. Under this mild, highly efficient condition, an α-hydroxy-β-N-biaryl sulfonamide (106) skeleton is rapidly assembled leading ultimately to a novel series of matrix metalloproteinase-9 inhibitors with single digit nanomolar activities (Yang and Murray, 2008). Miranda et al. (2009) reported the efficient microwave-assisted Au(I) catalyzed cleavage of the pyran ring of brevifloralactone. They described one of the first combination reactions of gold and microwave irradiation. They illustrated the synthesis of nine new ent-clerodane (107) diterpene derivatives obtained by partial synthesis of brevifloralactone, a naturally occurring clerodane-type diterpene isolated in large quantities from the aerial part of *Salvia breviflora*. The clerodane diterpenes have very interesting biological activities and the semisynthetic approach described in several papers represents an alternative to obtain them from other major diterpenes isolated from natural sources. Tabatabaeian et al. (2008) have reported the ruthenium-catalyzed regioselective ring opening of aliphatic and aryl epoxides under solvent-free conditions. It was found that $RuCl_3 \cdot nH_2O$ catalyzes the Friedel–Crafts alkylation of indoles, providing 3-alkylated derivatives (**108**) in good yields under mild reaction conditions. Nadir and Singh (2005) reported that *N*-tosylaziridines react efficiently with amines in the presence of montmorillonite K10 as catalyst under microwave irradiation in solvent-free conditions to yield the corresponding achiral and chiral diamines regio- and stereoselectively in a few minutes and in high yields (**109** and **110**).

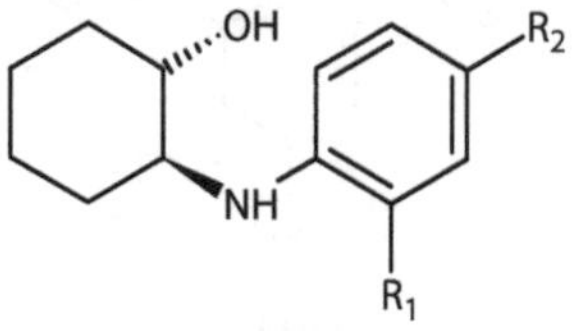

(**105**)

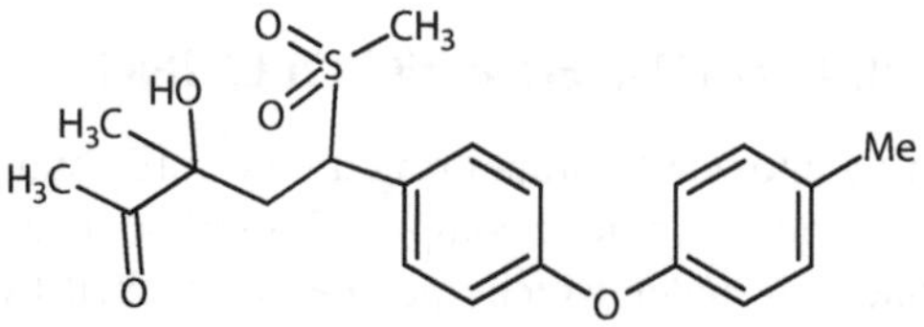

(**106**)

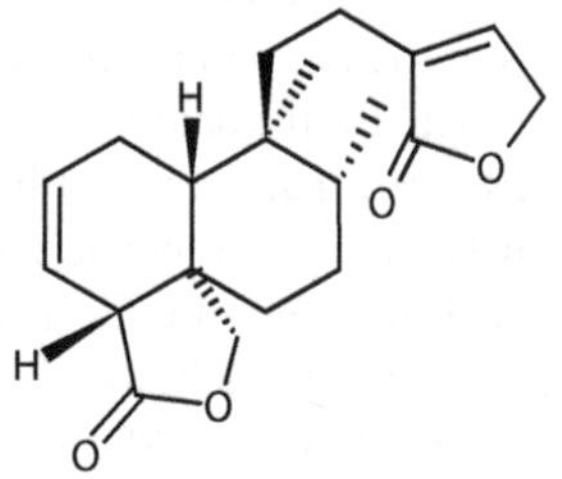

(**107**)

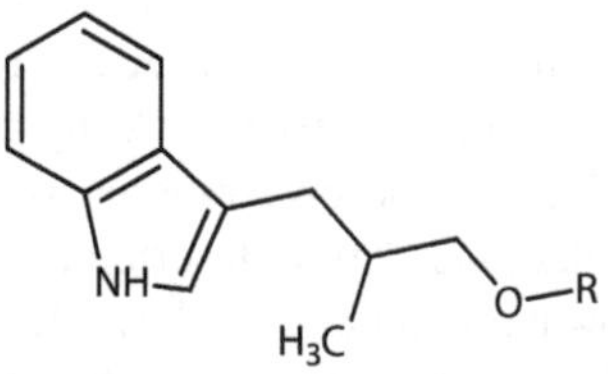

(**108**)

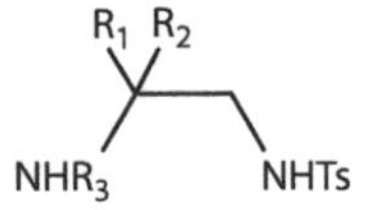

(109)

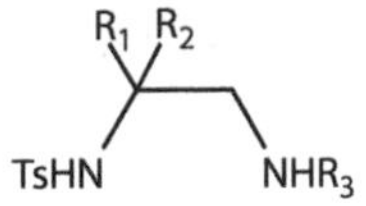

(110)

Ring-opening fluorination reactions of epoxides using tetrabutylammonium bifluoride (TBABF)-KHF$_2$ or Et$_3$N-3HF under microwave irradiation were applied for the introduction of a fluorine atom into the carbohydrate molecules (111 and 112). When TBABF-KHF$_2$ was used as the fluorination reagent, a fluorine atom was introduced regioselectively and various functional groups could tolerate the conditions. When Et$_3$N-3HF was used under microwave irradiation, the reaction time could be remarkably shortened compared with the conventional oil-bath heating (Akiyama et al., 2006). Cardillo et al. (2000) reported the microwave-assisted rearrangement of N-Boc-chiral aziridine-2-imides and esters to oxazolidin-2-ones (113 and 114) in the presence of different Lewis acids. The regio-selectivity of the reaction strongly depends upon the Lewis acids and the reaction conditions. Chakraborti et al. (2004) demonstrated the scope and limitations of montmorillonite K10 in ring opening of epoxides by amines. Montmorillonite K10 efficiently catalyzed the opening of epoxide rings by amines in high yields with excellent regio- and diastereo-selectivities under solvent-free conditions at room temperature affording an improved process for the synthesis of 2-amino alcohols. The reaction of cyclohexene oxide with aryl/alkyl amines leads to the formation of trans-2-aryl/alkylaminocyclohexanols (115). The ^{1}H and ^{13}C NMR analyses confirmed the trans stereochemistry of the product. For unsymmetrical epoxides, the regioselectivity is controlled by the electronic and steric factors associated with the epoxide and the amine. Carlos et al. (2006) reported the living cationic ring-opening polymerization of 2-ethyl-2-oxazoline to yield (116) in an ionic liquid under microwave irradiation. The reaction showed an enhanced polymerization rate in comparison to the reaction in common organic solvents. Also, an ionic liquid was efficiently recovered and reused in new reaction cycles.

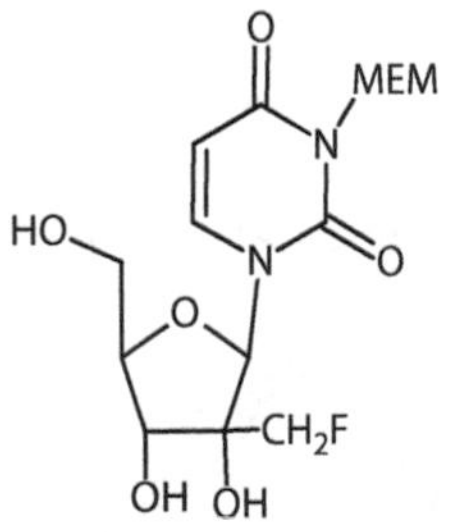

(111)

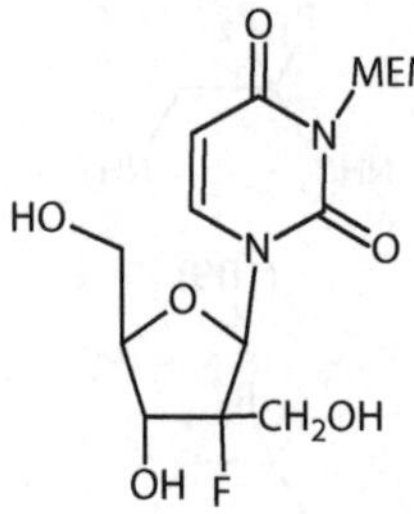

(112)

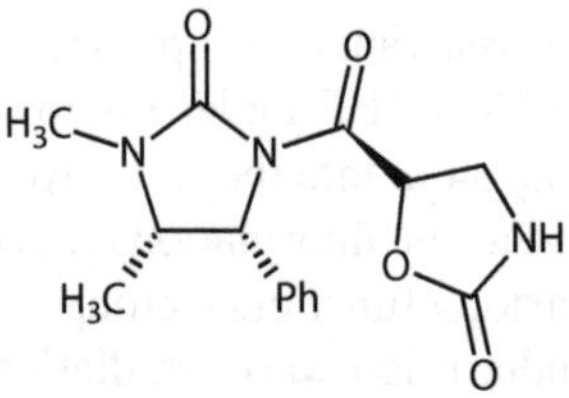

(113)

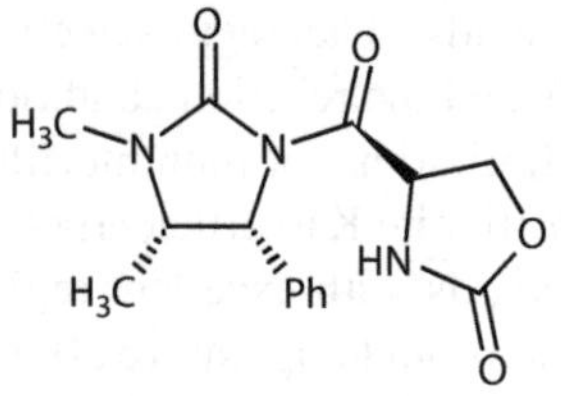

(114)

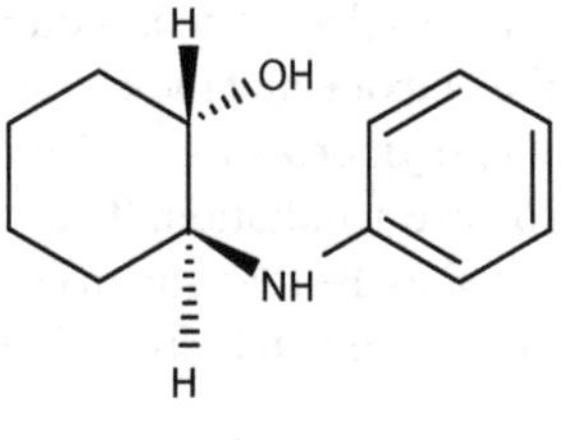

(115)

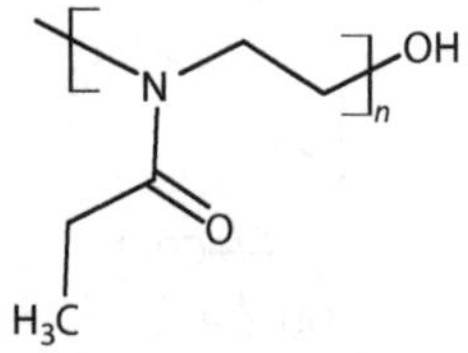

(116)

Microwave-assisted ring-opening polymerization of ε-caprolactone (ε-CL) initiated by benzoic acid was carried out to give (117). The mixtures of ε-CL/benzoic acid were heated under microwaves and the temperatures were self-regulated to equilibrium from 204 to 240°C (Yu and Liu, 2004). Mori et al. (2007) have worked on the ring-opening polymerization of a representative N-carboxy-α-amino acid anhydride (NCA) in ionic liquids. The polymerization of γ-benzyl-L-glutamate-N-carboxyanhydride (BLG-NCA) with n-butylamine as an initiator in an ionic liquid ([BMI][PF6]) proceeded as a milky white dispersion with no evidence of macroscopic precipitation. The polymerization rate was slightly affected by the nature of the anion and hydrophobicity of the ionic liquids, while poly-(BLG)s (118) having low polydispersities, were obtained regardless of the kind of the ionic liquids. Several parameters, such as the existence of organic solvent as a co-solvent and monomer concentration, had also clear effects on the polymerization rate and/or the polydispersity of the resulting poly-(BLG)s.

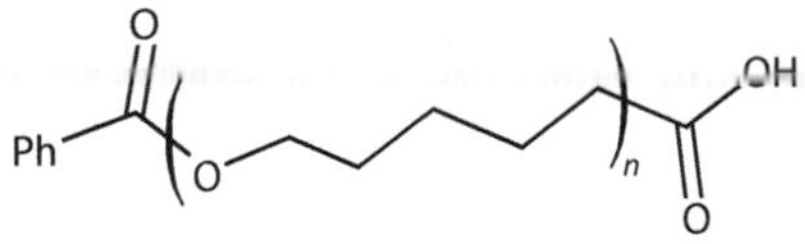

(117)

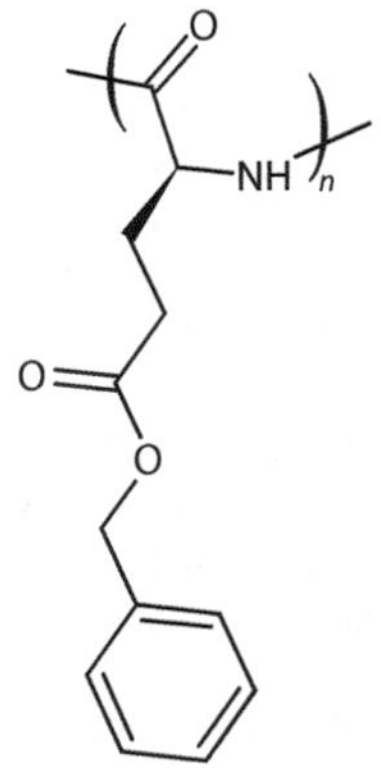

(118)

The effect of microwaves on the kinetics and the selectivity of the Cr (salen)-catalyzed asymmetric ring opening of epoxides was investigated by Dioos and Jacobs (2005). It was found that the reaction rate of the Cr (salen)-catalyzed kinetic resolution of terminal epoxides and the asymmetric ring opening (ARO) of meso-epoxides could be increased by three orders of magnitude without impairing the selectivity.

α,β-Dehydroamino amides are important intermediates for the synthesis of biologically active molecules. A rapid procedure has been reported for the synthesis of α,β-dehydroamino amides (119) from azalactones under microwave irradiation.

Azalactones were produced via the Erlenmeyer reaction and submitted to ring-opening reactions under a microwave during 15–30 min (Valdésa et al., 2007). Functional groups were selectively introduced at the C-2 position of isophorone via the epoxide ring-opening with several nucleophiles. Various behaviors were observed depending on the reaction conditions and the nature of nucleophilic reagents. Electronic and steric effects of the reactants were discussed. The best experimental systems involved LiClO$_4$ salt effect in acetonitrile, phase transfer catalysis or KF ± alumina under solvent-free conditions and under a microwave (Rissafi et al., 2001).

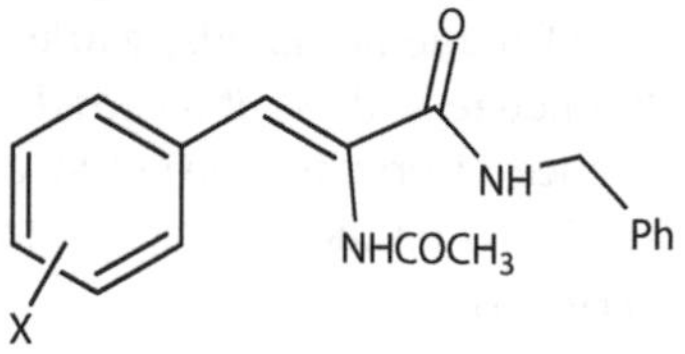

(119)

REFERENCES

Abramovitch, R.A. and Bulman, A. 1992. Fischer cyclizations by microwave heating. *Synlett*, 795–96.

Abramovitch, R.A., Shi, Q., and Bogdal, D. 1995. Microwave-assisted alkylations of activated methylene groups. *Synthetic Communications*, 25:1–8.

Adamczyk, M. and Rege, S. 1998. Microwave assisted sulfopropylation of *N*-heterocycles using 1,3-propane sultone. *Tetrahedron Letters*, 39:9587–88.

Akamatsu, H., Fukase, K., and Kusumoto, S. 2004. Solid phase synthesis of indol-2-ones by microwave-assisted radical cyclizations. *Synlett*, 1049–53.

Akiyama, Y., Hiramatsu, C., Fukuhara, T., and Hara, S. 2006. Selective introduction of a fluorine atom into carbohydrates and a nucleoside by ring-opening fluorination reaction of epoxides. *Journal of Fluorine Chemistry*, 127:920–23.

Alloum, A.B., Bakkas, S., Bougrin, K., and Soufiaoui, M. 1998. Synthèse de nouvelles spiro-rhodanine-pyrazolines par addition dipolaire-1,3 de la diphenylnitrilimine sur quelques 5-arylidènerhodanines en milieu sec et sous irradiation micro-onde. *New Journal of Chemistry*, 22:809–12.

Alloum, A.B., Labiad, B., and Villemin, D. 1989. Application of microwave heating techniques for dry organic reactions. *Journal of Chemical Society Chemical Communications*, 386–87.

Arrieta, A., Carrillo, J.R., Cossio, F.P., Diaz-Ortiz, A.,Gómez-Escalonilla, A., De la Hoz, A., Langa, F., and Moreno, A. 1998. Efficient tautomerization hydrazone-azomethine imine under microwave irradiation. Synthesis of [4,3'] and [5,3']bipyrazoles. *Tetrahedron*, 54:13167–80.

Avalos, M., Babiano, R., Cintas, P., Clemente, F.R., Jimenez, J.L., Palacios, J.C., and Sanchez, J.B. 1999. Hetero-Diels–Alder reactions of homochiral 1,2-diaza-1,3-butadienes with diethyl azodicarboxylate under microwave irradiation—Theoretical rationale of the stereochemical outcome. *Journal of Organic Chemistry*, 64:6297–6305.

Azizian, J., Asadi, A., and Jadidi, K. 2001. One-pot highly diastereo-selective synthesis of new 2-substituted 8-(spiro-3'indolino-2-one)-pyrrolo[3,4-a]-pyrrolizine-1,3-diones mediated by azomethine ylide induced by microwave irradiation. *Synthetic Communications*, 31:272–33.

Bagnell, L., Cablewski, T., and Strauss, C.R. 1999. A catalytic symmetrical etherification. *Chemical Communications*, 283–84.

Baruah, B., Boruah, A., Prajapati, D., and Sandhu, J.S., 1997a. BiCl$_3$ or CdI$_2$ catalyzed Michael addition of 1,3-dicarbonyl compounds under microwave irradiations. *Tetrahedron Letters*, 38:1449–50.

Baruah, B., Prajapati, D., Boruah, A., and Sandhu, J.S. 1997b. Microwave induced 1,3-dipolar cycloaddition reactions of nitrones. *Synthetic Communications*, 27:2563–67.

Bogdał, D., Pielichowski, J., and Boroń, A. 1996. Remarkable fast microwave-assisted *N*-alkylation of phthalimide in dry media. *Synlett*, 873–74.

Bogdal, D., Pielichowsk, J., and Jaskot, K. 1997. New synthesis method of *N*-alkylation of carbazole under microwave irradiation in dry media. *Synthetic Communications*, 27:1553–60.

Bogdal, D., Pielichowski, J., and Boroń, A. 1998. New synthetic method of aromatic ethers under microwave irradiation in dry media. *Synthetic Communications*, 28:3029–39.

Bougrin, K., Soufiaoui, M., Loupy, A., and Jacquault, P. 1995. Addition dipolaire 1,3-de la diphenylnitrile imine sur quelques dipolarophiles en "milteu sec" sous microonds. *New Journal of Chemistry*, 19:213.

Bourgeois, D., Craig, D., King, N.P., and Mountford, D.M. 2005. Synthesis of homoallylic sulfones through a decarboxylative Claisen rearrangement reaction. *Angewandte Chemie International Edition*, 44:618–21.

Bourgeois, D., Craig, D., Grellepois, F., Mountford, D.M., and Stewart, A.J.W. 2006. A [3,3]-sigmatropic process catalysed by acetate. The decarboxylative Claisen rearrangement. *Tetrahedron*, 62:483–95.

Cado, F., Jacquault, P., Dozias, M.J., Bazureau, J.P., and Hamelin, J. 1997. Tandem conjugate carbon addition-intermolecular hetero Diels–Alder reactions using ethyl 1H-perimidine-2-acetate as a ketene aminal with heating or microwave activation. *Journal of Chemical Research*, 176–77.

Calter, M., Hollis, T.K., Overman, L.E., Ziller, J., and Zipp, G.G. 1997. First enantioselective catalyst for the rearrangement of allylic imidates to allylic amides. *Journal of Organic Chemistry*, 62:1449–56.

Cardillo, G., Gentilucci, L., Gianotti, M., and Tolomelli, A. 2000. Influence of Lewis acids on the regioselectivity of *N*-Boc-aziridine-2-carboxylate microwave-assisted rearrangement. *Synlett*, 1309–11.

Carlos, G.S., Richard, H., and Ulrich, S.S. 2006. Fast and green living cationic ring opening polymerization of 2-ethyl-2-oxazoline in ionic liquids under microwave irradiation. *Chemical Communications*, 36:3797–99.

Castro, A.M. 2004. Claisen rearrangement over the past nine decades. *Chemical Review*, 104:2939–3002.

Chakraborti, A.S., Kondaskar, A., and Rudrawar, S. 2004. Scope and limitations of montmorillonite K10 catalysed opening of epoxide rings by amines. *Tetrahedron*, 60:9085–91.

Cheng, Q., Zhang, W., Tagami, Y., and Oritani, T. 2001. Microwave-induced 1,3-dipolar intramolecular cycloadditions of *N*-substituted oximes, nitrones, and azomethine ylides for the chiral synthesis of functionalized nitrogen heterocycles. *Journal of Chemical Society Perkin Transactions*, 1:452–56.

Choshi, T., Kumemura, T., Nobuhiro, J., and Hibino, S. 2008. Novel synthesis of the 2-azaanthraquinone alkaloid, scorpinone, based on two microwave-assisted pericyclic reactions. *Tetrahedron Letters*, 49:3725–28.

Craig, D. and Grellepois, F. 2005. Decarboxylative Claisen rearrangement reactions of allylic tosylmalonate esters. *Organic Letters*, 7:463–65.

Das, B., Madhusudhan, P., and Kashinatham, A. 2000. Unprecedented adducts formed by camptothecin and mappicine ketone with maleic anhydride under microwave radiation. *Indian Journal of Chemistry*, 39B:326.

De la Cruz, P., De la Hoz, A., Langa, F., Illescas, B., and Martin, N. 1997. Cycloadditions to [60] fullerene using microwave irradiation: A convenient and expeditious procedure. *Tetrahedron*, 53:2599–608.

De la Cruz, P., de la Hoz, A., Font, A.M., Langa, F., and Pérez-Rodríguez, M.C. 1998. Solvent-free phase transfer catalysis under microwaves in fullerene chemistry. A convenient preparation of N-alkylpyrrolidino [60] fullerenes. *Tetrahedron Letters*, 39:6053–56.

Deng, R., Wang, Y., and Jiang, Y. 1994. Solid–liquid phase transfer catalytic synthesis X: The rapid alkylation of ethyl acetoacetate under microwave irradiation. *Synthetic Communications*, 24:111–15.

Diaz-Ortiz, A., Diez-Barra, E., De la Hoz, A., Moreno, M., Goomez-Escalonilla, M.J., and Loupy, A. 1996. 1,3-Dipolar cycloaddition of nitriles under microwave irradiation in solvent-free conditions. *Heterocycles*, 43:1021–30.

Diaz-Ortiz, A., Carrillo, J.R., Cossio, F.P., Gomez-Escalonilla, M.J., De la Hoz, A., Moreno, A., and Prieto, P. 2000. Synthesis of pyrazolo[3,4-b]pyridines by cycloaddition reactions under microwave irradiation. *Tetrahedron*, 56:1569–77.

Dinica, R.M., Druta, I.I., and Pettinari, C. 2000. The synthesis of substituted 7,7'-bis-indolizines via 1,3-dipolar cycloaddition under microwave irradiation. *Synlett*, 1013–15.

Dioos, B.M.L. and Jacobs, P.A. 2005. Microwave-assisted Cr (salen)-catalysed asymmetric ring opening of epoxides. *Journal of Catalysis*, 235:428–30.

Dwek, R.A. 1996. Glycobiology: Toward understanding the function of sugars. *Chemical Review*, 683–720.

El-Shaaer, H., Foltinova, P., Lacova, M., Chovancová, J., and Stankovicova, H. 1998. Synthesis, antimicrobial activity and bleaching effect of some reaction products of 4-oxo-4H-benzopyran-3-carboxaldehydes with aminobenzothiazoles and hydrazides. *Il Farmaco*, 53:224–32.

Ferrier, R.J. and Prasad, N. 1969. Unsaturated carbohydrates. Part IX. Synthesis of 2,3-dideoxy-α-D-*erythro*-hex-2-enopyranosides from tri-O-acetyl-D-glucal. *Journal of Chemical Society C*, 570–74.

Ferrier, R.J. 2001. Substitution with allylic rearrangement reactions of glycal derivatives. *Topics in Current Chemistry*, 215:153–75.

Fraga-Drubreuil, J. and Cherouvrier, J.R. 2000. Clean solvent-free dipolar cycloaddition reactions assisted by focused microwave irradiations for the synthesis of new ethyl 4-cyano-2-oxazoline-4-carboxylates. *Green Chemistry*, 2:226–29.

Fringuelli, R., Schiaffella, F., Bistoni, F., Pitzurra, L., and Vecchiarelli, A. 1998. Azole derivatives of 1,4-benzothiazine as antifungal agents. *Bioorganic and Medicinal Chemistry*, 6:103–08.

Fujita, M., Ito, S., Ota, A., Kato, N., Yamamoto, K., Kawashima, Y., Yamauchi, H., and Iwao, J. 1990. Synthesis and calcium ion antagonistic activity of 2-[2-[(aminoalkyl)oxy]-5-methoxyphenyl]-3,4-dihydro-4-methyl-3-oxo-2H-1,4-benzothiazines. *Journal of Medicinal Chemistry*, 33:1898–1905.

Gaplovsky, A., Donovalová, J., Lacova, M., Mracnova, R., and El-Shaaer, H.M. 2000. The photochemical behaviour of 6-X-4H-3(bicyclo[2.2.1]-5-heptene-2,3-dicarboximidoiminomethyl)-4-chromones. Photochromism and thermochromism. *Journal of Photochemistry Photobiology A: Chemistry*, 136:61–65.

Gasparova, R., Lacova, M., and EL-Shaaer, H.M. 1997. Synthesis and antimycobacterial activity of some new 3-heterocyclic substituted chromones. *Il Farmaco*, 52:251–53.

Giguere, R.J., Bray, T.L., Duncan, S.M., and Majetich, G. 1986. Application of commercial microwave ovens to organic synthesis. *Tetrahedron Letters*, 27:4945–48.

Giguere, R.J., Namen, A.M., Lopez, B.O., Arcpally, A., Ramos, D.E., Majetich, G., and Defauw, J. 1987. Studies on tandem ene/intramolecular diels-alder reactions. *Tetrahedron Letters*, 28:6553–56.

Gonzalez, I., Bellas, I., Souto, A., Rodriguez, R., and Cruces, J. 2008. Microwave-assisted aza-Cope rearrangement of *N*-allylanilines. *Tetrahedron Letters*, 49:2002–4.

Guarda, V.L.M., Perrissin, M., Thomasson, F., Ximenes, E.A., Galdino, S.L., Ptta, I.R., Luu-Duc, C., and Barbe, J. 2003. Synthesis of 4-octyl-2*H*-1,4-benzo-thiazin-3-ones. *European Journal of Medicinal Chemistry*, 38:769–73.

Gonda, J., Martinkova, M., Zadrosova Monika Sotekova, A., Aschmanova, J., Conka, P., Gajdosikova, E., and Kappe, C.O. 2007. Microwave accelerated aza-Claisen rearrangements. *Tetrahedron Letters*, 48:6912–15.

Heber, D. and Stoyanov, E.V. 1999. The microwave assisted nucleophilic substitution of 4-hydroxy-6-methyl-2(1*H*)-pyridones. *Synlett*, 1747–48.

Hollis, T.K. and Overman, L.E. 1997. Cyclopalladated ferrocenyl amines as enantioselective catalysts for the rearrangement of allylic imidates to allylic amides. *Tetrahedron Letters*, 38:8837–40.

Hricovíniova, Z. 2008. Microwave-assisted stereospecific intramolecular rearrangement of (1 → 6)-linked disaccharides catalyzed by Mo (VI). *Tetrahedron: Asymmetry*, 19:1853–56.

Hu, Y., Yu, J., Yang, S., Wang, J., and Yin, Y. 1999. Substitution reaction of nitro group on α-nitrostyrene by organozinc halides under microwave irradiation. *Synthetic Communications*, 29:1157–64.

Jacob, R.G., Perin, G., Botteselle, G.V., and Lenardao, E.J. 2003. Clean and atom economic synthesis of octahydroacridines: Application to essential oil of citronella. *Tetrahedron Letters*, 44:6809–12.

Jaisinghani, H.G. and Khadilkar, B.M. 1997. Rapid and efficient synthesis of 1-arylpiperazines under microwave irradiation. *Tetrahedron Letters*, 38:6875–76.

Jaisinghani, H.G. and Khadilkar, B.M. 1999. Microwave assisted highly efficient solid state N- and S-alkylation. *Synthetic Communications*, 29:3693–98.

Jones, G.B., Huber, R.S., and Chau, S. 1993. The Claisen rearrangement in synthesis: Acceleration of the Johnson orthoester protocol *en route* to bicyclic lactones. *Tetrahedron*, 49:369–80.

Kajino, M., Mizuno, K., Tawada, H., Shibouta, Y., Nishikawa, K., and Meguro, K. 1991. Synthesis and biological activities of new 1,4-benzothiazine derivatives. *Chemical and Pharmaceutical Bulletin*, 39:2888–95.

Katritzky, A.R. and Singh, S.K. 2002. Synthesis of *c*-carbamoyl-1,2,3-triazoles by microwave-induced 1,3-dipolar cycloaddition of organic azides to acetylenic amides. *Journal of Organic Chemistry*, 67:9077–79.

Kerneur, G., Lerestif, J.M., Bazureau, J.P., and Hamelin, J. 1997. Convenient preparation of 4-alkylidene-1*H*-imidazol-5(4*H*)-one derivatives from imidate and aldehydes by a solvent-free cycloaddition under microwaves. *Synthesis*, 287–89.

Khadilkar, B.M. and Bendale, P.M. 1997. Microwave enhanced synthesis of epoxypropoxy-phenols. *Synthetic Communications*, 27:2051–56.

Khaddar, H., Hamelin, J., and Benhaoua, H. 1999. Microwave-assisted 1,3-dipolar cycloaddition reactions of nitrilimines and nitrile oxides. *Journal of Chemical Research*, 718–19. Also available at http://www.rsc.org/delivery/_ArticleLinking/DisplayArticleForFree.cfm?doi=a907010i&JournalCode=JC

Kidwai, M., Bhushan, K.R., Sapra, P., Saxena, R.K., and Gupta, R. 2000. Alumina-supported synthesis of antibacterial quinolines using microwaves. *Bioorganic and Medicinal Chemistry*, 8:69–72.

Krishnaiah, A. and Narsaiah, B. 2002. Studies on inverse electron demand hetero Diels–Alder reaction of perfluoroalkyl 2(1H) pyridones with different dienophiles under microwave irradiation. *Journal of Fluorine Chemistry*, 113:133–37.

Kumar, H.M.S., Subba Reddy, B.V., Reddy, E.J., and Yadav, J.S. 1999. Microwave-assisted eco-friendly synthesis of 2-alkylated hydroquinones in dry media. *Green Chemistry*, 1:141–42.

Kureshy, R.I., Singh, S., Khan, N.H., Abdi, S.H.R., Agrawal, S., Mayani, V.J., and Jasra, R.V. 2006. Microwave-assisted asymmetric ring opening of meso-epoxides with aromatic amines catalyzed by a Ti-S-(−)-BINOL complex. *Tetrahedron Letters*, 47:5277–79.

Lamberto, M., Corbett, D.F., and Kilburn, J.D. 2003. Microwave assisted free radical cyclisation of alkenyl and alkynyl isocyanides with thiols. *Tetrahedron Letters*, 44:1347–49.

Lei, B. and Fallis, A.G. 1990. Direct total synthesis of (+)-longifolene via an intramolecular Diels–Alder strategy. *Journal of the American Chemical Society*, 112:4609.

Lei, B. and Fallis, A.G. 1993. Cycloaddition routes to tricyclo[5.4.01,7.02,9] undecanes: A direct total synthesis of (+)-longifolene via an intramolecular Diels–Alder strategy. *Journal of Organic Chemistry*, 58:2186–95.

Lerestif, J.M., Perrocheau, J., Tonnard, F., Bazureau, J.P., and Hamelin, J. 1995. 1,3-Dipolar cycloaddition of imidate ylides on imino-alcohols: Synthesis of new imidazolones using solvent free conditions. *Tetrahedron*, 51:6757–74.

Linders, J.T.M., Kokje, J.P., Overand, M., Lie, T.S., and Maat, L. 1988. Chemistry of opium alkaloids. Part XXV. Diels-Alder reaction of 6-demethoxy-β-dihydrothebaine with methyl vinyl ketone using microwave heating; preparation and pharmacology of 3-hydroxy-α,α,17-trimethyl-6β,14β-enthenomorphinan-7β-methanol, a novel deoxygenated diprenorphine analog. *Rec. Trav. Chim Pays-Bas*, 107:449–54.

Louërat, F., Bougrin, K., Loupy, A., Retana, A.M.O., Pagaladay, J., and Palacios, F. 1998. Cycloaddition reactions of azidomethyl phosphonate with acetylenes and enamines. Synthesis of triazoles. *Heterocycles*, 48:161–70.

Loupy, A., Petit, A., and Bonnet-Delpon, D. 1995. Improvements in 1,3-dipolar cycloaddition of nitrones to fluorinated dipolarophiles under solvent-free microwave activation. *Journal of Fluorine Chemistry*, 75:215–16.

Margita, L., Agnieszka, P., Solcanyova, E., Lac, J., Kois, P., Chovancova, J., and Danuta, R. 2005. 3-Formylchromones IV. The rearrangement of 3-formylchromone enamines as a simple, facile route to novel pyrazolo[3,4-b]pyridines and the synthetic utility of the latter. *Molecules*, 10:809–821.

Matsui, T., Nakamura, Y., Ishikawa, H., Matsuura, A., and Kobayashi, F. 1994. Pharmacological profiles of a novel aldose reductase inhibitor, SPR-210, and its effects on streptozotocin-induced diabetic rats. *Japanese Journal of Pharmacology*, 64:115–24.

Mehmandoust, M., Petit, M., and Larcheveque, Y. 1992. Synthesis of (E)-β-γ-unsaturated α-aminoacids by rearrangement of allyltrichloracetimidates. *Tetrahedron Letters*, 33:4313–16.

Michaud, D., Texier-Boullet, F., and Hamelin, J. 1997a. Michael monoaddition of nitromethane on gem-diactivated alkenes in dry media coupled with microwave irradiation. *Tetrahedron Letters*, 38:7563–64.

Michaud, D., Abdallah-El Ayoubi, S., Dozias, M. J., Toupet, L., Texier-Boullet, F., and Hamelin, J. 1997b. New route of functionalized cyclohexenes from nitromethane and electrophilic alkenes without solvent under focused microwave irradiation. *Chemical Communications*, 1613–14.

Miranda, L.D., Marrero, J.G., Bautista, E., Maldonado, E., and Ortega, A. 2009. Microwave-assisted gold (I) catalyzed pyran ring opening in brevifloralactone: Synthesis of the hawtriwaic acid core. *Tetrahedron Letters*, 50:633–35.

Moghaddam, F.M., Mohammadi, M., and Hosseinnia, A. 2000. Water promoted Michael addition of secondary amines to unsaturated carbonyl compounds under microwave irradiation. *Synthetic Communications*, 30:643–50.

Mojtahedi, M.M., Sharifi, A., Mohsenzadeh, F., and Saidi, M.R. 2000. Microwave-assisted aminomethylation of electron-rich compounds under solvent-free condition. *Synthetic Communications*, 30:69–72.

Mori, H., Iwata, M., and Takeshi, S. 2007. Endo ring-opening polymerization of γ-benzyl-L-glutamate-N-carboxyanhydride in ionic liquids. *Polymer*, 48:5867–77.

Motorina, I.A., Parly, F., and Grierson, D.S. 1996. Selective *O*-alkylation of amidoalcohols on solid support. *Synlett*, 389–91.

Nadir, U.K. and Singh, A. 2005. Microwave-induced clay-catalyzed ring opening of *N*-tosylaziridines: A green approach to achiral and chiral diamines. *Tetrahedron Letters*, 46:2083–86.

Nagy, G., Filip, S.V., Surducan, E., and Surducan, V. 1997. Solvent-free synthesis of substituted phenoxyacetic acids under microwave irradiation. *Synthetic Communications*, 27:3729–36.

Nishikawa, T., Asai, M., Ohyabu, N., Yamamoto, N., Fukuda, Y., and Isobe, M. 2001. Synthesis of a common key intermediate for (−)-tetrodotoxin and its analogs. *Tetrahedron*, 57:3875–83.

Olofsson, K., Kim, S.Y., Larhed, M., Curran, D.P., and Hallberg, A. 1999. High-speed, highly fluorous organic reactions. *Journal of Organic Chemistry*, 64:4539–31.

Overman, L.E. and Zipp, G.G. 1997. Allylic transposition of alcohol and amine functionality by thermal or palladium (II)-catalyzed rearrangements of allylic *N*-benzoylbenzimidates. *Journal of Organic Chemistry*, 62:2288–91.

Pang, J., Xi, X., Cao, G., and Yuan, Y. 1996. Phase transfer catalyzed synthesis of *o*-ethoxyphenol under microwave irradiation. *Synthetic Communications*, 26:3425–29.

Quan., X., Zhang., Y., Chen., S., Zhao, Y., and Yang, F. 2007. Generation of hydroxyl radical in aqueous solution by microwave energy using activated carbon as catalyst and its potential in removal of persistent organic substances. *Journal of Molecular Catalysis A: Chemical*, 263:216–22.

Ramesh, N.G. and Balasubramanian, K.K. 1992. Reaction of 2,3-unsaturated aryl glycosides with Lewis acids: A convenient entry to C-aryl glycosides. *Tetrahedron Letters*, 33:3061–64.

Ranu, B.C., Saha, M., and Bha, S. 1997. Microwave assisted Michael addition of cycloalkenones and substituted enones on the surface of alumina in dry media. *Synthetic Communications*, 27:621–26.

Ranu, B.C., Guchhait, S.K., Ghosh, K., and Patra, A. 2000. Construction of bicyclo-[2.2.2] octanone systems by microwave-assisted solid phase Michael addition followed by Al_2O_3-mediated intramolecular aldolisation. An eco-friendly approach. *Green Chemistry*, 1:5–6.

Rissafi, B., Rachiqi, N., El Louzi, A., Loupy, A., Petit, A., and Teatouani, S.F. 2001. Epoxyisophorone ring-opening: an efficient route for the introduction of functional groups at position 2 of isophorone. *Tetrahedron*, 57:2761–68.

Rudd, P.M., Elliott, T., Cresswell, P., Wilson, I.A., and Dwek, R.A. 2001. Glycosylation and the immune system. *Science*, 291:2370–76.

Sabitha, G. 1996. 3-Formylchromone as a versatile synthon in heterocyclic chemistry. *Aldrichimica Acta*, 29:15–25.

Sagar, A.D., Patil, D.S., and Bandgar, B.P. 2000. Microwave assisted synthesis of triaryl cyanurates. *Synthetic Communications*, 30:1719–23.

Scharn, D., Wenschuh, H., Reineke, U., Mergener, J.S., and Germeroth L. 2002. Spatially addressed synthesis of amino- and amino-oxy-substituted 1,3,5-triazine arrays on polymeric membranes. *Journal of Combinatorial Chemistry*, 2:361–69.

Schulz, M.J., Coats, S.J., and Hlasta, D.J. 2004. Microwave assisted preparation of aryltetrazoleboronate esters. *Organic Letters*, 6:3265–68.

Shanmugasundaram, B., Bose, A.K., and Balasubramanian, K.K. 2002. Microwave-induced, montmorillonite K10-catalyzed Ferrier rearrangement of tri-*O*-acetyl-D-galactal: mild, eco-friendly, rapid glycosidation with allylic rearrangement. *Tetrahedron Letters*, 43:6795–98.

Soares, M.I.L., Teresa M.V.D., and Melo, P. 2008. Microwave-assisted generation and reactivity of aza- and diazafulvenium methides: Heterocycles via pericyclic reactions. *Tetrahedron Letters*, 49:4889–93.

Soriente, A., Spinella, A., De Rosa, M., Giordano, M., and Scettri, A. 1997. Solvent-free reaction under microwave irradiation: A new procedure for Eu^{+3}—Catalyzed Michael addition of 1,3-dicarbonyl compounds. *Tetrahedron Letters*, 38:289–90.

Srikrishna, A. and Nagaraju, S. 1992. Acceleration of ortho ester Claisen rearrangement by a commercial microwave oven. *Journal of Chemical Society, Perkin Transactions*, 1:311–12.

Stambouli, A., Chastrette, M., and Soufiaoui, M. 1991. Reactions de cycloaddition [4 + 2] sous micro-ondes des derives du glyoxal. *Tetrahedron Letters*, 32:1723–24.

Summa, V., Petrocchi, A., Matassa, V.G., Taliani, M., Laufer, R., Francessco, R.D., Altamura, S., and Pace, P. 2004. HCV NS5b RNA-dependent RNA polymerase inhibitors: From α,γ-diketoacids to 4,5-dihydroxypyrimidine- or 3-methyl-5-hydroxypyrimidinonecarboxylic acids. Design and synthesis. *Journal of Medicinal Chemistry*, 47:5336–39.

Syassi, B., Bougrin, K., Lamiri, M., and Soufiaoui, M. 1998. Synthese one-pot de quelques dérivés pyrazoliniques et pyrazoliques en milieu sec et sous irradiation micro-onde. *New Journal of Chemistry*, 22:1545.

Tabatabaeian, K., Mamaghani, M., Mahmoodi, N.O., and Khorshidi, A. 2008. Solvent-free, ruthenium-catalyzed, regioselective ring-opening of epoxides, an efficient route to various 3-alkylated indoles. *Tetrahedron Letters*, 49:1450–54.

Touaux, B., Klein, B., Texier-Boullet, F., and Hamelin, J. 1994. Synthesis in dry media coupled with microwave irradiation: Application to alkoxycarbonylformonitrile oxide generation and 1,3-dipolar cycloaddition. *Journal of Chemical Research*, 116.

Touaux, B., Texier-Boullet, F., and Hamelin, J. 1998. Synthesis of oximes, conversion to nitrile oxides and their subsequent 1,3-dipolar cycloaddition reactions under microwave irradiation and solvent-free reaction conditions. *Heteroatom Chemistry*, 9:351–54.

Valdésa, R.H., Aranda, D.A.G., Alvarez, H.M., and Antunes, O.A.C. 2007. Microwave-promoted ring opening reaction of azalactones. *Letters in Organic Chemistry*, 35–38.

Vanelle, P., Gellis, A., Kaafarani, M., Maldonado, J., and Crozet, M.P. 1999. Fast electron transfer *C*-alkylation of 2-nitropropane anion under microwave irradiation. *Tetrahedron Letters*, 40:4343–46.

Varki, A. 1993. Biological roles of oligosaccharides: All of the theories are correct. *Glycobiology*, 3:97–130.

Villemin, D., Jaffres, P.A.. and Simeon, F. 1997. Rapid and efficient phosphonation of aryl halides catalysed by palladium under microwave irradiation. *Phosphorus Sulphur and Silicon and the Related Elements*, 130:59–63.

Wang, J.X., Wang, C.H., Zhang, M., and Hu, Y. 1998. Synthesis of 8-quinolinyl ethers under microwave irradiation. *Synthetic Communications*, 28:2407–13.

Wang, W.B. and Roskamp, E.J. 1992. New technology for the construction of bicyclo [6.2.1] ring systems. *Tetrahedron Letters*, 33:7631–34.

Wetter, C. and Studer, A. 2004. Microwave-assisted free radical chemistry using the persistent radical effect. *Chemical Communications*, 174–75.

Williams, N.R. and Wander, J.D. 1980. *The Carbohydrates. Chemistry and Biochemistry*. New York: Academic Press, pp. 761–98.

Yadav, J.S. and Subba Reddy, B.V. 2000. CsF. Al$_2$O$_3$ mediated rapid condensation of phenols with aryl halides: Comparative study of conventional heating *vs.* microwave irradiation. *New Journal of Chemistry*, 24:489–91.

Yadav, J.S., Reddy, B.V.S., Rasheed, M.A., and Kumar, H.M.S. 2000. Zn^{2+} Montmorillonite catalyzed 3-aza-cope rearrangement under microwave irradiation. *Synlett*, 487–88.

Yang, S.M. and Murray, W.V. 2008. Microwave assisted ring-opening of epoxides with *N*-biaryl sulfonamides in the synthesis of matrix metalloproteinase-9 inhibitors. *Tetrahedron Letters*, 49:835–39.

Yu, Z.J. and Liu, L.J. 2004. Effect of microwave energy on chain propagation of poly (ε-caprolactone) in benzoic acid-initiated ring opening polymerization of ε-caprolactone. *European Polymer Journal*, 40:2213–20.

Zadmard, R., Aghapoor, K., Bolourtchian, M., and Saidi, M.R. 1998. Solid composite copper-copper chloride assisted alkylation of naphthols promoted by microwave irradiation. *Synthetic Communications*, 28:4495–96.

Zuo, H., Bo Li, Z., Ren, F.K., Falck, J.R., Lijuan, M., Ahn, C., and Shin, D.S. 2008. Microwave-assisted one-pot synthesis of benzo[b][1,4]thiazin-3(4H)-ones via Smiles rearrangement. *Tetrahedron*, 64:9669–74.

9 Composting as a Bioremediation Technique for Hazardous Organic Contaminants

Ackmez Mudhoo and Romeela Mohee

CONTENTS

9.1 INTRODUCTION—THE POLLUTION ISSUE

Hazardous organic wastes are generated widely in domestic, municipal, agricultural, industrial, and military activities (Chaudhry and Chapalamadugu, 1991). From agricultural, home, and industrial usage, pesticides and other petroleum wastes can enter into crop residues, municipal sludges, farm manures, and soils (Dao and Unger, 1995). Insect, disease, and weed control in farming may lead to accumulation of pesticides in soils, in which decomposition of these materials may occur slowly (Gevao et al., 2000; Barker and Bryson, 2000). Organic contaminants such as

215

polychlorinated biphenyls (PCBs), polycyclic aromatic hydrocarbons (PAHs), and pesticides (Fu et al., 2003) can enter into the soil from fuel combustion or from sewage sludge and other feedstocks (Edwards, 1983; Fries, 1982; Barker and Bryson, 2000). Persistent organic substances such as benzopyrene, heptachlor, chlordecone, chlordane, lindane, dieldrin, parathion, atrazine, picloram, and aldrin introduced through sludge additions to land may persist in soils or contaminate forage on which livestock graze. Spillage of fuel oil hydrocarbons can contaminate soils (Barker and Bryson, 2000). These hydrocarbons will inhibit seed germination and plant growth (Chaîneau et al., 1998). Hazards associated with these pollutants are their persistence in the environment, their bioaccumulation potential in the tissues of animals and humans through the food chain, and their toxic properties for humans and wildlife (Fu et al., 2003).

Chemical pollution of the environment has hence become a major source of concern (Lodolo et al., 2001; Semple et al., 2001). Studies on degradation of organic compounds have shown that some microorganisms are extremely versatile at catabolizing recalcitrant molecules. By harnessing this catabolic potential, it is possible to bioremediate some chemically contaminated environmental systems (Semple et al., 2001). Composting matrices and composts are rich sources of xenobiotic-degrading microorganisms including bacteria, actinomycetes, and lignolytic fungi, which can degrade pollutants to harmless compounds such as carbon dioxide and water (Ryckeboer et al., 2003). These microorganisms can also metabolize pollutants into less toxic substances and/or lock up pollutants within the organic matrix, thereby reducing the pollutant bioavailability to other vulnerable media. The success of a composting remediation strategy depends however on a number of factors and the most important of which are pollutant bioavailability and biodegradability (Semple et al., 2001). In this perspective, the present review has been performed to appraise the growing application of and research done on composting to promote the biodegradation of hazardous organic molecules that find their way into the environment through polluting discharges. With the intention of a balanced and reasonably comprehensive review approach to the use of composting as a bioremediation technique for harmful organic species, a variety of research conducted on the subject has been presented and discussed herein. This discussion cannot be complete and the author hopes to initiate further applied research and discussion in professional magazines, conferences, and scientific journals so that the body of knowledge and engineering application on this emerging green technology is expanded, implemented, and enriched.

9.2 BIOREMEDIATION FOR ENVIRONMENTAL CLEANUP

The intensification of agriculture and manufacturing industries has resulted in an increased release of a wide range of xenobiotic compounds to the environment (Pieper and Reineke, 2000), and consequently the excess organic loading of hazardous wastes into the various biospheres has led to scarcity of clean water, disturbances of soil thus limiting crop production (Kamaludeen et al., 2003), and serious health threats to the living biota. Although enactment of stringent regulations has led to less indiscriminate disposal of organic and inorganic wastes (Kamaludeen et al., 2003), many challenges remain that require other interventions on both the institutional and research and development fronts.

Bioremediation is an emerging, promising, and less expensive technique (Lynch and Moffat, 2005; Jeyasingh and Philip, 2005; Okoh and Trejo-Hernandez, 2006) for cleaning up contaminated soil and water (Vidali, 2001; Kamaludeen et al., 2003; Fu et al., 2003). This technology accelerates the naturally occurring biodegradation under optimized conditions such as oxygen supply, temperature, pH, presence or addition of suitable microbial population (bioaugmentation) and nutrients (biostimulation), water content, and mixing (Gentry et al., 2004; Okoh and Trejo-Hernandez, 2006). Bioremediation uses biological agents, mainly microorganisms, for example, yeast, fungi, or bacteria to cleanup contaminated soil and water (Watanabe, 2001; Lynch and Moffat, 2005; Strong and Burgess, 2008). This technology relies on promoting the growth of specific microflora or microbial consortia that are indigenous to the contaminated sites that are able to perform desired activities (Watanabe, 2001; Chatterjee et al., 2008). Establishment of such microbial consortia can be done in several ways. These are by promoting growth through addition of nutrients, by adding terminal electron acceptor, or by controlling the moisture and temperature conditions, among others (Smith et al., 1998). Microorganisms have a huge metabolic range that enables them to degrade a variety of organic pollutants and in many cases, the complex biochemistry and molecular biology of the catabolic pathways involved have been identified (Cerniglia, 1992; Knackmuss, 1996; Head, 1998). In bioremediation processes, microorganisms use the contaminants and/or pollutants as nutrients or energy sources (Tang et al., 2007; Okoh and Trejo-Hernandez, 2006). Despite valuable basic knowledge on the mechanisms of pollutant biodegradation, bioremediation has yet to be accepted as a routine treatment technology and the environmental industry is suspicious of applying bioremediation for the treatment of contaminated sites (Head, 1998). Furthermore, microorganisms offer the possibility that organic pollutants can be completely mineralized to inorganic materials, making bioremediation an attractive treatment strategy (Head, 1998). Establishment and maintenance of favorable conditions for microbial growth and process control are the basic requirements (Chatterjee et al., 2008). There are fundamentally two approaches to bioremediation (Vidali, 2001). The first one is *in situ* bioremediation, and this involves the treatment of the contaminants where they are actually located. In this case, the microorganisms come into direct contact with the dissolved and sorbed contaminants and use them as substrates for transformation. Since the *in situ* process is slow, it is not usually the best approach when immediate site cleanup is desired. *Ex situ* bioremediation is a different approach that utilizes specially constructed treatment facility, but it is more expensive than *in situ* bioremediation (Chatterjee et al., 2008). *Ex situ* bioremediation technologies (Vidali, 2001; Pavel and Gavrilescu, 2008) include slurry-phase remediation, where a water phase is added to enhance the physical mixing; treatment-bed remediation, where usually only nutrients are added and a bed of usually 0–1 m height is agitated mechanically at intervals by a mixing device. Biopiles refer to the piling of the material to be biotreated by adding nutrients and air into piles or windrows usually to a height of 2–4.5 m (Li et al., 2002; Pavel and Gavrilescu, 2008). Biopiles may be static with installed aeration piping or they may be turned or mixed by special devices for this purpose. Biopiles may be amended with a bulking agent, usually with straw, saw dust, bark or wood chips, or some other organic materials. If an organic material is added, the technology is termed composting. A more advanced type of composting is a drum composter, which is closed and has a continuous feeding and output. Many organic contaminants have successfully been bioremediated in biopiles. This technology has been demonstrated to function in field pilot or full scale especially for petroleum hydrocarbons (Koren et al., 2003), polyaromatic hydrocarbons (Prenafeta-Boldú et al., 2004), chlorophenols (Laine and Jørgensen, 1997, 1998), and nitro aromatics (Symons and Bruce, 2006).

Bioremediation can be relatively cheap, but methods to confirm its efficacy on a field scale have either been unavailable or not applied routinely (Fu et al., 2003). The unpredictability of bioremediation, meanwhile, stems from a lack of understanding of the behavior of microbial populations in natural environments and how physical, biological, and chemical factors interact to control their activity against environmental pollutants (Head, 1998). It is now clear that access of microorganisms to pollutants *in situ* is a critical factor in determining the success of bioremediation. Consequently, methods are being developed to enable predictions regarding the feasibility of bioremediation based on pollutant bioavailability and biodegradation. With the knowledge of the causes of unsuccessful bioremediation, methods can be formulated to overcome these limitations (Head, 1998). The prospects for predictable evaluation of bioremediation efficacy are gradually improving, but it is not clear that this knowledge will allow development of more effective means to treat organic pollutants biologically in contaminated soils and sediments (Head, 1998). A number of strategies for improving mass transfer have been suggested. Diffusion and desorption of pollutants are temperature dependent (Cornelissen et al., 1997) and this is the basis of physicochemical treatments such as thermal desorption and steam-stripping. In the context of bioremediation it has been suggested that the use of composting systems, where elevated temperatures are maintained, should improve mass transfer and hence bioremediation rates (Pignatello and Xing, 1996).

9.3 COMPOSTING AS A BIOREMEDIATION TECHNIQUE

There are many well-established bioremediation technologies applied commercially at contaminated sites (Lodolo et al., 2001; Khan and Anjaneyulu, 2006). One such technology is the use of compost material. Composting matrices and composts are rich sources of microorganisms, which can degrade contaminants to harmless compounds such as carbon dioxide and water. Composting is applied in bioremediation as a means of degrading toxic organic compounds and perhaps lessening the toxicity of metallic contaminants in organic residues, wastes, and by-products (Chaney et al., 2001; Keener et al., 2001; Khan and Anjaneyulu, 2006). In addition, composting stabilizes wastes for ultimate disposal in traditional manners in landfills or on farmland. Composting has been practiced to reduce volume and water content of feedstock, to destroy pathogens, and to destroy odor-producing nitrogenous and sulfurous compounds (Tiquia and Tam, 1998; Veeken et al., 2001). Composting also is considered as a remediation method for handling contaminated soil, sediment, and organic wastes (Williams and Keehan, 1993; Symons and Bruce,

2006). Mechanical treatment by grinding, mixing, and sieving out nondegradable or disturbing materials (metals, plastics, glass, stones, gravels) gives good conditions for biological treatment of compostable materials (Zach et al., 1999). The biological treatment builds up stable organic compounds through humification and reduces concentrations of organic pollutants. Composting can be used to lower the levels of chemical contaminants in residues or in soils to which polluted residues have been added (Khan and Anjaneyulu, 2006). The processes of remediation in compost are similar to those that occur biologically in soil. However, composting may accelerate the destruction of contaminants (Rao et al., 1995, 1996). Temperatures are generally higher in composts than in soils, resulting in increased solubility of contaminants and higher metabolic activity in composts. High levels of substrate in composts can lead to microbial cometabolism and degradation of organic contaminants (Horvath, 1972; Johnsen et al., 2005; Okpokwasili and Nweke, 2006). The microbiological population can be more numerous and diverse in composts than in soils. The nature of the organic contaminant, composting conditions and procedures, microbial communities, and time, all affect mechanisms of conversions in composts or soils (Rao et al., 1996).

9.4 THE COMPOSTING PROCESS

9.4.1 COMPOSTING DYNAMICS ESSENTIALS

Composting can be defined as the controlled biological decomposition of organic substrates carried out by successive microbial populations combining both mesophilic and thermophilic activities, leading to the production of a final product sufficiently stable for storage and application to land without adverse environmental effects (Haug, 1993; Petiot and de Guardia, 2004). Under optimal conditions, composting proceeds from the initial ambient state through three phases:

1. Mesophilic or moderate-temperature phase, which lasts for a few days
2. Thermophilic or high-temperature phase, which can last from a few days to several months
3. Cooling and maturation phase which lasts for several months (Ghaly et al., 2006)

This thermophilic stage is governed by the basic principles of heat and mass transfer and by the biological constraints of living microorganisms (Keener et al., 1993). The published data on the rate of heat production by organic material decomposition (Miller, 1984; Iwabuchi et al., 1995) suggest that there are wide variations in the rate of heat released by compost. This is because the rate of heat production is a function of chemical, physical, and biological properties of the composted material (Ghaly et al., 2006; Mohee and Mudhoo, 2005; Mudhoo and Mohee, 2006). Composting and other aerobic solid-state fermentation processes rely on gas transfer through a biologically active porous medium with evolving physical and chemical properties. Two of the physical parameters of particular importance to process control and analysis are air-filled porosity and permeability of the matrix (Richard et al., 2004).

Air-filled porosity is the volume fraction of air (often reported on a percentage basis) in a porous matrix. Permeability is a measure of the ability of fluids to flow through a multiphase material (Lange and Antohe, 2000). Composting is a solid-phase process that exploits the phenomenon of microbial self-heating (Finstein and Morris, 1975; Mudhoo and Mohee, 2007, 2008; Mohee et al., 2008). The material being composted serves as its own matrix, permitting gas exchange, and provides its own source of nutrients, water, and an indigenous, diverse inoculum. It also serves as its own waste sink and thermal insulation (Hogan et al., 1989). Hence, metabolically generated heat is conserved within the system, elevating its temperature.

The physical properties of compost materials play an important role in every stage of compost production as well as in the handling and utilization of the end product. From the mixing of various feedstocks and process monitoring and maintenance (Agnew and Leonard, 2003) to the packaging and shipping of the final product, parameters such as bulk density and porosity dictate the requirements for the optimum composting environment and the design of machinery and aeration equipment used in the system. The wet bulk density of compost is a measure of the mass of material (solids and water) within a given volume and is important in the determination of initial compost mixtures (Mohee and Mudhoo, 2005). The wet bulk density determines how much material can be placed at a certain site or hauled in a truck of a given size. The density of compost also influences the mechanical properties such as strength, porosity, and ease of compaction (Agnew and Leonard, 2003), and ultimately affects the rate of biodegradation. Air-filled porosity and permeability relationships (Richard et al., 2004) are different for different substrates and mix of these substrates, as each substrate and mix will have a different distribution of particle sizes and shapes, and these mixtures can be compacted to various wet and dry bulk densities (Bear, 1972). The relationship between air-filled porosity and biodegradation rate has also been examined by several investigators (Schulze, 1962; Jeris and Regan, 1973; Richard et al., 2002) and incorporated in kinetic models of the composting process (Haug, 1993; Hamelers, 2001).

Composting systems fall into two main categories: the "fully or partially open to air" systems and the "in-vessel" systems (Gajalakshmi and Abbasi, 2008). The first category comprises of systems ranging from the ones used from prehistoric times to the windrow, static pile, and "household" systems used in the present day. The second category consists of "tunnel" systems, the rotary drum composting systems, and other "in-vessel" or "reactor" systems of various designs. Depending on the physical location where the composting experiments are being carried out, the substrate, the scale of operation, and the skills and the machinery available, one or the other type of system is used. A more extensive review of the various composting systems has been recently performed by Gajalakshmi and Abbasi (2008).

9.4.2 VERMICOMPOSTING

One growing variant of the composting process is vermicomposting whereby earthworms are used for the degradation of the wastes substrates. Earthworm farming (vermiculture) is another biotechnique for converting the solid organic waste into compost (Aalok et al., 2008). An innovative discipline of vermiculture biotechnology, the breeding and propagation of earthworms and the use of castings has become an important tool of waste recycling across the globe. Essentially, the vermiculture provides for the use of earthworms as natural bioreactors for cost-effective and environmentally sound waste management (Aalok et al., 2008). Owing to their biological, chemical, and physical actions, earthworms can be directly employed within bioremediation strategies to promote biodegradation of organic contaminants (Hickman and Reid, 2008). Earthworms have been shown to aerate soils and improve their nutritional status and fertility, which are variables known to have limited bioremediation (Hickman and Reid, 2008). Earthworms have also been shown to retard the binding of organic contaminants to soils, release previously soil-bound contaminants for subsequent degradation, and promote and disperse organic contaminant degrading microorganisms (Hickman and Reid, 2008).

Vermicomposting is the term given to the process of conversion of biodegradable matter by earthworms into vermicompost (Tognetti et al., 2005; Garg et al., 2006, 2009; Yadav et al., 2009). Vermicomposts are finely divided mature peat-like materials with high porosity, aeration, drainage and water-holding capacity, and microbial activity, which are stabilized by interactions between earthworms and microorganisms in a nonthermophilic process (Edwards and Burrows, 1988; Aalok et al., 2008). Vermicomposts contain most nutrients in plant-available form such as nitrates, phosphates, and exchangeable calcium and soluble potassium. Vermicomposts have large particulate surface areas that provide many microsites for microbial activity and for the strong retention of nutrients. Vermicomposts are rich in microbial populations and diversity, particularly fungi, bacteria, and actinomycetes (Edwards, 1998). Vermicomposts consistently promote biological activity that can cause plants to germinate, flower, and grow and yield better than in commercial container media, independent of nutrient availability (Atiyeh et al., 2000; Aalok et al., 2008).

In the vermicomposting process, a major fraction of the nutrients contained in the organic matter (OM) is converted into more bioavailable forms. The first step in vermicomposting occurs when earthworms break the substrate down to small fragments as a prelude to ingesting the substrate (Gajalakshmi and Abbasi, 2008; Garg et al., 2009; Yadav et al., 2009). The earthworms possess a grinding gizzard that enables the mincing of the substrate. This increases the surface area of the substrate, facilitating microbial action. The substrate is then ingested and goes through a process of "digestion" brought about by numerous species of bacteria and enzymes present in the worm gut (Gajalakshmi and Abbasi, 2008). During this process, important plant nutrients such as nitrogen, potassium, phosphorus, and calcium present in the feed material are converted into forms that are much more water-soluble and bioavailable to the plants than those in the parent substrate (Martín-Gil et al., 2008; Ravikumar et al., 2008). The earthworms derive their nourishment from the microorganisms that grow upon the substrate particles. At the same time, they promote further microbial activity in the residuals so that the fecal material that they produce is much more fragmented and microbially active than what was ingested (Gajalakshmi and Abbasi, 2008). Worms can digest more than their own weight each day, and since the retention time of the waste in the earthworm is short, large quantities are passed through an average population of earthworms. In vermicomposting, the earthworms take over the roles of both turning and maintaining aeration

(Contreras-Ramos et al., 2006; Suthar, 2008) for the organics to be in an aerobic condition, thereby eliminating the need for mechanical or forced aeration. In the digestive system of these worms, microorganisms are responsible for transforming more than 35 (Pereira and Arruda, 2003) organic species (proteins, nucleic acids, fats, and carbohydrates) into a more stable product (vermicompost). This product presents a high cation exchange capacity, high humidity content, wide particle size distribution, high concentration of nutrients (calcium, magnesium, sodium, potassium, phosphorus, sulfur, and nitrogen), and a characteristic black color due to the presence of humic substances (Arancon et al., 2006). Vermicomposting is not an entirely exothermic process and, unlike conventional composting processes (Mason, 2006), does not lead to any perceptible rise in the vermireactor temperature (Gajalakshmi and Abbasi, 2008). To ensure that the earthworms remain maximally active, the vermireactor should be kept at conditions of temperature and soil moisture as close to the given earthworm species' naturally preferred habitat as possible. Earthworm species such as *Eisenia fetida*, *Lumbricus terrestris*, and *Lumbricus rubellus* have been successfully used in composting processes.

9.5 BIOREMEDIATION OF HAZARDOUS ORGANIC SPECIES BY COMPOSTING

Many anthropogenic organic contaminants entering the environment are not fully degraded during treatment and eventually accumulate in biosolids (Ang et al., 2005; Bhandari and Xia, 2005; Krauss and Wilcke, 2005). Owing to their relatively low water solubility and high lipophilicity (Bhandari and Xia, 2005), organic contaminants easily partition into biosolids resulting in their accumulation in biosolids at concentrations several orders of magnitude greater than influent concentrations (Govind et al., 1991). The following sections now present and discuss selected research findings for the application of composting in bioremediating PAHs, petroleum-based hydrocarbons, phenol derivatives, PCBs, phthalic acid esters (PAEs), pesticides, pharmaceutical compounds, and their metabolites.

9.5.1 BIOREMEDIATION OF PAHs

PAHs are a class of organic compounds that have accumulated in the natural environment mainly as a result of anthropogenic activities such as the combustion of fossil fuels (Juhasz and Naidu, 2000; Fu et al., 2003; Bamforth and Singleton, 2005; Johnsen et al., 2005). The increasing use of fossil fuels and their combustion products by human beings during the past two centuries raises several questions about PAHs hazards for living organisms. First, apart from accidental oil spills leading to massive pollution, the precise origin of trace PAHs, for example, natural versus anthropogenic, has rarely been clear traced. Second, the toxicity of PAHs, like other hazardous chemicals, requires their bioavailability. Since most PAHs are highly hydrophobic (Wild and Jones, 1992), their pathways of transfer through geological and biological media are far from being understood. Third, explicit correlations between PAH sources and carcinogenic effects have been reported only for intense exposure to PAHs, for example, for coal-mine workers. As for pesticides, PAHs and their metabolites are expelled from the aqueous phase to adsorb on the hydrophobic surfaces such as OM (Onken and Traina, 1997). They are either trapped in the pores, fixed with covalent or hydrogen bonds, or bound during humification processes (Hatzinger and Alexander, 1995). PAHs structure and stability stand in the way of their biodegradation by microorganisms (fungi and bacteria). Biodegradation is slow and is a function of environmental parameters such as oxygen, water, and nutriment contents. Migration of PAHs from the top soil is slow. PAHs seem to migrate bounded to particles (Onken and Traina, 1997). The major ways of entry of PAHs in plants seem to be through leaves, from the vapor phase, and by contact with contaminated soil particles. Interest has been continuously focused on the occurrence and distribution of PAHs for many decades due to their potentially harmful effects to human health (Juhasz and Naidu, 2000). Although various physicochemical methods have been used to remove these compounds from our environment, they have many limitations (Samanta et al., 2002). This concern has prompted researchers to address ways to detoxify and remove these organic compounds from the natural environment (Bamforth and Singleton, 2005). Bioremediation is one approach that has been used to remediate contaminated land and waters, and this has promoted the natural attenuation of the contaminants using the *in situ* microbial community of the site. PAHs are recalcitrant and can persist in the environment for long periods, but are conducive to biodegradation by certain enzymes found in bacteria and fungi (Hammel et al., 1986; Juhasz and Naidu, 2000; Ang et al., 2005). Over the past several years, several oxidoreductases such as laccases and cytochrome P450 monooxygenases have been exploited for the enzymatic degradation of PAHs.

Composting has been applied as a bioremediation technique for degrading toxic organic compounds and perhaps lowering their persistence and toxicity in organic residues and wastes (Barker and Bryson, 2002). The biochemical and physicochemical processes of remediation in composts are similar to those that usually occur biologically in soil. However, composting may accelerate the destruction of organic contaminants since metabolic temperatures developed are generally higher in composts than in soils. This remediation characteristic of composts and composting matrices has been successfully explored and exploited for the degradation of PAHs. Joyce et al. (1998) studied a laboratory-scale, batch-type, in-vessel composter, which was charged with simulated municipal solid waste deliberately contaminated with a mixture of 3- and 4-ringed PAHs. The extent of their biodegradation was monitored over a 2-month period consisting of 30 days of active composting and 30 days of compost curing. The analytical techniques used to quantify PAHs concentration were Soxhlet extraction and gas chromatography analysis. Joyce et al. (1998) deduced that most of the biodegradation of PAHs occurred in the active composting phase and very little occurred during the curing phase of the composting process. It was also found that fluorene was too volatile for this type of bioremediation, and benz[a]anthracene was persistent in the compost throughout both phases of the experiment while anthracene, phenanthrene, and pyrene showed excellent reduction in their respective levels in the actively composting phase, in contrast to their mercuric chloride-poisoned controls, indicating that biodegradation was the removal process favored by the composting mass. Later, Al-Daher et al. (2001) selected the bioremediation technique involving the use of composting soil piles from among the most appropriate methods and evaluated its performance to remediate PAHs on a pilot scale. Soil piles were con-structed from the contaminated soil after amendment with necessary soil additives and the piles were subjected to regular irrigation and turning, and a monitoring pro-gram was carried out, including monthly soil sample collection from each pile for the measurement of petroleum hydrocarbon PAHs, soil microbial counts, and mineral and metal concentrations. Al-Daher et al. (2001) found that the composting soil pile treatment resulted in the reduction of up to 59% total extractable matter of oil con-tamination within eight months of the composting process. More interestingly, Reid et al. (2002) studied the catabolism of phenanthrene within mushroom compost result-ing from its incubation with (1) phenanthrene, and (2) PAH-contaminated soil. Respirometers measuring mineralization of freshly added ^{14}C-9-phenanthrene were used to evaluate the induction of phenanthrene catabolism. Where pure phenan-threne spiked at a concentration of 400 mg kg^{-1} wet weight was used to induce phenanthrene catabolism in compost, induction was measurable, with maximal min-eralization observed after 7 weeks phenanthrene–compost contact time. Where PAH-contaminated soil was used to induce phenanthrene catabolism in uninduced compost, induction was observed after 5 weeks soil–compost contact time. Microcosm-scale amelioration of soil contaminated with ^{14}C-phenanthrene (aged in soil for 516 days prior to incubation with compost) indicated that both induced (using pure phenan-threne) and uninduced mushroom composts were equally able to promote degradation of this soil-associated contaminant. After 111 days incubation time, 42.7% loss of soil-associated phenanthrene was observed in the induced-compost soil mixture, while 36.7% loss of soil-associated phenanthrene was observed in the uninduced-compost soil mixture. Reid et al. (2002) concluded that these results are notable as they indicated that while preinduction of phenanthrene catabolism within compost is possible, it does not significantly increase the extent of degradation when the compost is used to ameliorate phenanthrene-contaminated soil. Thus, compost could be used directly in the amelioration of contaminated land and without preinduction of catabo-lism. In their research on phenanthrene and fluoranthene, Carlstrom and Tuovinen (2003) have studied the potential of phenanthrene and fluoranthene biodegradation in yard waste compost materials. Compost samples were incubated in biometers with ^{14}C-labeled phenanthrene and the evolution of ^{14}CO$_2$ was assessed as a measure of mineralization. The ^{14}CO$_2$ evolution varied widely among replicate biometers. Carlstrom and Tuovinen (2003) suggested that the variation could be the result of an uneven and patchy colonization of phenanthrene-degrading microorganisms on com-post particles and/or from a nonuniform dispersion of the labeled substrate spike into the yard waste microenvironment. Mineralization of phenanthrene reached about 40% extent of ^{14}CO$_2$ evolution at best before leveling off, but the maximum varied from sample to sample and could be as low as 1% after 3 months. Active minerali-zation occurred at mesophilic and thermophilic temperatures of the composting process. Carlstrom and Tuovinen (2003) also found that the low extraction yield and relatively low maximum mineralization (<40%) indicated that residual phenan-threne was sorbed and bound within the compost matrix in the biometer. Still on the use of compost to bioremediate PAHs, Lau et al. (2003) used spent mushroom compost (SMC) to promote the degradation of PAHs. The SMC of *Pleurotus pulmo-narius* immobilized laccase (0.88 mmol min^{-1} g^{-1}) and manganese peroxidase

(0.58 mmol min^{-1} g^{-1}) the optimal temperatures of which were 45°C and 75°C, respectively, were tested for the bioremediation performance. Lau et al. (2003) noted that in laboratory tests, complete degradative removal of individual naphthalene, phenanthrene, benzo[a]pyrene and benzo[g,h,i]perylene (200 mg PAH kg^{-1} sandy-loam soil) by 5% SMC was obtained in 2 days under continuous shaking at 80°C. The SMC-treated PAH samples had *significantly* reduced their toxicities as revealed by the Microtox® bioassay. These results were further confirmed by gas chromatography–mass spectrometry analysis on the breakdown products, and on the whole, demonstrated the potential in employing SMC in *ex situ* bioremediation of PAHs.

Antizar-Ladislao et al. (2005a) investigated the biodegradation of 16 United States Environmental Protection Agency (USEPA)-listed PAHs (Fu et al., 2003) present in contaminated soil from a manufactured gas plant site using laboratory-scale in-vessel composting-bioremediation reactors over 8 weeks. Antizar-Ladislao et al. (2005a) found that temperature and amendment ratio were important operating parameters for PAH removal for in-vessel composting bioremediation of aged coal tar-contaminated soil and thereafter recommended that when conventional composting processes using temperature profiles to meet regulatory requirements for pathogen control need to be used, these should be preferably started with a prolonged mesophilic stage followed by thermophilic, cooling, and maturation stages. To further understand the bioremediation of aged coal tar-contaminated soil Antizar-Ladislao et al. (2005b) conducted several other investigations over 16 weeks to determine the optimum soil composting temperature. Three new tests were performed; first, soil was composted with green waste, with a moisture content of 60%. Second, microbial activity was HgCl$_2$-inhibited in the soil green-waste mixture with a moisture content of 60%, to evaluate abiotic losses, while in the third experiment only soil was incubated at three different temperatures. Antizar-Ladislao et al. (2005b) determined that PAHs were lost from all treatments with 38°C being the optimum temperature for both PAH removal and microbial activity. They also calculated activation energy values for total PAHs and concluded that the main loss mechanism in the soil–green waste reactors was biological (most probably being microbially mediated degradation processes), whereas in the soil reactors it was chemical. Also, total PAH losses in the soil–green waste composting mixtures were by pseudo-first-order kinetics at 38°C. Additionally, Antizar-Ladislao et al. (2006) came to a similar conclusion that the optimal operational conditions for degradation of PAHs occurred at a moisture content of 60%, soil:green waste dry mass basis of 0.8:1 and 38°C when investigating the in-vessel composting of PAHs present in contaminated soil from a manufactured gas plant site over 98 days using laboratory-scale in-vessel composting reactors.

More recent studies on the application of composting to degrade PAHs have been conclusive and in concert with the findings of earlier studies. Cai et al. (2007) lately investigated the efficiency of four different composting processes to bioremediate PAHs-contaminated sewage sludge. Before composting, sewage sludge coming from the Datansha wastewater treatment plant (Guangzhou, Guangdong Province, China) was mixed with rice straw to obtain a carbon to nitrogen ratio (C/N) of 13:1. Cai et al. (2007) observed that after 56 days of composting, the total concentrations of 16 PAHs ranged from 1.8 to 10.2 mg kg^{-1} dry weight, decreasing in order of inoculated-manual exhibiting removal rates of 64%, 70%, 85%, and 94%, respectively.

Turned compost > Manual turned compost > Continuous aerated compost >
Intermittent aerated compost

The intermittent aerated compost treatment had a higher removal rate of high molecular weight and carcinogenic PAHs comparing to the other composting processes. On a fresh tone of research for PAH bioremediation using composting, Plaza et al. (2009) investigated the binding of phenanthrene and pyrene, by humic acids (HAs) isolated from an organic substrate at different stages of composting and a soil using a batch fluorescence quenching method and the modified Freundlich model. With respect to soil HA, the organic substrate HA fractions were characterized by larger binding affinities for both phenanthrene and pyrene. Further, Plaza et al. (2009) found that the isotherm deviation from linearity was larger for soil HA than for organic substrate HAs, indicating a larger heterogeneity of binding sites in the former. The composting process decreased the binding affinity and increased the heterogeneity of binding sites of HAs and hence Plaza et al. (2009) inferred that the changes undergone by the HA fraction during composting may be expected to contribute to facilitate microbial accessibility to PAHs. The results obtained also suggested that bioremediation of PAH-contaminated soils with matured compost, rather than with fresh organic amendments, may result in faster and more effective cleanup. The beneficial use of compost to bioremediate PAHs was further evidenced from the findings of Yuan et al. (2009). Yuan et al. (2009) studied the biodegradation of phenanthrene and pyrene in compost and compost-amended soil. The degradation rate of phenanthrene was found to be more than that of pyrene. The degradation of the PAHs was enhanced when the two species were present simultaneously in the soil, thereby suggesting some kind of mutually supported synergistic effect that favored their individual degradation rate since the addition of either of the two types of compost (straw and animal manure) individually enhanced PAH degradation. Further to analyze the effect of compost size, compost samples were separated into fractions with various particle size ranges, which spanned 2–50, 50–105, 105–500, and 500–2000 μm. Yuan et al. (2009) observed that the compost fractions with smaller particle sizes demonstrated higher PAH degradation rates but when the different compost fractions were added to soil, compost particle size had no *significant* effect on the rate of PAH degradation. Of the microorganisms isolated from the soil-compost mixtures, *Arthrobacter nicotianae*, *Pseudomonas fluorescens*, and *Bordetella petrii*, respectively, demonstrated the best degradation ability for the PAHs studied.

9.5.2 Bioremediation of Petroleum-Based Hydrocarbons

Total petroleum hydrocarbons (TPH) is a term used to describe a broad family of several hundred chemical compounds that originally come from crude oil.* Crude oils can vary depending on their chemical constituents, and so can the petroleum

---

* Agency for Toxic Substances and Disease Registry (ATSDR). 1999. *Toxicological Profile for Total Petroleum Hydrocarbons (TPH)*. Atlanta, GA: US Department of Health and Human Services, Public Health Service.

products that are made from crude oils. Some are clear or light-colored liquids that evaporate easily, and others are thick, dark liquids or semi-solids that do not evaporate. Many of these products have characteristic gasoline, kerosene, or oily odors. Because modern society uses so many petroleum-based products (gasoline, kerosene, fuel oil, mineral oil, and asphalt), contamination of the environment by them is potentially widespread. Contamination caused by petroleum products contains a variety of these hydrocarbons. Because they are found in a complex mixture, it is not usually practical to measure each one individually and treat them separately with complete remediation. The amount of TPH found in a sample is useful as a general indicator of petroleum contamination at that site. Several techniques have been employed to cleanup soil contaminated with TPH, and composting is one bioremediation method that is slowly gaining more research attention.

Composting of contaminated soil in biopiles is an *ex situ* technology, where OM such as bark chips are added to contaminated soil as a bulking agent. Composting of lubricating oil-contaminated soil was performed in field scale (5 × 40 m^3) using bark chips as the bulking agent, and two commercially available mixed microbial inocula as well as the effect of the level of added nutrients (nitrogen, potassium, and phosphorus) were tested by Jørgensen et al. (2000). Jørgensen et al. (2000) also performed the composting of diesel oil-contaminated soil at one level of nutrient addition and with no inoculum. Jørgensen et al. (2000) noted that the mineral oil degradation rate was most rapid during the first months of the composting process, and it followed a typical first-order degradation curve. During these 5 months, composting of the mineral oil had decreased in all piles with lubrication oil from approximately 2400 to 700 mg kg^{-1} dry weight, which was about 70% of the mineral oil content. Correspondingly, the mineral oil content in the pile with diesel oil-contaminated soil decreased with 71% from 700 to 200 mg kg^{-1} dry weight. In this type of treatment with the addition of a large amount of OM, the general microbial activity as measured by soil respiration was enhanced and no particular effect of added inocula was observed, thereby advocating the suitability of composting to bioremediate diesel oil-contaminated soil. Guerin (2001a) designed a remediation program and implemented it at a site in southeastern Australia that had become contaminated with nonvolatile, *n*-alkane TPH, and the remediation was conducted in two stages. The excavation, validation, and reinstatement of two contaminated areas on the site were first conducted, followed by the development of a composting treatment process, which was the actual bioremediation strategy being implemented. The total volume of contaminated soil (TPH concentration exceeding 1000 mg kg^{-1} C_{10}–C_{36}) was 4300 m^3 with a concentration of 3100 mg kg^{-1}. The soil was stockpiled into four windrows, on a compacted, bundled clay base. Approximately 35% (v/v) of raw materials (green tree waste, cow manure, gypsum, and nutrients) were added to initiate the composting process, and the piles were kept moist during the summer months, and no other maintenance was conducted. After 6 months treatment, the average TPH concentration (C_{10}–C_{36}) was 730 mg kg^{-1}, which met the relevant clean fill criteria applicable to the site. Besides the satisfactory cleanup performance of the composting process, there were no other contaminants of significance in the treated soil compost that could pose a risk to human health or the environment. Namkoong et al. (2002) conducted research to find the appropriate

mix ratio of organic amendments for enhancing diesel oil degradation during contaminated soil composting by adding sewage sludge or compost as an amendment
for supplementing OM for composting of contaminated soil. Namkoong et al. (2002)
thereafter found that the degradation of diesel oil was *significantly* enhanced by the
addition of these organic amendments relative to straight soil. The degradation rates
of TPH and *n*-alkanes were found to be greatest at the ratio of 2:1 of contaminated
soil to organic amendments on wet weight basis. Li et al. (2002) examined the field-
scale remediation of oil-contaminated soils from the Liaohe Oil Fields in China
using composting biopiles in windrow technology. Micronutrient-enriched chicken
excrement and rice husk were applied as nutrition and a bulking agent. Li et al.
(2002) found that the inoculum of indigenous fungi they had used increased both
the total colony-forming units and increased the rate of degradation of TPH in all
contaminated soils but at different rates. In sharp contrast to other studies, the introduction of exotic microorganisms did not improve the remediation, and suggested
that inoculation of oil-contaminated sites with nonindigenous species is likely to
fail. On the other hand, indigenous genera of microbes were found to be very effective in increasing the rate of degradation of TPH. According to the results of Li
et al. (2002), the degradation of TPH was mainly controlled by the compositions of
aromatic hydrocarbons and asphaltene and resin. Between 38% and 57% degradation of crude oils in contaminated soils was achieved after 53 days of the composting process, hence Li et al. (2002) concluded that the construction and operation of
field-scale composting biopiles in windrows with passive aeration is a cost-effective
bioremediation technology oil-contaminated soil.

The work of Marín et al. (2006) ascertains the efficacy of composting as a low-
cost technology bioremediation technique for reducing the hydrocarbon content of
oil refinery sludge with a large total hydrocarbon content of 250–300 g kg^{-1}, in semiarid conditions. The composting system designed by Marín et al. (2006), which
involved open air piles turned periodically over a period of 3 months, proved to be
inexpensive and reliable. The influence on hydrocarbon biodegradation of adding
wood shavings as bulking agent and inoculation of the composting piles with pig
slurry was also studied. Marín et al. (2006) determined that the most effective composting treatment was the one in which the bulking agent was added, where the initial hydrocarbon content was reduced by 60% in 3 months as compared with the 32%
reduction achieved without the bulking agent. Although spiking the piles with an
organic fertilizer did not significantly improve the degree of hydrocarbon degradation, Marín et al. (2006) concluded that the composting process without doubt led to
the biodegradation of toxic compounds. Mihial et al. (2006) determined that bioremediation by composting was a suitable alternative for the remediation of soil in and
around a pit contaminated with petroleum waste comprising used oil, gasoline, diesel fuel, and paint thinners. Mihial et al. (2006) conducted a bench-scale treatability
study to assess the potential for successful bioremediation of the site using composting. They set up two reactors each with ammonium phosphate fertilizer as the nutrient amendment using a mixture of grass clippings and sheep manure in one reactor
to determine whether the composting process could be accelerated by the addition of
these abundantly available waste materials. Based on the results of the treatability
study, the half-life of the petroleum hydrocarbons at the subject site was estimated to

be 36.3 and 121.6 days with and without the addition of grass clippings and sheep manure, respectively. It was estimated that it would take approximately 192 and 643 days to remediate the soil and reduce the TPH to 1000 mg L^{-1} using the amendments of the reactors, respectively. On the whole, Marín et al. (2006) concluded that the site could be remediated within acceptable cleanup criteria using composting. One more interesting study where composting has been successfully applied to bioremediate contaminated soil with more than 380,000 mg kg^{-1} TPH has been reported by Atagana (2008). In their work, Atagana (2008) inoculated the contaminated soil with sewage sludge and incubated the mix for 19 months. Compost heaps were set up in triplicates on wood pallets covered with double layers of nylon straw sheets and control experiments that contained contaminated soil and wood chips but without sewage sludge were set up in triplicate, and the concentrations of selected hydrocarbons in the contaminated soil were measured monthly during the incubation period. Atagana (2008) noted a typical composting performance with an increase in temperature to about 58°C in the sewage sludge compost within 60 days of incubation, while temperature in the control fluctuated between 15°C and 35°C throughout the incubation period. All the more, TPH was reduced by 17% in the control experiments and up to 99% in the sewage sludge compost at the end of the incubation period. Much promisingly as being a reliable bioremediation technique, the composting process reduced the concentrations of the TPH by up to 100% within the same period.

9.5.3 BIOREMEDIATION OF PHENOL DERIVATIVES

Phenol is both a synthetically and naturally produced aromatic compound. Microorganisms capable of degrading phenol are common and include both aerobes and anaerobes (van Schie and Young, 2000). Many aerobic phenol-degrading microorganisms have been isolated and the pathways for the aerobic degradation of phenol are now established (van Schie and Young, 2000). The first steps include oxygenation of phenol by phenol hydroxylase enzymes to form catechol, followed by ring cleavage adjacent to or in between the two hydroxyl groups of catechol. Phenol can also be degraded under anaerobic conditions, but this process is less well understood and documented, and only a few anaerobic phenol-degrading bacteria have been isolated to date (van Schie and Young, 2000). A number of practical applications exist for microbial phenol degradation and these comprise the exploitation of anaerobic phenol-degrading bacteria in the *in situ* bioremediation of creosote-contaminated subsurface environments, and the use of phenol as a cosubstrate for aerobic phenol-degrading bacteria to enhance *in situ* biodegradation of chlorinated solvents. Chlorophenols have been introduced into the environment through their use as biocides and as by-products of chlorine bleaching in the pulp and paper industry (Field and Sierra-Alvarez, 2008). Chlorophenols are subject to both anaerobic and aerobic metabolism (Antizar-Ladislao and Galil, 2003). Under anaerobic conditions, chlorinated phenols can undergo reductive dechlorination when suitable electron-donating substrates are available (Field and Sierra-Alvarez, 2008). Under aerobic conditions, both lower and higher chlorinated phenols can serve as sole electron and carbon sources supporting growth. Two main strategies are used by aerobic bacteria for the degradation of chlorophenols. Lower chlorinated phenols are initially attacked by monooxygenases yielding chlorocatechols as the first intermediates while polychlorinated phenols are converted into chlorohydroquinones as the initial intermediates. Fungi and some bacteria are additionally known to cometabolize chlorinated phenols (Field and Sierra-Alvarez, 2008). These microbial degradation mechanisms have gradually been put to use for the remediation of phenol- and phenol derivatives-contaminated soil and other strata through composting. Some of the most conclusive studies where composting has been applied to bioremediate phenol-contaminated strata are now discussed.

Valo and Salkinoja-Salonen (1986) studied the microbiological decontamination of technical chlorophenol-containing soil by composting. In two 50 m^3 windrows, the concentration of chlorophenols went down from 212 to 30 mg kg^{-1} in 4 summer months and after the second summer of composting it was only 15 mg kg^{-1}. Valo and Salkinoja-Salonen (1986) also reported that all the chlorophenol congeners present in the technical chlorophenol were degraded, but the main dimeric impurities, polychlorinated phenoxyphenols, had remained recalcitrant. In view to verify the degradation, Valo and Salkinoja-Salonen (1986) conducted laboratory experiments with samples from the windrow compost and demonstrated that the chlorophenols were truly degraded. Laboratory experiments also showed that degradation of chlorophenols was accelerated when sterilized contaminated soil was inoculated with *Rhodococcus chlorophenolicus* of the field composts. Laine et al. (1997) studied the fate of chlorophenols during the composting of sawmill soil and impregnated wood to see whether chlorophenols, in addition to mineralization, would form any harmful metabolites. The toxicity assessed by luminescent bacteria tests decreased during the composting, and it followed the chlorophenol concentrations in the compost piles. The threshold value for chlorophenol toxicity appeared to be 200 mg of total chlorophenols kg^{-1} dry weight. On the basis of the results obtained, Laine et al. (1997) deduced that the toxicity tests were a quick and promising tool for assessing the toxicity changes in chlorophenol-contaminated soil but were not sensitive enough to detect the concentrations that would meet the remediation criteria for cleanup of chlorophenol-contaminated soil, which in this case was 10 mg kg^{-1} dry weight total chlorophenols. In conclusion, Laine et al. (1997) found that no harmful metabolites were formed during composting of chlorophenol-contaminated soil, but the existing ones such as polychlorinated dibenzo-*p*-dioxins and dibenzofurans (PCDD/Fs) compounds were not removed during the biological treatment. The results of biotransformation studies suggested that 30–40% of the carbon in chlorophenols that disappeared but did not mineralize during the composting process most likely was built into the bacterial biomass.

Benoit and Barriuso (1997) carried out experiments to study the transformation of ^{14}C-ring-labeled 2,4-D, 4-chlorophenol (4-CP) and 2,4-dichlorophenol (4-DCP) during straw composting under controlled laboratory conditions. Incubation under sterile and nonsterile conditions was done to evaluate the relative importance of the biotic and abiotic processes. Precomposted straw was treated with three chemicals and the availability of the different chemicals was monitored during incubations as well as their degradation. Under nonsterile conditions, Benoit and Barriuso (1997) observed that the mineralization of both chlorophenols reached 20% of the applied compounds, whereas it was 52% for 2,4-D. Also, transitory water-soluble metabolites of 2,4-D and chlorophenols were formed, but they disappeared rapidly. After 21 days of composting, 21% of 2,4-D and 38% of 2,4-DCP were stabilized as bound residues under nonsterile conditions. Under the nonsterile conditions 71% of 4-CP was recovered as bound residues, whereas under sterile conditions 30% of the applied 4-CP formed bound residues after formaldehyde addition and only 8% with autoclaved straw. Benoit and Barriuso (1997) observed that the overall microbial activity decreased in the presence of chlorophenols probably due to toxicity effects. In conclusion, data from the study of Benoit and Barriuso (1997) indicated that the biological activity associated with straw transformation during composting had stimulated the depletion of 2,4-D and chlorophenols by mineralization and by formation of bound residues.

As the most commonly used detergents for this purpose, nonylphenol ethoxylates (NPE), are toxic to the environment, their fate must be carefully evaluated when disposal options for these sludges are considered. Jones and Westmoreland (1998) examined the fate of NPE and their metabolites produced during the composting of a mixture of these sludges and municipal green waste. Over 14 weeks of composting, the NPE residues were decreased by more than 96%. According to Jones and Westmoreland (1998), the main degradation pathway involved the oxidative hydrolytic shortening of the poly(ethylene oxide) chain of the hydrophile to produce low levels of the biorefractory metabolites nonylphenol, nonylphenol monoethoxylate, nonylphenol diethoxylate, nonylphenoxy acetic acid, and nonylphenoxyethoxy acetic acid. Still pursuing on nonylphenols, it is established that 4-nonylphenol (4-NP), a degradation intermediate of commercial surfactant and known endocrine disruptor, has been frequently detected at levels up to several thousand μg L^{-1} in surface waters and up to several hundred mg kg^{-1} dry weight in soil and sediment samples (Das and Xia, 2008). Large quantities of 4-NP can be quickly sorbed by the organic-rich solid phase during wastewater treatment and are concentrated in biosolids, a possible major source for 4-NP in the environment. Composting, as a process of solid matrix transformation where biological activity is enhanced by process control, has recently been used successfully in remediation of contaminated soils and sludges in a study by Das and Xia (2008). In their study, Das and Xia (2008) characterized the transformation kinetics of 4-NP and its isomers during biosolids composting. Five distinctive 4-NP isomer groups with structures relative to α- and β-carbons of the alkyl chain were identified in biosolids. Composting biosolids mixed with wood shaving at a dry weight percentage ratio of 43:57 (C:N ratio of 65:1) removed 80% of the total 4-NP within 2 weeks of the composting experiments. Das and Xia (2008) also found that isomers with α-methyl-α-propyl structure transformed significantly slower than those with less branched tertiary α-carbon and those with secondary α-carbon, suggesting isomer-specific degradation of 4-NP during biosolids composting.

9.5.4 BIOREMEDIATION OF PCBS

PCBs, which can be mixtures of up to 209 congeners, were first manufactured in 1929 (Bhandari and Xia, 2005) and these are among the most widely detected chemicals in wastewater residual biosolids. Although PCBs are no longer produced in the

United States because they build up in the environment and can cause harmful health effects, they are still in use in many other countries. Polychlorinated dibenzodioxins and polychlorinated dibenzofurans (dioxins) (Fu et al., 2003) consist of 210 different compounds that have similar chemical properties (Bhandari and Xia, 2005). This class of compounds is persistent, toxic, and bioaccumulative. They are generated as by-products during incomplete combustion of chlorine containing wastes such as municipal solid waste, sewage sludge, and hospital and hazardous wastes (Bhandari and Xia, 2005). PCBs were widely used in the past and now contaminate many industrial and natural areas.

PCBs can be degraded by microorganisms via a metacleavage pathway to yield tricarboxylic acid cycle intermediate and (chloro)benzoate (CBA). The initial step in the aerobic biodegradation of PCBs is the dioxygenation of PCB congeners by the biphenyl dioxygenase enzyme (Ang et al., 2005). In this step, the enzyme catalyzes the incorporation of two hydroxyl groups into the aromatic ring of a PCB congener, which increases the reactivity of the PCBs, rendering them more susceptible to enzymatic ring fission reactions (Bruhlmann and Chen, 1999). To the purpose of the present review, a single research paper was reported in the literature where composting has been applied for bioremediating PCBs. Michel Jr. et al. (2001) determined the effects of soil to amendment ratio on PCB degradation when a PCB-contaminated soil from a former paper mill was mixed with a yard trimmings amendment and composted in field-scale piles. Temperature, oxygen concentrations, and a number of other environmental parameters that usually influence microbial activity during composting were monitored. The PCBs in the contaminated soil had a concentration of 16 mg kg^{-1} dry weight and an average of four chlorines per biphenyl. The soil was composted with five levels of yard trimmings amendment (14–82% by weight) in pilot-scale compost piles of volume 25 m^3 and turned once every month. Michel Jr. et al. (2001) observed up to a 40% loss of PCBs with amendment levels of 60% and 82%. Also, congener-specific PCB analysis indicated that less chlorinated PCB congeners (1–3 chlorines per biphenyl) were preferentially degraded during the composting process. On the other hand, bench-scale studies indicated that less than 1% of the PCBs in the contaminated soil were volatilized from composts during incubation with forced aeration at 55°C. In conclusion, Michel Jr. et al. (2001) observed PCB loss during the composting of the PCB-contaminated soil and this appeared to be for the most part due to biodegradation, rather than volatilization.

9.5.5 Bioremediation of PAEs

Phthalate esters are manufactured in large quantities and have been used in the production of plastics. Di-(2-ethylhexyl) phthalate (DEHP), the most widely used phthalate ester, is persistent during sewage treatment and readily accumulates in sediments and lipid tissues in aquatic organisms (Bhandari and Xia, 2005). DEHP, a suspected endocrine disruptor (Hoyer, 2001) has been reported in a variety of media including water, atmospheric deposition, sediments, soil, biosolids, biota, and food products (Fu et al., 2003; Bluthgen, 2000). Among the PAEs targeted by the USEPA as priority pollutants, DEHP is the major pollutant identified at high concentrations level in lagooning sludge at about 28.67 mg kg^{-1} and in activated sludge at about 6.26 mg kg^{-1}.

Other PAEs, such as di-butyl phthalate (DBP) and di-methyl phthalate (DMP) show very low concentrations (Amir et al., 2005). Several studies have been carried out to assess the biodegradability and bioremediation of phthalate esters by composting and results have so far been promising. Some selected but hopefully representative findings of such studies are discussed hereafter.

Marttinen et al. (2004) studied the potential of composting and aeration to remove DEHP from municipal sewage sludge with raw sludge and anaerobically digested sludge. They found that composting removed 58% of the DEHP of the raw sludge and 34% of the anaerobically digested sludge during 85 days stabilization in compost bins, while a comparable removal for the anaerobically digested sludge was achieved in a rotary drum composter in 28 days. Although DEHP removal was greater from raw sludge compost than anaerobically digested sludge compost, the total and volatile solids removals were similar in the two composts. Moreover, Marttinen et al. (2004) determined that in the aeration process mode of raw sludge at 20°C, the DEHP removals were 33–41% and 50–62% in 7 and 28 days, respectively. The pool of results hence collected by Marttinen et al. (2004) indicated that both composting and aeration have the potential to reduce the DEHP contents typically found in sewage sludges to levels acceptable for agricultural use. As a result of their findings, Amir et al. (2005) also suggested that composting could be a detoxification process for the removal of PAEs, and mainly DEHP, from sludges after a sufficient time of treatment to provide a safe end product. During the sludge composting experiments monitored by Amir et al. (2005), after the composting stabilization phase, the subsequent appearance of di-ethyl phthalate and then DMP occurred indicating that microbial metabolism began by alkyl side-chain degradation before aromatic ring cleavage. The appearance and accumulation of PAEs with a short alkyl side chain in the last stages of activated sludge and lagooning sludge composting studied by Amir et al. (2005) has been suggested to originate from the degradation of phthalates with a much long side chain. The better DEHP biodegradation rate of 2.4×10^{-2} day^{-1} have been observed in the case of activated sludge composting compared to lagooning sludge compost evaluated at 1.53×10^{-2} day^{-1}. On a similar note, in assessing sludge composting as a bioremediation approach for DEHP, Gibson et al. (2007) investigated the impact of pilot-scale composting and drying of sludge on the physico-chemical characteristics and on the concentrations of some organic contaminants. During the 143-day composting experiments, the OM content fell by 22% and moisture by 50%. Concentrations of 4-nonylphenols fell by 88% and DEHP by 60%, and these losses continued throughout the procedure. The drying process was much shorter and lasted for only 40 days, yet OM content decreased by 27% and moisture by 85%. Losses of 4-NPs (39%) and DEHP (22%) were less than in composting and stopped when the moisture content quasi stabilized. Gibson et al. (2007) concluded that composting would be the method of choice for reducing organic contaminants but this bioremediation technique requires much longer times than drying. Cheng et al. (2008) also came to similar inferences as Amir et al. (2005) when investigating the potential degradation of DEHP and OM of sewage sludge by composting using laboratory reactors at different operating conditions. At the end of composting, Cheng et al. (2008) observed that the total DEHP degradation was more than 85% in all conditions and the total carbon reduction varied from 7.6% to 11.8%.

Cheng et al. (2008) deduced that the degradation kinetics of DEHP in thermophilic phase and the phase thereafter were modeled by first-order and fractional power kinetics, respectively. Recently, Chang et al. (2009) investigated the biodegradation of PAEs, DBP, and DEHP in compost and compost-amended soil. Both DBP and DEHP were added to straw and animal manure composts at a concentration of 50 mg kg^{-1} and subsequently added to the soil. Chang et al. (2009) noted that the addition of either compost individually improved the rate of PAE degradation in the compost-amended soil.

9.5.6 BIOREMEDIATION OF PESTICIDES

Chemical pesticides* have consistently demonstrated their merit by increasing the global agricultural productivity (Ecobichon, 2001), reducing insect-borne, endemic diseases and protecting plantations, forests, and harvested wood (Ecobichon, 2000). As of date, pesticides are more valued in developing countries, particularly those in tropical regions seeking to enter the global economy by providing off-season fresh fruits and vegetables to countries in more temperate climates (Ecobichon, 2001). However, the continuous use of pesticides has caused severe irreversible damage to the environment, caused human ill-health, negatively impacted on agricultural production, and reduced agricultural sustainability (Pimentel and Greiner, 1997; Wilson and Tisdell, 2001). Traditional methods of pesticide remediation that are however relatively costly include excavation and/or chemical oxidation processes (e.g., photocatalysis, ozonation, and iron-catalyzed Fenton's reaction) or thermal processes (e.g., low-temperature thermal desorption, incineration). On the other hand, bioremediation and phytoremediation are the biotic processes that are sometimes employed for the remediation of pesticide-contaminated sites (Lynch and Moffat, 2005). The use of phytotechnologies to remediate these more persistent pesticides is only emerging (Chaudhry et al., 2002; Henderson et al., 2006; Zhuang et al., 2007). Still, difficulties persist, including the potential phytotoxicity of some herbicides (Eullaffroy and Vernet, 2003; Van Eerd et al., 2003) that were originally developed but destroyed plant material. Typically, the mechanisms involved in pesticide phytoremediation are phytodegradation, rhizodegradation, and phytovolatilization. As a form of low-cost cleanup bioremediation option, composting and biobeds† are increasingly being assessed as an approach to remediate pesticides. Some studies have been carried out to this end and they unanimously are in favor of composting.

The outcome of the widely used lawn-care herbicide 2,4-dichlorophenoxyacetic acid (2,4-D) during the composting of yard trimmings consisting of primarily leaves and grass is an important unexplored question. In their study, Michel Jr. et al. (1995)

---

* According to the United States Environmental Protection Agency (US EPA) (1999), the term "pesticide" is a broad nonspecific term covering a large number of substances including, insecticides, herbicides, and fungicides, "though often misunderstood to refer only to insecticides."
† A biobed in its simplest form is a rectangular lined pit, 1–1.3 m deep, filled with a mixture of topsoil, peat-free compost, and straw in a ratio of 1:1:2, respectively and turfed over. The aim of the technique is to use the biobed in the sprayer filling/wash down area to minimize point source pollution, and contain and treat any potential pollutants. Biobeds filter out pesticides and use enhanced microbial activity to break them down.

determined the extent of 2,4-D mineralization, incorporation into humic matter, volatilization, and sorption during the composting of yard trimmings. Yard trimmings (2:1 [wt/wt] leaves–grass) were amended with ^{14}C-ring-labeled 2,4-D (17 mg kg^{-1} dry weight) and composted in a temperature-controlled laboratory-scale compost system. During composting, thermophilic microbes were numerically dominant, reaching a maximum of 2×10^{11} g^{-1}. At the end of composting, 46% of the OM present in the yard trimmings was lost and the compost was stable, with an oxygen uptake rate of 0.09 mg O_2 g^{-1} OM h^{-1}, and was well humified. Michel Jr. et al. (1995) also observed that the mineralization of the OM temporally paralleled the mineralization of 2,4-D. In the final compost, 47% of the added 2,4-D carbon was mineralized, about 23% was complexed with high-molecular-weight HAs, and about 20% remained bound. With very little volatilization of 2,4-D occurred during the composting process, Michel Jr. et al. (1995) noted with interest that their results indicated an active mineralization of 2,4-D at composting temperatures of 60°C. Pursuing research on the fate of pesticides during compositing, Michel Jr. et al. (1997) investigated the fate of diazinon (O,O-diethyl O-[2-isopropyl-6-methyl-4-pyrimidinyl] phosphorothioate) during the composting of a mixture of leaves and grass (2:1 w/w). The yard trimmings were amended with [Δ-2-^{14}C] labeled diazinon (10 mg kg^{-1} wet weight) and composted in a laboratory-scale compost system for 54 days. During the composting process, 48% of the initial total OM was lost as CO_2 and 11% of the ^{14}C-diazinon was mineralized to $^{14}CO_2$. Initially, 83% of the added ^{14}C-diazinon was ether extractable but less than 1% was ether extractable after composting. Advanced analysis using thin-layer chromatography and mass spectrometry indicated that the ^{14}C was in the form of 2-isopropyl-6-methyl 4-hydroxy pyrimidine (IMHP), which had been produced after the hydrolysis of diazinon. Once more, Michel Jr. et al. (1997) observed that volatilization of diazinon was negligible. However, their new set of results showed that during the composting of yard trimmings, a relatively small amount of diazinon was mineralized to CO_2, while a majority of the diazinon had been converted into potentially leachable, but less toxic IMHP, high-molecular-weight residues, and unextractable residues that are presumed to have low bioavailability.

To elucidate the hazard potential of compost application, Hartlieb et al. (2003) amended municipal biowaste with ^{14}C labeled pyrene and simazine, which they incubated in a pilot-scale composting simulation system. A mass balance incorporating the mineralization, metabolism, and sorption of the two model substances was then established over a period of 370 days. Hartlieb et al. (2003) found that the results were quite different for the two chemicals, thereby reflecting their intrinsic properties during their degradation in the composting environment. Hartlieb et al. (2003) determined that over 60% of the applied ^{14}C-simazine had resulted in nonextractable residues (NERs). Silylation experiments that were then conducted indicated that the formation of NER from simazine and its metabolites was due to both physical entrapment in the compost matrix and chemical binding. Earlier (in the same research group of Hartlieb et al., 2003), Ertunç et al. (2002) had composted ^{14}C-labeled simazine together with biowaste on a pilot scale where it was also found that the herbicide was quickly bound to the compost matrix. By aqueous extraction of 29- and 200-days-old compost, only 4.2% and 3.1%, respectively, of the radioactivity in the compost samples could be extracted with water. Further analysis of the extracts by

Ertunç et al. (2002) using high-performance size-exclusion chromatography (HPSEC) revealed that the dissolved organic matter (DOM) had molecular weights ranging between 2 and 28 kDa. The amount of DOM associated radioactivity increased from 53% (day 29) to 65% (day 200) of total extractable radioactivity. The type of binding of the ^{14}C-labeled residues and the DOM was elucidated by silylation of humic matter and subsequent HPSEC. Ertunç et al. (2002) interpreted that besides polar metabolites, intact simazine was also bound to the DOM. They also observed a distinct shift from rather weak interactions to strong covalent linkages of simazine and its metabolites with increasing age of the compost. Ertunç et al. (2002) concluded that the shift toward stable covalent linkages was equivalent to a detoxification of the contaminant in aged compost, and as a result the use of the analyzed compost in its mature stage should not pose an environmental risk to groundwater or the subsoil when applied as an agricultural amendment.

Ghaly et al. (2007) evaluated the effectiveness of in-vessel thermophilic composting on the destruction of pirimiphos-methyl (O-(2-diethylamine-6-methylpyrimidin-4-yl)O,O-dimethyl phosphorothioate). Pirimiphos-methyl is an insecticide with both contact and fumigant action and shows activity against a wide variety of insects including ants, beetles, caterpillars, cockroaches, fleas, flies, mites, mosquitoes, and moths. With a half-life of 117 days in water, 180–270 days on greens and seeds, pirimiphos-methyl has been reported to cause cholinesterase inhibition in humans, which at high-dose rates results in nausea, dizziness, and confusion and at high exposure due to accidents and major spills results in respiratory paralysis and death. The bioreactor for the composting process studied by Ghaly et al. (2007) was operated on a mixture of tomato plant residues, wood shavings, and municipal solid compost. Ghaly et al. (2007) found that the composting process successfully destroyed 81–89% of pirimiphos-methyl within the first 54 h of the composting process, while the complete destruction of the pesticide required approximately 440 h. Ghaly et al. (2007) also inferred that a number of physical, chemical, and biological mechanisms contribute to the degradation of pirimiphos-methyl in the environment and these consist of mineralization, abiotic transformations, adsorption, leaching, humification, and volatization. During composting of greenhouse wastes, in particular, the degradation of pirimiphos-methyl is accelerated by high temperatures developed during the thermophilic stage of the process, OM content, moisture of the compost matrix, and level of biological activity.

Delgado-Moreno and Peña (2009) amended a typical calcareous agricultural soil of the Mediterranean area contaminated with four triazine herbicides with olive cake, compost, and vermicompost of olive cake at rates 4 times higher than the agronomic dose in order to stimulate the biodegradation of simazine, terbuthylazine, cyanazine, and prometryn. Delgado-Moreno and Peña (2009) observed that the residual herbicide concentrations at the end of the degradation assay showed no *significant* differences between nonamended and amended soil. However, interestingly, Delgado-Moreno and Peña (2009) found that the addition of compost and vermicompost had enhanced the biological degradation rate of triazines during the first week of incubation, with half-lives ranging from 5 to 18 days for the amended soils. In contrast, the olive cake did not significantly modify the degradation of triazines in spite that the addition of this amendment to soil resulted in the highest dehydrogenase activity values. In all the substrates, it was deduced that the degradation of cyanazine and prometryn was faster than those of terbuthylazine and simazine. Last but not least, Fogg et al. (2003) assessed whether biobeds could degrade complex pesticide mixtures when applied repeatedly. A pesticide mixture containing isoproturon, pendimethalin, chlorpyrifos, chlorothalonil, epoxiconazole, and dimethoate was incubated in a biomix and topsoil at concentrations to simulate pesticide disposal. Although the data suggested that interactions between pesticides were possible, the effects were of less significance in the biomix than in topsoil. The same mixture was applied on three occasions at 30-day intervals. This time, Fogg et al. (2003) observed that the degradation was significantly quicker in the biomix than in topsoil. The rate of degradation, however, decreased with each additional treatment, possibly due to toxicity of the pesticide mixture to the microbial community. Fogg et al. (2003) concluded that biobeds could offer a viable means of treating pesticide waste.

9.5.7 BIOREMEDIATION OF PHARMACEUTICALS

Pharmaceutical compounds and their metabolites are emerging contaminants that have been frequently detected at low levels in environmental samples (Kolpin et al., 2002; Bhandari and Xia, 2005), biota, and human tissues. The persistence of pharmaceutical contaminants in the environment has been attributed to human consumption of drugs and subsequent discharges from sewage treatment plants, as well as veterinary use of drugs and nonpoint discharges from agricultural runoff. The use of some pesticides may have experienced a fall in recent years as new laws have been introduced to minimize their use (Daughton, 2002), but even if they should prove problematic, pharmaceuticals are unlikely to be restricted in this way, due to their beneficial human health effects and economic importance (Jones et al., 2005). Indeed, their use is expected to grow with the increasing average age of the population (Daughton and Ternes, 1999), and consequently, they and their metabolites are much likely to be detected in the environment (Sedlak et al., 2003). Drugs may be degraded during sewage treatment processes (Jones et al., 2005) but many pharmaceuticals are not thermally stable (Ternes, 2001) and so might be expected to break down during processes such as composting due to microbial heat generated during the process.

Guerin (2001b) investigated soil composting as an alternative to incineration for treatment of soils at a commercial facility that had been contaminated with the pharmaceutical chemical residues, Probenecid and Methaqualone. In laboratory trials, Guerin (2001b) used factorial experimental design to evaluate the OM amendment type and concentration, and incubation temperature. Guerin (2001b) determined that in pilot-scale trials, Probenecid was reduced from 5100 to less than 10 mg kg^{-1} within 20 weeks in mesophilic composting treatments, and an 8-ton pilot-scale treatment conducted subsequently confirmed that thermophilic composting was much effective under field conditions. In the full-scale treatment, 180 tons of soil was composted. The initial concentrations of the major contaminants in the full-scale composting treatment were 1160 and 210 mg kg^{-1}, for Probenecid and Methaqualone, respectively. Results were highly positive and encouraging since Guerin (2001b) noted that the concentration of Probenecid reached the target level of 100 mg kg^{-1} in 6 weeks, and the reduction of Methaqualone to less than 100 mg kg^{-1} was achieved after 14 weeks. On the basis of these set of results, Guerin (2001b) concluded that cocomposting was effective in reducing soil concentrations of Probenecid and Methaqualone residues to safe and acceptable values, and cocomposting is a technology that has a potentially wide application in the remediation of pharmaceutical contaminants in soil.

More recent studies on the application of composting to remediate pharmaceuticals have been reported by Dolliver et al. (2008) and Bao et al. (2009). Dolliver et al. (2008) quantified the degradation of chlortetracycline (CTC), monensin, sulfamethazine, and tylosin in spiked turkey litter during composting. Three manure composting treatments were evaluated: a control treatment with manure pile and no disturbance or adjustments after initial mixing, a managed compost pile with weekly mixing and moisture content adjustments, and in-vessel composting. Despite significant differences in temperature, mass, and nutrient losses between the composting treatments and the control, Dolliver et al. (2008) observed that there was no difference in antibiotic degradation among the treatments. CTC concentrations declined rapidly during composting, whereas monensin and tylosin concentrations declined gradually in all three treatments, while there was no degradation of sulfamethazine in any of the treatments. At the end of the composting period that lasted for 22–35 days, Dolliver et al. (2008) calculated that there was more than 99% reduction in CTC, whereas monensin and tylosin reduction ranged from 54% to 76% in all three treatments. The data of Dolliver et al. (2008) suggested that managed composting in a manure pile or in a vessel may not be better than the control treatment in degrading certain antibiotics in manure. Hence, Dolliver et al. (2008) proposed that stockpiling, after an initial adjustment of water content may be a more practical and economical option for livestock producers in reducing antibiotic levels in manure before land application. CTC is one of the most important pharmaceuticals occurring in the environment. Its recent increase in application as feed supplement for livestock and poultry in the world has led to a considerable CTC contamination of manures, because most of the CTC is excreted to manure. Bao et al. (2009) recently adopted a simulation experiment of aerobic composting to investigate CTC depletion in aged and spiked manure composting, and to evaluate the extent of CTC depletion during composting. The results of Bao et al. (2009) showed that the extractable CTC initial concentration was markedly different between the different manures, with 94.71 mg kg^{-1} in broiler manure and 879.6 mg kg^{-1} in hog manure. The concentration of extractable CTC also decreased rapidly at the initial stage of composting, and subsequently declined slowly during aged and spiked manure composting. At the end of composting, more than 90% of CTC in the manure composting process had been depleted.

9.6 CONCLUDING REMARKS

Over the past years, human activities have caused significant pollution in the natural environment. A number of organic pollutants such as PAHs, PCBs, and pesticides are relatively recalcitrant to degradation and consequently pose an ongoing toxicological threat to both wildlife and human beings. The industrial and manufacturing activities can no longer be denied and hampered as it remains a fact that the survival of the humankind depends crucially on the continuation of these economic activities.

Hence, a compromise has to be worked out and this need to strike the new balance that paves the way to sustainable development (Zechendorf, 1999). Sustainable development has become a priority for the world's policy makers, industrialists, researchers, engineers, and almost all alike. Among the broad range of technologies with the potential to reach the goal of sustainability, biotechnology could take a crucial place, especially in the fields of food production, renewable raw materials and energy, pollution prevention, and bioremediation (Zechendorf, 1999; Alcalde et al., 2006). The various technologies summarized under "biotechnology" could play a key role in most of these fields. Four fields have been selected in which biotechnological applications might have a major impact: food production, renewable materials, waste prevention, and bioremediation. Bioremediation is indeed an attractive alternative to traditional physicochemical techniques for the remediation of these pollutants at a contaminated site, as it can be more cost effective and it can selectively degrade the pollutants without damaging the site or its indigenous flora and fauna (Ang et al., 2005). Biotechnology has, to date, made relatively little contribution to recycling except in the area of paper recycling. On the other hand, the treatment of aqueous and solid wastes of industrial, agricultural, and domestic origin offers a number of opportunities to apply a wider range of biotechnological methods. The crux remains that the efficiency of these methods is based on the capacity of the microorganisms to degrade organic material and absorb hazardous substances (Zechendorf, 1999).

The present review has successfully demonstrated that composting, as an emerging green technology for bioremediation, exploits the biodegradative abilities of live organisms to support the long-term restoration of petroleum hydrocarbon, pesticides, PAHs, PCBs, phenol derivatives, PAEs, and pharmaceuticals polluted systems, with the added advantage of cost efficiency and environmental friendliness (Okoh and Trejo-Hernandez, 2006). Indeed, xenobiotic-degrading microorganisms have great potential for bioremediation, but "green" modifications are required to make such microorganisms more effective, efficient, and action specific in removing these compounds, which were initially thought to be refractory. At this point, the principles underlined by green chemistry (Anastas and Kirchhoff, 2002; Kirchhoff, 2003) may be applied in research and development for harnessing the bioremediative capacity of composting in detoxifying hazardous organic contaminants. Controlled chemotaxis such as metabolism-independent chemotaxis, metabolism-dependent chemotaxis, chemotaxis toward simple aromatic compounds, bacterial chemotaxis toward pollutants, and chemotaxis toward nitroaromatic compounds (Pandey and Jain, 2002) could be investigated and applied to enhance the biodegradation of pollutants. Additionally, genetic and biomolecular engineering can be successfully used to improve the capabilities of the enzymes and engineered biocatalysts in bioremediation systems (Pieper and Reineke, 2000; Ang et al., 2005) toward the biodegradation of pollutants such as lindane, atrazine, dibenzothiophenes, methyl-parathion, and chrysene (Alcalde et al., 2006). Nevertheless, prior to resorting to these more complex strategies for improving the efficiency of naturally occurring microorganisms, field bioremediation could be significantly improved in a first instance by optimizing certain factors such as bioavailability, adsorption, and mass transfer (Zechendorf, 1999).

ACKNOWLEDGMENTS

The authors wish to express their gratitude and appreciation to all the researchers whose valuable data, research findings, and discussions as reported in their respective publications and reports (cited below) have been of significance in crafting this review. Apologies go to the other researchers who are involved in extensive research on hazardous organic species remediation but whose work could not be reported presently due to lack of space. Finally, the authors are thankful to other colleagues and anonymous reviewers whose suggestions and criticisms have benefited the review.

REFERENCES

Aalok, A., Tripathi, A.K., and Soni, P. 2008. Vermicomposting: A better option for organic solid waste management. *Journal of Human Ecology*, 24:59–64.

Agnew, J.M. and Leonard, J.J. 2003. Literature review—The physical properties of compost. *Compost Science and Utilization*, 11:238–64.

Alcalde, M., Ferrer, M., Plou, F.J., and Ballesteros, A. 2006. Environmental biocatalysis: From remediation with enzymes to novel green processes. *TRENDS in Biotechnology*, 24:281–7.

Al-Daher, R., Al-Awadhi, N., Yateem, A., Balba, M.T., and ElNawawy, A. 2001. Compost soil piles for treatment of oil-contaminated soil. *Soil Sediment Contamination: An International Journal*, 10:197–209.

Amir, S., Hafidi, M., Merlina, G., Hamdi, H., Jouraiphy, A., El Gharous, M., and Revel, J.C. 2005. Fate of phthalic acid esters during composting of both lagooning and activated sludges. *Process Biochemistry*, 40:2183–90.

Anastas, P.T. and Kirchhoff, M.M. 2002. Origins, current status, and future challenges of green chemistry. *Accounts of Chemical Research*, 35:686–94.

Ang, E.L., Zhao, H., and Obbard, J.P. 2005. Recent advances in the bioremediation of persistent organic pollutants via biomolecular engineering. *Enzyme Microbiology and Technology*, 37:487–96.

Antizar-Ladislao, B. and Galil, N.I. 2003. Simulation of bioremediation of chlorophenols in a sandy aquifer. *Water Research*, 37:238–44.

Antizar-Ladislao, B., Lopez-Real, J., and Beck, A.J. 2005a. In-vessel composting-bioremediation of aged coal tar soil: Effect of temperature and soil/green waste amendment ratio. *Environment International*, 31:173–78.

Antizar-Ladislao, B., Lopez-Real, J., and Beck, A.J. 2005b. Laboratory studies of the remediation of polycyclic aromatic hydrocarbon contaminated soil by in-vessel composting. *Waste Management*, 25:281–89.

Antizar-Ladislao, B., Lopez-Real, J., and Beck, A.J. 2006. Degradation of polycyclic aromatic hydrocarbons (PAHs) in an aged coal tar contaminated soil under in-vessel composting conditions. *Environment Pollution*, 141:459–68.

Arancon, N.Q., Edwards, C.A., Lee, S., and Byrne, R. 2006. Effects of humic acids from vermicomposts on plant growth. *European Journal of Soil Biology*, 42:65–9.

Atagana, H.I. 2008. Compost bioremediation of hydrocarbon-contaminated soil inoculated with organic manure. *African Journal of Biotechnology*, 7:1516–25.

Atiyeh, R.M., Arancon, N.Q., Edwards, C.A., and Metzger, J.D. 2000. Influence of earthworm-processed pig manure on the growth and yield of greenhouse tomatoes. *Bioresource Technology*, 75:175–80.

Bamforth, S.M. and Singleton, I. 2005. Bioremediation of polycyclic aromatic hydrocarbons: Current knowledge and future directions. *Journal of Chemical Technology and Biotechnology*, 80:723–36.

Bao, Y., Zhou, Q., Guan, L., and Wang, Y. 2009. Depletion of chlortetracycline during composting of aged and spiked manures. *Waste Management*, 29:1416–23.

Barker, A.V. and Bryson, G.M. 2002. Bioremediation of heavy metals and organic toxicants by composting. *The Scientific World Journal*, 2:407–20.

Bear, J. 1972. *Dynamics of Fluids in Porous Media*. Elsevier, New York.

Benoit, P. and Barriuso, E. 1997. Fate of ^{14}C-ring-labeled 2,4-D, 2,4-dichlorophenol and 4-chlorophenol during straw composting. *Biology and Fertility of Soils*, 25:53–9.

Bhandari, A. and Xia, K. 2005. Hazardous organic chemicals in biosolids recycled as soil amendments. *Handbook of Environmental Chemistry Vol. 5, Part F* 1:217–39.

Bluthgen, A. 2000. Organic migration agents into milk at farm level (illustrated with diethylhexyl phthalate). *Bulletin-International Dairy Federation*, 356:39–42.

Bruhlmann, F. and Chen, W. 1999. Tuning biphenyl dioxygenase for extended substrate specificity. *Biotechnology and Bioengineering*, 63:544–51.

Cai, Q.-Y., Mo, C.-H., Wu, Q.-T., Zeng, Q.-Y., Katsoyiannis, A., and Férard, J.-F. 2007. Bioremediation of polycyclic aromatic hydrocarbons (PAHs)-contaminated sewage sludge by different composting processes. *Journal of Hazardous Materials*, 142:535–42.

Carlstrom, C.J. and Tuovinen, O.H. 2003. Mineralization of phenanthrene and fluoranthene in yardwaste compost. *Environmental Pollution*, 124:81–91.

Cerniglia, C.E. 1992. Biodegradation of polycyclic aromatic hydrocarbons. *Biodegradation*, 3:351–68.

Chaîneau, C.H., Morel, J.L., and Oudot, J. 1998. Phytotoxicity and plant uptake of fuel oil hydrocarbons. *Journal of Environmental Quality*, 26:1478–83.

Chaney, R.L., Ryan, J.A., Kukier, U., Brown, S.L., Siebielec, G., Malik, M., and Angle, J.S. 2001. Heavy metal aspects of compost use. In *Compost Utilization in Horticultural Cropping Systems*. Stoffella, P.J. and Kahn, B.A. (Eds). Lewis Publishers, Boca Raton, FL, pp. 323–59.

Chang, B.V., Lu, Y.S., Yuan, S.Y., Tsao, T.M., and Wang, M.K. 2009. Biodegradation of phthalate esters in compost-amended soil. *Chemosphere*, 74:873–77.

Chatterjee, S., Chattopadhyay, P., Roy, S., and Sen, S.K. 2008. Bioremediation: A tool for cleaning polluted environments. *Journal of Applied Biosciences*, 11:594–601.

Chaudhry, G.R. and Chapalamadugu, S. 1991. Biodegradation of halogenated organic compounds. *Microbiology and Molecular Biology Reviews*, 55:59–79.

Chaudhry, Q., Schröder, P., Werck-Reichhart, D., Grajek, W., and Marecik, R. 2002. Prospects and limitations of phytoremediation for the removal of persistent pesticides in the environment. *Environmental Science and Pollution Research*, 9:4–17.

Cheng, H.-F., Kumar, M., and Lin, J.-G. 2008. Degradation kinetics of di-(2-ethylhexyl) phthalate (DEHP) and organic matter of sewage sludge during composting. *Journal of Hazardous Materials*, 154:55–62.

Contreras-Ramos, S.M., Álvarez-Bernal, D., and Dendooven, L. 2006. *Eisenia fetida* increased removal of polycyclic aromatic hydrocarbons from soil. *Environmental Pollution*, 141:396–401.

Cornelissen, G., van Noott, P.C.M., Parsons, J.R., and Covers, H.A.J. 1997. Temperature dependence of slow adsorption and desorption kinetics of organic compounds in sediments. *Environmental Science and Technology*, 31:454–60.

Dao, T.H. and Unger, P.W. 1995. Agronomic practices in relation to soil amendments and pesticides. In *Soil Amendments and Environmental Quality*. Rechcigl, J.E. (Ed.). Lewis Publishers, Boca Raton, FL, pp. 427–70.

Das, K.C. and Xia, K. 2008. Transformation of 4-nonylphenol isomers during biosolids composting. *Chemosphere*, 70:761–68.

Daughton, C.G. 2002. Environmental stewardship and drugs as pollutants. *Lancet*, 360:1035–36.

Daughton, C.G. and Ternes, T.A. 1999. Pharmaceuticals and personal care products in the environment: Agents of subtle change? *Environmental Health Perspectives*, 107:907–42.

Delgado-Moreno, L. and Peña, A. 2009. Compost and vermicompost of olive cake to bioremediate triazines-contaminated soil. *Science of the Total Environment*, 407:1489–95.

Dolliver, H., Gupta, S., and Noll, S. 2008. Antibiotic degradation during manure composting. *Journal of Environmental Quality*, 37:1245–53.

Ecobichon, D.J. 2000. Our changing perspectives on benefit and risks of pesticides: A historical overview. *Neurotoxicology*, 21:211–18.

Ecobichon, D.J. 2001. Pesticide use in developing countries. *Toxicology*, 160:27–33.

Edwards, C.A. 1998. The use of earthworms in the breakdown and management of organic wastes. In *Earthworm Ecology*. Edwards C.A. (Ed.). CRC Press, Boca Raton, FL, pp. 327–354.

Edwards, C.A. and Burrows, I. 1988. The potential of earthworm composts as plant growth media. In *Earthworms in Environmental and Waste Management*. Edwards, C.A. and Neuhauser E. (Eds). SPB Academic Publ. B.V., the Netherlands, pp. 211–220.

Edwards, N.T. 1983. Polycyclic aromatic hydrocarbons (PAHs) in the terrestrial environment—A review. *Journal of Environmental Quality*, 12:427–41.

Ertunç, T., Hartlieb, N., Berns, A., Klein, W., and Schaeffer, A. 2002. Investigations on the binding mechanism of the herbicide simazine to dissolved organic matter in leachates of compost. *Chemosphere*, 49:597–604.

Eullaffroy, P. and Vernet, G. 2003. The F684/F735 chlorophyll fluorescence ratio: A potential tool for rapid detection and determination of herbicide phytotoxicity in algae. *Water Research*, 37:1983–90.

Field, J.A. and Sierra-Alvarez, R. 2008. Microbial degradation of chlorinated phenols. *Reviews in Environmental Science and Biotechnology*, 7:211–41.

Finstein, M.S. and Morris, M.L. 1975. Microbiology of municipal solid waste composting. *Advances in Applied Microbiology*, 19:113–51.

Fogg, P., Boxall, A.B.A., Walker, A., and Jukes, A.A. 2003. Pesticide degradation in a "biobed" composting substrate. *Pest Management and Science*, 59:527–37.

Fries, G.T. 1982. Potential polychlorinated biphenyl residues in animal products from application of contaminated sewage sludge to land. *Journal of Environmental Quality*, 11:14–20.

Fu, J., Mai, B., Sheng, G., Zhang, G., Wang, X., Peng, P., Xiao, X., et al. 2003. Persistent organic pollutants in environment of the Pearl River Delta, China: An overview. *Chemosphere*, 52:1411–22.

Gajalakshmi, S. and Abbasi, S.A. 2008. Solid waste management by composting: State of the art. *Critical Reviews in Environmental Science and Technology*, 38:311–400.

Garg, P., Gupta, A., and Satya, S. 2006. Vermicomposting of different types of waste using *Eisenia foetida*: A comparative study. *Bioresource Technology*, 97:391–5.

Garg, V.K., Gupta, R., and Kaushik, P. 2009. Vermicomposting of solid textile mill sludge spiked with cow dung and horse dung: A pilot-scale study. *International Journal of Environmental Pollution*, 38:385–96.

Gentry, T., Rensing, C., and Pepper, I. 2004. New approaches for bioaugmentation as a remediation technology. *Critical Reviews in Environmental Science and Technology*, 34:447–94.

Gevao, B., Semple, K.T., and Jones, K.C. 2000. Bound pesticide residues in soils: A review. *Environmental Pollution*, 108:3–14.

Ghaly, A.E., Alkoaik, F., and Snow, A. 2006. Thermal balance of invessel composting of tomato plant residues. *Canadian Biosystems Engineering*, 48:1–11.

Ghaly, A.E., Alkoaik, F., and Snow, A. 2007. Degradation of pirimiphos-methyl during thermophilic composting of greenhouse tomato plant residues. *Canadian Biosystems Engineering*, 49:1–11.

Gibson, R.W., Wang, M.-J., Padgett, E., Lopez-Real, J.M., and Beck, A.J. 2007. Impact of drying and composting procedures on the concentrations of 4-nonylphenols, di-(2-ethylhexyl)phthalate and polychlorinated biphenyls in anaerobically digested sewage sludge. *Chemosphere*, 68:1352–58.

Govind, R., Lai, L., and Dobbs, R. 1991. Integrated model for predicting the fate of organics in wastewater treatment plants. *Environmental Progress*, 10:13–23.

Guerin, T.F. 2001a. Co-composting of residual fuel contamination in soil. *Soil and Sediment Contamination: An International Journal*, 10:659–73.

Guerin, T.F. 2001b. Co-composting of pharmaceutical wastes in soil. *Letters in Applied Microbiology*, 33:256–63.

Hamelers, H.V.M. 2001. A mathematical model for composting kinetics. Dissertation, Wageningen University, Netherlands.

Hammel, K.E., Kalyanaraman, B., and Kirk, T.K. 1986. Oxidation of polycyclic aromatic-hydrocarbons and dibenzo P-dioxins by phanerochaete chrysosporium ligninase. *Journal of Biological Chemistry*, 261:6948–52.

Hartlieb, N., Ertunç, T., Schaeffer, A., and Klein, W. 2003. Mineralization, metabolism and formation of nonextractable residues of ^{14}C-labelled organic contaminants during pilot-scale composting of municipal biowaste. *Environmental Pollution*, 126:83–91.

Hatzinger, P.B. and Alexander, M. 1995. Effect of aging of chemicals in soil on their biodegradability and extractability. *Environmental Science and Technology*, 29:537–45.

Haug, R.T. 1993. *The Practical Handbook of Compost Engineering*. Lewis Publishers, Boca Raton, FL.

Head, I.M. 1998. Bioremediation: Towards a credible technology. *Microbiology*, 144:599–608.

Henderson, K.L.D., Belden, J.B., Zhao, S., and Coats, J.R. 2006. Phytoremediation of pesticide wastes in soil. Zeitschrift fur Naturforschung, 61:213–21.

Hickman, Z.A. and Reid, B.J. 2008. Earthworm assisted bioremediation of organic contaminants. *Environment International*, 34:1072–81.

Hogan, J.A., Miller, F.C., and Finstein, M.S. 1989. Physical modeling of the composting ecosystem. *Applied and Environmental Microbiology*, 55:1082–92.

Horvath, R.S. 1972. Microbial co-metabolism and the degradation of organic compounds in Nature. *Bacteriological Reviews*, 36:146–55.

Hoyer, P.B. 2001. Reproductive toxicology: Current and future directions. *Biochemical Pharmacology*, 62:1557–64.

Iwabuchi, K., Kimura, T., and Kamide, J. 1995. A study of the heat production rate during composting of dairy cattle feces. In *Symposium on Automation and Robotics in Bioproduction and Processing*. Kobe, Japan, pp. 432–439.

Jeris, J.S. and Regan, R.W. 1973. Controlling environmental parameters for optimum composting. Part II: Moisture, free air space and recycle. *Compost Science*, 14:8–15.

Jeyasingh, J. and Philip, L. 2005. Bioremediation of chromium contaminated soil: Optimization of operating parameters under laboratory conditions. *Journal of Hazardous Materials*, B118:113–20.

Johnsen, A.R., Wick, L.Y., and Harms, H. 2005. Principles of microbial PAH-degradation in soil. *Environmental Pollution*, 133:71–84.

Jones, F.W. and Westmoreland, D.J. 1998. Degradation of nonylphenol ethoxylates during the composting of sludges from wool scour effluents. *Environmental Science and Technology*, 32:2623–27.

Jones, O.A.H., Voulvoulis, N., and Lester, J.N. 2005. Human pharmaceuticals in wastewater treatment processes. *Critrical Reviews in Environmental Science and Technology*, 35:401–27.

Jørgensen, K.S., Puustinen, J., and Suortti, A.-M. 2000. Bioremediation of petroleum hydrocarbon-contaminated soil by composting in biopiles. *Environmental Pollution*, 107:245–54.

Joyce, J.F., Sato, C., Cardenas, R., and Surampalli, R.Y. 1998. Composting of polycyclic aromatic hydrocarbons in simulated municipal solid waste. *Water Environment Research*, 70:356–61.

Juhasz, A.L. and Naidu, R. 2000. Bioremediation of high molecular weight polycyclic aromatic hydrocarbons: A review of the microbial degradation of benzo[*a*]pyrene. *International Biodeterioration & Biodegradation*, 45:57–88.

Kamaludeen, S.P.B.K., Arunkumar, K.R., Avudainayagam, S., and Ramasamy, K. 2003. Bioremediation of chromium contaminated environments. *Indian Journal of Experimental Biology*, 41:972–85.

Keener, H.M., Dick, W.A., and Hoitink, H.A.J. 2001. Composting and beneficial utilization of composted by-product materials. In *Land Application of Agricultural, Industrial, and Municipal By-products*. Power J.F. and Dick W.A. (Eds). Soil Science Society of America, Madison, WI, pp. 315–41.

Keener, H.M., Marugg, C., Hansen, R.C., and Hoitink, H.A.J. 1993. Optimizing the efficiency of the composting process. In *Proceedings of the International Composting Research Symposium*. Hoitink, H.A.J. and Keener, H.M. (Eds). Renaissance Publications, Columbus, OH, pp. 59–94.

Khan, Z. and Anjaneyulu, Y. 2006. Bioremediation of contaminated soil and sediment by composting. *Remediation Journal*, 16:109–122.

Kirchhoff, M.M. 2003. Promoting green engineering through green chemistry. *Environmental Science and Technology*, 37:5349–53.

Knackmuss, H.-J. 1996. Basic knowledge and perspectives of bioelimination of xenobiotic compounds. *Journal of Biotechnology*, 51:287–95.

Kolpin, D.W., Furlong, E.T., Meyer, M.T., Thurman, E.M., Zaugg, S.D., Barber, L.B., and Buxton, H.T. 2002. Pharmaceuticals, hormones, and other organic wastewater contaminants in U.S. streams, 1999–2000: A national reconnaissance. *Environmental Science and Technology*, 36:1202–11.

Koren, O., Knezevic, V., Ron, E.Z., and Rosenberg, E. 2003. Petroleum pollution bioremediation using water-insoluble uric acid as the nitrogen source. *Applied Environmental Microbiology*, 69:6337–39.

Krauss, M. and Wilcke, W. 2005. Persistent organic pollutants in soil density fractions: Distribution and sorption strength. *Chemosphere*, 59:1507–15.

Laine, M.M., Ahtiainen, J., Wågman, N., Öberg, L.G., and Jørgensen, K.S. 1997. Fate and toxicity of chlorophenols, polychlorinated dibenzo-*p*-dioxins, and dibenzofurans during composting of contaminated sawmill soil. *Environmental Science and Technology*, 31:3244–50.

Laine, M.M. and Jørgensen, K.S. 1997. Effective and safe composting of chlorophenol-contaminated soil in pilot scale. *Environmental Science and Technology*, 31:371–8.

Laine, M.M. and Jørgensen, K.S. 1998. Bioremediation of chlorophenol-contaminated soil by composting in full scale. In *Bioremediation and Phytoremediation: Chlorinated and Recalcitrant Compounds*. Wickramanayake, G.B. and Hinchee, R.E. (Eds). Batelle Press, Columbus, OH, pp. 45–50.

Lange, J.L. and Antohe, B.V. 2000. Darcy's experiments and the deviation to nonlinear flow regime. *ASME Journal of Fluids Engineering*, 122:619–25.

Lau, K.L., Tsang, Y.Y., and Chiu, S.W. 2003. Use of spent mushroom compost to bioremediate PAH-contaminated samples. *Chemosphere*, 52:1539–46.

Li, P., Sun, T., Stagnitti, F., Zhang, C., Zhang, H., Xiong, X., Allinson, G., Ma, X., and Allinson, M. 2002. Field-scale bioremediation of soil contaminated with crude oil. *Environmental Engineering and Science*, 19:277–89.

Lodolo, A., Gonzales-Valencia, E., and Miertus, S. 2001. Remediation of persistent toxic substances. *Arh Hig Rada Toksikol*, 52:253–80.

Lynch, J.M. and Moffat, A.J. 2005. Bioremediation—prospects for the future application of innovative applied biological research. *Annals of Applied Biology*, 146:217–21.

Marín, J.A., Moreno, J.L., Hernández, T., and García, C. 2006. Bioremediation by composting of heavy oil refinery sludge in semiarid conditions. *Biodegradation*, 17:251–61.

Martín-Gil, J., Navas-Graci, L.M., Gómez-Sobrino, E., Correa-Guimaraes, A., Hernández-Navarro, S., Sánchez-Báscones, M., and del Carmen Ramos-Sánchez, M. 2008.

Composting and vermicomposting experiences in the treatment and bioconversion of asphaltens from the Prestige oil spill. *Bioresource Technology*, 99:1821–9.

Marttinen, S.K., Hänninen, K., and Rintala, J.A. 2004. Removal of DEHP in composting and aeration of sewage sludge. *Chemosphere*, 54(3):265–72.

Mason, I.G. 2006. Mathematical modeling of the composting process: A review. *Waste Management*, 26:3–21.

Michel Jr., F.C., Quensen, J., and Reddy, C.A. 2001. Bioremediation of a PCB-contaminated soil via composting. *Compost Science and Utilization*, 9:274–83.

Michel Jr., F.C., Reddy, C.A., and Forney, L.J. 1995. Microbial degradation and humification of the lawn care pesticide 2,4-dichlorophenoxyacetic acid during the composting of yard trimmings. *Applied Environmental Microbiology*, 61:2566–71.

Michel Jr., F.C., Reddy, A.C., and Forney, L.J. 1997. Fate of carbon-14 diazinon during the composting of yard trimmings. *Journal of Environmental Quality*, 26:200–5.

Mihial, D.J., Viraraghavan, T., and Jin, Y.-C. 2006. Bioremediation of petroleum-contaminated soil using composting. *Practice Periodical of Hazardous, Toxic and Radioactive Waste Management*, 10:108–15.

Miller, F. 1984. Thermodynamic and metric water potential analysis in field and laboratory scale composting ecosystem. Dissertation, Rutgers University, New Brunswick, NJ.

Mohee, R. and Mudhoo, A. 2005. Analysis of the physical properties of an in-vessel composting matrix. *Powder Technology*, 155:92–9.

Mohee, R., Mudhoo, A., and Unmar, G.D. 2008. Windrow co-composting of shredded office paper and broiler litter. Special Issue on Solid Waste Management—Part 1. *International Journal of Environmental Waste Management*, 2:3–23.

Mudhoo, A. and Mohee, R. 2006. Sensitivity analysis and parameter optimization of a heat loss model for a composting system. *Journal of Environmental Informatics*, 8:100–10.

Mudhoo, A. and Mohee, R. 2007. Overall heat transfer coefficients in organic substrates composting. *Journal of Environmental Informatics*, 9:87–99.

Mudhoo, A. and Mohee, R. 2008. Modeling heat loss during self-heating composting based on combined fluid film theory and boundary layer concepts. *Journal of Environmental Informatics*, 11:74–89.

Namkoong, W., Hwang, E.-Y., Park, J.-S., and Choi, J.-Y. 2002. Bioremediation of diesel-contaminated soil with composting. *Environmental Pollution*, 119:23–31.

Okoh, A.I. and Trejo-Hernandez, M.R. 2006. Remediation of petroleum hydrocarbon polluted systems: Exploiting the bioremediation strategies. *African Journal of Biotechnology*, 5:2520–5.

Okpokwasili, G.C. and Nweke, C.O. 2006. Microbial growth and substrate utilization kinetics. *African Journal of Biotechnology*, 5:305–17.

Onken, B.M. and Traina, S.J. 1997. The sorption of nonionic organic solutes to humic acid–mineral complexes: effect of cosolutes. *Journal of Environmental Quality*, 26:132–8.

Pandey, G. and Jain, R.K. 2002. Bacterial chemotaxis toward environmental pollutants: Role in bioremediation. *Applied Environmental Microbiology*, 68:5789–95.

Pavel, L.V. and Gavrilescu, M. 2008. Overview of *ex situ* decontamination techniques for soil cleanup. *Environmental Engineering and Management Journal*, 7:815–34.

Pereira, M.G. and Arruda, M.A.Z. 2003. Vermicompost as a natural adsorbent material: Characterization and potentialities for cadmium adsorption. *Journal of Brazilian Chemical Society*, 14:39–47.

Petiot, C. and De Guardia, A. 2004. Composting in a laboratory reactor, a review. *Compost Science and Utilization*, 12:69–79.

Pieper, D.H. and Reineke, W. 2000. Engineering bacteria for bioremediation. *Current Opinion in Biotechnology*, 11:262–70.

Pignatello, J.J. and Xing, B. 1996. Mechanisms of slow sorption of organic chemicals to natural particles. *Environmental Science and Technology*, 30:1–11.

Pimentel, D. and Greiner, A. 1997. Environmental and socioeconomic costs of pesticide use. In *Techniques for Reducing Pesticide Use: Economic and Environmental Benefits*. Pimentel, D. (Ed.). John Wiley and Sons, Chichester, pp. 51–78.

Plaza, C., Xing, B., Fernández, J.M., Senesi, N., and Polo, A. 2009. Binding of polycyclic aromatic hydrocarbons by humic acids formed during composting. *Environmental Pollution*, 157:257–63.

Prenafeta-Boldú, F.X., Ballerstedt, H., Gerritse, J., and Grotenhuis, J.T.C. 2004. Bioremediation of BTEX hydrocarbons: Effect of soil inoculation with the toluene-growing fungus *Cladophialophora* sp. Strain T1. *Biodegradation*, 15:59–65.

Rao, N., Grethlein, H.E., and Reddy, C.A. 1995. Mineralization of atrazine during composting with untreated and pretreated lignocellulosic substrates. *Compost Science and Utilization*, 3:38–46.

Rao, N., Grethlein, H.E., and Reddy, C.A. 1996. Effect of temperature on composting of atrazine-amended lignocellulosic substates. *Compost Science and Utilization*, 4(3):83–8.

Ravikumar, T.N., Yeledhalli, N.A., Ravi, M.V., and Narayana Rao, K. 2008. Physical, physico-chemcial and enzymes activities of vermiash compost. *Karnataka Journal of Agricultural Science*, 21:222–6.

Reid, B.J., Fermor, T.R., and Semple, K.T. 2002. Induction of PAH-catabolism in mushroom compost and its use in the biodegradation of soil-associated phenanthrene. *Environmental Pollution*, 118(1):65–73.

Richard, T.L., Veeken, A., De Wilde, V., and Hamelers, H.V.M. 2004. Air-filled porosity and permeability relationships during solid-state fermentation. *Biotechnology Progress*, 20:1372–81.

Richard, T.L., Hamelers, H.V.M., Veeken, A.H.M., and Silva, T. 2002. Moisture relationships in composting processes. *Compost Science and Utilization*, 10:286–302.

Ryckeboer, J., Mergaert, J., Vaes, K., Klammer, S., De Clercq, D., Coosemans, J., Insam, H., and Swings, J. 2003. A survey of bacteria and fungi occurring during composting and self-heating processes. *Annals of Microbiology*, 53:349–410.

Samanta, S.K., Singh, O.V., and Jain, R.K. 2002. Polycyclic aromatic hydrocarbons: Environmental pollution and bioremediation. *TRENDS in Biotechnology*, 20(6):243–8.

Schulze, K.L. 1962. Continuous thermophilic composting. *Applied Microbiology*, 10:108–22.

Sedlak, D.L., Pinkston, K.E., Gray, J.L., and Kolodziej, E.P. 2003. Approaches for quantifying the attenuation of wastewater-derived contaminants in the aquatic environment. *Chimia*, 57:567–9.

Semple, K.T., Reid, B.J., and Fermor, T.R. 2001. Impact of composting strategies on the treatment of soils contaminated with organic pollutants. *Environmental Pollution*, 112(2):269–83.

Smith, V.H., Graham, D.W., and Cleland, D.D. 1998. Application of resource-ratio theory to hydrocarbon biodegradation. *Environmental Science and Technology*, 32:3386–95.

Strong, P.J. and Burgess, J.E. 2008. Treatment methods for wine-related ad distillery wastewaters: A review. *Bioremediation Journal*, 12:70–87.

Suthar, S. 2008. Development of a novel epigeic–anecic-based polyculture vermireactor for efficient treatment of municipal sewage water sludge. *International Journal of Environment and Waste Management*, 2: 84–101.

Symons, Z.C. and Bruce, N.C. 2006. Bacterial pathways for degradation of nitroaromatics. *Natural Product Reports*, 23(6):845–50.

Tang, C.Y., Criddle, Q.S., Fu, C.S., and Leckie, J.O. 2007. Effect of flux (transmembrane pressure) and membranes properties on fouling and rejection of reverse osmosis and nanofiltration membranes treating perfluorooctane sulfonate containing waste water. *Environmental Science and Technology*, 41:2008–14.

Ternes, T.A. 2001. Analytical methods for the determination of pharmaceuticals in aqueous environmental samples. *Trends in Analytical Chemistry*, 20:419–34.

Tiquia, S.M. and Tam, N.F.Y. 1998. Composting of spent pig litter in turned and forced-aerated piles. *Environmental Pollution*, 99:329–37.

Tognetti, C., Laos, F., Mazzarino, M.J., and Hernandez, M.T. 2005. Composting vs. vermicomposting: A comparison of end product quality. *Compost Science and Utilization*, 13:6–13.

Valo, R. and Salkinoja-Salonen, M. 1986. Bioreclamation of chlorophenol-contaminated soil by composting. *Applied Microbiology and Biotechnology*, 25(1):68–75.

Van Eerd, L.L., Hoagland, R.E., Zablotowicz, R.M., and Hall, J.C. 2003. Pesticide metabolism in plants and microorganisms. *Weed Science*, 51(4):472–95.

van Schie, P.M. and Young, L.Y. 2000. Biodegradation of phenol: Mechanisms and applications. *Bioremediation Journal*, 4(1):1–18.

Veeken, A.M.H., Adani, F., Nierop, K.G.J., de Jager, P.A., and Hamelers, H.V.M. 2001. Degradation of biomacromolecules during high-rate composting of wheat straw-amended feces. *Journal of Environment Quality*, 30:1675–84.

Vidali, M. 2001. Bioremediation. An overview. *Pure and Applied Chemistry*, 73(7):1163–72.

Watanabe, K. 2001. Microorganisms relevant to bioremediation. *Current Opinion in Biotechnology*, 12:237–41.

Wild, S.R. and Jones, K.C. 1992. Organic chemicals entering agricultural soils in sewage sludges: Screening for their potential to transfer to crop plants and livestock. *Science of the Total Environment*, 119:85–119.

Williams, R.T. and Keehan, K.R. 1993. Hazardous and industrial waste composting. In *Science and Engineering of Composting*. Hoitink, H.A.J. and Keener, H.M. (Eds). Renaissance Press, Worthington, OH, pp. 363–82.

Wilson, C. and Tisdell, C. 2001. Why farmers continue to use pesticides despite environmental, health and sustainability costs. *Ecological Economics*, 39:449–62.

Yadav, A., Mudhoo, A., and Garg, V.K. 2009. Growth and fecundity of *Eisenia fetida* earthworm during vermicomposting of food industry sludge. *International Journal of Process Wastes Treatment*, 1(1):71–81.

Yuan, S.Y., Su, L.M., and Chang, B.V. 2009. Biodegradation of phenanthrene and pyrene in compost-amended soil. *Journal of Environmental Science and Health, Part A*, 44(7):648–53.

Zach, A., Latif, M., Binner, E., and Lechner, P. 1999. Influence of mechanical–biological pretreatment on the toxicity of municipal solid waste. *Compost Science and Utilization*, 7(4):25–33.

Zechendorf, B. 1999. Sustainable development: How can biotechnology contribute? *TIBTECH*, 17:219–25.

Zhuang, X., Chen, J., Shim, H., and Bai, Z. 2007. New advances in plant growth-promoting rhizobacteria for bioremediation. *Environment International*, 33:406–13.

10 Policies Promoting Green Chemistry
Substitution Efforts in Europe

Carl Dalhammar

CONTENTS

10.1 INTRODUCTION

Traditionally, public policies addressing chemicals have mainly dealt with the risks posed to health and the environment associated with the production, use and handling of these chemicals. There has been less focus on how public policies may stimulate research and development (R&D) with the aim to develop new, safer chemicals, and safer production processes, and to engage various actors in the substitution of chemicals when safer alternatives exist. Thus, a true "innovation" component is often missing from public policies. Whereas "traditional" policies that ban or restrict the manufacturing, use, and marketing of chemicals may trigger some R&D activities among industries, the effects have been limited. Public intervention is necessary in order to speed up these efforts, as incumbents on the market have made significant investments in existing chemicals and may have little to gain from expensive R&D into new alternatives, where the potential future profits are uncertain. When the chemicals industry has made voluntary commitments—most notably through the Responsible Care initiative—it is questionable if the results go beyond business-as-usual, and there is a lack of focus on chemical innovation and substitution efforts (Purvis and Bauler, 2004). There is thus a need to focus more on innovation in chemicals policy, and to engage various manufacturers and users of chemicals in substitution efforts.

The European Union (EU) has overtaken the role previously held by the United States (US) as the main leader in environmental policymaking (Vogel, 2003), not least in the field of chemical policy (Selin and VanDeveer, 2006; Dalhammar, 2007a). Since the 1960s, the EU has enacted a large amount of legislation that governs the manufacturing and use of chemicals, including rules on development and testing, import, production and storage, transport, packaging and labeling, marketing, use in preparations or other products/articles, and rules governing export/import of chemicals or chemicals in products (for an overview, please refer to Pallemaerts, 2006). Relevant provisions for dealing with chemicals include not only laws that directly govern the use of chemicals or information about chemicals, but also legislation that relates to the protection of workers, rules that set legally binding standards for water and air quality (the so-called environmental quality standards), and rules that regulate emissions from facilities through traditional licensing instruments, making use of binding individual emissions targets for large emitters. EU has also started to regulate chemicals in articles (Onida, 2004; Dalhammar and Nilsson, 2005; Selin and VanDeveer, 2006; Dalhammar, 2007a, 2007b), not least in the field of electrical and electronic products.

Generally, early EU legislation in the field of chemicals intended to promote the free circulation of chemicals in the EU and remove barriers to trade, whereas a second wave of regulations—often triggered by national policies in EU Member States (Onida, 2004; Dalhammar, 2007a)—addressed concerns about human health impacts associated with chemicals, often by effect-oriented approaches that often relied on end-of-pipe and waste management oriented measures (Pallemaerts, 2006). Before the enactment of the registration, evaluation, authorization, and restriction of chemicals (REACH) reform, an absence of a master plan for dealing with chemicals was noted (Pallemaerts, 2006, p. 231):

> It seems that the legislatures focus of attention in the reduction of chemical risks to health and the environment constantly shifted from one activity and type of risk to another, pursuing varying legislative rationales and responding to changing external pressures, often addressing the same substances or groups of substances from different angles and using different policy instruments with little or no coordination. Overall, the primary focus of regulation seems to have been almost exclusively on 'downstream', post-production aspects of chemicals' life cycles.

Also in the case of policies addressing chemicals in articles, the lack of coordinated efforts has been obvious (Dalhammar and Nilsson, 2005).

The EU chemicals policy has generally been poor at addressing the early stage of the life cycle of chemicals, and has not been very good at promoting innovation, for example, through enforcing substitution policies and initiating R&D activities. However, the REACH reform, enacted through Regulation (EC) No 1907/2006, lays the fundament for a more coordinated chemicals policy, with a stated intention to stimulate innovation and substitution efforts.

While REACH has many benefits, it also entails some risks. One such risk is that chemical legislation is becoming ever more harmonized in the EU, which limits the potential for EU Member States to enact national policies. A certain degree of harmonization is necessary as it replaces a plethora of national policies regulating chemicals and chemicals in articles, which may limit export/import; harmonization promotes the functioning of the EU Internal Market (the free flow of goods within the EU). However, legal harmonization is a complex topic and will not be dealt with presently (Dalhammar, 2007a). The principle is that when there are EU rules that regulate chemicals and their use, EU Member States cannot regulate the same chemicals for the same uses; EU rules take preference. This also means that when a chemical is allowed according to the EU rules, EU Member States may not ban it, unless they can provide very good motivations for why this is necessary to protect human health or the environment. There have been several legal battles between the European Commission and EU Member States, and battles between EU Member States and other EU states or enterprises, over the right to enact national rules.

While legal harmonization has many benefits, it may hinder the design of national policies, and this is problematic, for two main reasons. First of all, when the EU nations cannot pursue desired policies, it limits the likelihood that set national objective can be reached, as the policy options are constrained. Second, while the current discourse in environmental policy tend to stress the need for international cooperation to solve environmental problems, recent research also reveal *the importance of national initiatives in order to drive innovation and diffusion of environmental policy innovations* (Jänicke and Jacob, 2004; Jänicke, 2005; Jörgens and Tews, 2005; Dalhammar, 2007a). Environmental innovations originate mainly at the national level, in countries that have progressive environmental policies, and other countries tend to adopt the leading standards, for various reasons (Jänicke, 2005). Some standards are also adopted in international agreements.

The same mechanisms that work at the international level are relevant also within the EU. Many EU Member States are pushing for progressive product policies, for instance, producer responsibility and chemicals policies, which have forced the Commission to develop EU policies to avoid distortions on the Internal Market (e.g., in the field of chemical policy, see Onida, 2004). For some chemical standards, for instance, EU rules on chemicals in electronics, other countries are forced to adopt similar standards in order to make sure that their companies can export their products on the global market (Selin and VanDeveer, 2006; Dalhammar, 2007a).

This contribution reviews European policies for the promotion of green chemistry, focusing on how policies may stimulate substitution of chemicals for less hazardous ones, as this is a crucial aspect of a successful chemicals policy. We need to change the mindset from one based on the paradigm that all chemicals can be controlled, toward a mindset built more on prevention of an innovation, and substitution policies are key in this respect. Sweden has a long tradition of working progressively with substitution policies, and the REACH reform signals an intention to shift toward a stronger focus on substitution also in EU policies. Here the "substitution" element in REACH and the Swedish experiences with substitution policies have been reviewed. Hopefully, this chapter can provide more insight on the topic and possibly be an inspiration for other countries that wish to work more actively with substitution.

In the next section, the so-called Substitution Principle is discussed. This is followed by a brief outline of relevant policies and policy instruments that may be applied in order to stimulate substitution efforts. Then, the REACH reform is

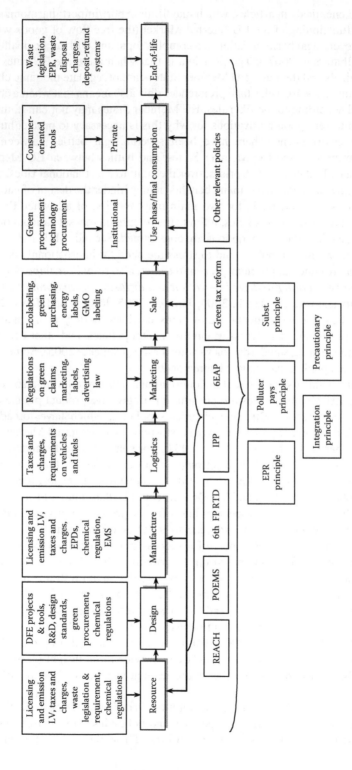

FIGURE 10.1 European policies for sustainable consumption and production. (Adapted from Mont, O. and Dalhammar, C. 2006. *International Journal of Sustainable Development*, 8(4):258–79.)

evaluated from the perspective of substitution, followed by a review of Swedish experiences from working with substitution policies. The chapter ends with some concluding remarks.

10.2 THE SUBSTITUTION PRINCIPLE AND ITS APPLICATIONS

The most well-known legal principle in the field of chemicals policy is the Precautionary Principle; it is so well-known that it hardly requires any further introduction here. It was introduced into EU environmental law through the Maastricht Treaty in 1992. The principle's value as a guide for decision making in different policy areas is often questioned, and some analyses of its impact on EU chemicals policy indicate that its influence is not that significant (Heyvaert, 2006), though it can be hard to determine exactly how influential the principle has been for important law cases from the Court of Justice of the European Communities.

The "Substitution Principle," sometimes referred to as the "Product Choice Principle," or as the "Principle of Substitution," is not so well-known (in Sweden, the term "Substitution Principle" was used until 1999, and then the principle was renamed the Product Choice Principle). It is an established legal principle in some countries, including Sweden where the principle was first established in law in 1985 (though the wording has changed over the years). The exact definition of the principle may vary between legal jurisdictions, but the essential content of the principle is:

1. Whenever possible, chemicals that might constitute a risk should be replaced with other, less harmful chemicals or with organic materials.
2. Goods that contain or have been treated with chemicals should be replaced with other, less harmful goods when this is possible.
3. The main limitation of the principle is that the costs for the substitution should not be unreasonable.

The principle might also be applicable on methods and processes, since methods and processes that involve toxic chemicals can sometimes be replaced with processes with less toxic chemicals, or even with organic or mechanical processes.

In Swedish law, the principle is expressed in the 2nd chapter of the Environmental Code, where it is one of the so-called rules of consideration (the following translations are taken from the official translation of the Environmental Code):

> Persons who pursue an activity or take a measure, or intend to do so, shall avoid using or selling chemical products or biotechnical organisms that may involve risks to human health or the environment if products or organisms that are assumed to be less dangerous can be used instead. The same requirement shall apply to goods that contain or are treated with a chemical product or a biotechnical organism.

The principle should be read in conjunction with the "proportionality rule" in the same chapter of the Code:

> The rules of consideration [. . .] shall be applicable where compliance cannot be deemed unreasonable. Particular importance shall be attached in this connection to the benefits

of protective measures and other precautions in relation to their cost. The cost–benefit relationship shall also be taken into account in assessments relating to total defence activities or where a total defence measure is necessary. [...]

In Sweden, the principle do not only constitute that chemicals/goods should be replaced by other chemicals/goods. The principle also means that *methods or techniques* should be used that minimizes or excludes the use of a chemical (Swedish Government Proposition, 1997/98, p. 45). It is not only the *toxicity* of the alternatives that are to be evaluated, but also the *risks* they pose to the environment and human health (Swedish Government Proposition, 1997/98, p. 45). Sometimes a more toxic chemical might be preferred to a less toxic one, since there are better methods developed to take care of it and reduce risks involved (SOU, 1996, p. 133). To make the substitution process easier for industries and other actors, authorities can provide information on different chemicals/products, provide information about chemicals that should be phased out or used to the least possible extent, and provide good examples of successful substitutions. More details about different considerations when applying the principle is provided in Nilsson (1997), Dalhammar (2000), Mont (2001), and Kemikalieinspektionen (2008).

The Substitution Principle is to be applied for weighing two alternatives in a specific situation. It is not to be used as a guideline for general bans on chemicals and products, although it might be used to ban the use of a chemical in a specific case. The Precautionary Principle is better suited as a guideline when the total ban of a substance is discussed. However, it is obvious that it is easier to ban a substance when a good alternative is available, and that the possibility to substitute is a very important factor to consider (Swedish Government Proposition, 1997/98, p. 45). In practice, it is very difficult to ban a substance if there are no good alternatives, and a ban is often preceded by an investigation of substitutes and the costs for replacement. Often the Precautionary Principle is very difficult to use by itself. It is (and has to be) expressed in a vague manner and the acceptable risks must be evaluated in every specific case. A Swedish Government proposition has stated that it is most useful when it is interpreted in combination with other principles, for example, the Substitution Principle (Swedish Government Proposition, 1997/98, p. 45). If there is a good substitute for a substance, the evidence that the substance poses a risk should not have to be as strong as in a situation where there is no suitable alternative.

The Swedish expression of the principle is that actors should replace chemicals with less harmful ones when the costs are not unreasonable. While this may appear to be a little more than common sense, the principle and its application are quite controversial. The principle is not mentioned in the Rio Declaration and is not considered to be an established principle in international environmental law, although Agenda 21 (19.41) refer to the substitution of hazardous substances for less hazardous substances as the classical example of risk reduction. So why is the principle not established in international environmental law? Several reasons exist. The application of the principle has several limitations. Apart from the cost issue, it can of course be difficult to evaluate whether a chemical is a better option than another one if none of them are properly tested, or if only one of them have been properly tested.

Another limitation is that sometimes it might be difficult to evaluate whether a substitution in fact leads to an environmental improvement since the effects in the whole technical system where the chemical is used might have to be examined using relevant life cycle assessment (LCA) methodologies. However, in Sweden this uncertainty is not seen as a main barrier, because the choice between alternatives is a *dynamic process*, not a one-time selection; as the knowledge on chemicals and techniques increase, new choices might have to be made (this point has been made by the Swedish government in Proposition, 1997/98, p. 45). Actors are thus mandated to act based on their best and latest knowledge.

Another problem concerns whether private citizens should have a duty to follow the principle, as these obviously have less knowledge about suitable alternatives than industries. But this has not been a main issue is Sweden either: whereas citizens should act upon their best knowledge, the principle is only enforced vis-à-vis corporations and other large users (how this can be done will be discussed later in this chapter) (Dalhammar, 2000).

Many corporations dislike the principle for several reasons. The principle puts a constant pressure upon users of chemicals to keep informed and act accordingly, and thus provides limited certainty regarding what is expected of them. The principle may therefore not work well in countries where the legal culture may mean that corporations may have to defend their substitution efforts (or lack thereof) in courts, or where authorities may have to defend why they have not done more to promote substitution if they have established the Substitution Principle in law. Similar problems exist when it comes to the potential use of the Precautionary Principle in the United States, and this in one reason as to why principles are easier to deal with in Civil law countries like Sweden than in the United States.

Although not an established principle in international environmental law, the Substitution Principle still has support in international law making. Agenda 21 (4.19) states that governments shall support the introduction of more environment-friendly products on their markets to promote sustainable consumption and production patterns. Substitution is also expressed as one of the best environmental options in some international agreements, for instance, the Baltic Sea Convention from 1992 (Nilsson, 1997). The EU has started to promote the concept of substitution in several documents and directives, although the Substitution Principle has not become an established principle in EU law. Thus, several of the EU's directives refer to, or actively promote, substitution. In the IPPC (Integrated Pollution Prevention and Control) Directive (Dir. 96/61/EC), substitution of chemicals is mentioned as one example of Best Available Techniques (BAT), and several directives dealing with chemicals,— for example, the VOC Solvents Directive (Directive 1999/13/EC)—promote substitution, stating that the environmental gains shall be compared to the costs.

10.3 POLICIES PROMOTING SUBSTITUTION

Policy principles act as "compasses" for sustainability, but they must be supported by different policies and action programs, and ultimately policy instruments. Figure 10.1 represents a conceptual image of different policy levels for the promotion of sustainable consumption and production in Europe.

TABLE 10.1

Policy Instruments Providing Different Actors with Incentives for Change

	Mandatory Instruments	Voluntary Instruments
Administrative	Bans, licenses, requirement on EHS information, EPR, recycling and recovery quotas, material and quality requirements, emission levels, chemicals regulation	Responsible care and similar initiatives, POEMS, application of product standards, product panels, EMS, functionality panels, agreements between government and industry
Economic	Deposit-refund systems, taxes and charges, liability rules	Green public procurement, technology procurement, R&D investments
Informative	Requirement on EHS information, emission registers, material and quality requirements, chemicals regulation on information for professional and private users, energy labeling, marketing regulations	Eco-labeling ISO type I, EPDs, green claims, energy labeling, organic labeling of food, for example, certification schemes of hotels, consumer advice, consumer campaigns, education

Source: Mont, O. and Dalhammar, C. 2006. *International Journal of Sustainable Development*, 8(4):258–79.

Ultimately, it is the policy instruments that provide different actors with incentives for change. Some examples of policy instruments are given in Table 10.1. A carefully planned environmental policy should provide for a positive interaction of different policy instruments, which may happen in several ways (Bemelmans-Videc et al., 1997; Gunningham and Grabosky, 1998; Dalhammar, 2007a).

In the case of green chemistry, and more especially chemical substitution, a number of policy instruments are relevant. The traditional approach is to ban certain toxic chemicals in order to induce substitution efforts. Such bans are usually preceded by examples of successful substitutions, as it is controversial to ban chemicals when no alternatives exist at reasonable cost. Otherwise, industry is often granted generous phase-in periods, in order to develop substitutes. A third way is to grant derogations when it is hard or very costly to develop substitutes. The latter approach has been applied in the context of the EU Restriction of Hazardous Substances (RoHS) Directive (Directive 2002/95/EC), which bans six substances in electrical and electronic products. A slightly less interventionist approach is to put restrictions on uses of certain chemicals. Other administrative approaches include the ban of chemicals, or restrictions in use, in individual operations when these apply for permits.

Informative instruments play an increasingly important role in chemicals policy. Information provided to different actors, such as consumers, professional purchasers, and nongovernmental organizations (NGOs), is often a prerequisite for change, as the information may be used for design, purchase, in recycling, and in recovery operations, or in order to put pressure on chemical manufactures or users to develop and/or use better alternatives.

There are several countries around the world that have enacted policies that provide relevant information to chemicals users (e.g., safety data sheets), and NGOs and other actors (e.g., the Toxic Release Inventory [TRI] program in the United States and corresponding EU policies), although the accuracy of information provided in those schemes have often been questioned. There is however a lack of schemes that brings relevant information to private citizens (Dalhammar and Nilsson, 2005). Similarly, few schemes are well adapted to the needs of professional purchasers.

Also many ecolabeling schemes tend to set restrictions on chemicals, and this is of course more relevant in some product groups than others.

The Swedish Chemicals Agency has set up lists of chemicals that should preferably be substituted (and sunset lists), and these lists have often been used by different authorities and enterprises when engaging in substitution work and in their procurement practices (Dalhammar, 2000). In Sweden, previous lists are now replaced with a new tool, PRIO, which is also available in an English version at www.kemi.se.

With regard to voluntary initiatives without its basis in regulation, there are examples of relevant industry initiatives that deal with chemicals in articles. An interesting case is the BASTA system, set up by the Swedish construction industry to phase-out the use of particularly hazardous chemical substances from chemical products and building products (available at www.bastaonline.se in English). In the BASTA system, suppliers of construction materials can register articles that fulfill the criteria concerning chemical content, and independent auditors will make tests to ensure credibility. Globally, textile, electronics, and car industries are developing information systems to provide environmental data on components and articles in international product chains (Kemikalieinspektionen, 2005).

Another instrument of voluntary character without formal basis in regulation is public procurement. Here, it is quite common to include criteria for chemicals in articles. Many Swedish authorities do this on a regular basis (Naturvårdsverket, 2005), often using the tools developed by the Swedish Chemicals Agency to aid government and enterprises in their work with chemicals substitution. Statskontoret, the authority responsible for coordination of government procurement in Sweden, has included criteria on chemicals content in relevant product groups, for example, personal computers (Dalhammar and Nilsson, 2005).

In order to more directly induce innovation for substitution, authorities have often created projects in cooperation with progressive manufacturers, main users, and universities, in order to develop proper substitutes. Examples include the Grön Kemi project in Southwest Sweden and which may be visited at http://www.gronkemi.nu/start_eng.html. Also, various initiatives to phase out lead and other substances from various product groups in the Danish Renere Produkter program may be consulted on OXFORD AS/IIIEE 2006. Several countries have initiated research programs that aim to phase out chemicals from medical products and other applications.

Technology procurement is also a very interesting approach (Edquist et al., 2000). Technology procurement is public procurement that intends to stimulate innovation through invitation to tenders that demand innovative solutions that are not yet available on the market. If run by the government, the governmental authorities can collect desired product characteristics from potential buyers, and these buyers will also commit to purchase a large number of products from the winner; that is, the manufacturer that comes up with the best product conforming to the criteria. This ensures that there is a market potential for the technology developed, and thus decreases the economic risks associated with product development. Technology procurement can aid both the market introduction and commercialization of new products. Technology procurement has mostly been successfully applied in order to promote more energy-efficient products, and there is significant potential to use it also for substitution of chemicals. In Sweden, there are plans to make more use of the instrument in order to promote substitution of chemicals. For more information, the reader is directed to the website of the Swedish Environmental Management Council available at http://www.msr.se/en/.

10.4 THE REACH REFORM

In EU law, a requirement for premarket testing of chemicals was introduced by the famous 6th amendment of Directive 67/548/EEC, adopted in 1980. While the new requirement of testing of chemicals before market introduction was indeed very reasonable, a problem was that the *existing* substances on the market (which had been introduced before 1981) did not require testing. Later (actually not until 1993), rules on testing of existing substances were introduced (through Regulation 793/93), but the obligation to test chemicals for their effects on human health and the environment was not imposed on the industry but on public authorities. Owing to the long and cumbersome process of testing, the rules had little effect on existing substances, and few substances were banned. The major bulk of chemicals on the market were introduced before 1980, and therefore most substances on the market were not properly tested. Further, the fact that chemicals introduced after 1980 had to be tested (which is costly) before being introduced on the market while substances that were already on the market in 1980 did not require testing, which led to a situation where new chemicals were put at a disadvantage. This posed a hindrance for innovative activities, most especially for the development of new and less dangerous alternatives. This state of affairs was considered deeply unsatisfactory by many actors. Another problem in EU chemicals policy was the "patchwork" character of policies; there was a need for integration of the different pieces of legislation.

In 1998 it was decided that the EU chemicals regulatory system needed reform. A long process of analysis and discussion resulted in a proposal for a new system, called REACH. The REACH Regulation was adopted in 2006 (through Regulation (EC) No 1907/2006) and came into force in June 2007. Through REACH, a new agency, the European Chemicals Agency (ECHA), was initiated.

In this chapter, only a brief description of some of the main elements of REACH is provided, and its effect on substitution discussed. Somewhat simplified, the main elements of REACH are as follows:

1. A substance must be registered before it is put on the market, if used in quantities above 1 ton year^{-1}. Substance manufacturers and importers (into the EU) are obliged to send a registration dossier containing safety data to the ECHA. Information about the substance and potential uses should be included.
2. A chemical safety report is required when manufacturers and importers use 10 tons or more of a substance per year. The report must outline the

hazard and risk assessment of the substance for specified ("identified") uses—including its inclusion in articles—and how the risks posed by the chemical can be "adequately controlled" for these uses. One of the outputs of the assessment is an "exposure scenario."

3. The registration dossier is sent to ECHA, which reviews the registrations and, if necessary, demands supplementary information. ECHA will however not have the resources to review the quality and accuracy of data of all registrations.

4. When the registration process is complete, the manufacturer/importer may manufacture/import the substance (within the limits of what is included in the registration). Downstream users may use the substance in any way covered by the registration. For additional uses, they may have to make their own registration.

5. ECHA will check the compliance of a minimum of 5% of registration dossiers.

6. ECHA, in cooperation with EU Member State authorities, may request further information from industry on particular substances, when there are suspicions of risks for humans and animals. The costs for these procedures should be shared among registrants of the substance. Substance evaluation will be prioritized on the basis of risk following guidance developed by the ECHA.

7. Evaluation may lead to the conclusion that the use of certain substances must be put under restrictions or that authorization is considered necessary for some substances.

8. For substances of very high concern (SVHC), the so-called authorization is required for their use and their placing on the market. These substances are considered to have hazardous properties of such high concern that it is necessary to regulate them centrally, through a mechanism to ensure that the risks related to their actual uses are assessed, considered, and then decided on an EU-wide basis. The main justification is that the effects on humans and the environment of these substances can be serious and irreversible. Substances that fall into these categories will come into the authorization system as resources allow. Thus, existing resources are a main constraint for the functioning of the system; the uses of potentially dangerous substances will not be banned by default. An authorization can be based solely on "adequate control" (if demonstrated in the chemical safety report) of the substance for the use in question. Still, "adequate control" cannot apply to PBTs, vPvPs, and nonthreshold CMRs as it is not possible to determine a "safe" exposure level. *All applications for an authorization must be accompanied by an analysis of possible alternatives considering their risks and the technical and economic feasibility of substitution. All authorizations will be subject to a time-limited review. This would enable further consideration of the availability of alternatives at some point in the future.*

While there are several deficiencies in REACH, it is without doubt a landmark in modern chemicals policy. The fact that the reform was adopted despite heavy lobbying from, among others, European industries and the Bush government (for more details on the political controversies and lobbying efforts, refer to Schörling, 2004; The Waxman Report, 2004; Selin, 2007) it is quite an achievement. REACH lays the foundations for a more effective European regime—at the European and national levels—for better chemicals management as it will make sure that information is available and that it can be used for bans and restrictions. REACH also serves as an example for other jurisdictions, and there are examples of how countries and federal agencies outside of the EU are planning similar reforms, most notably California.

A stated intention of REACH is to promote innovation and substitution (European Commission, 2009). Recital 12 of the REACH Regulation states thus:

> *'An important objective [...] is to encourage and in certain cases to ensure that substances of high concern are eventually replaced by less dangerous substances or technologies where suitable economically and technically viable alternatives are available'*, while Recital 70 states : *'Adverse effects on human health and the environment from substances of very high concern should be prevented through the application of appropriate risk management measures to ensure that any risks from the uses of a substance are adequately controlled, and with a view to progressively substituting these substances with a suitable safer substance.'*

In the REACH Regulation, there are several articles that promote substitution practices, not least the rules on authorization and restrictions. In connection to the Authorization procedure, Article 55 states:

Aim of Authorization and Considerations for Substitution
The aim of this Article is to ensure the good functioning of the internal market while assuring that the risks from SVHC are properly controlled and that these substances are progressively replaced by suitable alternative substances or technologies where these are economically and technically viable. To this end, all manufacturers, importers, and downstream users applying for authorizations shall analyze the availability of alternatives and consider their risks, and the technical and economic feasibility of substitution.

In the authorization process, the importer, user, or manufacturer shall submit a review of possible substitutes, and if relevant develop a substitution plan, and is, *inter alia*, required to submit (Article 62):

> (e) an analysis of the alternatives considering their risks and the technical and economic feasibility of substitution and including, if appropriate information about any relevant research and development activities by the applicant;
> (f) where the analysis referred to in point (e) shows that suitable alternatives are available, taking into account the elements in Article 60(5), a substitution plan including a timetable for proposed actions by the applicant.

In terms of restrictions put on substances of high concern, Article 68 states:

> When there is an unacceptable risk to human health or the environment, arising from the manufacture, use or placing on the market of substances, which needs to be addressed on a Community-wide basis, Annex XVII shall be amended in accordance with the

procedure referred to in Article 133(4) by adopting new restrictions, or amending current restrictions in Annex XVII, for the manufacture, use or placing on the market of substances on their own, in preparations or in articles, pursuant to the procedure set out in Articles 69 to 73. Any such decision shall take into account the socio-economic impact of the restriction, including the availability of alternatives.

Thus, the availability of substitutes is crucial when authorities decide on whether a substance may be "Authorized" for certain uses, and also when authorities decide upon necessary restrictions in the use of a substance.

REACH will ensure that more information about substances is developed, and ECHA will make much of this information publicly available (although there are restrictions due to trade secrets and other parameters). Implementing REACH will however be difficult and many issues need to be resolved (Forbes, 2009; Peters, 2009).

While it is clear that REACH may lay the foundation for a regime that focuses more on substitution and thus promotes innovation, the extent to which REACH will promote substitution depends on several factors. These include the following:

1. The quality of investigations made by the users, and the quality of data received by authorities; without proper data it is hard to make decisions on substitution possibilities.
2. The resources that authorities devote to conduct proper investigations regarding: (a) if the information supplied by applicants are correct, and (b) if substitution opportunities exist, which the applicants have not accounted for. Information exchanges between the ECHA and EU Member State authorities will be crucial in this respect.
3. There is a need for efforts within the EU Member States in order to provide incentives for different actors to engage in substitution activities; the Member States cannot rely on the ECHA.
4. Despite the "progressiveness," REACH is still a "permissive" regime; it originates from the fundamental belief that risks can be "adequately controlled" (Clean Production Action, 2003; Dalhammar and Nilsson, 2005). The ECHA as well as national bodies must therefore work in a more preventive manner; otherwise the substitution incentives might be lax.

If substitution is progressively promoted through the REACH regime, at both the EU and national levels, it has the potential to drive R&D activities, and help create markets for firms that want to develop new, safer alternatives. But such effects are quite uncertain as it is hard to evaluate how progressively substitution will be enforced, and long time lags in the system may be expected. Typically, companies that identify a suitable substitute for a chemical process need a couple of years to complete the switch (Mont, 2001). Therefore, an effective substitution regime requires that EU Member States and ECHA work progressively with identifying and enforcing substitution possibilities; otherwise little will happen in effect. In many cases, it can be expected that chemical users have little interest in identifying substitutes when such substitutes exist, and therefore public authorities must pressure them to investigate the appropriateness of substitutes. Without such efforts, REACH will mainly generate data about risk, but not lead to the exploitation of risk reduction opportunities and substitution (Koch and Ashford, 2006). Authorities also need to identify areas where they can support R&D activities through various measures such as technology procurement. But an effective policy will require political will and resources. It is also valuable to look at previous successful efforts for substitution, including the Massachusetts Toxics Use Reduction Act (TURA) (Koch and Ashford, 2006) and the Swedish experiences.

10.5 SUBSTITUTION POLICIES IN SWEDEN

While some experts may doubt the effectiveness of the Substitution Principle, the Swedish experiences have been positive. This section will account for chemical policies in Sweden, how public authorities promote the Substitution principle, and some of the success factors.

First it needs to be pointed out that Sweden is considered to be a very progressive nation in the area of chemical policy. The country initiated a comprehensive Chemicals Act already in 1985, and has a separate Chemicals Agency with considerable resources and with extensive powers at its disposal. Sweden has been a pioneer in setting strict standards for heavy metals and other chemicals, and has often been an early mover when it comes to regulation of chemicals in articles. Sweden has also been a strong promoter of policies that would promote more and improved information about different substances. Sweden has also set its own independent national targets related to chemicals in the system of national Environmental Quality Objectives, which may be consulted at http://www.miljomal.nu/Environmental-Objectives-Portal/.

The Substitution Principle was introduced as a legal principle in Sweden through the enactment of the 1985 Chemicals Act. When the Environmental Code replaced the Chemicals Act and several other major environmental laws in 1999, the principle was expressed in the 2nd chapter of the Code, now named the Product Choice Principle.

There are a number of authorities that have the responsibility for the supervision and control of the use of chemicals in Sweden. The different authorities have three main tasks:

1. Issue more specific rules that supplement and clarify the chemical legislation (mainly the central authorities)
2. Work for improving the knowledge about chemicals and the risks involved and spread this information
3. Supervise that the laws, regulations, and rules on the handling of chemicals are being followed

The Swedish EPA and the Chemicals Agency have the responsibility at the central level. The County Administrative Boards in each region have the responsibility at the regional level and the municipalities at the local level. The Chemicals Agency has the right to issue general rules on how the supervision and control in accordance with the Environmental Code shall be carried out, and provides information and guidelines to the regional and municipal authorities. Furthermore, it evaluates and makes up lists of dangerous substances (among other things the Agency have set up the PRIO tool discussed previously in this chapter). The substitution principle has been enforced through various measures. For more details, we refer to Nilsson (1997), Dalhammar (2000), Mont (2001), Dalhammar (2007a), and Kemikalieinspektionen (2008). Past and current efforts include the following:

Pesticide approval: In the system applied in Sweden before it became a member of the EU, the Substitution Principle was applied when pesticides were reviewed. An application to import and sell a certain type of pesticide on the Swedish market was to be denied if a less dangerous pesticide that could be used for the same purpose already existed on the market.

Licensing and inspections: It is only the national authorities that can issue *general norms*, but the regional and municipal authorities can use prohibitions and injunctions toward single manufacturers, sellers, and users. Chemical use can also be regulated in individual permits, when companies seek a license to start up or expand its operations. In some cases the authorities have prohibited actors to use or sell a certain chemical with the motivation that good alternatives exist. Examples include process chemicals, fertilizers, and pesticides. The authorities have also mandated chemical users to examine the substitution possibilities in cooperation with the authorities. In some cases, large manufacturers who use large amounts of chemicals have also been asked to provide more information about chemicals used than what is already provided in the safety data sheets by manufacturers/ importers. This has however been quite controversial as it may be expensive to perform own tests. The fact that the regional authorities cannot issue general norms has been a main limitation for them. For instance, they cannot issue a general prohibition to sell a certain type of chemical aimed at all sellers in an area, but must address all the sellers individually.

Information campaigns and voluntary initiatives: The different authorities often launch information campaigns to inform the different actors about the obligation to substitute chemicals and give examples of methods to achieve this. This includes a "good example," which often stresses the savings than can be made. Sometimes the authorities cooperate with the industry associations to find good solutions and voluntary initiatives and information campaigns have often been launched by the industry on a voluntary basis. Kemikontoret, the association for the Swedish chemical industry, has issued written recommendations for how the Substitution Principle can be used in the Chemical industry. The labor unions, which have a very strong position in Sweden, strive for the replacement of chemicals that might pose a threat to workers' health. Examples of campaigns concern the substitution of certain flame retardants and nonyl phenol ethoxylates (Swedish Trade Union Confederation, 1999). Many big corporations work actively for the substitution of chemicals and have achieved significant results. While information campaigns put no formal pressure on the companies, pressure comes from other actors, such as the strong Swedish labor unions and the consumers. Many big Swedish companies are very concerned about their environmental image (and environmental stewardship) and are therefore inclined to take efforts when put under scrutiny.

Procurement: It is quite common to include criteria for chemicals in articles in public procurement. Many Swedish authorities do this on a regular basis, often using the tools developed by the Chemicals Agency to aid government and enterprises in their work with chemicals substitution. Statskontoret, the authority responsible for coordination of government procurement, has included criteria on chemicals content in relevant product groups, for example, personal computers. There are a great number of tools supporting green public procurement, developed by both state agencies and regional authorities.

Agreements: There is at least one example of how a formal agreement has been made between authorities and an industry sector to promote substitution. In 1994, Swedish EPA and the Swedish Petroleum Institute made an agreement were the industry would introduce a "green" gasoline (Miljöklass 2) in the whole of Sweden if the Swedish EPA could arrange a different tax rate on this type of gasoline.

Sometimes coercive measures have been considered necessary, when authorities have regarded the process of change as too slow in some industries. Often however, information campaigns can be a less controversial way to induce change, compared to the use of other instruments. Cooperation between different policy departments can lead to significant progress. As an example, a number of Swedish authorities were concerned about the increased use of antibacterial products in the late 1990s. In 2000, five Swedish authorities issued a joint press release, asking producers, retailers, and consumers not to market or buy unnecessary antibacterial products. The press release got massive response and set off a lot of actions, including a decision by a majority of the retailers not to sell consumer products that were marketed as antibacterial or that contained triclosan. This example shows how a joint initiative from a number of authorities may be very effective, and that cooperation is important in understanding and handling complex issues that involve the competence of many authorities.

The main question is of course: how successful have these efforts been in promoting substitution? A problem when the effectiveness of the principle is examined is of course that it is hard to say what the different actors would have done even if they were not legally obliged to consider the substitution of chemicals. Different studies and other bodies have however showed that the substitution policies aimed at certain industries and substances have been successful and achieved significant results (IVL, 1993; Kemikalieinspektionen, 1997; Swedish Trade Union Federation, 1999; Mont, 2001). Many types of products have undergone rapid changes, and this is to a large extent due to substitution policies from both the authorities and the industry. Product changes are clearly visible, for example, in different types of detergents and paints (Kemikalieinspektionen, 1997). For instance, substitution policies definitely speeded up the replacement of products based on solvents for products based on water.

Although significant results have been made, substitution is not the best choice in all situations. The following are some interim inferences:

- The complexity of the production system must be taken into account. If the substitution of a chemical causes large changes in the whole system there might be other negative impacts. An integrated approach is crucial to avoid replacing one risk with another.
- An LCA approach is sometimes needed to make a good choice.
- A substitution can lead to an increase in energy consumption.
- The gain of a substitution might vary in different workplaces.
- Sometimes the more toxic alternative can be the best one, if there are good methods to deal with it. Not only the toxicity but also the *risk* is to be evaluated, but this can sometimes be forgotten.

The Swedish experiences show that certain measures can improve the effectiveness of substitution policies (Dalhammar, 2000; Mont, 2001). These include the following:

- The regional and local authorities need guidelines from central authorities on how to promote substitution, and what areas to prioritize. The same goes for sellers and users of chemicals; in many cases, they might not know how dangerous the different substances are and how they can be substituted, and therefore need good guidelines.
- It is important to spread information about successful substitution efforts ("good examples").
- Information should not only be directed to professional users but also consumers need information about different chemicals and how these can be substituted.
- Public purchasing should further favor substitution by including demands on chemicals in tenders.

Why has Sweden managed to successfully promote substitution? Success factors include the following:

- The companies in Sweden are in general quite aware of environmental issues and strive to have a good environmental performance, and have a strong tradition of good legal compliance. Many large Swedish corporations support the Substitution Principle.
- The central environmental authorities have been able to make up lists of chemicals that should be phased out and provide information on how the substitution can be made. This is vital since the regional/local authorities and the users/sellers often need this information to know what to substitute and what alternatives are available.
- The regional and local authorities have an independent position, with considerate powers at their disposal; they have often been willing to challenge industries on substitution matters even when this has been controversial.

- The Swedish labor unions have a very strong position within the Swedish society and have a lot of power. They are very interested in the substitution of chemicals since these are often a serious threat toward the health and welfare of workers. Health and safety issues have a very long tradition in Sweden and especially big corporations are working very hard with these issues.
- Sweden does not have a great number of large chemical manufacturers: this means that there has been limited resistance, and antilobbying efforts, against substitution policies.

Criminal enforcement of the Substitution Principle is a contentious issue. While the principle can be criminally enforced according to the Environmental Code, criminal enforcement is difficult to use in practice, and so far no one has been charged because he did not substitute chemicals. There has however been at least one case where the environmental authorities have reported a suspected crime based on the fact that a substitution was not made (Kemikalieinspektionen, 1997). Criminal enforcement is still useful as a potential threat. It can be used in a situation where it is easy to prove that the user was aware of the risk of a substance and that there were alternatives. This would typically be the case when the authorities have informed the user of the toxicity of the chemicals used and that substitution possibilities do exist.

10.6 CONCLUDING REMARKS

There is an ongoing paradigm shift in the EU, the United States, and many other states, toward knowledge-driven policies where the burden-of-proof is placed primarily on the main manufacturers and users of chemicals, rather than on the authorities. However, as discussed in the chapter, authorities can do more to drive technological innovation.

Successful substitution policies have been pursued in Sweden and elsewhere, and it is important that past experiences are taken into account when REACH is implemented, otherwise its stated objective to promote substitution will fail, or at least be limited. The EU Member States must devote political will and resources for effective substitution policies, and information exchanges between relevant authorities and ECHA will be crucial. However, REACH will in any case mainly compel substitution when relevant substitutes are available on the market. Therefore, policies that make advance R&D are needed as well. Such policies can involve cooperation between policymakers, industries, universities, research institutes, and other actors as well, and could make use of technology procurement and other relevant policies in order to ensure that companies will be rewarded for green and innovative efforts.

In this era of international cooperation, there is a need for progressive national policies as well. Successful national policies will eventually influence the EU and international policies. In the EU chemicals policy, there is a drive toward harmonized legal standards. This means that the EU Member States are restricted in their national lawmaking, and should therefore rely also on other instruments, such as procurement policies, in order to drive innovations and promote chemical substitution.

REFERENCES

Bemelmans-Videc, M.-L., Rist, R., and Vedung, E. (Eds). 1997. *Carrots, Sticks & Sermons: Policy Instruments and their Evaluation.* Transaction Publishers. New Brunswick, N.J. & London.

Clean Production Action. 2003. Safer chemicals within REACH. Using the substitution principle to drive green chemistry. Report prepared for Green Peace Environmental Trust.

Dalhammar, C. 2000. The Swedish Product Choice Principle. Substitution of chemicals as a pollution prevention strategy. Working paper, IIIEE, Lund University.

Dalhammar, C. 2007a. An emerging product approach in environmental law: Incorporating the life cycle perspective. Doctoral dissertation, IIIEE, Lund University.

Dalhammar, C. 2007b. Product and life cycle issues in European environmental law: A review of recent developments. In *Yearbook of European Environmental Law Vol. 7.* Oxford University Press, New York.

Dalhammar, C. and Nilsson, A. 2005. Chemicals in articles—A look at current policy and law and recent trends. *The 7th Nordic Environmental Social Science Research Conference (NESS),* Gothenburg, June 15–17, 2005.

Edquist, C., Hommen, L., and Tsipouri, L. (Eds). 2000. *Public Technology Procurement and Innovation.* Boston/Dordrecht/London: Kluwer.

European Commission. 2009. Reviewing community innovation policy in a changing world. COM(2009) 442 final.

Forbes, R. 2009. The long arm of REACH: How to navigate through the compliance process. *European Energy and Environmental Law Review,* 18(1):34–49.

Gunningham, N. and Grabosky, P. 1998. *Smart Regulation. Designing Environmental Policy.* Oxford: Clarendon Press.

Heyvaert, H. 2006. Guidance without constraint: Assessing the impact of the precautionary principle on the European Community's chemicals policy. In *The Yearbook of European Environmental Law Vol. 6.* Etty, T. and Somsen, H. (Eds). Oxford: Oxford University Press.

Institutet för Vatten–och luftvårdsforskning (IVL) [Institute for Water and Air Research]. 1993. *Substitution av Farliga kemikalier—lösningen på miljöproblemen.* IVL, Stockholm.

Jänicke, M. 2005. Trend-setters in environmental policy: The character and role of pioneer countries. *European Environment,* 15:129–42.

Jänicke, M. and Jacob, K. 2004. Lead markets for environmental innovations: A new role for the nation state. *Global Environmental Politics,* 4(1):29–46.

Jörgens, H. and Tews, K. 2005. The global diffusion of regulatory instruments: The making of a new international environmental regime. *Annals of the American Academy of Political and Social Science,* 598(1):146–67.

Kemikalieinspektionen [Swedish Chemicals Agency]. 1997. *Förändringar i kemikalievalet.* PM nr 1/97.

Kemikalieinspektionen [Swedish Chemicals Agency]. 2005. *Information om varors innehåll av farliga kemiska ämnen.* Rapport 6/04, Kemikalieinspektionen, Stockholm (with English summary).

Kemikalieinspektionen [Swedish Chemicals Agency]. 2008. *Produktval, substitution och tillsyn.* Rapport 01/08, Kemikalieinspektionen, Stockholm (with English summary).

Koch, L. and Ashford, N. 2006. Rethinking the role of information in chemicals policy: Implications for TSCA and REACH. *Journal of Cleaner Production,* 14:31–46.

Mont, O. 2001. The Swedish product choice principle. *European Environmental Law Review,* 10(12):351–63.

Mont, O. and Dalhammar, C. 2006. Sustainable consumption: At the cross-road of environmental and consumer policies. *International Journal of Sustainable Development,* 8(4):258–79.

Naturvårdsverket [Swedish Environmental protection Agency]. 2005. *Miljöanpassad offentlig upphandling*. Rapport 5445, Naturvårdsverket, Stockholm.

Nilsson, A. 1997. *Att Byta Ut Skadliga Kemikalier. Substitutionsprincipen—En Miljörättslig Analys*. Nerenius and Santerus förlag, Göteborg.

Onida, M. 2004. Environmental protection by product policy: Focus on dangerous substances. ELNI Review No. 2/2004.

OXFORD Research A/S/IIIEE. 2006. *Evaluing of Program for renere Produkter*. Report to the Danish Ministry of Environment.

Pallemaerts, M. 2006. The EU and sustainable development: An ambiguous relationship. In *The European Union and Sustainable Development: Internal and External Dimensions*. M. Pallemaerts and A. Azmanova (Eds). Brussels: VUB Press.

Peters, P. 2009. SIEFs and dispute resolution. *European Energy and Environmental Law Review*, 18(4):204–208.

Purvis, M. and Bauler, J. 2004. *Irresponsible Care: The Failure of the Chemical Industry to Protect the Public from Chemical Accidents*. US Public Interest Research Group Education Fund, Washington, DC.

Schörling, I. 2004. *REACH What Happened and Why?* The Greens/European Free Alliance in the European Parliament.

Selin, H. 2007. Coalition politics and chemicals management in a regulatory ambitious Europe. *Global Environmental Politics*, 7(3):63–93.

Selin, H. and VanDeveer, S. 2006. Raising global standards. Hazardous substances and e-waste management in the European Union. *Environment*, 48(10):6–17.

Statens Offentliga Utredningar (SOU) 1996:103. [Swedish Public Inquiries].

Swedish Government proposition [Regeringens proposition om Miljöbalken]1997/98, p. 45.

Swedish Trade Union Federation. 1999. *Towards Safe Chemicals*. MB Communicate AB, Stockholm.

The Waxman Report. 2004. *A Special Interest Case Study: The Chemical Industry, The Bush Administration, and European Efforts to Regulate Chemicals*. Congressman Waxman, Special Investigations Division, United States House of Representatives Committee on Government Reform.

Vogel, D. 2003. The hare and the tortoise revisited: The new politics of consumer and environmental regulation in Europe. *British Journal of Political Science*, 33:557–80.

11 Integrated Numerical Modeling Software Tools for the Efficient Project Management of Sustainable Development Works in Solid Waste

Telemachus C. Koliopoulos

CONTENTS

11.1 INTRODUCTION

At the present time, complicated socioeconomic and environmental systems begin to cause problems when their environmental effects become an environmental public health risk. Environmental management is the discipline that is concerned with resources once society requires them. It is necessary to manage green chemistry's solid waste emissions in a sustainable way by minimizing the environmental impacts related to the operation of environmental systems. In an effort to meet growing environmental awareness, most companies include investments in their plans that are related to the production of eco-design, risk management, and the sustainable development and protection of the environment. However, the principles of signal processing and image processing can assist in the determination of the parameters required for the solution of particular geoinformation and environmental problems. Thus, improved monitoring and proper quality management tools of environmental systems are necessary [Alexander, 1996; Brimicombe, 2003; Brocks et al., 1994; Canter,

1996; Department of Environment (DOE), 1995; Gonzalez and Woods, 2002; Jain, 1989; Koliopoulos and Fleming, 2002; Wolf, 2000]. The following paragraphs discuss useful applications of numerical modeling software, which can be used in quality assurance and operational risk project management control of green chemistry's manufactures.

Solid waste green chemistry includes the International Standards Organization (ISO), which has published a series of certified systems including ISO 9001, 14001, and 18001 for the protection and certification of quality, environmental management, and health and safety, respectively (Fellows et al., 2002; Konstantinidis, 1998; Persse, 2006; Rothery, 1995). Monitoring schemes should be designed and inspections should be made frequently, utilizing proper systems analysis software tools especially in emergencies, not only to protect the optimum operation of solid waste green chemistry's eco-designs but also to support the sustainability of complicated environmental systems (Georgiadis, 2006; Gregoriadis et al., 2004; Koliopoulos, 2008a, 2008b, 2008c, 2009b, 2009c; Koliopoulos and Koliopoulou, 2007a, 2007b, 2009a). The continuous life cycle analysis of an environmental system is necessary.

11.2   PROJECT MANAGEMENT PROCESS AREAS

First, the project management process areas in landfill green chemistry need to be defined. These include the following stages:

- Project planning
- Project monitoring and control
- Supplier agreement management
- Integrated project management
- Risk management
- Quantitative project management

Project management applies to system analysis engineering, software engineering, and integrated product and process development (Anastasiou, 2005; Efraimidi, 2001; Fellows et al., 2002; Persse, 2006; Koliopoulos et al., 2007c; Ossenbruggen, 1984; Tersine, 1979). However, three goals should be established for project planning: creating formal estimates for the scope, size, duration, and resources needed for the project; developing a comprehensive project plan based on the estimates and management expectations; and reviewing the plan and obtaining commitment to it from team members, management, and stakeholders.

Two goals are defined for project monitoring and control: monitoring the actual performance and progress of the project against what has been documented in the project plan, and taking corrective actions when the project's performance or results deviate significantly from the plan, and managing these actions to closure. Moreover, the supplier agreement management has two goals: agreements with suppliers are established and maintained, and agreements with the suppliers that have to be satisfied by both the project and the supplier. Also, three goals govern the integrated project management: conducting the project using a set of defined processes that are tailored from the organization's set of standard processes, coordinating and collaborating with relevant stakeholders across the full life cycle of the project. Furthermore, for the risk management, the next stages for the identification of risks, risk quantification, risk allocation, and response should be identified. The source, pathway, and receptor should be analyzed so as to identify the risks from solid waste emissions at a landfill site.

11.3 SOLUTIONS TO THE INVESTIGATED PROBLEM

For the realization of the above-mentioned particular project management, several databases that control the life cycle of a technical, economical project in time taking the relative measures in order to achieve the successful completion of a technical green chemistry's project should be developed. SimGasRisk numerical modeling software could be used so as to analyze the life cycle analysis of biogas emissions (Koliopoulos and Koliopoulou, 2007a; Koliopoulos et al., 2007d). Moreover, the quantified landfill gas emissions can be manipulated properly, utilizing them so as to quantify leachate emissions and associated risks (Canter, 1996; Papatheodorou, 2001; Tchobanoglous et al., 1993; Koliopoulos, 2009b; Schnepp and Ganntt, 1999). Both landfill gas and leachate emissions could be compared with landfill settlements so as to establish an integrated database system for the life cycle analysis of landfill chemical emissions (Koliopoulos, 2009b; Koliopoulos et al., 2007d). Efficient landfill eco-designs should be preferred instead of uncontrolled dumps so as to achieve green chemistry's goals for the establishment of environmental sustainability (Koliopoulos et al., 2007d; Koliopoulos and Koliopoulou, 2007e; Tchobanoglous et al., 1993). Moreover, quality assurance computational software tools should be used for the control and safety of particular manufactures in the green chemistry's industry field, for example, reservoirs' stability located on slopes for leachates' collection treatment (Koliopoulos, 2009b, 2009c). All the related field data could be recorded on line using the proper software programming languages for the safe data transfer and lift in the internet so as to save time in data access of relative databases (Aitken, 2000; Deitel and Deitel, 2005; Gaddis and Irvine, 2009; McCarty, 1998).

In addition to the above edge detection, digital image processing algorithms should be used for the quick quality assurance of green chemistry's manufactures (Koliopoulos, 2008b, 2008c, 2009b, 2009c; Panorios, 2009). Some useful finite difference numerical solutions to the Laplace differential equation, mainly for educative purposes to the readers of this chapter are described by the following mathematical equation:

$$G(x,y) = \frac{\partial^2 f(x,y)}{\partial x^2} + \frac{\partial^2 f(x,y)}{\partial y^2}, \quad 0 < x, 0 < y, \tag{11.1}$$

where x and y depend on the number of pixels of the examining digital image.

The investigated problem has several applications in edge detection of structures and manufactures, and it can be combined with other risk assessment simulation model, that is, waste management risk analysis (Koliopoulos, 2008a, 2008b, 2008c,

2009b, 2009c). According to the finite difference numerical solution principle, a forward finite difference is applied for the term $\partial(x, y)/\partial x$ for the solution of Equation 11.1, which is as follows:

$$\frac{\partial f(x,y)}{\partial x} \rightarrow f_x(x,y) = f(x+1,y) - f(x,y). \tag{11.2}$$

Moreover, applying the backward finite difference, yields

$$\frac{\partial^2 f(x,y)}{\partial x^2} \rightarrow f_{xx}(x,y) = f(x,y) - f(x-1,y). \tag{11.3}$$

Equations 11.2 and 11.3 yield

$$\frac{\partial^2 f(x,y)}{\partial x^2} \rightarrow f_{xx}(x,y) = f(x+1,y) - 2f(x,y) + f(x-1,y). \tag{11.4}$$

In the same way, applying the proper finite difference schemes for the term $\partial^2 f(x, y)/\partial y^2$ yields

$$\frac{\partial^2 f(x,y)}{\partial x^2} \rightarrow f_{yy}(x,y) = f(x,y+1) - 2f(x,y) + f(x,y-1). \tag{11.5}$$

According to the finite difference numerical solution principle, the solution of Equation 11.1 is as follows:

$$\nabla^2 f(x,y) = f_{xx}(x,y) + f_{yy}(x,y), \tag{11.6}$$

or

$$\nabla^2 f(x,y) = f(x+1,y) + f(x-1,y) + f(x,y+1) + f(x,y-1) - 4f(x,y). \tag{11.7}$$

Equation 11.7 can be used for the weak filter representation of a given digital image.

Working in the same way, with different backward–forward finite difference schemes for the second-order finite differences, the solution of Equation 11.6 is as follows:

$$\nabla^2 f(x,y) = f(x+1,y) + f(x-1,y) + f(x,y+1) + f(x,y-1) \\ - 8f(x,y) + f(x-1,y-1) + f(x-1,y+1) \\ + f(x+1,y-1) + f(x-1,y+1), \tag{11.8}$$

or

$$\nabla^2 f(x,y) = f(x-1,y-1) + 2f(x-1,y) + f(x-1,y+1) + 2f(x,y-1)$$
$$- 12f(x,y) + 2f(x,y+1) + f(x+1,y-1) + f(x+1,y+1)$$
$$+ 2f(x+1,y), \tag{11.9}$$

where Equation 11.8 represents a strong Laplacian image processing filter and Equation 11.9 a very strong one. The above filters could be selected for a digital image filtering depending on the visual photoconstant points' identification according to the quality of recorded orthrophotos (Jain, 1989; Koliopoulos, 2008b, 2008c, 2009b, 2009c). The above-mentioned edge-detection image processing applications could be applied to any available digital datum spatial databases of specific reservoirs for the proper quality assurance and operational risk management control of manufactures under flood risk, for example, operation of flood safety pumps and proper waste management units' operation.

Moreover, another problem which is commonly met in project management and green chemistry's industry is the accurate numerical prediction of chemical indexes and spatial project management indexes in time. A solution to the latter problem can be given by proper time-series analysis based on the method of least-squares numerical modeling calculations. According to this methodology, an accurate prediction of trends in time series is made based on the particular numerical data and utilizing them properly taking into account the first-order and higher-order polynomial fitting-curve equations (Koliopoulos, 2009b, 2009c; Panorios, 2009).

In order to see the accuracy of the latter least-squares numerical modeling methodology and predictions for time series, the method of least squares based on surveying field data for the right monitoring of several surface compact units, which are located on slopes (i.e., leachate ponds and cylindrical reservoirs) can be used. Below is presented a methodology based on an experimental case study for the prediction of the movement of a cylindrical reservoir located on a given experimental slope by beans, lentils, and garbanzos (Koliopoulos, 2009c). The following formulae analyze the cylindrical reservoir's movement, for examining the experimental case study, describing the coordinates of a surveyed specific point on the cylindrical reservoir's cap due to several reasons (e.g., heavy traffic load). The movement of the cylindrical reservoir was monitored every 6 s putting a weight of 1 kg on the investigated slope. The field data include the first 36 s.

$$X(t) = 0.2086 + 8.085 \times 10^{-2} t + 2.206 \times 10^{-4} t^2, \tag{11.10}$$

$$Y(t) = 0.9943 + 2.54 \times 10^{-3} t + 2.646 \times 10^{-4} t^2, \tag{11.11}$$

where $X(t)$ and $Y(t)$ are the examining coordinates in time t seconds, the sum of squares equaling to 0.57 and 0.07 for $X(t)$ and $Y(t)$, respectively, and the standard error equaling to 0.38 and 0.13 for $X(t)$ and $Y(t)$, respectively, and with a mean algebraic sum of errors equal to 0.

On the basis of Equations 11.10 and 11.11, a prediction is made for the cylindrical reservoir's coordinates in the first 42s.

$$X(42) = 3.99 \text{ cm} \quad \text{movement on } x - \text{axis from a specific base point,}$$

$$Y(42) = 1.57 \text{ cm movement on } y - \text{axis from a specific base point,}$$

where the above results present an overestimation, as the real measured values were $X(42) = 3.95$ cm and $Y(42) = 1.25$ cm. The latter situation is good so as to take relative safety measures in time.

A prediction is also made by a back propagation artificial neural network (ANN), applying the relative Neurosolutions software, with one input being the time t, two outputs the x and y coordinates, and one hidden layer and the results are $X(42) = 3.26$ cm and $Y(42) = 1.45$ cm. Hence, by applying and examining the ANN architecture, there is an underestimation for x coordinate and an overestimation for y coordinate based on the real measured field data. Therefore, the method of least squares should be applied in terms of safety. Additional field data and ANN architectures should be examined by applying them properly either on real field data or on experimental ones. It can be seen that they are close to the ANN's results and the least-squares' ones. However, more field spatial data and photoconstant points are necessary in order to increase the accuracy in spatial analysis of moved manufactures on given slopes' morphologies. This methodology could be applied properly for the risk analysis either of manufactures or of slopes' stability under mobile heavy loads on given slopes (Koliopoulos and Koliopoulou, 2009a).

11.4 CONCLUSIONS

In this chapter, an overview of some useful software tools for several landfill green chemistry's project management applications in space have been presented. These tools provide accurate solutions in environmental sustainable development. Any numerical modeling software should include efficient numerical manipulation of schemes in their modules, and should be combined with any available dynamic simulation risk assessment tools, as these are necessary for evaluating the life cycle of a particular manufactured design and the associated operational project management, and also for demonstrating efficient sustainable designs that minimize any environmental impact on the receptors. Field data are of great importance, not only for making estimates, comparisons, and time-series predictions, but also for calibrating field data used in the mathematical numerical models to develop useful environmental assessment software tools and to take the right project management solutions and decisions for a given problem in time.

ACKNOWLEDGMENTS

The author acknowledges the UK Energy Technology Support Unit (ETSU, UK), Department of Trade and Industry (DTI) and UK Environment Agency (EA) for their support of the Mid Auchencarroch experimental project. The author also acknowledges the Office of the Surveyor in Northwestern Region in Nigeria for the collaboration extended. The conclusions expressed herein represent the findings of the author and are based on his expertise and experience in this area and his findings in the professional literature. It does not necessarily represent the views of EA or of the Office of the Surveyor.

REFERENCES

Aitken, P. 2000. *Programming with Visual Basic 6*. Arizona: Coriolis Publisher Group.

Alexander, N.A. 1996. *Engineering Mathematics*. London, UK: University of East London Press.

Anastasiou, T. 2005. *Economic and Technical Projects*. Athens, Greece: Ellin Publisher.

Brimicombe, A. 2003. *GIS, Environmental Modeling and Engineering*. Boston: Taylor & Francis.

Brocks, G., Yorks, L., and Ranney, G. 1994. *Beyond Total Quality Management*. New York: McGraw-Hill.

Canter, L. 1996. *Environmental Impact Assessment*. New York: McGraw-Hill.

Deitel, H. and Deitel, P. 2005. *Java how to Program*. New Jersey: Pearson Education Publisher.

Department of Environment (DOE). 1995. *A Guide to Risk Assessment and Risk Management for Environmental Protection*. London: HMSO.

Efraimidi, C. 2001. *Project Management*. Athens, Greece: Symmetria Publisher.

Fellows, R., Langford, D., Newcombe, R., and Urry, S. 2002. *Construction Management in Practice*. Oxford, UK: Blackwell Publishing.

Gaddis, T. and Irvine, K. 2009. *Visual Basic 2008*. New Jersey: Pearson Education Publisher.

Georgiadis, P. 2006. *Theory of Dynamic System Analysis*. Thessaloniki, Greece: Sofia Publisher.

Gonzalez, R.C. and Woods, R.E. 2002. *Digital Image Processing*. New Jersey: Prentice-Hall.

Gregoriadis, E., Doumpos, M., Zopounidis, K., and Matsatsinis, N. 2004. *Multicreterial Decision Making Analysis: Methodological Approaches and Applications*. Athens, Greece: New Technologies Publisher.

Jain, A.K. 1989. *Fundamentals of Digital Image Processing*. New Jersey: Prentice-Hall.

Koliopoulos, T.C. and Fleming, G. 2002. Application of life cycle analysis tools to sustainable waste management strategies. In: *Proceedings of International Solid Waste Association (ISWA) Congress*, Kocasoy, G., Atabarut, T., and Nuhoglu, I. (eds.), Vol. 1, pp. 355–61. Istanbul, Turkey: Bogazici University.

Koliopoulos, T.C. and Koliopoulou, G. 2007a. The use of input–output system analysis for sustainable development of multivariable systems. In: *American Institute of Physics Conference Proceedings*, Todorov, M. (eds.), Vol. 946, pp. 256–60. New York: American Institute of Physics Publisher.

Koliopoulos, T.C. and Koliopoulou, G. 2007b. Efficient numerical solution schemes combined with spatial analysis simulation models—diffusion and heat transfer problem. In: *American Institute of Physics Conference Proceedings*, Todorov, M. (eds.), Vol. 946, pp. 171–75. New York: American Institute of Physics Publisher.

Koliopoulos, T.C., Koliopoulou, G., and Axinte, C. 2007c. The use of efficient lining methods combined with numerical models for optimum project management of manufactures. *MOCM Journal*, 13: 379–84.

Koliopoulos, T.C., Kollias, V., Kollias, P., and Koliopoulou, G. 2007d. Evaluation of geotechnical parameters for effective landfill design and risk assessment. In: *Geotechnics Related to the Environment*, Sarsby, R.W. and Felton, A.J. (eds.), pp. 49–57. UK: Taylor & Francis Group.

Koliopoulos, T.C. and Koliopoulou, G. 2007e. Evaluation of optimum landfill design: Mid Auchencarroch experimental landfill emissions. In: *Wessex Institute of Technology Transactions on Computer Aided Optimum Design in Engineering (OPTI)*, Hernandez, S. and Brebbia, C.A. (eds.), pp. 231–39. Southampton, UK: W.I.T. Press.

Koliopoulos, T.C. 2008a. Landfill bioreactor's chemical balance and assessment of its biodegradation—Mid Auchencarroch experimental project. *RASAYAN Journal of Chemistry*, 1(1): 171–78.

Koliopoulos, T.C. 2008b. Carbon dioxide emissions' at Mid Auchencarroch experimental site and environmental impact assessment—Utilization of remote sensing and digital image processing software for an integrated Landfill Gas Risk assessment. *RASAYAN Journal of Chemistry*, 1(4): 766–73.

Koliopoulos, T.C. 2008c. Leachate emissions at Mid Auchencarroch experimental site & environmental impact assessment—Efficient spatial analysis utilizing remote sensing and digital image processing software for leachate monitoring. *RASAYAN Journal of Chemistry*, 1(4): 788–94.

Koliopoulos, T.C. and Koliopoulou, G. 2009a. *Asian Journal of Chemistry*, 21(4): 2989–3000.

Koliopoulos, T.C. 2009b. An efficient image processing software in geoinformatics and quality assurance of sustainable development projects. In: *Remote Engineering and Virtual Instrumentation International Conference Proceedings*, Gupta, N. and Sobh, T. (eds.), pp. 142–47. CT, USA: Bridgeport University Publishers.

Koliopoulos, T.C. 2009c. Meeting the challenges of digital image processing software in modern bioinformatics and geoinformatics engineering demands in the new global economy. In: *A.S.E.E. International Conference Proceedings*. CT, USA: Bridgeport University Publishers, paper 2, Session 29.

Konstantinidis, P. 1998. *Project Management*. Athens, Greece: Ministry of Education Publisher.

McCarty, B. 1998. *SQL Database Programming with Java*. Arizona: Coriolis Publisher.

Ossenbruggen, P. 1984. *Systems Analysis for Civil Engineers*. New York: Wiley.

Panorios, N. 2009. *Using Software for Digital Image Processing of Orthophotomaps and Geographical Analysis of Mountainous Regional Protection in Attica Prefecture*. Thesis, Technological Educational Institute of Athens, Greece.

Papatheodorou, T. 2001. *Algorithms*. Greece: University of Patras.

Persse, J. 2006. *Process Improvement Essentials*. California: O'Reilly Publisher.

Rothery, B. 1995. *ISO 14000 and ISO 9000 Series of Quality Standards*. London, UK: Gower Publishers.

Schnepp, R. and Gantt, P. 1999. *Hazardous Materials, Regulations, Response & Site Operations*. Kentucky: Delmar Publisher.

Tchobanoglous, G., Theisen, H., and Vigil, S.A. 1993. *Integrated Solid Waste Management, Engineering Principles and Management Issues*. New York: McGraw-Hill.

Tersine, P.J. 1979. *Management of Materials and Logistics*. New York: Elsevier Publishers.

Wolf, P. 2000. *Elements of Photogrammetry and Elements of G.I.S.* New York: McGraw-Hill.

12 Necessity of Risk-Assessment Tools in Solid Waste Emissions in Analyzing Environment and Public Health Protection

Georgia Koliopoulou

CONTENTS

12.1 INTRODUCTION

At the present time, the sustainability of an environmental system is heavily dependent on hazards' mitigation, energy savings, recovery of waste emissions, and their exploitation, taking into account the particular systems' characteristics as to the management of the necessities to better natural resources (Aldrich and Griffith, 1993; Canter, 1996; Daszak et al., 2000; Elliott et al., 2001; Friis and Sellers, 2004; Koliopoulos and Koliopoulou, 2007a, 2007b, 2007c; Koliopoulos and Koliopoulou, 2009; Rothery, 1995; Schnepp and Gantt, 1999; Tchobanoglous et al., 1993). Moreover, the specific flow of goods or waste-management stream characteristics over time in a given geographical area could be studied using signal processing theory and input–output theory to determine useful socioeconomic parameters. Image processing can determine the demands of resources for a particular input–output problem, based on the number of consumers of the resources per pixel or unit area (Koliopoulos, 2009a, 2009b; Koliopoulos and Koliopoulou, 2006; 2007d; Wolf, 2000). Hence, the principles of signal processing and image processing can assist in the determination

277

of the parameters required for the solution of particular geoinformation and environmental problems.

Risk-assessment tools are necessary not only to protect building properties, manufactures, infrastructure works, and environmental sustainable projects but also to set up efficient monitoring control systems for preventing and avoiding any emerging hazards and ecoterrorism acts (Canter, 1996; Koliopoulos et al., 2007, 2009a; Koliopoulos and Koliopoulou, 2007a, 2007c, 2007e, 2007f, 2007g, 2008; Lawrence, 2003). Spatial numerical modeling software diagnostic tools for the life cycle analysis of solid waste green chemistry's emissions should be used for the maintenance and control of particular ecodesign environmental systems, developing proper databases, solving effectively associated project management engineering problems and protecting public health indexes, and monitoring food chain quality from associated risks (Brimicombe, 2003; Brocks et al., 1994; Canter, 1996; DOE, 1995; Koliopoulos, 2000; Koliopoulos, 2008a, 2008b; Koliopoulos and Koliopoulou, 2007a, 2007f; Koliopoulos et al., 2003; Kortepeter, 2001).

12.2   UTILIZING RISK-ASSESSMENT TOOLS TO PREVENT PUBLIC HEALTH HAZARDS

At present times, many efforts in the green chemistry industry are made in order to increase the monitoring, surveillance, identification, life cycle analysis of goods, and reporting of disease agents, and to better understand the potential dynamics of disease transmission within human and animal populations in both industrialized and developing environments. Both landfill gas and leachate emissions could be compared with landfill settlements so as to establish an integrated monitoring database for landfill quality management and the life cycle analysis of landfill chemical emissions (Koliopoulos et al., 2007; Koliopoulos and Koliopoulou, 2007f; Koliopoulos, 2008a, 2008b). Efficient landfill ecodesigns should be preferred instead of uncontrolled dumps so as to achieve green chemistry's goals for the establishment of environment sustainability (Koliopoulos et al., 2007). Also, quality assurance computational software tools should be used for the control and safety of particular manufactures in the green chemistry's industry field, for example, reservoirs' stability located on slopes for leachates' collection treatment (Koliopoulos, 2009a, 2009b). Moreover, genetically or cultivated engineered biological organisms are ranked by some analysts as the most potentially dangerous of all existing weapons technologies, nuclear weapons notwithstanding (DOE, 1995; Elliott et al., 2001; Friis and Sellers, 2004; Koliopoulos and Koliopoulos, 2008).

In the latter cases, a proper simulation risk-assessment software should be designed, taking into account the particular constraints, physicochemical properties of a multivariable environmental system, installing a proper rehabilitation network for the prevention and right confrontation of the toxic emissions. The use of a proper risk-assessment methodology is seen as the principle of achieving the goal of biosecurity against bioterrorism acts, thereby minimizing any insecurity to the sustainable development of our society and taking the right civil defense measures in time (Koliopoulos and Koliopoulou, 2007d, 2007e, 2007f; Koliopoulos et al., 2007), and avoiding any long-term biogas or leachate hazardous thresholds in landfill management that would otherwise infringe with environmental sustainability. Proper spatial system analysis and combinations of robust, quick, and efficient lining methods for emergency monitoring and set up of biosecurity control systems, and the installation of proper sampling devices of particular landfill emissions on critical locations should take place on bioterrorism's probable spatial target zones for avoiding any forms of asymmetric chaotic treatment (Canter, 1996; Koliopoulos and Koliopoulos, 2008; Koliopoulos, 2008a, 2008b; Lawrence, 2003).

12.3 USE OF SPATIAL SIMULATION TOOLS FOR POTENTIAL SAFETY MANAGEMENT OF CHEMICAL HAZARDS

Volatile organic carbon (VOC) pollutants' emissions and leachate emissions, which can be found in gaseous and in liquid phases, respectively, at landfill sites, should be monitored and managed properly (DOE, 1995; Tchobanoglous et al., 1993; Koliopoulos, 2000; Rothery, 1995; Schnepp and Gantt, 1999). However, the prevention of environmental accidents and bioterrorism acts and the realization of disaster management measures are becoming a necessity for the avoidance of any associated public health's threat. Hence, a proper numerical modeling software should be used for the risk analysis, risk management, and prevention of any associated risks (Koliopoulos and Koliopoulou, 2007a, 2007b, 2007c, 2007d, 2007f, 2007g; Koliopoulos, 2008a, 2008b).

For air pollutants associated with public health problems, Equation 12.1 can be used taking into account a landfill gas concentration C of an air pollutant in the atmosphere along the central line of axis x from its source location, bearing in mind the source of the air pollutant on height H from the ground surface. In this way, the right safety measures for examining the spatial analysis of air pollutant's concentration on top soil could be taken (Davis and Cornwell, 1998; Koliopoulos and Koliopoulou, 2007f).

$$C(x,0,0,H) = \frac{Q}{\pi u \sigma_y \sigma_z} \exp\left(-\frac{1}{2}\frac{H^2}{\sigma_z^2}\right), \tag{12.1}$$

where σ_y and σ_z are dispersion coefficients dependent on x, and their relation is presented below. Also, the concentration C of an examining air pollutant, for example, carbon dioxide, VOC's emissions along the cross section to the central line at location (x, y) from landfill boundary, could be calculated by Equation 12.2, in order to take the right public health protection measures in time, taking the source of air pollutant on a height difference H related to the location (x, y) of the receptor (Davis and Cornwell et al., 1998; Koliopoulos and Koliopoulou, 2007f).

$$C(x,y,0,H) = \left[\exp\left(-\frac{1}{2}\frac{y^2}{\sigma_y^2}\right)\right]\frac{Q}{\pi u \sigma_y \sigma_z}\exp\left(-\frac{1}{2}\frac{H^2}{\sigma_z^2}\right), \tag{12.2}$$

where Q is the gas emission from the source (kg s^{-1}), C the air pollutant concentration at location x, y from a height H (kg m^{-3}), u the wind velocity (m s^{-1}), and x is defined by the respective x distance on the x-axis from the source of the air pollutant to a nearby civic, industrial, or agricultural land-use receptor area, based on the particular spatial data of the examining proposed landfill sites' areas. The selected x distance is applied on relative graphs so as to determine the respective dispersion coefficients in relation to atmospheric stability conditions (Davis and Cornwell, 1998; Koliopoulos and Koliopoulou, 2007f). y is defined as the transverse distance above the selected x location for which the air pollutant concentration is calculated. The y distance is based on particular map data of nearby land uses next to the proposed landfill sites' areas under examination. The selected y distance is applied on a relative graph of σ_z dispersion coefficient versus distance so as to determine the respective value of dispersion σ_z coefficient in relation to atmospheric stability conditions (Davis and Cornwell, 1998; Koliopoulos and Koliopoulou, 2007f). σ_y and σ_z dispersion coefficients are calculated based on relative graphs for the atmospheric stability categories that are selected based on meteorological conditions.

The establishment of an associated risk-assessment spatial analysis software using proper modules could provide a right solution to an examining public health problem for analyzing the particular factors and parameters that affect environmental sustainability by solid waste chemical emissions of green chemistry's ecodesigns (Koliopoulos, 2000, 2008a, 2008b, 2009a, 2009b; Koliopoulos and Koliopoulou, 2007a, 2007e, 2007f, 2008; Koliopoulos et al., 2007). Hence, the thresholds of a migrating explosive gas concentration could be calculated by the solution of the diffusion–advection gas flux in a porous medium problem, which is described in Equation 12.3.

$$R\frac{\partial C}{\partial t} = D\frac{\partial^2 C}{\partial x^2} - V\frac{\partial C}{\partial x} \qquad (12.3)$$

where C is the gas concentration by volume, x the distance along the migration pathway (m), t the time (s), D the diffusion coefficient (m^2 s^{-1}), V the velocity (m s^{-1}), and R the retardation factor. Moreover, according to the literature and field data measurements, the fundamental principles of gas migration in porous media could be studied and a relative estimation could be realized for the calculation of lateral distance of a gas migration and probable explosions of it. Indicatively, assuming that the threshold of 0.1 biogas (50% methane + 50% carbon dioxide) concentration by volume will take place in time t since the gas has started migrating from its source boundaries and taking the values of the advection velocity and diffusion coefficient constant, then the lateral distance of gas migration can be calculated (Koliopoulos, 2000; Koliopoulos and Koliopoulou, 2007c). Sometimes due to the anisotropic complex nature of the geological strata adjacent to gas migrated source boundaries, there are cases that are difficult to monitor by gas probes. Hence, in such cases it is preferable for safety reasons to investigate firstly the calculated half threshold of 0.5 gas concentration by volume, as in many cases, the respective measured real threshold value is equal to half or more than half of the calculated one. Taking into account all the above, any sustainable development land uses of agricultural products or other anthropogenic resources next to landfill boundaries could be initiated in a safe manner and with proper planning for ensuring this safety. Also, quality assurance of landfill settlements, time-series analysis, signal processing tools, and prediction of probable leachate migration in time in case of landfill design accidents should be simulated in advance using proper software tools so as to take the right measures in time and avoid any associated risks and chemical pollution threats (Canter 1995; Koliopoulos, 2008a, 2008b, 2009a, 2009b; Lawrence, 2003).

According to the above discussion, the proper lining of an efficient integrated safety project management system should take place so as to provide an efficient environmental sustainability of solid waste green chemistry's ecodesigns and confrontation of any associated risks to public health at landfill sites and for nearby areas. Several scientific teams from different disciplines should collaborate in order to give the right confrontation solutions for the several and different risk scenarios (including study source, pathway, and receptor routes) on given topographies with particular multivariable characteristics. The main steps that should be followed in the case of estimating landfill emissions for sustainable development works and project management of remote engineering life cycle analysis in structures for a landfill site include the following:

- Use proper remote engineering tools and online web applications
- Use numerical modeling software of produced landfill emissions for monitoring and quality management of landfill sites
- Develop and use properly spatial analysis and image processing software tools for the right project management and maintenance of ecodesigns
- Evaluate life cycle analysis of chemical emissions, utilize proper system analysis tools—numerical models, and propose the right economic development solutions for environmental sustainability

Moreover, the proper image processing can be used as a risk-assessment utility combining its results with efficient risk-assessment numerical models such as SimGasRisk 1 for the assessment of landfill emissions in time and the project management of relative sustainable development works (Koliopoulos, 2009a, 2009b; Koliopoulos and Koliopoulou, 2007b, 2007c). Therefore, landfill gas migration and leachate leakage on a given topography could be assessed accurately so as to protect flora and fauna, public health, agrotourism production works, and ecotourism development works, all with a common objective as to achieve a sustainable economic development. The next steps that should be followed for an efficient sustainable development on a given geography and its economic resources are the following:

- Utilize proper software so as to estimate waste emissions and landfill settlements
- Time-series analysis of chemical and spatial indexes for good timing in taking strategic management acts and mitigation of chemical hazards
- Utilize proper software so as to determine landfill sitting and to propose sustainable economic management solutions
- Efficient project management of technical works at a landfill site for public health protection
- Project management of landfill emissions' monitoring works (e.g., biogas production, leachates, and settlements) and register online chemical indexes

and surveying indexes. Compare the *in situ* field data with the numerical results of relative software utilities so as to realize an efficient environmental impact assessment and to take the right acts on time

- Utilize image processing tools for monitoring quality assurance of sustainable development projects (e.g., monitoring landfill gas migration, leachate leakage, settlements, drainage canals, and maintenance), application of ISO 14001, and slope stability control on given topographies

Thus, the use of an efficient landfill emissions' assessment software for a given waste production is necessary for the proper management of sustainable development works. However, artificial neural networks (ANNs) can be utilized for landfill site monitoring or other related brownfield sites to develop threshold maps and to determine risk zones for the proper design of further reclamation works. The ANN results are commonly satisfactory, while they should be combined with other available numerical models for better environmental impact assessments. ANN effectiveness depends not only on their architecture but also on the reliability of investigated parameters and the quantum of data for training the ANN. Also, efficient quality assurance software should be used for the proper project management of sustainable development works. However, more field spatial data and photoconstant points are necessary in order to increase the accuracy in spatial analysis of displaced structures and deformed manufactures on given slopes' morphologies. Efficient stereometric methodologies could be applied properly for the risk analysis either of manufactures (particular structures), trusses' quality assurance, or slopes' stability under mobile heavy loads on given slopes (Koliopoulos, 2008a, 2008b, 2009a, 2009b).

12.4 CONCLUSIONS

Field data are of great importance both in bioinformatics and geoinformatics software applications not only for making estimations, comparisons, and predictions, but also for calibrating field data in mathematical numerical models in order to develop useful simulation numerical spatial tools, which ultimately help take the right measures for a given topography on time.

Proper use of software is necessary for the simulation of solid wastes emissions' in green chemistry's ecodesigns to evaluate the life cycle of particular manufactured ecodesigns and proper project management of sustainable resources that will be able to minimize any environmental impacts to a variety of receptors. Special care should be given during the system analysis in the proper input–output simulation, right determination of chemical and waste biodegradation indexes, and constraints of the related problem taking fully into account the particular topographic characteristics of the area under investigation.

REFERENCES

Aldrich, T. and Griffith, J. 1993. *Environmental Epidemiology and Risk Assessment*. New York: Van Nostrand Reinhold.
Brimicombe, A. 2003. *GIS, Environmental Modeling and Engineering*. Boston: Taylor & Francis.

Brocks, G., Yorks, L., and Ranney, G. 1994. *Beyond Total Quality Management*. New York: McGraw-Hill.

Canter, L. 1996. *Environmental Impact Assessment*. New York: McGraw-Hill.

Daszak, P., Cunningham, A.A., and Hyatt, A.D. 2000. Emerging infectious diseases of wildlife—threats to biodiversity and human health. *Science*, 287:443–49.

Davis, M. and Cornwell, D. 1998. *Introduction to Environmental Engineering*. New York: McGraw-Hill.

Department of Environment (DOE). 1995. *A Guide to Risk Assessment and Risk Management for Environmental Protection*. London: HMSO.

Elliott, P., Briggs, D., Morris, S., Hoogh, C., Hurt, C., Jensen, T.K., Maitland, I., Richardson, S., Wakefield, J., and Jarup, L. 2001. Risk of adverse birth outcomes in populations living near landfill sites. *BMJ*, 323:363–68.

Friis, R.H. and Sellers, T.A. 2004. *Epidemiology for Public Health Practice*. Boston, Massachusetts: Jones and Bartlett Publishers.

Koliopoulos, T.C. 2000. Numerical modelling of landfill gas and associated risk assessment. Thesis, Department of Civil Engineering, University of Strathclyde, Glasgow, Scotland, UK.

Koliopoulos, T.C. 2008a. Carbon dioxide emissions' at Mid Auchencarroch experimental site and environmental impact assessment—Utilization of remote sensing and digital image processing software for an integrated landfill gas risk assessment. *RASAYAN Journal of Chemistry*, 1(4):766–73.

Koliopoulos, T.C. 2008b. Leachate emissions at Mid Auchencarroch experimental site & environmental impact assessment—efficient spatial analysis utilizing remote sensing and digital image processing software for leachate monitoring. *RASAYAN Journal of Chemistry*, 1(4):788–94.

Koliopoulos, T.C. 2009a. An efficient image processing software in geoinformatics and quality assurance of sustainable development projects. In: *Remote Engineering and Virtual Instrumentation International Conference Proceedings*, pp. 142–47. Connecticut: Bridgeport University.

Koliopoulos, T.C. 2009b. Meeting the challenges of digital image processing software in modern bioinformatics and geoinformatics engineering demands in the new global economy. In: *A.S.E.E. International Conference Proceedings*. Connecticut: Bridgeport University, paper 2, Session 29.

Koliopoulos, T.C. and Koliopoulou, G. 2006. Controlling landfill emissions for environmental protection: Mid Auchencarroch experimental project. *Asian Journal of Experimental Science*, 20(2):233–42.

Koliopoulos, T.C. and Koliopoulou, G. 2007a. Evaluating landfill chemical emissions–Mid Auchencarroch experimental design. *Asian Journal of Chemistry*, 19(5):3911–17.

Koliopoulos, T.C. and Koliopoulou, G. 2007b. Evaluating biomass temperature vs biodegradation for environmental impact minimization: Mid Auchencarroch experimental landfill. *Asian Journal of Experimental Science*, 20(2):43–54.

Koliopoulos, T.C. and Koliopoulou, G. 2007c. Risk analysis of landfill gas emissions: A report on Mid Auchencarroch project. *Asian Journal of Experimental Science*, 21(2): 215–26.

Koliopoulos, T.C. and Koliopoulou, G. 2007d. The use of input–output system analysis for sustainable development of multivariable systems. In: *American Institute of Physics Conference Proceedings,* S. Hernandez, C.A. Brebbia (eds), 946, pp. 256–60. Southampton, UK: WIT Press.

Koliopoulos, T.C. and Koliopoulou, G. 2007e. Evaluation of optimum landfill design: Mid Auchencarroch experimental landfill emissions. In: *Wessex Institute of Technology Transactions on Computer Aided Optimum Design in Engineering (OPTI)*, S. Hernantez and C.A. Brebbia (eds), pp. 231–39. Southampton, UK: WIT Press.

Koliopoulos, T.C. and Koliopoulou, G. 2007f. A diagnostic model for MSW landfill's operation and the protection of ecosystems with a spatial multiple criteria analysis–Zakynthos Island Greece. In: *Wessex Institute of Technology Transactions on Ecosystems and Sustainable Development*, E. Tiezzi, J.C. Marques, and C.A. Brebbia (eds), pp. 449–62. Jorgensen S.E., UK: WIT Press.

Koliopoulos, T.C. and Koliopoulou, G. 2007g. Efficient numerical solution schemes combined with spatial analysis simulation models—Diffusion and heat transfer problem. In: *American Institute of Physics Conference Proceedings*, Michail Todorov (ed.), 946, pp. 171–75. New York: American Institute of Physics Publisher.

Koliopoulos, T.C. and Koliopoulou, G. 2008. Prevention acts to bioweapons & bioterrorism— The use of spatial risk assessment tools for control systems' Disaster Management, A.L. Bhatia (ed.), pp. 297–307. Jaipur, India: Kulshrestha Pointer Publishers (http://www.vedamsbooks.com/no59575.htm).

Koliopoulos, T.C. and Koliopoulou, G. 2009. Biodegradation of iso-valeric acid in relation to other chemical indexes and spatial liner risk assessment at landfill topographies—mid auchencarroch experimental site. *Asian Journal of Chemistry*, 21(4):2989–3000.

Koliopoulos, T.C., Kollias, V., and Kollias, P. 2003. Modelling the risk assessment of groundwater pollution by leachates and landfill gases, U.K. Wessex Institute of Technology, In: *Wessex Institute Transactions on Water Pollution VII, Modelling, Measuring and Prediction.* C.A. Brebbia, D. Almorza, and D. Sales (Eds), pp. 159–69. UK: WIT Press.

Koliopoulos, T.C., Kollias, V., Kollias, P., and Koliopoulou, G. 2007. Evaluation of geotechnical parameters for effective landfill design and risk assessment. In *Geotechnics Related to the Environment.* Sarsby and Felton (eds), pp. 49–57. UK: Taylor & Francis Group.

Kortepeter, M., Christopher, G., Cielsak, T., Culpeper, R., Darling, R., Pavlin, J., Rowe, J., McKee, K.J., and Eitzen, E.J. 2001. *Medical Management of Biological Casualties Handbook* (4th edition). United States Army Medical Research Institute of Infectious Diseases, Fort Detrick, Frederick, Maryland.

Lawrence, D. 2003. *Environmental Impact Assessment.* New Jersey: John Wiley & Sons.

Rothery, B. 1995. *ISO 14000 and ISO 9000 Series of Quality Standards.* London, UK: Gower.

Schnepp, R. and Gantt, P. 1999. *Hazardous Materials, Regulations, Response & Site Operations.* Florence, Kentucky: Delmar Publishers.

Tchobanoglous, G., Theisen, H., and Vigil, S.A. 1993. *Integrated Solid Waste Management, Engineering Principles and Management Issues.* New York: McGraw-Hill.

Wolf, P. 2000. *Elements of Photogrammetry and Elements of G.I.S.* New York: McGraw-Hill.

13 Green Eluents in Chromatography

Ali Mohammad and Nazrul Haq

CONTENTS

13.1 INTRODUCTION

Solvents are important components of nature providing one or more liquid phases for chemical reactions and processes. Solvents are used to aid in mass and heat transfer and to facilitate separations and purifications. While some solvents are available from nature even in large quantities, most of the solvents are man-made. Solvents have been used extensively to provide one or more liquid phases for chemical reactions, regulate temperatures, moderate exothermic reactions, clean equipment and clothing, isolate and purify compounds by recrystallization or extractions, generate azeotropes for separation and assist structural and/or analytical characterization of chemicals (Reichardt, 2007). Solvents can have significant effects on the outcome of chemical reactions and physicochemical processes including extractions and crystallizations. Both the macroscopic (boiling point density) and microscopic (dipole moment and hydrogen bonding ability) properties of the solvent affect its influence on such processes and the choice of solvent for a chemical system. Many of the macroscopic physical properties of the solvent are dependent also on the molecular structure of the solvent molecules.

Solvents are often volatile organic compounds (VOCs) and are therefore a major environmental concern as they are able to form low-level ozone and smog through free radical air-oxidation processes (Lancaster, 2002). Also, they are often highly flammable and can cause a number of adverse health effects including eye irritation, headaches, and allergic skin reactions. Some VOCs are suspected carcinogens. One of the 12 principles of green chemistry stipulates the use of safer solvents and auxiliaries (Clark and Macquarrie, 2002). It is, therefore not surprising that over the last 10 years, chemistry research regarding the use of greener and alternative solvents has grown enormously (Nelson, 2003; Sheldon, 2005).

In most of the cases, it is not possible to perform a reaction without a solvent. In these cases, therefore, use of VOCs should be minimized. According to Hofer and Bigorra (2007), the demand for hydrocarbon and chlorinated solvents during the last decade is on a downward trend as a result of environmental regulations and public interests, with oxygenated and green solvents replacing them to a large extent. But, it is unrealistic to think that all VOCs can be replaced in every application and therefore there is a growing role for VOCs derived from renewable sources in the alternative solvent field. Green solvents are environment-friendly solvents or biosolvents, which are derived from processing of agricultural crops. Green solvents were developed as a more environment-friendly alternative to petrochemical solvents, for example, ethyl lactate, an ester of lactic acid, is a green solvent derived from processing cane. There is no perfect green solvent that can be applicable to all situations and therefore judicious decisions have to be taken in this regard.

Just because the solvent is nontoxic, it does not make the process "green." There is also a need to consider the energy required for processes such as evaporation, pressurization, and stirring for the solvent recycling, solubility of the solute, toxicity of any by-products, atom efficiency of the process, product separation, and whether the solvent affects the product/package. In recent years, efforts have been made to quantify the greenness of a wide range of solvents. Both green and common organic media were considered. A computer-aided method of organic solvent selection for reactions has been developed by Capello et al. (2007) using solvent reaction properties and solvent environment properties. Among the solvent environment properties, the two most important considerations were environment, health, and safety (EHS), and the life cycle assessment (LCA). EHS properties of a solvent include its ozone depletion potential, biodegradability, toxicity, and flammability. Fischer and Hungerbühler (2000) have developed a solvent assessment method based on EHS criteria, while the LCA approach evaluates the environmental burdens of a product, process, or activity; quantifies the resource use and emissions; assesses the environmental and human health impacts; and evaluates and implements opportunities for improvements. All alternative solvents have advantages and disadvantages. In all areas, there is a need to balance the technical advantages of a particular solvent with any environmental costs or other disadvantages. Manufactures and consumers need to decide if the advantages outweigh the disadvantages. Some of the important green solvents are discussed in the sections that follow.

13.2 POLYETHYLENE GLYCOL

The low solubility of organic molecules has been the main reason for a restricted use of water as a solvent in chemical analysis. It has been found that aqueous polyethylene glycol (PEG) solutions are a better alternative for use as green solvents. PEG $(HO-(CH_2CH_2O)n-H)$ is available in a variety of molecular weights (200 to tens of thousands). At ambient temperature, a PEG–water solution and hygroscopic polymer (molecular weight <600) are colorless, nonvolatile, and viscous liquids. Liquid PEG is miscible with water in all proportions and solid PEG is highly soluble in water. PEG has a number of benign characteristics, which are useful in bioseparations (Albertsson, 1986). Aqueous solutions of PEG are biocompatible and are used in tissue culture media for the preservation of organs. Some other attractive features of PEGs include (a) low flammability, (b) biodegradability, (c) reasonable resistance to certain acids and bases, (d) thermal stability, (e) resistance to oxidizing (O_2, H_2O_2) and reducing ($NaBH_4$) agents, and (f) low toxicity.

The unique character of PEG's chain, with its interaction properties, capacity of dissolving salts, and forming electrolyte complexes, has drawn interest about the use of PEGs in several frontal areas of research. PEG has been successfully applied as a green solvent in activator generated by electron transfer–atom transfer radical polymerization (AGET–ATRP) of methyl methacrylate (Hu et al., 2009). In the field of chromatography, PEG has been used as (a) general purpose polar stationary phase and (b) as an additive of mobile phase. PEG have been widely used as a general polar stationary phase in gas chromatography (GC), high-performance liquid chromatography (HPLC), and electrophoresis for the separation of volatile polar organic compounds such as flavors and fragrances, metal cations, and inorganic anions. Polymeric packing materials used in aqueous HPLC are found to be superior stationary phases to silica-based packing materials with respect to the chemical resistance and the reproducibility (Kurosu et al., 1984). With a PEG packing, characterized by other groups, the total analysis time is reduced due to the combination of increased retention for polar compounds and decreased retention of nonpolar compounds.

Ogino et al. (1995) have prepared oligo (ethylene glycol) (OEG)-based monodisperse hydrophilic polymer gel beads for their use as packing materials for aqueous size-exclusion chromatography (SEC) columns. In another study, Rong and Takeuchi (2004) examined PEG as the stationary phase for the separation of iodide. Guo et al. (2008) introduced click OEG as a good reversed-phase stationary phase. Though the structure of click OEG is identical to C18 phases, the presence of four ether bonds, a hydroxyl group, and a triazole ring on the click OEG phase make it susceptible for polar interactions such as hydrogen bonding and dipole–dipole interaction. In another study, Valiokas et al. (1999) demonstrated the ability of OEG molecules to form helix-like conformation in organic–aqueous media. Because of the multiple interactions and uniqueness of structure, OEG may bring about different separation mechanisms compared to that of C18. Furthermore, the PEG-based stationary phase was found suitable for the analysis of the active compounds in *H. perforatum* extracts and natural products (Pellati et al., 2005). PEG-based systems have been frequently used in the past for the analysis of cations by electrophoresis. Simunicova et al. (1994) demonstrated the separation of analyte cations by isotachophoresis using water–PEG (60:40 v/v) and 3 mM maleic acid at pH 5.0. According to Kaniansky et al. (1990), a volume fraction of 40–45% v/v of PEG in water was effective for the separation of alkali and alkaline earth metal cations. As PEG is a linear analog of cyclic polyether and is flexible in structure, the complex formation with the cations is not specific and is weak compared to that of their corresponding cyclic polyether. Ito and Hirokawa (1996) have studied the separation behavior of alkali and alkaline earth metal and NH_4^+ cations by capillary electrophoresis using PEG and tartaric acid as electrolyte systems. It was observed that the mobility of cations complexed with PEG was influenced by the changes in cation size and dielectric constant as well as by the viscosity of water–PEG mixtures. For alkaline earth metal cations, the addition of PEG retarded the migration in the order of $Mg^{2+} > Ca^{2+} > Sr^{2+} > Ba^{2+}$. The retardation effect by PEG was larger for alkaline metals than that for the alkali metal cations, suggesting a stronger complication by PEG for alkaline earth metal cations with divalent charge.

Aqueous solutions of polymers have proved to be novel solvents by virtue of their ability to solubilize otherwise poorly soluble species without the involvement of organic solvents (Huddleston et al., 1999). These wholly aqueous solvent extraction systems (WASE) represent environmentally benign alternatives to the conventional organic solvents in liquid–liquid extraction processes such as aqueous biphasic extraction chromatography (ABEC). Aqueous two-phase systems consisting of mixtures of two structurally different polymeric materials have been most suitable as environment-friendly solvents for the separation of biomaterials due to high water content in both phases (Walter and Johansson, 1994). The attractive feature of aqueous two-phase extraction is that cell separation, product purification, and concentration process can be performed in a single step. For industrial and large-scale enzyme extractions, PEG-salt systems have been mainly used and these have been successful. PEG has also found application as a mobile-phase additive in the chromatographic analysis of polar compounds. Kazoka and Shatz (1996) have showed that chromatographic systems with unmodified silica and binary mobile phases consisting of ethyl acetate and the weakly soluble polar components (ethylene glycol and/or formamide) have been useful in the chromatography of purine and pyrimidine derivatives.

To improve the chromatographic performance of unmodified silica in the analysis of highly polar solutes, solvent generated liquid–liquid systems are often employed. In general, two or three immiscible solvents are equilibrated and after the separation of layers, the less polar layer is used as a mobile phase. The polar component of this layer wets the silica surface forming a thin film of liquid that acts as a stationary phase in liquid–liquid partition chromatography. In an interesting study, Shatz and Kazoka (1991) have tested a mixture of ethylene glycol in ethyl acetate as a mixed partition-adsorption system (MPA) as a mobile phase for the first time. Addition of methanol to such binary mobile phase resulted in the enhancement of elution strength of the MPA (Kazoka and Shatz, 1996). With such systems, as the concentration of polar component decreases, retention of solutes increases. A later study by Tauc et al. (1998) revealed that the addition of ethylene glycol to a mobile phase containing 10–20 mM NaCl in 5 mM sodium acetate buffer, pH 5.4 does not appreciably change the chromatographic performance of buffer in the analysis of proteins.

13.3 SURFACTANTS

At the present time, the scientific society needs the development of new environment-friendly analytical methods. Thus, chemists are supposed to develop analytical methods that are free from the use of hazardous reagents and leave minimal amount of chemical waste. Surfactants play an important role in our everyday life because they are applied in households as well as in many industrial processes. Most surfactants are susceptible to biodegradation, metabolic, and other breakdown reactions that may lead to metabolites with significant environmentally benign chemical properties. Generally, most surfactants appear to be less toxic in the environment than would be inferred from laboratory tests. The quantities of each class of surfactant used are difficult to estimate. As an approximation, anionic surfactants are the most important, representing 60–70% of surfactants currently in use. Nonionic compounds constitute around 30% but their use is increasing, while cationic and amphoteric products make up the smallest proportion. In surfactant-based organized media, all changes in the solution properties occur locally near the solubilized particle rather than in the bulk of the solvent. Ionic surfactants are used, in particular, for the modification of physicochemical properties and reactivity of organic reagents. The modification is brought about by the formation of neutral reagent–surfactant ion pairs stabilized by both electrostatic and hydrophobic interactions.

An area of growing interest in green chemistry is the application of surfactants for controlling reactions based on different phases, such as liquid (water)–liquid (oil) in a nonorganic solvent reaction system. As a result, it has been possible to achieve several objectives including a nonorganic water solvent system, clean reaction conditions, high yields of halide-free products, water reduction, simplicity of operation (product separation), and cost effectiveness (Anastas and Warner, 1998). The effect of surfactant ions on the analytical processes and systems is based on different principles. As a rule, it is connected with the formation of hydrophobically hydrated neutral reagent (chelate)–surfactant ion pairs (Savvin et al., 1991). At present, surfactants are widely used in molecular and atomic absorption and emission (luminescence) spectroscopy, potentiometry, voltammetry, titrimetry, various versions of chromatography [thin-layer chromatography (TLC), HPLC, ion chromatography, affinity chromatography, supercritical fluid chromatography, and gel chromatography], extraction and floatation, ultracentrifugation, capillary and gel electrophoreses, and other methods for determination and separation of organic and inorganic substances. In addition, surfactants are used for modifying the proteolytic, tautomeric, and complexing properties of organic reagents and simplifying sample preparation for a number of samples. Figure 13.1 gives a break down of the application of surfactants in various branches of analytical chemistry.

Below the critical micellar concentration (CMC), ionic surfactants have been added to the mobile phase in reversed-phase liquid chromatography (RPLC) since long in the so-called ion-pair chromatography. In reversed-phase ion-pair HPLC and TLC, the use of surfactant ion makes it possible to control the eluting capacity of the water–organic mobile phase and the selectivity of separation of organic and inorganic ions. The effect of surfactant ions is due to their adsorption on the stationary phase and the electrostatic interaction with the counter ions to be separated in solution. An increase in the concentration of the surfactant ions in the mobile phase results in the increase in retention time of the counter ions along with extended elution intervals, thus improving the selectivity of their separation (Kord and Khaledi, 1992). Surfactant micelles are more widely used in analysis than ions. This is due to greater opportunities for varying the properties of the microscopic environment of the reagents and their solubility and the protective action of the interface produced by the surfactant molecule (ion) assembly. Organized media, especially normal micelles, are widespread in various separation and preconcentration methods. Micellar extraction is an example of the successful application of micelles in separation and preconcentration. Surfactant-based micellar mobile phases have also been used in gel chromatography (Armstrong and Fendler, 1977).

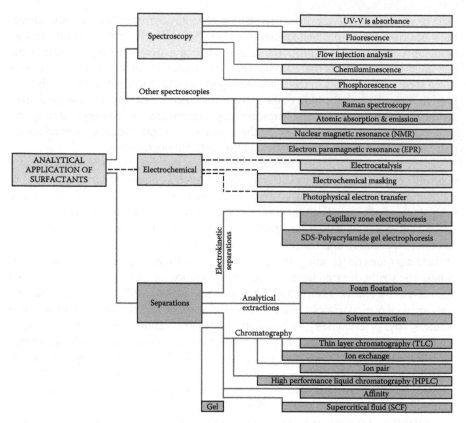

FIGURE 13.1   Application of surfactants in various branches of analytical chemistry.

The developed micellar liquid chromatography (MLC) made it possible to simultaneously separate hydrophilic and hydrophobic substances, charged and neutral
species, and optically active molecules. MLC is an efficient alternative to conventional RPLC with hydro-organic mobile phases. Current concern about the environment also reveals MLC as an interesting technique for "green" chemistry, because it
uses mobile phases containing 90% or more water. These micellar mobile phases
have a low toxicity and are not producing hazardous wastes (Ruiz-Angel et al., 2009).
For practical purposes, a suitable surfactant for MLC should have low CMC and
aggregation number and for ionic surfactants, a low-craft point. The unique capability of micellar mobile phases is attributed to the ability of micelles to selectively
compartmentalize and organize solvents at the molecular level. Solutes are separated
on the basis of their differential partitioning between the bulk aqueous phase and the
micellar aggregates in the mobile phase, and between the bulk aqueous phase and
the surfactant-coated stationary phase. In the first reports on MLC, the mobile phase
contained only a surfactant and occasionally, a buffer compound. In 1983, Dorsey
et al. recommended the addition of a short-chain alcohol, 1-propanol or butanol to
the micellar eluent to enhance the efficiency (Dorsey et al., 1983). The concentration
of organic solvent needed in the preparation of hybrid micellar mobile phases is

appreciably smaller than in conventional RPLC with hydro-organic mixtures (even for highly hydrophobic steroids). This is translated in a lower cost and toxicity, and the reduction of the environmentally important act of hazardous wastes. The organic additive (i) lowers the polarity of the aqueous solution, (ii) alters the micellar structure, and (iii) acts on the stationary phase changing the amount of adsorbed surfactant. Also, the organic solvent molecules wet the bonded phases, changing their physicochemical structure (rigidity) and hydrophobicity.

13.4 IONIC LIQUIDS

Several articles related to ionic liquids (IL) appearing in leading international journals over the last decade is a clear indication of the increasing interest of industry, research, and academia in exploiting the advantages of these existing and unique solvents as separating agents especially in chromatographic techniques as described by Koel (2005) and Pandey (2006). An IL generally consists of a large nitrogen-containing organic cation and a smaller inorganic anion. The asymmetry reduces the lattice energy of the crystalline structure and results in a salt of low melting point. The components of ILs are constrained by high columbic forces and thus exert practically no vapor pressure above the liquid surface. Not only are ILs more environmentally attractive than organic solvents, but also possess many unique physico-chemical properties, such as having very good dissolution properties for most organic and inorganic compounds, being nonflammable, fire resistant, having high thermal stability, and wide liquid temperature range.

Room temperature ionic liquids (RTILs) have been used in a variety of chromatographic methods as novel materials. Armstrong et al. (1999, 2001) observed that the wetting ability and the viscosity of RTILs made them ideal coating stationary phases in various GC applications. They discovered that RTILs act as low-polarity stationary phases to nonpolar compounds. Ding et al. (2004) and Berthod et al. (2001) were able to achieve first enantiomeric separations using chiral stationary phases in GC. Most of the investigations related to GC were concerned with the synthesis of IL-based stationary phases possessing unique IL properties such as chirality, high-thermal stability and efficiency on the GC column (Anderson and Armstrong, 2003, 2005; Berthod et al., 2001; Qi and Armstrong, 2007). Stalcup and Cabouska (2004) presented some important work on the applications of RTILs in liquid chromatography (LC) while Poole et al. (1986) were the first to demonstrate that liquid alkyl ammonium nitrate and thiocyanate salts could be used as mobile phases in reverse-phase (RP) LC when mixed with a second solvent of low viscosity. Further studies carried out by Waichigo et al. (2005) revealed that ILs behave like organic solvents with polarity similar to that of methanol. ILs at different pH and concentration levels have also been found useful as eluents with C18 and RP columns to achieve improved resolution and reduced band broadening (He et al., 2003). Zhang et al. (2003) have utilized RTILs as residual sianol-masking agents and also as separating agents for polar compounds, whereas Xiaohua et al. (2004) used amines and phthalic acids with 1-alkyl-3-methylimidazolium ILs as mobile-phase additives. When ILs were used as a mobile-phase additive at millimolar concentration, all their unique properties were lost and the substances became dissociated salts with hydrophobic cations and chaotropic anions. In addition to the advantages provided by the use of ILs such as enhanced separation efficiency and easy to handle, another important feature associated with their use is their negligible effect on pH of the mobile phase unlike triethylamine and other amines. Recently, utility of ILs have also been explored for stationary phase modification in LC by Sun et al. (2005) and Wang et al. (2006). Kaliszan et al. (2004) found that addition of 0.5–1.5% v/v of imidazolium BF_4 RTILs blocked silica group of silica gel and provided efficient separations of strongly basic drugs using TLC. Santos et al. (2004) have studied the fast screening of low-molecular weight compounds by TLC followed by direct on-spot MALDI-ToF mass spectrometry (MS) identification with nearly matrix free mass spectra using an UV-absorbing RTIL matrix, triethylamine 1λ-cyano-4-hydroxycinnamic acid. RTIL have also found application in counter-current chromatography both as solvent (Berthod and Carda-Broch, 2003) and also as coating material (Berthod and Carda-Broch, 2004). The use of RTILs have also been explored in capillary electrophoresis (Vaher et al., 2001), micellar electrokinetic chromatography (Mwongela et al., 2003; Tian et al., 2005), electrochemistry (Martiz et al., 2004), sensors (Wang et al., 2004), and nanotechnology (Zhao et al., 2004).

Although there has been intense interest in the use of ILs as green solvents as depicted by Figure 13.2, relatively very little is known about their biodegradability and toxicity. The "green character" of ILs has usually been justified with their negligible vapor pressure, but even if ILs do not evaporate and do not contribute to air pollution, most of these are water soluble and might enter into the environment by accidental spills or as effluents. When the ILs have served their operational use, their disposal becomes an issue. Within the field of green chemistry, it is unacceptable to produce large quantities of wastes that have high ecotoxicity or biological activity.

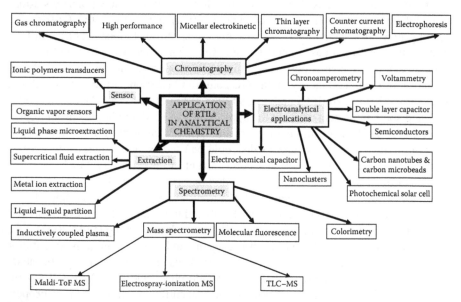

FIGURE 13.2   Fields of application of RTILs in analytical chemistry.

According to Gathergood et al. (2004), commonly used dialkylimidazolium ILs (bmix) have shown negligible toxicity. Structural similarities among certain ILs, herbicides, and plant growth regulators have been noted. These similarities raise significant environmental safety and health concerns and give ILs a potentially large unknown risk factor. New health and safety concerns could also result from ILs residues in polymers, and particularly in those used for packaging food and personal case products. Broad commercialization of ILs will therefore require a sound science-based understanding of their environmental safety and health impacts.

13.5 SUPERCRITICAL FLUIDS

A supercritical fluid (SCF) is defined as an element or compound above its critical pressure and critical temperature. Because of an increasing demand for environmental friendly processes, trends to banish some classes of solvents responsible of ozone depletion and to prevent or reduce the direct and indirect emissions of VOCs in the environment, the use of SCFs has become important. The first reported use of SCF reactions was by Daubrée in 1857, who conducted hydrothermal experiments to produce feldspar and quartz by heating kaolin with water for 2 days in a steel bomb. The first acknowledgment of the analytical potential for SCF probably came in a note made by James Lovelock in 1958. The last 30 years have seen an intense interest in the use of SCFs in separation science. The first supercritical fluid chromatographic separation was reported in 1962 by Klesper et al., who separated nickel etiporphyrin II from nickel mesoporphyrin IX dimethyl ester using dichlorodi-fluoromethane and monochlorodifluoromethane. After the pioneering work of Klesper et al. (1962), the main applications of SCFs focused on analytical chromatography. The main objective was to use the specific properties of the SCFs to improve peak resolution (by improvement of mass transfer) and to speed up analysis. Supercritical fluid chromatographic is essentially a normal-phase separation mode in which a primary retention mechanism is the interaction of the analyte with the stationary phase. The ability of SCF to dissolve low vapor pressure solids was first reported by Hannay and Hogarth (1879) when they placed several inorganic salts [cobalt(II) chloride and iron(III)chloride] into a glass tube and observed the solubility changes in pressure with ethanol above its critical temperature (240.8°C). They found that increasing pressure induced dissolution in the fluid phase whilst a decrease in pressure promoted precipitation.

Safety considerations led to the choice of carbon dioxide as the unique extraction medium. Besides the nonproblematic critical data (working above 32°C and 80 bar), carbon dioxide can be purchased with high purity at low cost and can be easily distilled (Figure 13.3). It is chemically inert, nonflammable and nonexplosive, has good solubility properties for nonpolar solutes. The removal of carbon dioxide from the extract and its disposal are easy. Another important advantage of carbon dioxide in food processing is that, as a naturally occurring gas that leaves no residues in the product.

One advantage of supercritical solvents when compared to conventional organic normal-phase solvents is that the equilibration rate after changing conditions is much more rapid even on interactive surfaces, such as silica (Steuer et al., 1988). The solvent power of an SCF is strongly linked to its density (controlled by pressure and temperature) and can also be adjusted by addition of an organic solvent (referred to

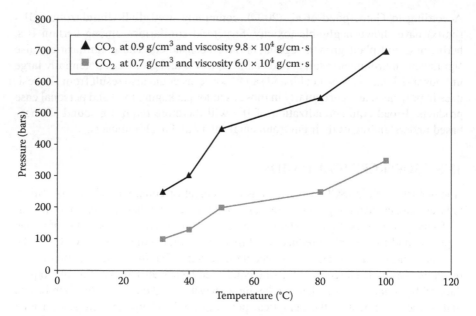

FIGURE 13.3 Variation of pressure with CO_2 with varying temperature at different viscosity and density.

as cosolvent) such as methanol and acetonitrile, and additives such as acids and bases. One of the attractions of supercritical fluid chromatography has been that it can use both GC- and LC-like detectors. Thus, it may make use of the universal FID instrument for nonvolatile and volatile analytes after separation on both capillary (Fjeldsted et al., 1983) and packed columns (Novotny, 1986). Nitrous oxide has been the widely examined alternative to carbon dioxide and was claimed to have stronger eluent strength for chromatography (Wright et al., 1985). There was particular interest in its application for the separation of amines (Khorassani et al., 1990) as there was a concern that basic amines would form carbamates in carbon dioxide. However, the oxidizing strength of nitrous oxide has proved a hazard with oxidizable analytes or with organic modifiers (Raynie, 1993). Although some other compounds such as Freon-22 and its fluoroform (Ong et al., 1990), the more eco-friendly 134a-Freon (Blackwell and Stringham, 1997), superheated water (Smith and Burgess, 1997), and supercritical ammonia (Giddings et al., 1968) have in the past been used as SCF, only carbon dioxide, with methanol or acetonitrile as a modifier, has reached anything like routine acceptability for extraction or chromatography. None of the other supercritical materials has shown sufficient advantages for warranting their general use, when compared with the ready availability, low cost, low toxicity, and readily obtained critical conditions offered by carbon dioxide.

Supercritical fluid chromatography coupled with MS has been one of the most successful applications of SCF (Coombs et al., 1997). It is much easier to evaporate a supercritical mobile phase into the MS source than most LC solvents. A number of workers have also examined the application of supercritical fluid chromatography coupled to nuclear magnetic resonance (NMR) spectroscopy for the separation and identification of complex mixtures (Albert, 1997). Calvey et al. (1991) have analyzed hydrogenated soybean oil by supercritical fluid chromatography/FT–IR with capillary columns while Borch-Jensen and Mollerup (1996) used a combination of hydrogenation and supercritical fluid chromatography or preparative TLC and supercritical fluid chromatography to analyze fish, shark, and seal oils. The Soxtec and supercritical fluid extraction (SFE)–TLC method (Rahman, 1996) was used as a rapid screening method for the qualitative determination of 2-dodecylcyclobutanone (DCB) in irradiated chicken. The high-performance TLC plate was used to detect DCB quantitatively. The important applications of SCF in chromatography are summarized in Table 13.1.

TABLE 13.1
The Important Applications of Supercritical Fluid in Chromatography

Analyte	Remark	References
Fatty acids	Pure CO_2 is perfectly suited for the separation	Alkio et al. (2000)
	Separation by preparative-supercritical fluid chromatography	Jusforgues and Shaimi (1998)
	Underivatized fatty acids and methyl esters of fatty acids are easily eluted using a bonded silica phase packed column and pure carbon dioxide	Via et al. (1994)
	Use of deltabond octyl column with water saturated carbon dioxide for the analysis of fatty acids, di- and triglycerides in abused vegetable oils	France et al. (1991)
	First report on the separation of underivatized fatty acids using water in carbon dioxide with FID	Geiser et al. (1988)
Petrochemicals	Simulated distillations for the characterization of oil fractions	Venter et al. (1999)
Metals	Study of metal ions related to nuclear industry	Lin et al. (1998)
	Investigations on separation and spectroscopic behavior of metal chelates in SCFs	Laintz et al. (1992)
	In spite of limited solubility metal chelates of dithiocarbamates are well separated in supercritical fluid chromatography	Laintz et al. (1991)
Pesticides	Faster analyses, lower detection limits, and greater injection to injection reproducibility were obtained with packed column supercritical fluid chromatography	Wheeler and McNally (1987)
Polymers	Describes the use of online SFE/supercritical fluid chromatography/FT-IR to extract, separate, and identify the extractables in nylon and polystyrene	Jordan et al. (1997)
Environmental samples	Supercritical fluid chromatography was used to rapidly determine phenols and nitrophenols in water	Pocurull et al. (1996)
Oligomers	Results are useful for grading oils	Campbell et al. (1988)
	Chromatographed by means of a pressure gradient on a capillary column coated with poly(dimethyl siloxane)	White and Houck (1986)
	Oligo- and polysaccharides as well as polyglycerol esters were studied using capillary columns and pure carbon dioxide	Anton (1989)

continued

TABLE 13.1   (continued)
The Important Applications of Supercritical Fluid in Chromatography

Analyte	Remark	References
	A polar stationary phases such as ODS and TMS and polar such as CN, diol, or NO_2 have been successfully applied to the separation of monosaccharides	Herbreteau et al. (1990)
	Separation of maltodextrin derivatized with trifluoroacetic anhydride using FID and a CN packed column	Oudsema and Poole (1992)
	Capillary supercritical fluid chromatography-MS is well suited to confirm the identity of the TMS derivative of inositol triphosphate	Pinkston et al. (1989)
Chiral analytes	Enantioseparation	Williams and Sander (1997)
Drugs	Quantification of five different drugs in pigs kidney	Ramsey et al. (1989)
	Separation of carboxylic polyether antibiotics (monensin, salinomycin, and narasin) with an amino bonded phase and 15% methanol in carbon dioxide using light-scattering detector	Berry et al. (1996)
	Analysis of hydroxylated metabolites of dialkyldithio-carbamates	Evans and Smith (1994)
	Separation of dihydroxypyridin enantiomers and their three major metabolites, chiral antimalarial compounds, β-blockers, and separation of eight members of the "profen" family	Peytavin et al. (1993), Biermanns et al. (1993), Terfloth et al. (1995)
Biomolecules	Micelles were formed in carbon dioxide through use of a fluoroether surfactant that supported the transport of proteins in an internal aqueous environment	Johnston et al. (1996)
Organic compounds	Carbobenzyloxy derivatized product demonstrated enhanced chiral	Kraml et al. (2005), Macaudiere et al. (1989)
	Enantiomeric separation of a wide range of secondary amides using binary mobile phases	
Food	Applications of supercritical fluid chromatography, particularly in the food and drug industry	Anklam et al. (1998)

13.6   CONCLUSIONS

Owing to the negative impact of VOCs on the environment, the demand for use of environment-friendly solvents to facilitate separations and purifications is increasing. As a result, green solvents are being developed as environment-friendly alternative solvents to petrochemical solvents. The low solubility of organic molecules has been the main reason for a restricted use of water as a solvent in chemical analysis. Aqueous solutions of polymers have proved to be novel solvents by virtue of their ability to solubilize otherwise poorly soluble species without the involvement of organic solvents. Aqueous two-phase systems comprising of mixtures of two different polymeric materials have been found to be environment-friendly solvent for the separation of biomaterials due to high water content in both phases. PEGs have found application as mobile phase additive in the chromatographic analysis of polar compounds, in addition to their use as general polar stationary phase. Surfactant micelles have been more widely used in chemical analysis compared to their use as electrolyte. The developed MLC, made it possible to simultaneously separate hydrophilic and hydrophobic substances, charged and neutral species, and optically active molecules.

Advantages of IL as unique solvents in separation science especially in chromatographic techniques are evident by the appearance of large number of publications on their use as mobile-phase systems. Not only are ILs more environmentally attractive than organic solvents because of their negligible vapor pressure, but also possess many unique physicochemical properties, such as good dissolution properties for most organic and inorganic compounds, nonflammability high thermal stability, negligible effect on pH of the mobile phase, and wide liquid temperature range. The wetting ability of ILs makes them ideal for coating stationary phases for GC. SCFs have found application in both GC and HPLC. Supercritical fluid chromatography in combination with MS, NMR, and IR has been the choice of most of the chromatographers for the rapid screening of a variety of organic compounds. Various favorable properties of CO_2 (a naturally occurring gas) such as critical data (working above 32°C and 80 bars), high purity, low cost, easy distillation, chemical inertness, nonflammability, and nonexplosiveness, good solubility properties for nonpolar solutes, have been the main reasons for the use of CO_2 as the unique extraction medium.

There is no perfect green solvent that can be applicable to all situations and therefore judicious decisions have to be taken by the user in this regard. All alternative solvents have advantages and disadvantages. In all areas, there is a need to balance the technical advantages of a particular solvent with any environmental cost or other disadvantages.

REFERENCES

Albert, K. 1997. Supercritical fluid chromatography—proton nuclear magnetic resonance spectroscopy coupling. *Journal of Chromatography A*, 785:65–83.

Albertsson, P.A. 1986. *Partition of Cell Particles and Macromolecules*. New York: Wiley.

Anastas, P.T. and Warner, T.C. 1998. *Green Chemistry: Theory and Practice*. London: Oxford University Press.

Alkio, M., Gonzales, C., Jântii, M., and Aaltonen, O. 2000. Purification of polyunsaturated fatty acid esters from tuna oil with supercritical fluid chromatography. *Journal of the American Oil Chemists' Society*, 77:315–21.

Anderson, J.L. and Armstrong, D.W. 2003. High-stability ionic liquids. A new class of stationary phases for gas chromatography. *Analytical Chemistry*, 75:4851–8.

Anderson, J.L. and Armstrong, D.W. 2005. Immobilized ionic liquids as high-selectivity/ high-temperature/high-stability gas chromatography stationary phases. *Analytical Chemistry*, 77:6453–62.

Anklam, E., Berg, H., Mathiasson, L., Sharman, M., and Ulberth, F. 1998. Supercritical fluid extraction (SFE) in food analysis: A review. *Food Additives & Contaminants: Part A*, 15:729–50.

Anton, K. 1989. In K.E. Markides and M.L. Lee (eds), SFC applications. *The 1989 Workshop on Supercritical Fluid Chromatography*. Utah: Brigham Young University Press, p. 346.

Armstrong, D.W. and Fendler, J.H. 1977. Differential partitioning of tRNA's between micellar and aqueous phases: A convenient gel-filtration method for separation of tRNA's. *Biochimica et Biophysica Acta*, 478:75–80.

Armstrong, D.W., He, L.F., and Liu, Y.S. 1999. Examination of ionic liquids and their interaction with molecules, when used as stationary phases in gas chromatography. *Analytical Chemistry*, 71:3873–6.

Berry, A.J., Ramsey, E.D., Newby, M., and Games, D.E. 1996. Applications of packed-column SFC using light-scattering detection. *Journal of Chromatography Science*, 34:245–53.

Berthod, A. and Carda-Broch, S. 2003. A new class of solvents for CCC: The room temperature ionic liquids. *Journal of Liquid Chromatography & Related Technologies*, 26:1493–1508.

Berthod, A. and Carda-Broch, S. 2004. Use of the ionic liquid 1-butyl-3-methylimidazolium hexafluorophosphate in countercurrent chromatography. *Analytical and Bioanalytical Chemistry*, 380:168–77.

Berthod, A., He, L., and Armstrong, D.W. 2001. Ionic liquids as stationary phase solvents for methylated cyclodextrins in gas chromatography. *Chromatographia*, 53:63–8.

Biermanns, E., Miller, C., Lyons, V., and Wilson, W.H. 1993. Chiral resolution of β-blockers by packed column supercritical fluid chromatography. *Liquid Chromatography Gas Chromatography*, 10:744–7.

Blackwell, J.A. and Stringham, R.W. 1997. Temperature effects for chiral separations using various bulk fluids in near-critical mobile phases. *Chirality*, 9:693–8.

Borch-Jensen, C. and Mollerup, J. 1996. Supercritical fluid chromatography of fish, shark and seal oils. *Chromatographia*, 42:252–8.

Calvey, E.M., McDonald, R.E., Page, S.W., Mossoba, M.M., and Taylor, L.T. 1991. Evaluation of SFC/FT–IR for examination of hydrogenated soybean oil. *Journal of Agricultural and Food Chemistry*, 39:542–8.

Campbell, R.M., Djordjevic, N.M., Markides, K.E., and Lee, M.L. 1988. Supercritical fluid chromatographic determination of hydrocarbon groups in gasolines and middle distillate fuels. *Analytical Chemistry*, 60:356–62.

Capello, C., Fischer, U., and Hungerbuhler, K. 2007. What is a green solvent? A comprehensive framework for the environmental assessment of solvents. *Green Chemistry*, 9:927–34.

Clark, J.H. and Macquarrie, D.J. 2002. *Handbook of Green Chemistry and Technology*. London: Blackwell Science.

Coombs, M.T., Khorassani, M.A., and Taylor, L.T. 1997. Packed column supercritical fluid chromatography–mass spectroscopy: A review. *Journal of Chromatography A*, 785:85–100.

Ding, J., Welton, T., and Armstrong, D.W. 2004. Chiral ionic liquids as stationary phases in gas chromatography. *Analytical Chemistry*, 76:6819–22.

Dorsey, J.G., DeEchegaray, M.T., and Landy, J.S. 1983. Liquid chromatographic phosphorescence detection with micellar chromatography and postcolumn reaction modes. *Analytical Chemistry*, 54:1552–8.

Evans, M.B. and Smith, M.S. 1994. A comparative study of the chromatography of hydroxylated dialkyldithiocarbamates as models for drug metabolites. *Chromatographia*, 39:569–76.

Fischer, U. and Hungerbühler, K. 2000. Application of indicators for assessing environmental aspects of chemical processes to case studies from pharmaceutical production. *CHIMIA International Journal for Chemistry*, 54(9):494–500.

Fjeldsted, J.C., Kong, R.C., and Lee, M.L. 1983. Capillary supercritical-fluid chromatography with conventional flame detectors. *Journal of Chromatography*, 279:449–55.

France, J.E., Snyder, J.M., and King, J.W. 1991. Packed-microbe supercritical fluid chromatography with flame ionization detection of abused vegetable oils. *Journal of Chromatography*, 540:271–8.

Gathergood, N., Garcia, M.T., and Scammells, P.J. 2004. Biodegradable ionic liquids: Part I. Concept, preliminary targets and evaluation. *Green Chemistry*, 6:166–75.

Geiser, F.O., Yocklovich, S.G., Lurcott, S.M., Guthrie, J.W., Guthrie, E.J., and Levy, E.J. 1988. Water as a stationary phase modifier in packed-column supercritical fluid chromatography: I. Separation of free fatty acids. *Journal of Chromatography*, 459:173–81.

Giddings, J.C., Myers, M.N., McLaren, L., and Keller, R.A. 1968. High pressure gas chromatography of non volatile species. *Science*, 162:67–73.

Guo, Z.M., Liu, Y.F., Xu, J.Y., Xu, Q., Xue, X., Zhang, F., Ke, Y., Liang, X., and Lei, A. 2008. Novel reversed-phase high-performance liquid chromatography stationary phase with oligo(ethylene glycol) "click" to silica. *Journal of Chromatography A*, 1191:78–82.

Hannay, J.B. and Hogarth, J. 1879. On the solubility of gases. *Proceedings of the Royal Society of London*, 29:324–6.

He, L.J., Zhang, W.Z., Zhao, L., Liu, X., and Jiang, S.X. 2003. Effect of 1-alkyl-3-methylimidazolium-based ionic liquids as the eluent on the separation of ephedrines by liquid chromatography. *Journal of Chromatography A*, 1007:39–45.

Herbreteau, B., Lafosse, M., Morin-Allory, L., and Dreux, M. 1990. Analysis of sugars by supercritical fluid chromatography using polar packed columns and light-scattering detection. *Journal of Chromatography*, 505:299–305.

Hofer, R. and Bigorra, J. 2007. Green chemistry a sustainable solution for industrial specialties applications. *Green Chemistry*, 9:203–12.

Hu, Z., Shen, X., Qiu, H., Lai, G., Wu, J., and Li, W. 2009. AGET ATRP of methyl methacrylate with poly(ethylene glycol) (PEG) as solvent and TMEDA as both ligand and reducing agent. *European Polymer Journal*, 45:2313–8.

Huddleston, J.G., Willauer, H.D., Griffin, S.T., and Rogers, R.D. 1999. Aqueous polymeric solutions as environmentally benign liquid/liquid extraction media. *Industrial and Engineering Chemistry Research*, 38:2523–39.

Ito, K. and Hirokawa, T. 1996. Separation of alkali and alkaline-earth metal and ammonium cations by capillary electrophoresis using poly(ethylene glycol) and tartaric acid. *Journal of Chromatography A*, 742:281–8.

Johnston, K.R., Harrison, K.L., Clarke, M., Howdle, S.M., Heitz, M.P., Bright, F.V., Carlier, C., and Randolph, T.W. 1996. Water-in-carbon dioxide microemulsions: An environment for hydrophiles including proteins. *Science*, 271:624–6.

Jordan, S.L., Taylor, L.T., Seemuth, P.D., and Miller, R.J. 1997. Analysis of additives and monomers in nylon and polystyrene. *Textile Chemist and Colorist*, 29:25–32.

Jusforgues, P. and Shaimi, M. 1998. Preparative supercritical fluid chromatography. *Analysis*, 26:55–60.

Kaliszan, R., Marszall, M.P., Markuszewski, M.J., Baczek, T., and Pernak, J. 2004. Suppression of deleterious effects of free silanols in liquid chromatography by imidazolium tetrafluoroborate ionic liquids. *Journal of Chromatography A*, 1030:263–71.

Kaniansky, D., Zelensky, I., Valaskova, I., Marak, J., and Zelenska, V. 1990. Isotachophoretic separation of alkali and alkaline earth metal cations in water–polyethylene glycol mixtures. *Journal of Chromatography A*, 502:143–53.

Kazoka, H. and Shatz, V. 1996. Dynamically generated stationary liquid phase systems with silica and ternary mobile phases containing ethylene glycol and formamide. *Journal of Chromatography A*, 732:231–8.

Khorassani, M.A., Taylor, L.T., and Zimmerman, P. 1990. Nitrous oxide versus carbon dioxide for supercritical-fluid extraction and chromatography of amines. *Analytical Chemistry*, 62:1177–80.

Klesper, E., Corwin, A.H., and Turner, D.A. 1962. High pressure gas chromatography above critical temperature. *Journal of Organic Chemistry*, 27:700.

Koel, M. 2005. Ionic liquids in chemical analysis. *Critical Reviews in Analytical Chemistry*, 35:177–92.

Kord, A.S. and Khaledi, M.G. 1992. Chromatographic characteristics of surfactant-mediated separations: Micellar liquid chromatography *vs* ion pair chromatography. *Analytical Chemistry*, 64(17):1901–7.

Kraml, C.M., Zhou, D., Byrne, N., and McConnell, O. 2005. Enhanced chromatographic resolution of amine enantiomers as carbobenzyloxy derivatives in high-performance liquid chromatography and supercritical fluid chromatography. *Journal of Chromatography A*, 1100:108–15.

Kurosu, Y., Kawasaki, H., Chen, X., Amano, Y., Yan-Il, F., Isobe, T., and Okuyama, T. 1984. Comparison of retention times of polypeptides in reversed phase high performance liquid chromatography of polystyrene resin and on alkyl bonded silica. *Bunseki Kagaku*, 33:301–8.

Laintz, K.E., Yu, J.J., and Wai, C.M. 1992. Separation of metal ions with sodium bis(trifluoroethyl) dithiocarbamate chelation and supercritical fluid chromatography. *Analytical Chemistry*, 64:311–5.

Laintz, K.E., Wai, C.M., Yonker, C.R., and Smith, R.D. 1991. Solubility of fluorinated metal diethyldithiocarbamates in supercritical carbon dioxide. *Journal of Supercritical Fluids*, 4:194–8.

Lancaster, M. 2002. *Green Chemistry: An Introductory Text*. Cambridge: Royal Society of Chemistry.

Lin, Y.H., Wu, H., Smart, N.G., and Wai, C.M. 1998. Investigation of adducts of lanthanide and uranium β-diketonates with organophosphorus Lewis bases by supercritical fluid chromatography. *Journal of Chromatography A*, 793:107–13.

Macaudiere, E., Lienne, M., Caude, M., Rosset, R., and Tambute, A. 1989. Resolution of π-acid racemates on π-acid chiral stationary phases in normal-phase liquid and subcritical fluid chromatographic modes: A unique reversal of elution order on changing the nature of the achiral modifier. *Journal of Chromatography*, 467:357–72.

Martiz, B., Keyrouz, R., Gmouh, S., Vaultier, M., and Jouikov, V. 2004. Superoxide-stable ionic liquids: New and efficient media for electrosynthesis of functional siloxanes. *Chemical Communications*, 6:674–5.

Mwongela, S.M., Numan, A., Gill, N.L., Agbaria, R.A., and Warner, I.M. 2003. Separation of achiral and chiral analytes using polymeric surfactants with ionic liquids as modifiers in micellar electrokinetic chromatography. *Analytical Chemistry*, 75:6089–96.

Nelson, W.M. 2003. *Green Solvents for Chemistry: Perspective and Practice*. Oxford: Oxford University Press.

Novotny, M. 1986. New detection strategies through supercritical fluid chromatography. *Journal of High Resolution Chromatography*, 9:137–44.

Ogino, K., Sato, H., Aihara, Y., Suzuki, H., and Moriguchi, S. 1995. Preparation and characterization of monodisperse oligo(ethylene glycol) dimethacrylate polymer beads for aqueous high-performance liquid chromatography. *Journal of Chromatography A*, 699:67–72.

Ong, C.P., Lee, H.K., and Li, S.F.Y. 1990. Chlorodifluoromethane as the mobile phase in supercritical fluid chromatography of selected phenols. *Analytical Chemistry*, 62:1389–91.

Oudsema, J. and Poole, C. 1992. Some practical experiences in the use of a solventless injection system for packed column supercritical fluid chromatography. *Journal of High Resolution Chromatography*, 15:65–70.

Pandey, S. 2006. Analytical applications of room-temperature ionic liquids: A review of recent efforts. *Analytica Chimica Acta*, 556:38–45.

Pellati, F., Benvenuti, S., and Melegari, M. 2005. Chromatographic performance of a new polar poly(ethylene glycol) bonded phase for the phytochemical analysis of *Hypericum perforatum* L. *Journal of Chromatography A*, 1088:205–17.

Peytavin, G., Gimenez, F., Genissel, B., Gillotin, C., Baillet, A., Wainer, I.W., and Farinotti, R. 1993. Chiral resolution of some antimalarial agents by sub- and supercritical fluid chromatography on an (*S*)-naphthylurea stationary phase. *Chirality*, 5:173–80.

Pinkston, J.D., Bowling, D.J., and Delaney, T.E. 1989. Industrial applications of supercritical-fluid chromatography–mass spectrometry involving oligometric materials of low volatility and thermally labile materials. *Journal of Chromatography*, 474:97–111.

Pocurull, E., Marce, R.M., Borrull, F., Bernal, J.L., Toribio, L., and Serna, M.L. 1996. On-line solid-phase extraction coupled to supercritical fluid chromatography to determine phenol and nitrophenols in water. *Journal of Chromatography*, 755:67–74.

Poole, C.F., Kersten, B.R., Ho, S.S.J., Coddens, M.E., and Furton, K.G. 1986. Organic salts, liquid at room temperature, as mobile phases in liquid chromatography. *Journal of Chromatography*, 352:407–25.

Qi, M. and Armstrong, D.W. 2007. Dicationic ionic liquid stationary phase for GC–MS analysis of volatile compounds in herbal plants. *Analytical and Bioanalytical Chemistry*, 388:889–99.

Rahman, R., Matabudall, D., Haque, A.K., and Sumar, S. 1996. A rapid method (SFE–TLC) for the identification of irradiated chicken. *Food Research International*, 29:301–7.

Ramsey, E.D., Perkins, J.R., Games, D.E., and Startin, J.R. 1989. Analysis of drug residues in tissue by combined supercritical-fluid extraction–supercritical-fluid chromatography–mass spectrometry–mass spectrometry. *Journal of Chromatography*, 464:353–64.

Raynie, D.E. 1993. Warning concerning the use of nitrous oxide in supercritical fluid extractions. *Analytical Chemistry*, 65:3127–8.

Reichardt, C. 2007. Organic process research & development solvents and solvent effects: An introduction. *Organic Process Research & Development*, 11:105–13.

Rong, L. and Takeuchi, T. 2004. Determination of iodide in seawater and edible salt by microcolumn liquid chromatography with poly(ethylene glycol) stationary phase. *Journal of Chromatography A*, 1042:131–5.

Ruiz-Angel, M.J., Garcia-Alvarez-Coque, M., and Berthod, A. 2009. New insights and recent developments in micellar liquid chromatography. *Separation and Purification Reviews*, 38:45–96.

Santos, L.S., Haddad, R., Hoehr, N.F., Pilli, R.A., and Eberlin, M.N. 2004. Fast screening of low molecular weight compounds by thin-layer chromatography and "on-spot" MALDI–TOF mass spectrometry. *Analytical Chemistry*, 76:2144–7.

Savvin, S.B., Chernova, R.K., and Shytkov, S.N. 1991. *Poverkhnostno–aktivnye veshchestva (Surfactants)*. Moscow: Nauka.

Shatz, V.D. and Kazoka, H. 1991. Adsorption and partition mode in high-performance liquid chromatography of highly polar solutes on silica. *Journal of Chromatography*, 552:23–9.

Sheldon, R.A. 2005. Green solvents for sustainable organic synthesis: State of the art. *Green Chemistry*, 7:267–78.

Simunicova, E., Kaniansky, D., and Loksikova, K. 1994. Separation of alkali and alkaline earth metal and ammonium cations by capillary zone electrophoresis with indirect UV absorbance detection. *Journal of Chromatography A*, 665:203–9.

Smith, R.M. and Burgess, R.J. 1997. Superheated water as an eluent for reversed-phase high-performance liquid chromatography. *Journal of Chromatography A*, 785:49–55.

Stalcup, A.M. and Cabouska, B. 2004. Ionic liquids in chromatography and capillary electrophoresis. *Journal of Liquid Chromatography & Related Technologies*, 27:1443–59.

Steuer, W., Schidnler, M., and Erni, F. 1988. Gradient elution with normal phases on silica: A comparison between high-performance liquid and supercritical fluid chromatography. *Journal of Chromatography*, 454:253–9.

Sun, Y., Cabovska, B., Evans, C.E., Ridgway, T.H., and Stalcup, A.M. 2005. Retention characteristics of a new butylimidazolium-based stationary phase. *Analytical and Bioanalytical Chemistry*, 382:728–34.

Tauc, P., Cochet, S., Algiman, E., Callebaut, I., Cartron, J.-P., Brochon, J.C., and Bertrand, O. 1998. Ion-exchange chromatography of proteins: Modulation of selectivity by addition of organic solvents to mobile phase: Application to single-step purification of a proteinase inhibitor from corn and study of the mechanism of selectivity modulation. *Journal of Chromatography A*, 825:17–27.

Terfloth, G.J., Pirkle, W.H., Lynam, K.G., and Nicolas, E.C. 1995. Broadly applicable polysiloxane-based chiral stationary phase for high-performance liquid chromatography and supercritical fluid chromatography. *Journal of Chromatography A*, 705:185–94.

Tian, K., Qi, S., Cheng, Y., Chen, X., and Hu, Z. 2005. Separation and determination of lignans from seeds of *Schisandra* species by micellar electrokinetic capillary chromatography using ionic liquid as modifier. *Journal of Chromatography A*, 1078:181–7.

Vaher, M., Koel, M., and Kaljurand, M. 2001. Nonaqueous capillary electrophoresis in acetonitrile using ionic-liquid buffer electrolytes. *Chromatographia*, 53:302–6.

Valiokas, R., Svedhem, S., Svensson, S.C.T., and Liedberg, B. 1999. Self-assembled monolayers of oligo(ethylene glycol)-terminated and amide group containing alkanethiolates on gold. *Langmuir*, 15:3390–4.

Venter, A., Rohwer, E.R., and Laubscher, A.E. 1999. Analysis of alkane, alkene, aromatic and oxygenated groups in petrochemical mixtures by supercritical fluid chromatography on silica gel. *Journal of Chromatography A*, 847:309–21.

Via, J., Taylor, L.T., and Schweighardt, F.K. 1994. Experimental determination of changes in methanol modifier concentration in premixed carbon dioxide cylinders. *Analytical Chemistry*, 66:1459–61.

Waichigo, M.M., Riechel, T.L., and Danielson, N.D., 2005. Ethylammonium acetate as a mobile phase modifier for reversed phase liquid chromatography. *Chromatographia*, 61:17–23.

Walter, H. and Johansson, G. 1994. *Methods in Enzymology*. New York: Academic Press.

Wang, Q., Baker, G.A., Baker, S.N., and Colon, L.A. 2006. Surface confined ionic liquid as a stationary phase for HPLC. *Analyst*, 131:1000–5.

Wang, R., Hoyano, S., and Ohsaka, T. 2004. O_2 gas sensor using supported hydrophobic room-temperature ionic liquid membrane-coated electrode. *Electrode; Chemistry Letters*, 33:6–7.

Wheeler, J.R. and McNally, M.E. 1987. Comparison of packed column and capillary column supercritical fluid chromatography and high-performance liquid chromatography using representative herbicides and pesticides as typical moderate polarity and molecular weight range molecules. *Journal of Chromatography*, 410:343–53.

White, C.M. and Houck, R.K. 1986. Supercritical fluid chromatography and some of its applications: A review. *Journal of High Resolution Chromatography*, 9:4–17.

Williams, K.L. and Sander, L.C. 1997. Enantiomer separations on chiral stationary phases in supercritical fluid chromatography. *Journal of Chromatography A*, 785:149–58.

Wright, B.W., Kalinoski, H.T., and Smith, R.D., 1985. Investigation of retention and selectivity effects using various mobile phases in capillary supercritical fluid chromatography. *Analytical Chemistry*, 57:2823–9.

Xiaohua, X., Liang, Z., Xia, L., and Shengxiang, J. 2004. Ionic liquids as additives in high performance liquid chromatography: Analysis of amines and the interaction mechanism of ionic liquids. *Analytica Chimica Acta*, 519:207–11.

Zhao, F., Wu, X., Wang, M., Liu, Y., Gao, L., and Dong, S. 2004. Electrochemical and bioelectrochemistry properties of room-temperature ionic liquids and carbon composite materials. *Analytical Chemistry*, 76:4960–7.

Zhang, W.Z., He, L.J., Gu, Y.L., Liu, X., and Jiang, S.X. 2003. Liquids as mobile phase additives on retention of catecholamines in reversed-phase high-performance liquid chromatography. *Analytical Letters*, 36:827–38.

14 Ultrasound Applications in Biorenewables for Enhanced Bioenergy and Biofuel Production

Samir Kumar Khanal, Devin Takara,
Saoharit Nitayavardhana, Buddhi P. Lamsal,
and Prachand Shrestha

CONTENTS

14.1 INTRODUCTION

The sky-rocketing crude oil prices in recent years and the continuous exploitation of fossil fuels demand that we make serious efforts toward sustainable biofuel and bioenergy production. Renewable energy derived from plant-based feedstocks, organic residues, and biowastes is expected to reduce our dependency on fossil fuels, reduce greenhouse gas emissions, and enhance the rural economy (Schmer et al., 2008). In the United States, liquid biofuels such as ethanol and biodiesel are primarily derived from corn starch and soybeans, respectively. Lignocellulosic biomass is an alternative abundant renewable resource for biofuel production. Research on bioconversion of lignocellulosic biomass into biofuel has been vigorously pursued in the United States (Lynd et al., 1991; Anex et al., 2007). There is also a considerable amount of focus placed on producing gaseous forms of energy, such as methane and hydrogen, from biowastes. These high solids feedstocks require various forms of pretreatment to enhance biofuel and bioenergy production.

Biomass pretreatments—size reduction and cell wall disruption, have been widely adopted to enhance enzymatic activity for higher biofuel/bioenergy yield. Ultrasound technology is a promising pretreatment method in biorenewables. Ultrasound is a sound wave with a frequency above the normal hearing range of humans (>15–20 kHz). When the ultrasound wave propagates in a liquid medium, it generates a repeating pattern of compressions and rarefactions. In the rarefaction zone, liquid molecules are unduly stretched in the medium where microbubbles are formed. The microbubbles oscillate under the influence of variable pressure and their size begins to increase before violently collapsing (Khanal et al., 2007a). As the bubbles collapse, a localized temperature of up to 5000 K and pressures of up to 180 MPa are produced (Suslick, 1990). This collapsing phenomenon, known as cavitation, generates enough energy to mechanically shear the particles in aqueous phase surrounding the bubbles.

Several studies examined the applications of ultrasound in biofuel/bioenergy production. Wood et al. (1997) reported nearly a 20% increase in ethanol production from mixed waste office paper after 96 h of fermentation by intermittent exposure to ultrasonic energy. High-power ultrasound was extremely effective in delignifying wheat straw, and the residual lignin content for pulp and paper production was reduced by more than 75% (Csóka et al., 2008). Khanal et al. (2007b) examined the effect of ultrasound on particle size reduction and subsequent sugar release from corn slurry. The authors reported a 20-fold reduction in particle size during 40 s of sonication, with about 30% higher sugar release with respect to nonsonicated samples. In another study, Nitayavardhana et al. (2008) examined the effect of ultrasound on sugar release from cassava chips, the results of which showed an increase in reducing sugar release by as much as 180% with respect to the nonsonicated samples. Fermentation studies also showed significantly higher ethanol yields for sonicated samples compared to nonsonicated samples. The ethanol yields from sonicated samples were over 250% higher than yields from control samples (Nitayavardhana et al., 2010). The fermentation time for sonicated samples could be reduced by nearly 24 h to achieve the same ethanol yield with respect to the controls.

Lignocellulosic biomass is one of the abundant feedstocks for biofuel production. The United States has the potential of producing 1.3 billion dry tons of plant biomass annually and over 900 million dry tons of this biomass is available as agricultural residues (USDA and USDOE Joint Report, 2005). Pretreatment of the biomass is required to make the cellulose/hemi-cellulose more accessible to the enzymes. The single most important factor affecting enzymatic activity is the available surface area of the substrate and the opening of the pore volume. High-power ultrasonication has the potential to break down the ground lignocellulosic biomass into finer particles. Sonication of ammonia-steeped switchgrass followed by enzymatic hydrolysis increased glucose yield by nearly 10% in an exploratory study (Montalbo-Lomboy et al., 2007).

Application of high-power ultrasound in soybean oil extraction with hexane-isopropanol as a solvent was studied by Li et al. (2004). The sonication process resulted in over 60% oil yield, whereas untreated soybean flakes had only 40% oil yield with the same solvent mixture. Transesterification of the vegetable oil was possible in a short reaction time with a small amount of base catalyst in the presence of ultrasonic treatment at a frequency of 28 kHz (Stavarache et al., 2005). The total reaction time varied between 10 and 40 min for a 91–98% methyl esters yield at 0.5–1% (w/w) NaOH concentration. Ji et al. (2006) also reported the shortest reaction time and highest yield of biodiesel following ultrasonication (250 Wh kg^{-1} soybean oil) during the transesterification of soybean oil to produce biodiesel. Hielscher Ultrasonics (2008) conducted a pilot-scale testing of ultrasound technology in biodiesel production. Sonication reduced the processing time significantly with a reduction in the amount of catalyst required by up to 50%.

Ultrasound technology has also been widely adopted for the disruption of biological cells from wastewater residuals. The exposure of microbial cells to ultrasound breaks the cell wall and membrane and releases the intracellular organics in the bulk solution, enhancing the overall digestibility and methane yield (Khanal et al., 2007a). Ultrasound pretreatment alone, however, may require a high-energy input for effective cell disruption. One approach of improving ultrasound efficiency is to sonicate the microbial biomass prior to chemical pretreatment (acid or alkali) to weaken the cell wall. This chapter summarizes the major findings of ultrasound applications in biofuel/bioenergy production.

14.2 ULTRASOUND APPLICATIONS IN BIOETHANOL PRODUCTION

14.2.1 CORN SLURRY

14.2.1.1 Scanning Electron Microscopy Examination

The scanning electron micrographs (SEM) of raw and cooked corn slurry samples before and after sonication at high-power level are shown in Figure 14.1. Figures 14.1a and c show cells that appear almost intact. With ultrasonic treatment for 40 s, complete disintegration of cells was observed with large numbers of fragmented cell materials (Figures 14.1b and d). Several micropores were also visible within the disintegrated corn particles. The SEM images clearly demonstrated the changes in the structures of corn particles following ultrasound pretreatment.

14.2.1.2 Particle Size Distribution

The effect of ultrasonic treatment on corn particles was examined by sonicating both cooked and raw corn slurry samples and the resulting particle size was compared with nonsonicated samples (controls). The peak of the particle size distribution curve shifted from 800 μm to around 80 μm following sonication at high-power levels for cooked corn slurry samples. In addition, the particle size reduction was directly related to the power level and sonication time. The particle size reduction at the

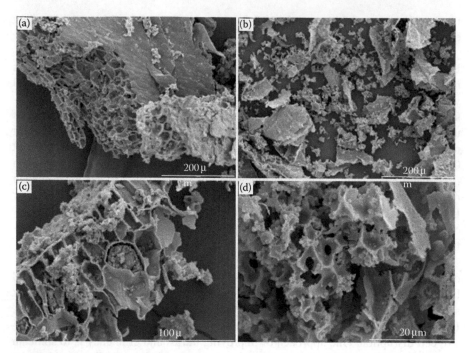

FIGURE 14.1 SEM images of corn slurry: (a) raw corn (control); (b) raw corn sonicated (40 s); (c) cooked corn (control); (d) cooked corn sonicated (40 s).

higher power level and longer sonication time is in close agreement with glucose yield under similar conditions as discussed later. Similar results were also obtained for raw corn slurry (data not shown here).

14.2.1.3 Glucose Release

The concentrations of glucose release from raw corn slurry samples are presented in Figure 14.2. The highest glucose release increases of 32 and 27% with respect to the controls were obtained for SD_{40} (enzyme addition during 40 s of sonication) at both low and medium power inputs, respectively. The glucose release, however, dropped by nearly 22% at the high-power setting for SD_{40}. A similar trend was also observed for SD_{20}, in which the glucose released dropped by 11%. These tests were conducted without temperature control, and the final temperature of the ultrasound treated samples increased with the power level and treatment time. Therefore, these decreases in glucose concentration could be attributed to the excessive ultrasonic treatment, which may have resulted in the degradation or denaturation of the enzymes at high-power settings. This finding is in close agreement with the glucose released data without enzymes (SA_{20} and SA_{40}), which showed an improvement in the glucose released following sonication, irrespective of power levels. This suggests that the higher power level did not cause the gelatinization of starch due to sonication. The additional sugar yield from ultrasound-treated samples could also be due to the release of starch that was bound to lipids and did not have access to the hydrolyzing enzyme.

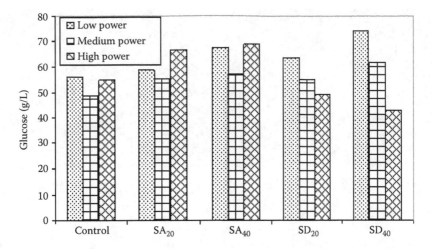

FIGURE 14.2 Glucose release of raw corn slurry at varying power input levels.

14.2.2 CASSAVA SLURRY

14.2.2.1 Particle Size Distribution and SEM Examination

Particle size distribution and SEM images were examined in order to evaluate the efficacy of ultrasound to disintegrate cassava chip slurries. The cassava slurries were sonicated at high-power levels under temperature-controlled condition in order to eliminate the effect of temperature on particle disintegration. There were three peaks of 600, 200, and 15 μm in the particle size distribution curves. The particle size was reduced following sonication. The peak shifted from 600 to 200 and 15 μm. The original peak (600 μm) is expected to represent milled cassava particles that were not affected by ultrasound treatment. The 200 μm peak corresponds to cell morphologies of the cassava chips affected by sonication. The 15 μm peak essentially represents the individual starch granules. This was further substantiated by SEM as discussed later.

The SEM images of cassava slurry samples after sonication for 30 s at high power and the control are presented in Figure 14.3. As seen in Figures 14.3a and b, the cassava fibrous structure was disintegrated by sonication. The destruction of cassava cell structures resulted in the release of more individual starch granules confined within the fibrous structure in the slurry, which enhanced the enzymatic hydrolysis as discussed later. The starch granules appeared to be unaffected by ultrasound pretreatment (Figures 14.3c and d). The average size of cassava starch granules was around 15 μm. This finding was in close agreement with the result of particle size distribution.

14.2.2.2 Reducing Sugar Release and Ethanol Fermentation

The reducing sugar and ethanol concentration profiles of samples sonicated at high power for 10 and 30 s, and the controls are presented in Figure 14.4. As seen in the figure, reducing sugar release increased with ultrasound pretreatment. The cassava cell structure disintegrated during sonication released more starch granules in the

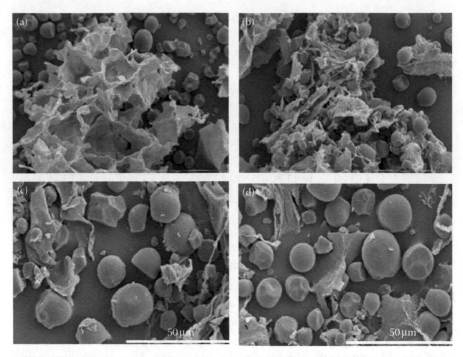

FIGURE 14.3 SEM images of cassava starch granules: (a) and (c) control; (b) and (d) high power (30 s).

aqueous phase, thereby exposing a much larger surface area to enzymes, subsequently enhancing the enzymatic hydrolysis reaction. With respect to ethanol yields, control samples showed a continuous increase during fermentation. However, for both 10 and 30 s sonicated samples, the ethanol yield reached a plateau by 48 h of incubation. Consequently, sonicated samples resulted in significantly higher ethanol production rates than the controls. The ultrasound pretreatment reduced the fermentation time by nearly 24 h with respect to the control. Furthermore, the maximum ethanol yield of sonicated samples was much higher than the control. The percentage in ethanol yield improvement with respect to the controls was as much as 266% for samples sonicated at high power for 30 s. Thus, sonication not only reduced the fermentation time but also enhanced ethanol production.

14.2.3 Lignocellulosic Biomass

14.2.3.1 SEM Examination
High-power ultrasound has the potential to increase the pore volume and to reduce the crystallinity of cellulosic biomass. The effect of ultrasound pretreatment on the microstructure of alkaline-pretreated corn stover is shown in Figures 14.5a and b. The pretreated stover sample showed significant disruption of the cell wall (Figure 14.5b). Thermal and chemical pretreatments (acid) make the cellulose more accessible to enzyme by opening up cleavage sites. Such pretreatments however, also

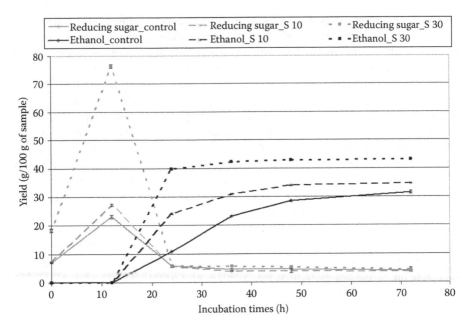

FIGURE 14.4 The percentage of reducing sugar and ethanol yield profiles for sonicated samples at high-power level for 10 s (S 10) and 30 s (S 30) and the control (nonsonicated). (Adapted from Nitayavardhana, S. et al. 2010. *Bioresource Technology*, 10:2741–2747.)

produce lignin that obscure the access of cellulase enzymes to cellulose through surface redeposition. The use of high-power ultrasonics before enzyme hydrolysis may disrupt lignin–cellulose hydrophobic interactions and enhances the enzymatic conversion of cellulose into fermentable sugars.

14.3 ULTRASOUND APPLICATION IN BIODIESEL PRODUCTION

The application of ultrasound in biodiesel production is gaining considerable attention lately. Improvements in mixing efficiency due to a streaming effect have the

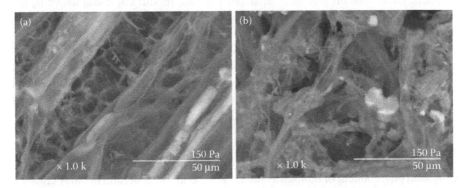

FIGURE 14.5 Scanning electron micrographs of alkaline-pretreated corn stover: (a) control (nonsonicated); (b) sonicated for 40 s.

potential to improve biodiesel yields, curtail processing times, and reduce the alcohol-base catalysts ratio. Rokhina et al. (2009) reviewed the application of low-frequency ultrasound in biofuel. The ultrasonic mixing has been studied using different catalysts and alcohols (Stavarache et al., 2005, 2006, 2007).

A significant reduction in the reaction time of mono-alcohols (e.g., methanol) and fatty acids in the presence of a base catalyst (e.g., NaOH or KOH) has been reported in comparison to conventional mechanical stirring (Stavarache et al., 2005). The authors also reported that 28 kHz ultrasonication for 60 min resulted in higher yields (99%) of fatty acid methyl esters—FAME (biodiesel) compared to mechanical stirring. In similar studies by Colucci et al. (2005), 20 kHz ultrasonication resulted in a FAME yield of 99.4% at a methanol to soybean oil ratio of 6:1. The improvement in mass transfer and reaction kinetics were attributed to an increase in interfacial area created by macro and micro bubbles during sonication. The increase in interfacial area is also due to a reduction in the droplet size of the reaction mixture (over 40%, following sonication) compared to mechanical agitation (Wu et al., 2007). There are several studies that reported a reduced time for FAME production at a 1:6 ratio of vegetable oil to alcohol–base mixture. Singh et al. (2007) and Chand et al. (2008) reported higher biodiesel yield in 5 and 1.5 min, respectively, via sonication under different laboratory conditions.

14.4   ULTRASOUND APPLICATION IN ALGAE CELL DISRUPTION

Microbial cell disruption by ultrasound is extensively employed in laboratory experiments for intracellular and metabolic products (Lambert, 1983; Chisti and Moo-Young, 1986). Lee et al. (2001) subjected blue-green algae samples to low-frequency sonication (28 kHz). The authors reported sonication as a promising method to control blue-green algal bloom. Disruption of algal cells and the collapse of gas vacuoles were confirmed via electron microscopy examination. In another study, Bosma et al. (2003) reported the application of ultrasound to harvest microalgae via acoustically induced aggregation and sedimentation in laboratory-scale experiments.

A Los Angeles-based company OriginOil, reported an economic way of extracting lipid by breaking algal cell walls via the combined effect of ultrasound and electromagnetic pulse. The oil was then separated from the mixture of disrupted cells, oil, and water mixture by lowering pH using carbon dioxide (Heger, 2009) and flotation. The skimmed bio-oil can be processed further to produce biodiesel using ultrasound.

The principles of sonochemistry can also be applied to disrupt different species of oil-bearing microalgae cells. Detailed experimental results are necessary to support the cost-effectiveness and industrial scale applicability of ultrasound for microbial lipid extraction, and subsequent biodiesel production in comparison to conventional methods (Mata et al., 2010).

14.5   ULTRASOUND APPLICATION IN MUNICIPAL SLUDGE

Chemical pretreatment of biological sludge using acid, base, or enzyme may enhance the sonication effect. Such chemicals weaken the cell wall, effectively disrupting the biological cells with less energy input. Bases such as NaOH and KOH are widely

TABLE 14.1

Methane Production, Methane Yield, and Methane Content Under Different Conditions

	Methane Production (mL day^{-1})	Gas Yield (m^3 CH$_4$ g^{-1} VS$_{removed}$)	Methane Content (%)
SRT: 25 days			
Control	0.56	358	58
Ultrasonic	0.56	404	57
Chemical-ultrasonic	0.52	420	58
SRT: 15 days			
Control	0.52	478	58
Ultrasonic	0.54	589	58
Chemical-ultrasonic	0.49	626	58
SRT: 10 days			
Control	0.34	485	58
Ultrasonic	0.4	636	59
Chemical-ultrasonic	0.38	688	59

Source: Data from Bunrith, S. 2008. Anaerobic digestibility of ultrasound and chemically pretreated waste activated sludge. Master's thesis, Asian Institute of Technology.

used to improve cell hydrolysis. Literature describing the effect of such pretreatment on ultrasound disintegration is not available currently. In a collaborative research project conducted at the Asian Institute of Technology (Thailand), the authors found significantly higher methane yield of chemical-ultrasound pretreated sludge with respect to control and sonicated samples as illustrated in Table 14.1.

14.6   SUMMARY

With rising fuel price, there is a growing interest toward maximizing biofuel/bioenergy production from all available renewable feedstocks. Most feedstocks with high solids contents, for example, corn, cassava chips, cellulosic biomass, biological sludge, animal manure, and food wastes, are ideal feedstocks for renewable energy production. One major challenge of these feedstocks is their slow digestibility due to a rate-limiting hydrolysis step. Pretreatment of these feedstocks is essential to enhance their hydrolysis/digestibility for biofuel/bioenergy generation. Pretreatment is still an emerging field and process comparisons are difficult with limited data. The focus of bench-scale or pilot-scale studies for most of the emerging pretreatment technologies is often narrow, and the reported conclusions must be synthesized from different sources to obtain a complete and balanced perspective. Information is also lacking on factors related to the operating parameters and costs involved to quantify the economic benefits from pretreatment. One of the key considerations in ultrasound pretreatment is net energy gain from the improved biofuel/bioenergy production.

REFERENCES

Anex, R.P., Lynd, L.R., Laser, M.S., Heggenstaller, A.H., and Liebman, M. 2007. Potential for enhanced nutrient cycling through coupling of agricultural and bioenergy systems. *Crop Science*, 47:1327–1335.

Bosma, R., van Spronsen, W.A., Tramper, J., and Wijffels, R.H. 2003. Ultrasound, a new separation technique to harvest microalgae. *Journal of Applied Phycology*, 15(2–3):143–153.

Bunrith, S. 2008. Anaerobic digestibility of ultrasound and chemically pretreated waste activated sludge. Master's thesis, Asian Institute of Technology.

Chand, P., Reddy, C.V., Verkade, J.G., and Grewell, D. 2008. Enhancing biodiesel production from soybean oil using ultrasonics. American Society of Agricultural and Biological Engineers (ASABE)——Annual International Meeting (AIM), Rhode Island, June 29 to July 2, 2008, Paper no. 084374.

Chisti, Y. and Moo-Young, M. 1986. Disruption of microbial cells for intracellular products. *Enzyme Microbiology and Technology*, 8:194–204.

Colucci, J.A., Borrero, E.E., and Alape, F. 2005. Biodiesel from an alkaline transesterification reaction of soybean oil using ultrasonic mixing. *Journal of the American Oil Chemists' Society*, 82:525–530.

Csóka, L., Lorincz, A., and Winkler, A. 2008. Sonochemically modified wheat straw for pulp and papermaking to increase its economical performance and reduce environmental issues. *BioResources*, 3:91–97.

Heger, M. 2009. A new processing scheme for algae biofuels in the World Wide Web. Technology Review. http://www.technologyreview.com/energy/22572/?a=f (accessed October 23, 2009).

Hielscher Ultrasonics. 2008. Ultrasonic transesterification of oil to biodiesel in the World Wide Web. Hielsher-Ultrasound Technology. http://www.hielscher.com/ultrasonics/biodiesel_transesterification_01.htm (accessed May 11, 2008).

Ji, J., Wang, J., Li, Y., Yu, Y., and Xu, Z. 2006. Preparation of biodiesel with the help of ultrasonic and hydrodynamic cavitation. *Ultrasonics*, 44:411–414.

Khanal, S.K., Grewell, D., Sung, S., and van Leeuwen, J. 2007a. Ultrasound applications in wastewater sludge pretreatment: A review. *Critical Reviews in Environmental Science and Technology*, 37:1–37.

Khanal, S.K., Montalbo, M., van Leeuwen, J., Srinivasan, G., and Grewell, D. 2007b. Ultrasound enhanced glucose release from corn in ethanol plants. *Biotechnology and Bioengineering*, 98:978–985.

Lambert, P.W. 1983. Industrial enzyme production and recovery from filamentous fungi. In J.E. Smith, D.R. Berry, and B. Kristiansen (Eds), *The Filamentous Fungi—Vol. IV (Fungal Technology)*, pp. 210–237. London: Edward Arnold.

Lee, T.J., Nakano, K., and Matsumara, M. 2001. Ultrasonic irradiation for blue-green algae bloom control. *Environmental Technology*, 22(4):383–390.

Li, H., Pordesimo, L., and Weiss, J. 2004. High intensity ultrasound-assisted extraction of oil from soybeans. *Food Research International*, 37:731–738.

Lynd, L.R., Cushman, J.H., Nichols, R.J., and Wyman, C.E. 1991. Fuel ethanol from cellulosic biomass. *Science*, 251:1318–1323.

Mata, T.M., Martins, A.A., and Caetano, N.S. 2010. Microalgae for biodiesel production and other applications: A review. *Renewable and Sustainable Energy Reviews*, 14:217–232.

Montalbo-Lomboy, M., Srinivasan, G., Raman, R.D., Anex, R.P., and Grewell, D. 2007. Influence of ultrasonics in ammonia steeped switchgrass for enzymatic hydrolysis. American Society of Agricultural and Biological Engineers (ASABE)—Annual International Meeting (AIM), St. Joseph, Michigan, 2007, Paper no. 076231.

Nitayavardhana, S., Rakshit, S.K., Grewell, D., Pometto III, A.L., van Leeuwen, J., and Khanal, S.K. 2008. Ultrasound pretreatment of cassava chip slurry to enhance sugar release for subsequent ethanol production. *Biotechnology and Bioengineering*, 101(3):487–496.

Nitayavardhana, S., Shrestha, P., Rasmussen, M.L., Lamsal, B.P, van Leeuwen, J., and Khanal, S.K. 2010. Ultrasound improved ethanol fermentation from cassava chips in cassava-based ethanol plants. *Bioresource Technology*, 10:2741–2747.

Rokhina, E.V., Lens, P., and Virkutyte, J. 2009. Low-frequency ultrasound in biotechnology: State of the art. *Trends in Biotechnology*, 27(5):298–306.

Schmer, M.R., Vogel, K.P., Mitchell, R.B., and Perrin, R.K. 2008. Net energy of cellulosic ethanol from switchgrass. *Proceedings of the National Academy of Sciences USA*, 105:464–469.

Singh, A.K., Fernando, S.D., and Hernandez, R. 2007. Base-catalyzed fast transesterification of soybean oil using ultrasonication. *Energy and Fuels*, 21:1161–1164.

Stavarache, C., Vinatoru, M., and Maeda, Y. 2006. Ultrasonic versus silent methylation of vegetable oils. *Ultrasonics Sonochemistry*, 13:401–407.

Stavarache, C., Vinatoru, M., and Maeda, Y. 2007. Aspects of ultrasonically assisted transesterification of various vegetable oils with methanol. *Ultrasonics Sonochemistry*, 14:380–386.

Stavarache, C., Vinatoru, M., Nishimura, R., and Maeda, Y. 2005. Fatty acids methyl esters from vegetable oil by means of ultrasonic energy. *Ultrasonics Sonochemistry*, 12:367–372.

Suslick, K.S. 1990. Sonochemistry. *Science*, 247:1439–1445.

USDA and USDOE Joint Report. 2005. A Billion-Ton Feed Stock Supply for Bioenergy and Bioproducts Industry: Technical Feasibility of Annually Supplying 1 Billion Dry Tons of Biomass. Joint Report—U.S. Department of Agriculture and U.S. Department of Energy, February 2005.

Wood, B.E., Aldrich, H.C., and Ingram, L.O. 1997. Ultrasound stimulates ethanol production during the simultaneous saccharification and fermentation of mixed waste office paper. *Biotechnology Progress*, 13(13):232–237.

Wu, P., Yang, Y., Colucci, J.A., and Grulke, E.A. 2007. Effect of ultrasonication on droplet size in biodiesel mixtures. *Journal of the American Oil Chemists' Society*, 84:877–884.

15 Microbes as Green and Eco-Friendly Nanofactories

Rashmi Sanghi and Preeti Verma

CONTENTS

15.1 INTRODUCTION

Nanotechnology is enabling technology at a leading edge that deals with nanometer-sized objects (Feynman, 1991) at several levels: materials, devices, and systems. At present, the nanomaterials level is the most advanced, both in scientific knowledge and in commercial applications. A decade ago, nanoparticles were studied because of their size-dependent physical and chemical properties (Murray et al., 2000). However now, they have entered a commercial exploration period (Mazzola, 2003; Paull et al., 2003). The development of synthetic procedures for uniform nanometer-sized particles is essential for many advanced applications (Schmid et al., 2004) because the monodispersity of metal nanoparticles induces their precise size and shape-dependent properties. From this point of view, a current research interest has been focused on the synthesis and properties of metal nanoparticles. Therefore, a large number of methods have been developed for the synthesis of metal nanoparticles over the last decades. A wide variety of physical, chemical, and biological processes results in the synthesis of nanoparticles, some of these are novel and others

315

are quite common. Normally, the size, shape, and surface modification of metal nanoparticles are some of the most important factors that may dramatically affect their physical/chemical properties. For example, the silver nanoparticles can be prepared by the chemical reduction of a homogeneous solution containing silver salt and a large amount of stabilizers (Bradley and Schmid, 2004), such as reverse micelles (Taleb et al., 1997), surfactants (Liz-Marzan and Tourinō, 1996), dendrimers (Balogh et al., 2001), alkanethiol (Pileni, 2001), alkylamine (Bunge et al., 2003), and carboxylic acid (Lin et al., 2003). Despite numerous reports, the synthetic methods based on chemical reduction still have difficulty in removing halides, by-products, and excess stabilizers such as surfactants.

Metal nanoparticles are generally prepared by reduction of metal salts either by borohydride, alcohols, citrates, or alkyl sulfates. The use of a strong reductant, such as borohydride, resulted in small particles that were somewhat monodispersed, but controlling the generation of the larger particles became difficult. The use of citrate, a weaker reductant, resulted in a slower reduction rate, but the size distribution was far from narrow. Metal nanoparticles are also produced by ligand exchange with a suitable organic molecule/biomolecule containing an SH or NH_2 group as a stabilizer, and in some cases, nanoparticles have also been prepared by reducing the metal ions in the presence of a stabilizer (Gole et al., 2001). One of the disadvantages of this method is that there is always a chance of particle aggregation during the process of ligand exchange and centrifugation for the removal of the undesired product (Hayat, 1991). Also, the chemical reduction method generally yields metal nanoparticles with wide size distributions (Nickel et al., 2000).

Therefore, the development of reliable experimental protocols for the synthesis of nanomaterials over a range of chemical compositions, sizes, and high monodispersity is one of the challenging issues in current nanotechnology. There is a need to develop an environment-friendly approach for nanomaterial synthesis that should not use toxic chemicals in the synthesis protocol. Several manufacturing techniques that usually employ atomistic, molecular, and particulate processing in a vacuum or in a liquid medium are in use (Daniel and Astruc, 2004). Most of the techniques are capital intensive, as well as inefficient in materials and energy use. Hence, there is an ever-growing need to develop clean, nontoxic, and environmentally benign synthesis procedures. Synthesis of metal nanoparticles of various size and shape, and their colloidal stabilization through biomolecule immobilization is very essential due to their usefulness in many biotechnological applications, such as sensors (Kim et al., 2006), optical (Krolikowska et al., 2003), catalysis (Gole et al., 2001), chemical (Kumar et al., 2003), photoelectrochemical (Chandrasekharan and Kamat, 2000), and electronic (Peto et al., 2002). In addition, these nanoparticles can be used as optical probes as a result of their color change upon the coupling of surface plasmon resonances of adjacent nanoparticles.

Currently, there is a growing need to develop environmentally benign nanoparticle synthesis processes that do not use toxic chemicals in the synthesis protocol. As a result, researchers in the field of nanoparticles synthesis and assembly have turned to biological systems for inspiration. The secrets gleaned from nature have led to the development of biomimetic approaches to the growth of advanced nanomaterials (Sastry et al., 2003). Besides this, the utilization of nontoxic chemicals, environmentally benign solvents and renewable materials are some of the key issues that merit a green synthesis protocol for nanoparticles. Increased industrialization and urbanization have damaged the environment by introducing a number of harmful and unwanted substances. Microorganisms have been exposed to a variety of such pollutants in the environment (including water and soil). Of which, metal ion, being nonbiodegradable and persistent in nature often causes toxicity and inhibits microbial growth. However, even at high metal ion concentration, microorganisms can survive and grow due to their ability to fight the metal stress. The mechanisms include efflux systems; alteration of solubility and toxicity via reduction or oxidation; biosorption; bioaccumulation; extracellular complexation or precipitation of metals; and lack of specific metal transport systems (Beveridge et al., 1997; Bruins et al., 2000). Metal ions in solution are generally adsorbed to the microbial surface by chemical functional groups on the cell wall biopolymers and then *in situ* reduced to metal atoms by reducing sugars from hydrolysates of polysaccharides of the biomass (Lin et al., 2005). Also, responsible for the reduction of metals are enzymes such as reductases in various microbes (Duran et al., 2005; Rosen, 2002). The nucleation and growth of the inorganic structures are mostly controlled by proteins and other biomacromolecules (Aizenberg, 1996; Cha et al., 1999; Klaus et al., 1999). These metal–microbe interactions have important roles in several biotechnological applications including the fields of bioremediation, biomineralization, bioleaching, and microbial corrosion. However, it is only recently that microorganisms have been explored as potential biofactory for synthesis of metallic nanoparticles, such as cadmium sulfide, gold, and silver (Sastry et al., 2003). In general, the synthesis of nanoparticles can be covered under three important research themes:

1. The chemical reduction of metal precursors by a reducing agent such as $NaBH_4$, citrate, and ascorbate (Setua et al., 2007).
2. The irradiation of the solution containing metal ions with ultraviolet or visible light, and microwave and ultrasound irradiation (Rocha et al., 2007).
3. The utilization of biomolecules and bio-organisms, such as bacteria, proteins, and polysaccharide, for the synthesis of Ag and gold nanoparticles (Chandran et al., 2006).

Recently, more and more researchers have turned to theme-3, now that the other two themes may be associated with environmental toxicity or biological hazards, while the biosynthesis could be easily controlled under mild and eco-friendly conditions. The use of microorganisms in the deliberate and controlled synthesis of nanoparticles is a relatively new and exciting area of research with considerable potential for development. Also, these methods involve the use of environmental-friendly aqueous solvents at biological pH. Furthermore, due to slower kinetics, they offer better manipulation and control over crystal growth and their stabilization. The work on the use of microorganisms to synthesize advanced nanoscale materials is sporadic. Among these microorganisms, only a few groups had been confirmed to be able to selectively reduce certain metal ions (Kamilo et al., 1991). Hence, both unicellular and multicellular organisms are known to produce inorganic materials either intra- or extracellularly (Mann, 1996).

Using metal-accumulating microorganisms as a tool for the production of nano-particles, and their assembly for the construction of new advanced materials, is a completely new technological approach. This chapter reviews the work reported by various researchers over the last one decade on the use of microorganisms for pro-duction of nanoparticles starting from the very primitive microorganisms the prokaryotes (bacteria and actinomycetes) to the more developed eukaryotes (algae, yeast, and fungi). The possible mechanisms postulated by various researchers have also been discussed.

15.2 BIOLOGICAL SYNTHESIS OF NANOPARTICLES

In modern nanoscience and technology (Zhu et al., 2000), the interaction between inorganic nanoparticles and biological structures are one of the most exciting areas of research. In this chapter, some of the known microbes being used in the formation of nanoparticles have been discussed and are summarized in Table 15.1.

TABLE 15.1

Use of Microbes in the Synthesis of Nanoparticles

Microbes	Nanoparticles	Extracellular/ Intracellular	References
Bacillus subtilis (bacterium)	Au		Beveridge and Murray (1980)
Candida glabrata (yeast)	CdS	Intracellular	Dameron et al. (1989)
Schizosaccharomyces pombe (yeast)	CdS	Intracellular	Dameron et al. (1989)
Klebsiella aerogenes (bacterium)	CdS	Intracellular	Holmes and coworkers (1995)
Pseudomonas stutzeri (bacterium)	Ag		Klaus et al. (1999); Joerger et al. (2000)
Verticillium (fungus)	Ag	Intracellular	Mukherjee et al. (2001)
Verticillium (fungus)	Au	Intracellular	Mukherjee et al. (2001a)
Magnetotactic bacteria	Magnetic (Fe_3O_4), greigite (Fe_3S_4)		Roh et al. (2001)
Desulfovibrio desulfuricans (bacterium)	Palladium		Yong et al. (2002)
Schizosaccharomyces pombe (yeast)	CdS	Intracellular	Kowshik et al. (2002)
Torulopsis sp. (yeast)	PbS	Intracellular	Kowshik and coworkers (2002)
Lactobacillus strains (bacterium)	Ag, Au, Ag–Au		Nair and Pradeep (2002)
Fusarium oxysporium (fungus)	Au		Mukherjee et al. (2002)
Yeast strain MKY3	Ag		Kowshik et al. (2003)
Thermomonospora sp. (actinomycete)	Au	Extracellular	Sastry et al. (2003)

TABLE 15.1 (continued)
Use of Microbes in the Synthesis of Nanoparticles

Microbes	Nanoparticles	Extracellular/ Intracellular	References
Verticillium sp. and *Fusarium oxysporum* (fungus)	Au, Ag		Sastry et al. (2003)
Fusarium oxysporum (fungus)	Ag		Ahmad et al. (2003b)
Rhodococcus sp. (actinomycete)	Au	Intracellular	Ahmad et al. (2003b)
Colletotrichum sp. (fungus)	Au	Extracellular	Shankar et al. (2003)
Thermomonospora sp. (actinomycetes)	Au	Extracellular	Ahmad et al. (2003c)
Verticillium (fungus)	Ag	Intracellular	Senapati et al. (2004)
Trichothecium sp. (fungus)	Au	Extracellular and intracellular	Ahmad et al. (2005)
Fusarium oxysporum (fungus)	Silica and titanium particles (SiF_6^{2-} and TiF_6^{2-})	Extracellular	Bansal et al. (2005)
Aspergillus fumigatus (fungus)	Ag	Extracellular	Bhainsa and D'Souza (2006)
Fusarium oxysporum (fungus)	Barium titanate	Extracellular	Bansal et al. (2006)
Fusarium oxysporum and *Verticillium* sp. (fungus)	Magnetite	Extracellular	Bharde et al. (2006)
Phaenerochaete chrysosporium (fungus)	Ag	Extracellular	Vigneshwaran et al. (2006)
P. jadinii (yeast)	Au	Intracellular	Gericke and Pinches (2006b)
V. luteoalbum (fungus)	Au	Intracellular	Gericke and Pinches (2006b)
Plectonema boryanum UTEX 485 (cyanobacterium)	Octahedral Au platlets	At the cell wall	Lengke et al. (2006a, 2006b)
Pseudomonas aeruginosa (bacterium)	Au	Extracellular	Husseiny et al. (2007)
Aspergillus flavus (fungus)	Ag		Vigneshwaran et al. (2007)
Nitrate reductases from *Fusarium oxysporum* (fungus)	Ag	Extracellular	Kumar et al. (2007a)
Fusarium oxysporum (fungus)	CdSe quantum dots	Extracellular	Kumar et al. (2007b)
Sargassum wightii (marine alga)	Au		Singaravelu et al. (2007)
Rhodopseudomonas capsulate (bacterium)	Au		He et al. (2007, 2008)
Helminthosporum solani (bacterium)	Au		Kumar et al. (2008)
Bacillus licheniformis (bacterium)	Ag	Extracellular	Kalishwaralal et al. (2008)
Fusarium solani (fungus)	Ag	Extracellular	Ingle et al. (2008)
Fusarium semitectum (fungus)	Au and Au–Ag alloy	Extracellular	Sawle et al. (2008)

continued

TABLE 15.1 (continued)
Use of Microbes in the Synthesis of Nanoparticles

Microbes	Nanoparticles	Extracellular/ Intracellular	References
Fusarium semitectum (fungus)	Ag	Extracellular	Basavaraja et al. (2008)
Trichoderma asperellum (fungus)	Ag		Mukherjee et al. (2008)
Bacillus licheniformis (bacterium)	Au		Kalishwarala et al. (2009)
Lactobacillus sp. and *Saccharomyces cerevisiae* (bacterium)	TiO$_2$ nanoparticles		Jha et al. (2009)
T. viride	Ag		Fayaz et al. (2009)
Pathogenic fungus Phytophthora infestans	Ag	Extracellular	Thirumurugan et al. (2009)
Hormoconis resinae (fungus)	Ag	Extracellular	Varshney et al. (2009)
Penicillium (fungus)	Ag	Extracellular	Maliszewska et al. (2009)
Penicillium brevicompactum WA 2315 (fungus)	Ag	Extracellular	Shaligram et al. (2009)
Penicillium fellutanum (fungus)	Ag		Kathiresan et al. (2009)
Coriolus versicolor (fungus)	Ag		Sanghi and Verma (2009a, 2009b)
Coriolus versicolor (fungus)	CdS	Extracellular	Sanghi and Verma (2009a, 2009b)

15.2.1 SYNTHESIS OF NANOPARTICLES USING BACTERIA

Although, silver in general is highly toxic to most microbial cells, most of the work reported on silver nanoparticles production is by the use of bacterial cells. Several bacterial strains are reported as silver resistant (Silver, 2003) and can accumulate silver at the cell wall to as much as 25% of the dry weight biomass. The silver-resistant bacterial strain *Pseudomonas stutzeri* AG259 accumulates silver nanoparticles, along with some silver sulfide, in the cell where particle size ranges from 35 to 46 nm (Slawson et al., 1992). The bacteria *Ps. stutzeri* AG259 isolated from a silver mine, when placed in a concentrated aqueous solution of AgNO$_3$, resulted in the formation of the silver-based single crystals with well-defined compositions and shapes up to 200 nm at different cellular binding sites of the bacterium (Joerger et al., 2000; Klaus et al., 1999). The bacterial cells could accumulate the triangular, hexagonal, and spheroidal-shaped silver crystals in large quantities and frequently, the cell poles serve as preferred accumulation sites. Large particles with distinct shapes as well as small colloidal particles were found all over the cell wall.

Nair and Pradeep (2002) have reported that common *Lactobacillus* strains found in buttermilk, when challenged with silver and gold ions, assisted the growth of microscopic gold, silver, and gold silver alloy crystals of well-defined morphology within the bacterial cells. However, the exact reaction mechanism leading to the formation of silver nanoparticles by this species of silver-resistant bacteria was not elucidated. Kalishwaralal et al. (2008) reported the synthesis of silver nanoparticles by reduction of aqueous Ag^+ ions with the culture supernatant of *Bacillus licheniformis*. The synthesis of nanoparticle circles around enzyme nitrate reductase, which exists in *B. licheniformis*. This enzyme has been previously used for *in vitro* synthesis of silver nanoparticle under anaerobic conditions. Nitrate reductase is known to shuttle electron from nitrate to the metal group and may be involved in the synthesis of silver nanoparticles in the range of 50 nm. The silver nanoparticles synthesized are highly stable and this method has advantages over other methods as the organism used here is a nonpathogenic bacterium leading to an easy procedure for producing silver nanoparticles with the added advantage of biosafety.

Beveridge et al. (1997) have demonstrated that gold particles of nanoscale dimensions may be readily precipitated within bacterial cells by incubation of the cells with Au^{3+} ions. Beveridge and Murray (1980) first demonstrated that the exposure of *Bacillus subtilis* treated with gold chloride resulted in the synthesis of gold nanoparticles. Biosynthesis of gold nanoparticles has been carried out using *Rhodopseudomonas capsulate* (He et al., 2007, 2008) and *Helminthosporum solani* (Kumar et al., 2008).

Pseudomonas aeruginosa is a Gram-negative bacterium that is capable of existing in multiple environmental niches and is an opportunistic pathogen, meaning that it exploits some break in the host defenses to initiate an infection. Husseiny et al. (2007) have reported the extracellular synthesis of gold nanoparticles using cell supernatant of the bacterium *P. aeruginosa*. The particle size and distribution was of the order of 40–15 nm and as the particle size increases, the color was found to be shifted from pink to blue due to the surface plasmon of Au NPs. The extracellular formation of particles offered a great advantage over an intracellular process of synthesis from the application point of view. The extracellular synthesis would make the process simpler and easier for downstream processing since the nanoparticles formed inside the bacteria would have required additional step of processing for the release of the nanoparticles from the bacteria by ultrasound treatment or by reaction with suitable detergents.

Morphological control over the shape of gold nanoparticles has been achieved by using *Plectonema boryanum* UTEX 485, a filamentous cyanobacterium by Lengke et al. (2006a). When it was reacted with aqueous $Au(S_2O_3)_2^{3-}$ and $AuCl_4^-$ solutions at 25–100°C for up to 1 month and at 200°C for 1 day resulted in the precipitation of cubic gold nanoparticles and octahedral gold platelets, respectively. Lengke et al. (2006b) further demonstrated the mechanisms of gold bioaccumulation by cyanobacteria (*P. boryanum* UTEX 485) from gold (III)–chloride solutions. The interaction of cyanobacteria with aqueous gold (III)–chloride initially promoted the precipitation of nanoparticles of amorphous gold (I)–sulfide at the cell walls, and finally deposited metallic gold in the form of octahedral (III) platelets near cell surfaces and in solution. Adding further to the mechanism, a sulfate-reducing bacterial enrichment was used to destabilize gold (I)–thiosulfate complex to elemental gold. It was proposed by Lengke and Southam (2006) that this could occur by three possible mechanisms involving iron sulfide, localized reducing conditions, and metabolism. In future, it would be important to understand the biochemical and molecular mechanism of nanoparticles synthesis by the cell filtrate in order to achieve better control over size and polydispersity of the nanoparticles.

Kalishwarala et al. (2009) have reported the synthesis of stable gold nanocubes by the reduction of aqueous $AuCl_4^-$ by *B. licheniformis*, at room temperature. They have reported a single-step process without the requirement of toxic chemicals and stringent conditions. The aqueous chloroaurate ions were reduced to metallic gold (Au^0) on exposure to the bacterial biomass. The color of the reaction solution turned from pale yellow to dark purple, which indicated the formation of gold nanoparticles (Kumar et al., 2008). The UV–vis spectra results indicated that the reaction solution has an absorption maximum at about 540 nm attributable to the surface plasmon resonance band (SPR) of the gold nanoparticles. Observation of this peak, assigned to a surface plasmon, is well documented for various metal nanoparticles with sizes ranging from 2 to 100 nm (Sastry et al., 1998). Metabolic products secreted by the bacteria, such as proteins, organic acids, and polysaccharides, are expected to interact with the crystal faces, thereby changing the surface energies of the latter in due course. In this process, the bacteria functions as a cellular efflux pumping system and a periplasmic protein that binds gold specifically at the cell surface (Beveridge and Murray, 1980; Matias and Beveridge, 2008) bringing about alteration of solubility and toxicity via reduction, biosorption, bioaccumulation, and lack of specific metal transport systems. The mechanism of reduction of Au nanoparticles in the bacterium is still not well understood. Previous reports have claimed that an enzyme belonging to the NADH reductase family is involved (He et al., 2007; Kumar et al., 2008). The reduction of $AuCl_4^-$ ions to Au^0 ions facilitates the synthesis of nanocubes. The primarily formed gold nanoparticles are thermodynamically unstable in aqueous solution because of insufficient capping agent. At this point, they would tend to form linear assemblies driven by Brownian motion and short-range interaction. It is probable that proteins acted as the major biomolecules involved in the bioreduction and facilitated the synthesis of gold nanoparticles. It was shown that the concentration of gold ions played an important role in forming and stabilizing the shape of gold nanocubes (Delapierre et al., 2008; He et al., 2008). Even other anisotropic metal nanostructures are expected using the biosynthetic method.

Similarly, in the presence of exogenous electron donor, sulfate reducing bacterium *Desulfovibrio desulfuricans* NCIMB 8307 has been shown to be synthesizing palladium nanoparticles (Yong et al., 2002). Magnetotactic bacteria (Roh et al., 2001) such as *Magnetospirillum magneticum* produces two types of particles; magnetite nanoparticles in chains and/or greigite nanoparticles. Some well-known examples of bio-organisms synthesizing inorganic materials include diatoms synthesizing siliceous materials (Mann, 2001) and S-layer bacteria producing gypsum and calcium carbonate layers (Pum and Sleytr, 1999).

Jha et al. (2009) reported a low-cost green and reproducible microbes (*Lactobacillus* sp. and *Saccharomyces cerevisiae*)-mediated biosynthesis of TiO_2 nanoparticles at room temperature in the laboratory ambience. The synthesis of n-TiO_2 might have resulted due to pH-sensitive membrane-bound oxido-reductases and carbon source-dependent rH2 in the culture solution. Holmes et al. (1995) have shown that the bacterium, *Klebsiella aerogenes* when exposed to Cd^{2+} ions resulted in the intracellular formation of CdS nanoparticles in the size range 20–200 nm. They also showed that the composition of the nanoparticles formed was a strong function of buffered growth medium for the bacteria.

15.2.2 SYNTHESIS OF NANOPARTICLES USING YEAST

After an extensive screening program, Kowshik et al. (2002) identified the yeast, *Torulopsis* sp. as being capable of intracellular synthesis of nanoscale PbS crystallites when exposed to aqueous Pb^{2+} ions. The PbS nanoparticles were extracted from the biomass by freeze-thawing and analyzed by a variety of techniques. Dameron et al. (1989) used *Candida glabrata* and *Schizosaccharomyces pombe* for the first time in the biosynthesis of cadmium sulfide (CdS) nanocrystals. These nanocrystals were produced using cadmium salts and are now used in quantum semiconductor crystallites. Further experiments were conducted to improve the quantity of semiconductor CdS nanocrystals production that was achieved by using *S. pombe* cells. When these cells were incubated with 1 mM Cd during their mid-log phase of growth, maximal nanocrystals were obtained (Kowshik et al., 2002) suggesting that the formation of CdS nanocrystals was dependent on the growth phase of yeast. When Cd was added during stationary phase, its uptake as well as production of CdS nanocrystals was decreased or resulted in no CdS formation. Upon adding Cd during early exponential phase of yeast growth, CdS nanocrystals were formed but this time it was affecting the cellular metabolism of the yeast and resulted in efflux of Cd from the cells (Williams et al., 1996). The possible mechanism of decrease in CdS nanocrystals formation could be that upon exposure to Cd as a stress, a series of biochemical reactions were triggered to overcome the toxic effects of this metal. First, an enzyme phytochelatin synthase was activated to synthesize phytochelatins (PC) that chelated the cytoplasmic Cd to form a low molecular weight PC–Cd complex and ultimately transport them across the vacuolar membrane by an ATP-binding cassette-type vacuolar membrane protein (HMT1). In addition to Cd, sulfide could also be added to this complex in the membrane resulting in the formation of high molecular weight $PCCdS^{2-}$ complex that also allowed sequestering into vacuole (Ortiz et al., 1995).

Conditions have also been standardized for the synthesis of large quantities of silver nanoparticles by using silver-tolerant yeast strain MKY3. On the basis of differential thawing of the samples, the procedure for separation of these silver particles has also been documented (Kowshik et al., 2003). Yeast strains have been identified for their ability to produce gold nanoparticles, whereby controlled size and shape of the nanoparticles could be achieved by controlling the growth and other cellular activities (Gericke and Pinches, 2006a). Gericke and Pinches (2006b) have reported the intracellular synthesis of gold nanoparticles using yeast *P. jadinii*.

15.2.3 SYNTHESIS OF NANOPARTICLES USING ALGAE

The reduction of aqueous $AuCl_4^-$ by the extract of marine alga *Sargassum wightii* resulted in the formation of high density, extremely stable gold nanoparticles in the size ranging from 8 to 12 nm with an average size ca. 11 nm (Singaravelu et al., 2007).

15.2.4 SYNTHESIS OF NANOPARTICLES USING ACTINOMYCETES

Sastry et al. (2003) observed that the extremophilic actinomycete, *Thermomonospora* sp. when exposed to gold ions reduced the metal ions extracellularly, yielding gold nanoparticles with a much improved polydispersity. In an effort toward elucidating mechanism or conditions favoring the formation of nanoparticles with desired features, Ahmad et al. (2003c) also carried out the reduction of $AuCl_4^-$ ions by using extremophilic *Thermomonospora* sp. biomass resulting in efficient synthesis of monodisperse gold nanoparticles. It was concluded by them that the reduction of metal ions and stabilization of the gold nanoparticles occurred by an enzymatic process (Ahmad et al., 2003b, 2003c). The nanoparticles formation could be due to extreme biological conditions such as alkaline and slightly elevated temperature conditions used for the synthesis of nanoparticles. On the basis of the hypothesis, alkalotolerant *Rhodococcus* sp. was used for the intracellular synthesis of good quality monodisperse gold nanoparticles (Ahmad et al., 2003b). It was observed that the concentration of nanoparticles were more on the cytoplasmic membrane than on the cell wall. This could be due to reduction of the metal ions by enzymes present in the cell wall and on the cytoplasmic membrane but not in the cytosol. These metal ions were not toxic to the cells, which were produced by them, and continued to multiply even after the biosynthesis of gold nanoparticles (Ahmad et al., 2003b).

15.2.5 SYNTHESIS OF NANOPARTICLES USING FUNGUS

Since fungi are known to secrete much higher amounts of proteins, they might have significantly higher productivity of nanoparticles in biosynthetic approach. Compared to bacteria, fungi could be a source for the production of a larger amount of nanoparticles. Filamentous fungi are very good candidates for such processes and furthermore, these biomasses are easy to handle.

The exposure of *Verticillium* sp. to silver ions resulted in a similar intracellular growth of silver nanoparticles (Mukherjee et al., 2001a). Mukherjee et al. (2001b) have reported the intracellular synthesis of gold nanoparticles using the fungus *Verticillium* sp. They have reported that the trapping of gold ions on the surface of fungal cells could occur by electrostatic interaction with positively charged groups (e.g., lysine residues) in enzymes that are present in the cell wall of the mycelia. It can be concluded from their study that fungi could be a source for the production of a large amount of nanoparticles. Mukherjee et al. (2002) have elucidated the mechanism of nanoparticles formation, as *in vitro* approach was followed where species-specific NADH-dependent reductase, released by the *Fusarium oxysporium*, was successfully used to carry out the reduction of gold ions to gold nanoparticles. Sastry et al. (2003) have reported that fungus *Verticillium* sp. and *F. oxysporum*, when exposed to gold and silver ions, reduced the metal ion rather rapidly and formed respective metallic nanoparticles. Endophytic fungus *Colletotrichum* sp. growing in the leaves of geranium was used for the synthesis of stable and various shaped gold nanoparticles. The reducing agents in this fungus were also polypeptides/enzymes (Shankar et al., 2003). Senapati et al. (2004) reported a novel biological method for the intra- and extracellular synthesis of silver nanoparticles using the fungi, *Verticillium* and *F. oxysporum*, respectively. This has opened up an exciting possibility wherein the nanoparticles may be entrapped in the biomass in the form of a film or produced in solution, both having interesting commercial potential.

Growth conditions also play an important role during the synthesis of nanoparticles while using the fungi cultures. When gold ions were incubated with the *Trichothecium* sp. biomass under stationary conditions, led to the formation of extracellular nanoparticles. While under shaking conditions, this resulted in the formation of intracellular gold nanoparticles. The possible reason for this could be the enzymes and proteins responsible for the synthesis of nanoparticles. These proteins were released into the medium under stationary conditions and did not release under shaking conditions (Ahmad et al., 2005).

Bharde et al. (2006) reported the synthesis of magnetic nanoparticles by using *F. oxysporum* and *Verticillium* sp. at room temperature. Both fungi secreted proteins, which were capable of hydrolyzing iron precursors extracellularly to form iron oxides predominantly in the magnetite (Fe_3O_4) phase. Tetragonal barium titanate ($BaTiO_3$) nanoparticles of sub-10 nm dimensions produced by *F. oxysporum* under ambient conditions have been observed to be eco-friendly and economically viable methods for the synthesis of complex oxide nanomaterials of technological interest (Bansal et al., 2006). Also, the synthesis of highly luminescent CdSe quantum dots at room temperature, reported recently by the fungus, *F. oxysporum* when incubated with a mixture of $CdCl_2$ and $SeCl_4$ would be of great importance (Kumar et al., 2007b). Bansal et al. (2005) have reported the formation of silica and titania nanoparticles from aqueous anionic complexes SiF_6^{2-} and TiF_6^{2-}, respectively using fungus *F. oxysporum*. Extracellular protein-mediated hydrolysis of the anionic complexes results in the facile room temperature synthesis of crystalline titania particles whereas calcination at 300°C is required for the crystallization of silica. Gericke and Pinches (2006b) have also reported the intracellular synthesis of gold nanoparticles using fungus *V. luteoalbum*.

The fungus, *Aspergillus flavus* when challenged with silver nitrate solution accumulated silver nanoparticles on the surface of its cell wall in 72 h. These nanoparticles dislodged by ultrasonication showed an absorption peak at 420 nm in UV–vis spectrum corresponding to the plasmon resonance of silver nanoparticles (Vigneshwaran et al., 2007). Synthesis of nanoparticles was found to be intracellular in all the examples given above except for *F. oxysporum*. Fungi are found to be capable of reducing the metal ions into their corresponding nanometals and the process can be either intracellular or extracellular depending on the position of the reduction enzymes. The intracellular production of nanoparticles makes the job of downstream processing difficult and beats the purpose of developing a simple and cheap process. Therefore, in recent times, research is focused on the development of an extracellular process.

Bhainsa and D'Souza (2006) investigated extracellular biosynthesis of silver nanoparticles using filamentous fungus *A. fumigatus*. The synthesis process was quite fast and silver nanoparticles were formed within minutes of silver ion coming in contact with the cell filtrate. *A. fumigatus* showed potential for extracellular synthesis of fairly monodispersed, silver nanoparticles in the range of 5–25 nm. The kinetics of silver nanoparticles synthesis using the cell filtrate indicated that the rapid synthesis of nanoparticles would be suitable for developing a biological process for mass scale production. Furthermore, the extracellular synthesis would make the process simpler and easier for downstream processing. Vigneshwaran et al. (2006) have reported the extracellular biosynthesis of silver nanoparticles using fungus *Phaenerochaete chrys-osporium*. The mycelial mat of *P. chrysosporium* when challenged with silver nitrate reduced the toxicity of silver ions by reducing them to silver metals in 24 h. They have confirmed that the protein is responsible for the stabilization of silver nanoparticle. The synthesized silver nanoparticles absorbed maximum at 470 nm in the visible region. IR spectroscopic study has confirmed that the carbonyl groups from amino acid residues and peptides of proteins have a stronger ability to bind metal, so that the proteins could most possibly form a coat covering the metal nanoparticle to prevent gathering of the particles. This evidence suggests that the biological molecules could possibly perform the function for the stabilization of the silver nanoparticles.

Achievement of the biosyntheses of metallic and bimetallic nanoparticles contributes to an increase in the efficiency of biosynthetic procedures using environment-benign resources as an alternative to chemical synthesis protocols. Crystallized and spherical-shaped Au and Au–Ag alloy nanoparticles have been synthesized and stabilized using a fungus, *Fusarium semitectum* in an aqueous system. Aqueous solutions of chloroaurate ions for Au and chloroaurate and Ag^+ ions (1:1 ratio) for Au–Ag alloy were treated with an extracellular filtrate of *F. semitectum* biomass for the formation of Au nanoparticles (AuNP) and Au–Ag alloy nanoparticles (Au–AgNP). Reduction of the equimolar mixture of Ag^+ and $AuCl_4^-$ ions by employing an extracellular solution of the biomass of *F. semitectum* was monitored as a function of reaction time by UV–vis spectroscopy measurement, and the spectra obtained at 420 nm. In addition to the peak at 420 nm, another peak at 378 nm is also seen that increases in intensity with time and appears as a shoulder in the UV–vis spectra after 3 h of the reaction (Sawle et al., 2008).

Thus, filamentous fungi are preferred over bacteria and unicellular organisms as they are capable of extracellular synthesis of metal nanoparticles, making downstream processing and biomass handling easier. But to date, all filamentous fungi such as *F. oxysporum* (Ahmad et al., 2003a) and *A. fumigatus* (Bhainsa and D'Souza, 2006) that have been reportedly used for the purpose of extracellular biomass-free synthesis of silver nanoparticles are pathogenic to either plants (Kolombet et al., 2004) and/or humans (Casadevall and Pirofski, 1999). This makes handling and disposal of the biomass a major inconvenience toward commercialization of the process and it would not be advisable to use these fungi for the aforesaid purpose. Hence, a newer approach of testing a nonpathogenic fungus for the successful synthesis and capping of nanosized silver grains was necessary. Ingle et al. (2008) have selected *Fusarium solani* for the synthesis of silver nanoparticles within range of 5–35 nm because it is a well-known plant pathogen and causes infection in many economically important plants. The exploitation of such pathogenic fungi is beneficial because it helps to study and detect the infections in plants as well as useful in nanoparticle synthesis. Up till now only *F. oxysporum* have been used for the synthesis of nanoparticles while other species are not well studied but from the present investigation it is clear that the other *Fusarium* species such as *F. solani* also have potential to synthesize the silver nanoparticles, which is an eco-friendly, safe, and reliable approach. The extracellularly formed silver nanoparticles are quite stable in solution due to capping of silver nanoparticles by proteins secreted by the fungus. This is an efficient, eco-friendly, and simple process.

Keeping this in view, Mukherjee et al. (2008) have developed a methodology to prepare metal nanoparticles using *Trichoderma asperellum*. Continuing the author's efforts in that direction, several agriculturally important fungi were investigated for the synthesis of silver nanoparticles in the range of 13–18 nm under optimized culture conditions. However, *T. asperellum* was found to yield the most reproducible results with respect to stability and size of the silver nanoparticles.

Basavaraja et al. (2008) reported the extracellular biosynthesis of silver nanoparticles using the filtrate of the fungus *F. semitectum*. Highly-stable and crystalline silver nanoparticles are produced in solution by treating them with the aqueous silver nitrate solution. Their studies suggest that the protein might have played an important role in the stabilization of silver nanoparticles through coating of protein moiety on the silver nanoparticles. Thirumurugan et al. (2009) have investigated the biological synthesis of silver nanoparticles by potato plant disease causing fungus *Phytophthora infestans* and the synthesized silver nanoparticles were found to be most active against the clinically isolated human pathogenic bacteria. The results proved that silver nanoparticles showed maximum activity at a least concentration, which revealed silver nanoparticles as novel antibacterial agents. Fayaz et al. (2009) have demonstrated simple, stable, and efficient biological method for synthesis of silver nanoparticles using fungus, *T. viride*. The synthesis is carried out at various temperature conditions and at lower reaction temperature the sizes of the nanoparticles were increased, whereas increases in temperature results decrease in size of the nanoparticles. An important potential benefit of this research is the development of eco-friendly protocol for biosynthesis of silver nanoparticles ranging from 2 to 4 nm, by controlling the reaction temperature.

Maliszewska et al. (2009) have reported the biological process for formation of silver nanoparticles using *Penicillium* sp. J3 strain isolated from the soil. The reduction of the metal ions occurs on the surface of the cells leading to the formation of nanosilver. The results of transmission electron microscopy (TEM) and scanning electron microscopy (SEM) suggested that the protein might have played an important role in the formation and stabilization of silver nanoparticles. Shaligram et al. (2009) have demonstrated an eco-friendly process for the extracellular synthesis of stable silver nanoparticles using the fungus, *Penicillium brevicompactum* WA 2315. The fungus has been previously utilized to study fermentative production of compactin in submerged and solid-state fermentation (Shaligram et al., 2008). The supernatant of the seed media obtained after separating the cells has been used for the synthesis of silver nanoparticles. After 72 h of treatment, silver nanoparticles obtained were in the range of 23–105 nm.

The mechanistic aspect of the fungal reduction of metal ions led by colloidal suspension is still an open question. However, in the fungal case, this process occurs probably either by reductase action or by electron shuttle quinines, or both. To elucidate the mechanism of nanoparticles formation, a novel fungal/enzyme-based *in vitro* approach was for the first time explained by Mukherjee et al. (2002). They successfully used species-specific NADH-dependent reductase, released by the *F. oxysporum*, to carry out the reduction of $AuCl_4^-$ ions to gold nanoparticles. Duran et al. (2005) later reported that the reduction of the metal ions occurs by a nitrate-dependent reductase and a shuttle quinone extracellular process. The same observation was reported with another strain of *F. oxysporum* and it was pointed out that this reductase was specific to *F. oxysporum*. However, *Fusarium moniliforme*, did not result in the formation of silver nanoparticles, neither intracellularly nor extracellularly but contained intra and extracellular reductases in a similar fashion as *F. oxysporum* (Ahmad et al., 2003a). This is an indication that probably the reductases in *F. moniliforme* are important for Fe (III) to Fe (II) but not to Ag (I) to Ag (0). Moreover, in *F. moniliforme* anthraquinone derivatives were not detected unlike the case of *F. oxysporum*. Both *Fusarium* were alike in the production of naphthaquinones (Medentsev and Alimenko, 1998) but differed in the production of anthraquinones. The UV–vis spectra from the *F. oxysporum* 07SD strain reaction with silver nitrate at different times of reaction showed strong surface plasmon resonance centered at ca. 415–420 nm, which clearly increased in intensity with time. The solution was extremely stable, with no evidence of flocculation of the particles even several weeks after reaction. The UV–vis spectra in low wavelength region recorded from the reaction medium exhibited an absorption band at ca. 265 nm and it was attributed to aromatic amino acids of proteins. It is well-known that the absorption band at ca. 265 nm arises due to electronic excitations in tryptophan and tyrosine residues in the proteins. This observation indicates the release of proteins into solution by *F. oxysporum* and suggests a possible mechanism for the reduction of the metal ions present in the solution. The FTIR analysis indicated the strong possibility of tryptophan, tyrosine, cysteine, and methionine as the main proteins involved in this reduction process (Ahmad et al., 2002). Earlier studies on enzyme-mediated synthesis of nanoparticles by the fungus have also indicated the presence of four bands in SDS-PAGE studies. Enzymes such as silver reductase (Rosen, 2002) and the amino acids (present in proteins/enzymes) such as aspartic acid and tryptophan (Mandal et al., 2005) have been reported in various microbial systems responsible for the reduction of metal ions. Toward elucidating the mechanism of synthesis of nanoparticles, a NADPH-dependent nitrate reductase and phytochelatin isolated from *F. oxysporum* has been used for *in vitro* silver nanoparticle production (Kumar et al., 2007a).

Minimum time, miniaturization, and nonhazardous process are key parameters for any kind of technology acceptance. The process should be such that it is suitable for speedy mass scale production of nanoparticles. Varshney et al. (2009) reported the extracellular synthesis of silver nanoparticles by previously unexploited non-pathogenic fungus *Hormoconis resinae* within 6 h in the range of 20–80 nm. On treatment of aqueous solutions of silver with fungus, silver nanoparticles could be rapidly fabricated within an hour. These nanoparticles were not in direct contact even within the aggregates indicating stabilization of the nanoparticles by a capping agent. These are surrounded by a faint thin layer of other material, which could be the capping organic material from *H. resinae* biomass. These may be tryptophan and tyrosine residues present in the proteins/enzymes found in the fungus.

Kathiresan et al. (2009) reported the biosynthesis of silver nanoparticles using $AgNO_3$ as a substrate by *Penicillium fellutanum* isolated from coastal mangrove sediment. The biosynthesis was faster within minutes of silver ion coming in contact with the cell filtrate and was maximum when the culture filtrate was treated with

1.0 mM AgNO$_3$, maintained at 0.3% NaCl and pH 6.0, incubated at 5°C for 24 h. The culture filtrate, precipitated with ammonium sulfate, showed a single protein band with a molecular weight of 70 kDa using PAGE.

15.2.6 Synthesis of Ag Nanoparticles Using White Rot Fungus, Coriolus Versicolor

Working toward an eco-friendly, simple yet speedy approach Sanghi and Verma (2009a) developed a one-step, easy, cheap, and convenient method for the biosynthesis of silver colloid nanoparticles using white rot fungus, *C. versicolor*. Initially, the fungus when challenged with silver nitrate solution accumulated silver nanoparticles on the surface of its cell wall in 72 h. On optimization of experimental parameters, the nanoparticle formation started within 5–10 min under alkaline conditions. The production was even faster compared to the reported physical and chemical processes of nanoparticles synthesis. The resulting Ag nanoparticles displayed controllable structural and optical properties depending on the experimental parameters, such as reactive temperatures, and concentrations of silver salt as suggested by different colors of Ag colloids. The washed fungal mycelium and the growth medium in which the fungus was harvested were separately challenged with 1 mM of AgNO$_3$ and incubated in a shaker at 37°C under dark conditions. Dark-brown deposits were seen on the fungal mycelium on the fourth day, which further darkened with time. Under normal pH conditions, the characteristic surface plasmon absorption band at 440 nm was observed on the second day that attained the maximum intensity on the third day in the case of media. With mycelium the time taken was much longer almost a difference of one day and the band was blue shifted to 430 nm with respect to Ag/media solution (Figure 15.1).

To optimize the reduction process, variations in the parameters such as pH, concentration of AgNO$_3$, and temperature were made. Under alkaline conditions and at room temperature, the color changes with time from colorless to light pink, reddish brown, and finally to dark brown (Figure 15.2) with increasing intensity and were indicative of the formation of Ag nanoparticles both intracellularly on fungal mycelium and extracellularly in media.

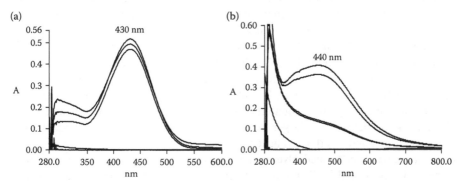

FIGURE 15.1 UV–vis spectra of the (a) Ag/media and (b) Ag/mycelium solutions at normal pH.

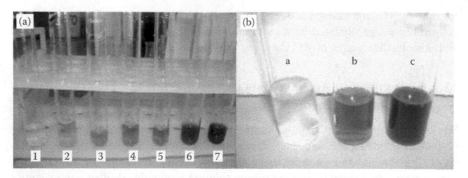

FIGURE 15.2 1 and 2 at normal pH on 2nd and 3rd day, respectively; 3 and 4 are Ag/mycelium at 30 min and 1 h, respectively; 5 is Ag/mycelium (1 h) at pH 10; 6 and 7 are Ag/medium and Ag/mycelium at pH 12 (1 h) (a), and Ag/medium at pH 10 with 0.2, 1, and 3 mM concentrations of AgNO₃, respectively (b).

The syntheses performed at various Ag concentrations and pH conditions produced a wide range of particle sizes (25–450 nm) with narrow size distributions. At pH 10, the absorbance band was blue shifted to 420 nm in Ag/fungal mycelium solution as well as Ag/media with respect to normal conditions and was much broader and asymmetric, and expected to contain double peaks as compared to normal conditions indicating decreased particle size. The absorbance at around 220 and 280 nm maintained the same intensity throughout the incubation time under normal pH conditions. With increase in pH to 10, both in mycelium and in media experiments, the peak at 220 and the hump at 280 was much intensified showing the significant role played by the aromatic proteins tryptophan and tyrosine in silver reduction. The average size, size distribution, morphology, and structure of particles were determined by atomic force microscopy (AFM), TEM, and UV–vis absorption spectrophotometry. Fourier transform infrared (FTIR) study disclosed that the amino groups were bound to the particles, which was accountable for the stability of nanoparticles. It further confirmed the presence of protein as the stabilizing and capping agent surrounding the silver nanoparticles. The coordination behaviors between amino groups in fungus and Ag⁺ ions are responsible for the stabilization of the silver nanoparticles.

The mechanistic route followed for the reduction process was different in the case of Ag/mycelium and Ag/media. In media, glucose was found mainly responsible for the reduction whereas in the case of mycelium, it was mainly the S-H group responsible for the same. The photoluminescence spectrum of these protein-stabilized silver nanoparticles also showed much enhanced fluorescence emission intensity.

AFM study was carried out to investigate any morphology change in formed silver nanoparticles that occurs from 1 and 3 mM concentration of silver nitrate solution in both fungal mycelium solution and media under alkaline conditions at pH 10. Clearly, the particles are in larger numbers, much denser, and closely packed and hence agglomerated together in clusters (Figure 15.3) as compared to that at normal pH. The shape of the formed silver nanoparticles is spherical in both cases and their particle size is around 25–75 nm in Ag/media and 444–491 nm in Ag/fungal mycelium. The particle size of nanoparticles of Ag/media is much lesser as compared to Ag/fungal mycelium, which could be due to the effect of the presence of reducing

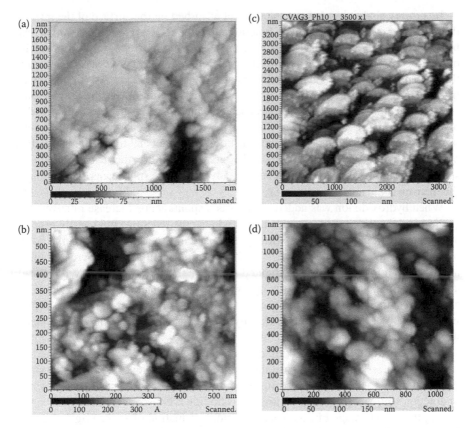

FIGURE 15.3 AFM of Ag/fungus solution at pH 10. (a) and (b) are Ag/media samples at 1 and 3 mM, respectively; (c) and (d) are Ag/mycelium samples at 1 and 3 mM, respectively.

agent glucose. With increase in the concentration of silver nitrate in both fungal media as well as mycelium solution from 1 to 3 mM, the shape of the formed silver nanoparticles though remains more or less the same but their particle size is reduced to around 18–48 nm in Ag/media and 107–213 nm in Ag/mycelium solution and arranged in monodisperse form. The ordered layers seen in the case of mycelium solution get disordered and are now seen as clusters in mycelium with increase in concentration. Both TEM and AFM data confirmed that the use of reducing agents yields smaller particle sizes at the same concentration of silver nitrate.

In general, the reduction reaction of metallic ions is sensitive to the solution's pH as it may affect the product's morphology via the formation of certain species. It has been mentioned by many researchers that the addition of alkaline ion is necessary to carry out the reduction reaction of metallic ions. Under normal conditions, when no hydroxide ion was added to the system, the time taken for the reduction of Ag ions was 3 days for media and 4 days for fungal mycelium, indicating the necessity of OH$^-$ to this reduction reaction. The absence of glucose in the fungal mycelium experiments is the possible reason for delayed reduction, whereby the S-H of the protein plays a key role in the reduction process.

In a rotary shaker, Ag ions in solution get adsorbed on the surface of the mycelia through interactions with chemical functional groups such as carboxylate anion, carboxyl, and peptide bond of proteins. The mycelia being more immobile are capable of binding Ag^+ more than that of the external cellular substances that distributed in the inter-mycelial space, thus reducing most of the Ag^+ *in situ* to Ag^0. In the meantime, stronger adsorptive groups such as the carbonyl group on the extracellular substances could further adsorb the particles located on the surface of the mycelia, resulting in capping these nanoparticles, while rocking. When other Ag^+ in the solution was rocked onto this overlay and was bound and reduced to Ag^0 on the surface of the layer and these Ag^0 might be further coated with the other extracellular substances; this process was repeated continuously until these substances distributing in the inter-mycelial space was used up.

When hydroxide ion was added, there was a spontaneous increase of silver conversion and the time taken was much less and also the reaction could easily proceed at room temperature without any stirring. This is mainly because the reducing power of the responsible protein, which acts as the reducing agent is much increased under alkaline conditions. Both cases exhibited abrupt changes initially followed by gradual and continuous changes, suggesting a very fast reaction at the beginning and then a slower reaction at the later stage for this process. As observed in the first 15 min, when $AgNO_3$ was mixed with NaOH, only Ag_2O was obtained in the case of media whereas in the case of fungal mycelium it was Ag_2S as well.

Most of the initially added alkaline ions reacted with Ag^+ to form an intermediate Ag_2O/Ag_2S, most of which was instantly reduced to Ag^0 after 30 min. At this stage, the alkaline ion may be involved in reduction of silver ion, water hydrolysis, and interaction with protein functionalities. This finding is consistent with the notion that during the second stage of this reaction system, reduction of silver might occur on the surface of existing colloids. More investigation is however required to further elucidate this aspect of reaction mechanism.

15.3 SYNTHESIS OF CdS NANOPARTICLES USING IMMOBILIZED FUNGUS, *C. VERSICOLOR*

Sanghi and Verma (2009b) have also reported for the first time a green route to prepare autocapped CdS nanoparticles under ambient aqueous conditions using immobilized fungus *C. versicolor* in continuous column mode. The average size of the particles estimated from the TEM image is about 5–9 nm (Figure 15.4a). The particles seem to be in groups of different sizes surrounded by a thin outer, not so dense layer, which is perhaps the protein covering. The CdS nanoparticles visualized by SEM showed micro to nanospheres in the 100–200 nm range (Figure 15.4b). CdS nanoparticles exhibited a uniform morphology, the size estimation by SEM cannot be relied upon as large variations in particle size were observed from a few to approximately 100 nm in diameter.

The CdS nanoparticles were autocapped by the fungal proteins, which were responsible for imparting them high stability. The colloidal solution of as-prepared CdS nanoparticles was extremely stable with no evidence for aggregation even after 10 months of continuously running the column. The long-term stability of the as-prepared CdS

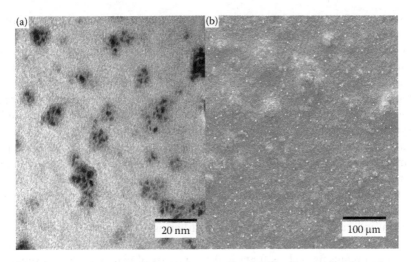

FIGURE 15.4 Transmission electron micrographs of CdS nanoparticles (a) and scanning electron micrographs of CdS nanoparticles (b).

nanoparticle solution is due to the presence of the proteins in the nanoparticle solution that bind to the surface of the nanoparticles and prevent aggregation.

The biocatalytic and stability characteristics of the enzymes secreted by the fungus on exposure to cadmium ions clearly highlighted the potential of this approach both in bioremediation as well as large-scale mineral growth of CdS NPs. Interestingly, no external source of sulfur is required and the thiol group of the fungal protein was found responsible for the transformation of toxic Cd to nontoxic CdS, which is well-known for its excellent optical properties. The thiol (SH) group of Cys in *C. versicolor* plays a critical role for binding to and sequestration of Cd^{2+} as well as reduction to CdS (Figure 15.5).

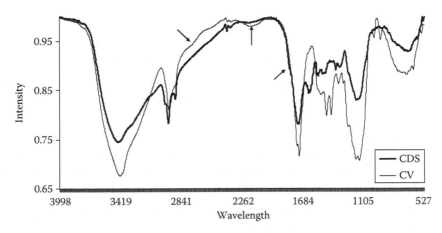

FIGURE 15.5 Shows FTIR spectra of as-prepared CdS nanoparticles as well as spectra of plain fungus (CV).

This is a safe, low cost, and more convenient approach as it does not involve special instrumentation, poisonous intermediates, and the growing rate can also be easily controlled. The utility of this process is underlined by the fact that even after continuous exposure to the toxic metal ions, the fungus readily grows and transforms the toxic conditions to nontoxic by reducing Cd to CdS without the use of any external source of sulfur. Another important, potential benefit of the process described is the fact that the semiconductor CdS nanoparticles, which are quite stable in solution, are synthesized extracellularly in large quantities. This is therefore, a very important advantage over other biosynthetic methods where the nanoparticles are entrapped within the cell matrix in limited quantity whereby an additional processing is required to release them from the matrix.

Extracellular secretion of enzymes offers the advantage of obtaining large quantities in a relatively pure state, free from other cellular proteins associated with the organism and can be easily processed by filtering of the cells and isolating the enzyme for nanoparticles synthesis from cell-free filtrate. The use of specific enzymes secreted by organisms such as fungi in the synthesis of nanoparticles is exciting for the following reasons. The process can be extended to the synthesis of nanoparticles of different chemical compositions and indeed, different shapes and sizes by suitable identification of enzymes secreted by the fungi (Sastry et al., 2003). Understanding the surface chemistry of the biogenic nanoparticles (i.e., nature of capping surfactants/peptides/proteins) would be equally important.

15.4 CONCLUSIONS AND FUTURE DIRECTIONS

In summary, an overview of the use of microorganisms such as bacteria, yeasts, algae, fungi, and actinomycetes in the biosynthesis of metal nanoparticles has been described. As can be seen from the above, the use of microorganisms in the synthesis of nanoparticles is a new and exciting area of research with considerable potential for green development.

REFERENCES

Ahmad, A., Mukherjee, P., and Mandal, D. 2002. Enzyme mediated extracellular synthesis of CdS nanoparticles by the fungus, *Fusarium oxysporum*. *Journal of the American Chemical Society*, 124:12108–9.

Ahmad, A., Mukherjee, P., and Senapati, S. 2003a. Extracellular biosynthesis of silver nanoparticles using the fungus *Fusarium oxysporum*. *Colloids and Surfaces B: Biointerfaces*, 28:313–8.

Ahmad, A., Senapati, S., and Khan, M.I. 2003b. Intracellular synthesis of gold nanoparticles by a novel alkalotolerant actinomycete, *Rhodococcus* species. *Nanotechnology*, 14:824–8.

Ahmad, A., Senapati, S., Khan, M.I., Kumar, R., and Sastry, M. 2003c. Extracellular biosynthesis of monodisperse gold nanoparticles by a novel extremophilic actinomycete, *Thermomonospora* sp. *Langmuir*, 19:3550–3.

Ahmad, A., Senapati, S., Khan, M.I., Kumar, R., and Sastry, M. 2005. Extra-/intracellular, biosynthesis of gold nanoparticles by an alkalotolerant fungus, *Trichothecium* sp. *Journal of Biomedical Nanotechnology*, 1:47–53.

Aizenberg, J. 1996. Stabilization of amorphous calcium carbonate by specialized macromolecules in biological and synthetic precipitates. *Advanced Materials*, 8:222–5.

Balogh, L., Swanson, D.R., Tomalia, D.A., Hagnauer, G.L., and McManus, A.T. 2001. Dendrimer–silver complexes and nanocomposites as antimicrobial agents. *Nano Letters*, 1:18–21.

Bansal, V., Poddar, P., Ahmad, A., and Sastry, M. 2006. Room-temperature biosynthesis of ferroelectric barium titanate nanoparticles. *Journal of the American Chemical Society*, 128:11958–63.

Bansal, V., Rautaray, D., Bharde, A., Ahire, K., Sanyal, A., Ahmad, A., and Sastry, M. 2005. Fungus-mediated biosynthesis of silica and titania particles. *Journal of Material Chemistry*, 15:2583–9.

Basavaraja, S., Balaji, S.D., Lagashetty, A., Rajasab, A.H., and Venkataraman, A. 2008. Extracellular biosynthesis of silver nanoparticles using the fungus *Fusarium semitectum*. *Materials Research Bulletin*, 43:1164–70.

Beveridge, T.J. and Murray, R.G.E. 1980. Sites of metal deposition in the cell wall of *Bacillus subtilis*. *Journal of Bacteriology*, 41:876–87.

Beveridge, T.J., Hughes, M.N., Lee, H., Leung, K.T., Poole, R.K., Savvaidis, I., Silver, S., and Trevors, J.T. 1997. Metal–microbe interactions: Contemporary approaches. *Advances in Microbial Physiology*, 38:177–243.

Bhainsa, K.C. and D'Souza, S.F. 2006. Extracellular biosynthesis of silver nanoparticles using the fungus *Aspergillus fumigatus*. *Colloids and Surfaces B: Biointerfaces*, 47:160–4.

Bharde, A., Rautaray, D., Bansal, V., Ahmad, A., Sarkar, I., Yusuf, S.M., Sanyal, M., and Sastry, M. 2006. Extracellular biosynthesis of magnetite using fungi. *Small*, 2:135–41.

Bradley, J.S. and Schmid, G. 2004. Clusters and colloids. In: *Nanoparticles: From Theory to Applications*. Schmid, G. (ed.). Chapter 3.2.1, pp. 186–99. Weinheim, Germany: VCH.

Bruins, R.M., Kapil, S., and Oehme, S.W. 2000. Microbial resistance to metal in the environment. *Ecotoxicology and Environmental Safety*, 45:198–207.

Bunge, S.D., Boyle, T.J., and Headley, T.J. 2003. Synthesis of coinage-metal nanoparticles from mesityl precursors. *Nano Letters*, 3:901–5.

Casadevall, A. and Pirofski, L.A. 1999. Host–pathogen interactions: Redefining the basic concepts of virulence and pathogenicity. *Infection and Immunity*, 67:3703–13.

Cha, J.N., Shimizu, K., Zhou, Y., Christiansen, S.C., Chmelka, B.F., Stucky, G.D., and Morse, D.E. 1999. Silicatein filaments and subunits from a marine sponge direct the polymerization of silica and silicones *in vitro*. *Proceedings of the National Academy of Sciences USA*, 96:361–35.

Chandran, S.P., Chaudhary, M., Pasricha, R., Ahmad, A., and Sastry, M. 2006. Synthesis of gold nanotriangles and silver nanoparticles using *Aloe vera* plant extract. *Biotechnology Progress*, 22:577–83.

Chandrasekharan, N. and Kamat, P.V. 2000. Improving the photoelectrochemical performance of nanostructured TiO_2 films by adsorption of gold nanoparticles. *Journal of Physical Chemistry B*, 104:10851–7.

Dameron, C.T., Reeser, R.N., Mehra, R.K., Kortan, A.R., Carroll, P.J., Steigerwald, M.L., Brus, E., and Winge, D.R. 1989. Biosynthesis of cadmium sulphide quantum semiconductor crystallites. *Nature*, 338:596–7.

Daniel, M.-C. and Astruc, D. 2004. Gold nanoparticles: Assembly, supramolecular chemistry, quantum-size-related properties, and applications toward biology, catalysis, and nanotechnology. *Chemical Reviews*, 104:293–346.

Delapierre, T.M., Majimel, J., Mornet, S., Duguet, E., and Ravaine, S. 2008. Synthesis of non-spherical gold nanoparticles. *Gold Bulletin*, 41:195–207.

Duran, N., Marcato, P.D., Alves, O.S., Souza, G.D., and Esposito, E. 2005. Mechanistic aspects of biosynthesis of silver nanoparticles by several *Fusarium oxysporum* strains. *Journal of Nanobiotechnology*, 3:1–7.

Fayaz, M.A., Balaji, K., Kalaichelvan, P.T., and Venkatesan R. 2009. Fungal based synthesis of silver nanoparticles—An effect of Temperature on the size of particles. *Colloids and Surfaces B: Biointerfaces*, 74(1):126–36.

Feynman, R. 1991. There's plenty of room at the bottom. *Science*, 254:1300–1.

Gericke, M. and Pinches, A. 2006a. Biological synthesis of metal nanoparticles. *Hydrometallurgy*, 83:132–40.

Gericke, M. and Pinches, A. 2006b. Microbial production of gold nanoparticles. *Gold Bulletin*, 39:22–8.

Gole, A., Dash, C., Ramakrishnan, V., Sainkar, S.R., Mandale, A.B., Rao, M., and Sastry, M. 2001. Pepsin–gold colloid conjugates: Preparation, characterization, and enzymatic activity. *Langmuir*, 17:1674–9.

Hayat, M.A. 1991. *Colloidal Gold: Principles, Methods, and Applications*. San Diego, CA: Academic Press.

He, S., Guo, Z., Zhang, Y., Zhang, S., Wang, J., and Gu, N. 2007. Biosynthesis of gold nanoparticles using the bacteria *Rhodopseudomonas capsulate*. *Materials Letters*, 61(18):3984–7.

He, S., Zhang, Y., Guo, Z., and Gu, N. 2008. Biological synthesis of gold nanowires using extract of *Rhodopseudomonas capsulate*. *Biotechnology Progress*, 24(2):476–80.

Holmes, J.D., Smith, P.R., Evans-Gowing, R., Richardson, D.J., Russell, D.A., and Sodeau, J.R. 1995. Energy-dispersive-X-ray analysis of the extracellular cadmium sulfide crystallites of *Klebsiella aerogenes*. *Archives of Microbiology*, 163:143–7.

Husseiny, M.I., Abd El-Aziz, M., Badr, Y., and Mahmoud, M.A. 2007. Biosynthesis of gold nanoparticles using *Pseudomonas aeruginosa*. *Spectrochimica Acta Part A*, 67:1003–6.

Ingle, A., Rai, M., Gade, A., and Bawaskar, M. 2009. *Fusarium solani*: A novel biological agent for the extracellular synthesis of silver nanoparticles. *Journal of Nanoparticle Research*, 11(8):2079–85.

Jha, A.K., Prasad, K., and Kulkarni, A.R. 2009. Synthesis of TiO$_2$ nanoparticles using microorganisms. *Colloids and Surfaces B: Biointerfaces*, 71(2):226–9.

Joerger, R., Klaus, T., and Granqvist, C.G. 2000. Biologically produced silver-carbon composite materials for optically functional thin film coatings. *Advanced Materials*, 12:407–9.

Kalishwaralal, K., Deepak, V., Ramkumarpandian, S., Nellaiah, H., and Sangiliyandi, G. 2008. Extracellular biosynthesis of silver nanoparticles by the culture supernatant of *Bacillus licheniformis*. *Materials Letters*, 62:4411–3.

Kalishwaralal, K., Deepak, V., Ramkumarpandian, S., and Sangiliyandi G. 2009. Biological synthesis of gold nanocubes from *Bacillus licheniformis*. *Bioresource Technology*, 100:5356–8.

Kamilo, M., Suzuki, T., and Kawai, K. 1991. Accumulation of rare-earth elements by microorganisms. *Bio Industry (Japan)*, 16:36–42.

Kathiresan, K., Manivannan, S., Nabeel, M.A., and Dhivya, B. 2009. Studies on silver nanoparticles synthesized by a marine fungus, *Penicillium fellutanum* isolated from coastal mangrove sediment. *Colloids and Surfaces B: Biointerfaces*, 71:133–7.

Kim, C.K., Kalluru, R.R., Singh, J.P., Fortner, A., Griffin, J., Darbha, G.K., and Ray, P.C. 2006. Gold nanoparticle based miniaturized laser induced fluorescence probe for specific DNA hybridization detection: Studies on size dependent optical properties. *Nanotechnology*, 17:3085–93.

Klaus, T., Joerger, R., Olsson, E., and Granqvist, C.G. 1999. Silver-based crystalline nanoparticles, microbially fabricated. *Proceedings of National Academy of Science, USA*, 96:13611–4.

Kolombet, L.V., Sokolov, M.S., Chuprina, V.P., Schisler, D.A., and Samuels, G.J. 2004. *8th International Workshop on Trichoderma and Gliocladium, Trichoderma, and the Environment*. Hangzhou, 394–5.

Kowshik, M., Ashtaputre, S., Kharrazi, S., Vogel, W., Urban, J., Kulkarni, S.K., and Paknikar, K.M. 2003. Extracellular synthesis of silver nanoparticles by a silver-tolerant yeast strain MKY3. *Nanotechnology*, 14:95–100.

Kowshik, M., Deshmukh, N., Vogel, W., Urban, J., Kulkarni, S.K., and Paknikar, K.M. 2002. Microbial synthesis of semiconductor CdS nanoparticles, their characterization, and their use in the fabrication of an ideal diode. *Biotechnology and Bioengineering*, 78:583–8.

Krolikowska, A., Kudelski, A., Michota, A., and Bukowska, J. 2003. SERS studies on the structure of thioglycolic acid monolayers on silver and gold. *Surface Science*, 532:227–32.

Kumar, A., Mandal, S., Selvakannan, P.R., Parischa, R., Mandale, A.B., and Sastry, M. 2003. Investigation into the interaction between surface-bound alkylamines and gold nanoparticles. *Langmuir*, 19:6277–82.

Kumar, S.A., Abyaneh, M.K., Gosavi, S.W. et al. 2007a. Nitrate reductase-mediated synthesis of silver nanoparticles from AgNO$_3$. *Biotechnology Letters*, 29:439–45.

Kumar, S.A., Ansary, A.A., Abroad, A., and Khan, M.I. 2007b. Extracellular biosynthesis of CdSe quantum dots by the fungus, *Fusarium oxysporum*. *Journal of Biomedical Nanotechnology*, 3:190–4.

Kumar, S.A., Peter, Y.A., and Nadeau, J.L. 2008. Facile biosynthesis, separation, and conjugation of gold nanoparticles to doxorubicin. *Nanotechnology*, 19:495101.

Lengke, M., Fleet, M.E., and Southam, G. 2006a. Morphology of gold nanoparticles synthesized by filamentous cyanobacteria from gold (I) thiosulfate and gold(III)–chloride complexes. *Langmuir*, 22:2780–7.

Lengke, M., Ravel, B., Fleet, M.E., Wanger, G., Gordon, R.A., and Southam, G. 2006b. Mechanisms of gold bioaccumulation by filamentous cyanobacteria from gold (III)–chloride complex. *Environmental Science & Technology*, 40:6304–9.

Lengke, M. and Southam, G. 2006. Bioaccumulation of gold by sulfate-reducing bacteria cultured in the presence of gold (I)-thiosulfate complex. *Geochimica et Cosmochimica Acta*, 70:3646–1.

Lin, X.Z. Teng, X., and Yang, H. 2003. Direct synthesis of narrowly dispersed silver nanoparticles using a single-source precursor. *Langmuir*, 19:10081–5.

Lin, Z., Wu, J., Xue, R., and Yong, Y. 2005. Spectroscopic characterization of Au^{3+} biosorption by waste biomass of *Saccharomyces cerevisiae*. *Spectrochimica Acta Part A*, 61:761–5.

Liz-Marzan, L.M.L. and Tourinõ, I. 1996. Reduction and stabilization of silver nanoparticles in ethanol by nonionic surfactants. *Langmuir*, 12:3585–9.

Maliszewska, I., Szewczyk, K., and Waszak, K. 2009. Biological synthesis of silver nanopartilces. *2nd National Conference on Nanotechnology 'NANO 2008' IOP Publishing Journal of Physics Conference Series*, 146:012025.

Mandal, S., Phadtare, S., and Sastry, M. 2005. Interfacing biology with nanoparticles. *Current Applied Physics*, 5:118–27.

Mann, S. 2001. *Biomineralization: Principles and Concepts in Bioinorganic Materials Chemistry*. Oxford: Oxford University Press.

Mann, S. (ed.) 1996. *Biomimetic Materials Chemistry*, pp. 1–40. New York: VCH Publishers.

Matias, V.R.F. and Beveridge, T.J. 2008. Lipoteichoic acid is a major component of the *Bacillus subtilis* periplasm. *Journal of Bacteriology*, 190:7414–8.

Mazzola, L. 2003. Commercializing nanotechnology. *Nature Biotechnology*, 21:1137–43.

Medentsev, A.G. and Alimenko, V.K. 1998. Naphthoquinone metabolites of the fungi. *Phytochemistry*, 47:935–59.

Mukherjee, P., Ahmad, A., Mandal, D., Sainkar, S.R., Khan, M.I., Parischa, R., Ajayakumar, P.V., Alam, M., Kumar, R., and Sastry, M. 2001a. Fungus mediated synthesis of silver nanoparticles and their immobilization in the mycelial matrix: A novel biological approach to nanoparticle synthesis. *Nano Letters*, 1:515–9.

Mukherjee, P., Ahmad, A., Mandal, D., Senapati, S., Sainkar, S.R., Khan, M.I., Ramani, R., Parischa, R., Ajayakumar, P.V., Alam, M., Sastry, M., and Kumar, R. 2001b. Bioreduction of AuCl₄⁻ ions by the fungus, *Verticillium* sp. and surface trapping of the gold nanoparticles formed. *Angewandte Chemie International Edition*, 40:3585–8.

Mukherjee, P., Roy, M., Dey, G.K., Mukherjee, P.K., Ghatak, J., Tyagi, A.K., and Kale, S.P. 2008. Green synthesis of highly stabilized nanocrystalline silver particles by a non-pathogenic and agriculturally important fungus *T. asperellum. Nanotechnology*, 19:art. no. 0751031.

Mukherjee, P., Senapati, S., Mandal, D., Ahmad, A., Khan, M.I., Kumar, R., and Sastry, M. 2002. Extracellular synthesis of gold nanoparticles by the fungus *Fusarium oxysporum. ChemBioChem*, 3:461–3.

Murray, C.B., Kagan, C.R., and Bawendi, M.G. 2000. Synthesis and characterization of monodisperse nanocrystals and close-packed nanocrystal-assemblies. *Annual Review of Material Science*, 30:545–610.

Nair, B. and Pradeep, T. 2002. Coalescence of nanoclusters and formation of submicron crystallites assisted by lactobacillus strains. *Crystal Growth and Design*, 2(4):293–8.

Nickel, U., Castell, A.Z., Pöppl, K., and Schneider, S. 2000. Silver colloid produced by reduction with hydrazine as support for highly sensitive surface-enhanced Raman spectroscopy. *Langmuir*, 16:9087.

Ortiz, D.F., Ruscitti, T., McCue, K.F., and Ow, D.M. 1995. Transport of metal binding peptides by HMT–1, a fission yeast ABC type vacuolar membrane protein. *Journal of Biological Chemistry*, 270:4721–8.

Paull, R., Wolfe, J., Hebert, P., and Sinkula, M. 2003. Investing in nanotechnology. *Nature Biotechnology*, 21:1134–47.

Peto, G., Molnar, G.L., Paszti, Z., Geszti, O., Beck, A., and Guczi, L. 2002. Electronic structure of gold nanoparticles deposited on SiOₓ/Si. *Materials Science and Engineering C*, 19:95–9.

Pileni, M.P. 2001. Nanocrystal self-assemblies: Fabrication and collective properties. *Journal of Physical Chemistry, B*, 105:3358–71.

Pum, D. and Sleytr, U.B. 1999. The application of bacterial S-layers in molecular nanotechnology. *Trends in Biotechnology*, 17:8–12.

Rocha, T.C.R., Winnischofer, H., Westphal, E., and Zanchet, D. 2007. Formation kinetics of silver triangular nanoplates. *Journal of Physical Chemistry, C*, 111:2885.

Roh, Y., Lauf, R.J., McMillan, A.D., Zhang, C., Rawn, C.J., Bai, J., and Phelps, T.J. 2001. Microbial synthesis and the characterization of metal-substituted magnetites. *Solid State Communications*, 118:529–34.

Rosen, B.P. 2002. Transport and detoxification systems for transition metals, heavy metals and metalloids in eukaryotic and prokaryotic microbes. *Comparative Biochemistry and Physiology—A Molecular and Integrative Physiology*, 133:689–93.

Sanghi, R. and Verma, P. 2009a. Biomimetic synthesis and characterization of protein capped silver nanoparticles. *Bioresource Technology*, 100(1):501–4.

Sanghi, R. and Verma, P. 2009b. A facile green extracellular biosynthesis of CdS nanoparticles by immobilized fungus. *Chemical Engineering Journal*, in press.

Sastry, M., Ahmad, A., Khan, M.I., and Kumar, R. 2003. Biosynthesis of metal nanoparticles using fungi and actinomycete. *Current Science*, 85:162–70.

Sastry, M., Patil, V., and Sainkar, S.R.J. 1998. Electrostatically controlled diffusion of carboxylic acid derivatized silver colloidal particles in thermally evaporated fatty amine films. *Journal of Physical Chemistry B*, 102:1404–10.

Sawle, B.D., Salimath, B., Deshpande, R., Bedre, M.D., Prabhakar, B.K., and Venkataraman, A. 2008. Biosynthesis and stabilization of Au and Au–Ag alloy nanoparticles by fungus, *Fusarium semitectum. Science and Technology of Advanced Materials*, 9: art. no. 035012.

Schmid, G., Talapin, D.V., and Shevchenko, E.V. 2004. In: Schmid, G. (ed.) *Nanoparticles: From Theory to Applications*, Chapter 4.2, pp. 251–98. Weinheim, Germany: VCH.

Senapati, S., Mandal, D., Ahmad, A., Khan, M.I., Sastry, M., and Kumar, R. 2004. Fungus mediated synthesis of silver nanoparticles: A novel biological approach. *Indian Journal of Physics*, 78A:101–5.

Setua, P., Chakraborty, A., Seth, D., Bhatta, M.U., Satyam, P.V., and Sarkar, N. 2007. Synthesis, optical properties, and surface enhanced Raman scattering of silver nanoparticles in nonaqueous methanol reverse micelles. *Journal of Physical Chemistry C*, 111:3901.

Shaligram, N.S., Bule, M., Bhambure, R., Singhal, R.S., Singh, S.K., Szakacs, G., and Pandey, A. 2009. Biosynthesis of silver nanoparticles using aqueous extract from the compactin producing fungal strain. *Process Biochemistry*, 44:939–43.

Shaligram, N.S., Singh, S.K., Singhal, R.S., Pandey, A., and Szakacs, G. 2008. Compactin production studies using *Penicillium brevicompactum* under solid-state fermentation conditions. *Applied Biochemistry and Biotechnology*, 159(2):1–16.

Shankar, S.S., Ahmad, A., Pasrichaa, R., and Sastry, M. 2003. Bioreduction of chloroaurate ions by geranium leaves and its endophytic fungus yields gold nanoparticles of different shapes. *Journal of Material Chemistry*, 13:1822–6.

Silver, S. 2003. Bacterial silver resistance: Molecular biology and uses and misuses of silver compounds. *FEMS Microbiology Reviews*, 27:341–53.

Singaravelu, G., Arockiamary, J.S., Ganesh K.V., and Govindaraju K.A. 2007. Novel extracellular synthesis of monodisperse gold nanoparticles using marine alga, *Sargassum wightii* Greville. *Colloids and Surfaces B: Biointerfaces*, 57:97–101.

Slawson, R.M., Van Dyke, M.I., Lee, H., and Trevors, J.T. 1992. Germanium and silver resistance, accumulation, and toxicity in microorganisms. *Plasmid*, 27(1):72–9.

Taleb, A., Petit, C., and Pileni, M.P. 1997. Synthesis of highly monodisperse silver nanoparticles from AOT reverse micelles: A way to 2D and 3D self-organization. *Chemistry of Materials*, 9:950–9.

Thirumurugan, G., Shaheedha, S.M., and Dhanaraju, M.D. 2009. *In-vitro* evaluation of antibacterial activity of silver nanoparticles synthesised by using *Phytophthora infestans*. *International Journal of ChemTech Research*, 1(3):714–6.

Varshney, R., Mishra, A.N., Bhadauria, S., and Gaur, M.S. 2009. A novel microbial route to synthesize silver nanopartilces using fungus *Hormoconis resinae*. *Digest Journal of Nanomaterials and Biostructures*, 4:349–55.

Vigneshwaran, N., Ashtaputre, N.M., Varadarajan, P.V., Nachane, R.P., Paralikar, K.M., and Balasubramanya, R.H. 2007. Biological synthesis of silver nanoparticles using the fungus *Aspergillus flavus*. *Materials Letters*, 61(6):1413–8.

Vigneshwaran, N., Kathe, A.A., Varadarajan, P.V., Nachane, R.P., and Balasubramanya, R.H. 2006. Biomimetics of silver nanoparticles by white rot fungus, *Phaenerochaete chrysosporium*. *Colloids and Surfaces B: Biointerfaces*, 53:55–9.

Williams, P., Keshavarz-Moore, E., and Dunnil, P. 1996. Production of cadmium sulphide microcrystallites in batch cultivation by *Saccharomyces pombe*. *Journal of Biotechnology*, 48:259–67.

Yong, P., Rowsen, N.A., Farr, J.P.G., Harris, I.R., and Macaskie, L.E. 2002. Bioreduction and biocrystallization of palladium by *Desulfovibrio desulfuricans* NCIMB 8307. *Biotechnology and Bioengineering*, 80:369–79.

Zhu, J., Liu, S., Palchik, O., Koltypin, Y., and Gedanken, A. 2000. Shape-controlled synthesis of silver nanoparticles by pulse sonoelectrochemical methods. *Langmuir*, 16:6396–9.

16 Plant-Based Biofortification

From Phytoremediation to Selenium-Enriched Agricultural Products

Liu Ying, Fei Li, Xuebin Yin, and Zhi-Qing Lin

CONTENTS

16.1 INTRODUCTION

Selenium (Se) was discovered by the Swedish chemist, Jakob Berzelius, in 1817. Its name comes from the Greek word for moon, *selene*. It is a group VIA metalloid with an atomic weight of 78.96. Se shares many similar chemical properties with sulfur (S), although the Se atom is slightly larger. Similar to S, Se can exist in five valence states, selenide (2–), elemental Se (0), thioselenate (2+), selenite (4+), and selenate (6+). Se was originally identified as a potentially toxic element long before it was recognized to be an essential micronutrient for animals and humans. Se salts are toxic in large amounts, but trace amounts of the element are necessary for cellular function. In 1973, Se was shown to form part of the important antioxidant enzyme, glutathione (GSH) peroxidase. It takes part in forming the active center of the enzymes, glutathione peroxidase and thioredoxin reductase, which indirectly reduce certain oxidized molecules and three known deiodinase enzymes that convert one thyroid hormone to another in most, if not all, animals.

341

The US National Institutes of Health (Office of Dietary Supplements) has issued a recommended dietary allowance (RDA) of 55 µg day^{-1} for adults (http://ods.od.nih.gov/factsheets/selenium.asp). However, the average daily intakes in various portions of many countries are often much lower than the RDA in the world. In trace amounts, Se has important benefits for animal and human nutrition. At high dosages, however, it may be toxic to animals and humans. The concentration range from nutrient requirement to being lethal is quite narrow. The minimal nutritional level for animals is about 0.05–0.10 mg Se kg^{-1} dry forage feed while the toxic level is 2–5 mg Se kg^{-1} dry forage.

The toxicity of Se has been known for many years. Se toxicity is generally encountered in arid and semiarid regions of the world that have seleniferous, alkaline soils derived from the weathering of seleniferous rocks and shale. Se is also released into the environment by various industrial activities, such as oil refineries and electric utilities. The narrow margin between the beneficial and harmful levels of Se has important implications for human health. For example, in the western parts of the United States, the leaching of soluble oxidized forms of Se, especially selenate, from local seleniferous soils is accelerated by intensive agricultural irrigation. Consequently, selenate has accumulated in high concentrations in drainage water.

The Se concentrations in plants differ among species; with some plants apparently accumulating very little. Although most of the studies show that Se should not be included as one of the essential micronutrients, recent studies suggest that Se could play a very important role in plant nutrient metabolisms (Zhu et al., in press). However, plants can play a pivotal role in Se utilization. Plants that accumulate Se may be useful as a "Se-biofortification system" (in forage or crops) to supplement Se in mammalian diet in many Se-deficient areas. On the other hand, the abilities of plants to absorb and sequester Se can also be harnessed to manage environmental Se contamination by phytoremediation as green plants and plant-associated microbes can be used to remove pollutants from contaminated soil or water. One special attribute of plants that has potential benefits for Se phytoremediation is their ability to convert inorganic Se into volatile compounds predominantly dimethylselenide (DMSe), through an ecological process called phytovolatilization (Lin et al., 1999, 2000, 2002; Lin and Terry, 2003).

In this chapter, the basic information on Se and its research history as well as its allowance and toxicity have been discussed. The role that plants can play in the phytoremediation and biofortification of Se has also been addressed.

16.2   NATURAL Se DISTRIBUTION IN CHINA

Se has a greatly uneven distribution in the world. For example, there are Se-deficient places with the Keshan disease due to scant Se intake in Keshan (China) while by contrast, in other places of China, the levels of Se concentration are toxic to humans and animals. The Se accumulation and distribution in plants hence varies among species, and as a function of chemical forms and concentrations of Se supplied. In general, with an elemental abundance of 0.08 µg g^{-1} in the Earth's crust and 13 µg g^{-1} in the Earth's interior, the existence and speciation of Se in nature is so dispersive that it commonly does not reach a threshold abundance warranting industrial exploitation. Owing to the scarcity and dispersion of Se in nature, Se is known as one of the scattered elements. Proved to be in 7.1×10^4 tons of reserves, the United States has 52.7% of the world's reserves of Se, followed by Asia (15.4%), and Africa (15.4%), while Oceania accounted for 4.4% only. Chile, the United States, Canada, China, Zambia, Zaire, Peru, Philippines, Australia, and Papua New Guinea collectively have 76.9% of the world's total reserves of Se, while more than 40 countries lack Se resources. According to previous reports, China ranked fourth in the world in Se reserves, after Canada, the United States, and Belgium. In spite of a relatively rich reserve, the distribution of Se in China is quite uneven, having 71.2% of the reserve in the Northwest and the South. For instance, in the Enshi region of Hubei province, the carbonaceous shale and stone coal contain approximately 30–30,000 mg Se kg^{-1} (Yin et al., unpublished data), and the Se mine storage reaches nearly 5 billion tons within a 850 km^2 area. Accordingly, Se reserves in the area was estimated to be 250,000–500,000 tons. Therefore, according to these new findings, the Se storage in the Enshi area accounts for a significant proportion of global Se storage. Indeed, Enshi can be entitled as "the World Capital of Selenium."

16.3   HIGHEST Se-RICH AREA, ENSHI

A survey has shown that the city of Enshi, Hubei Province, China, has three major Se-enriched deposition periods including the early Cambrian, Late Ordovician, and Permian periods. The stone coal formed during the first two periods has a low Se content of 30 µg g^{-1}, which is similar to the stone coal formed in Early Paleozoic in Southern Shaanxi. However, the black shale series formed in the Permian period have a Se concentration that is at least threefold higher than that of the former two. There is a cross-cutting low-Se area with brown soil series from northeastern to the southwestern China including more than 10 provinces. Approximately 72% Chinese are deficient in Se, while there are about 1.5 billion people living in Se-deficient areas worldwide (Bañuelos, 2009).

The Se distribution in the local environment is also related to human activities. The combustion of coal from the black shale has caused secondary Se contamination. Se migrates and ultimately gets transported to soils from the nearby combustion of Se-rich stone coal. Concentrations of Se in agricultural food grown in those soils were found to be highly toxic to the local residents. Yin and Wang (in press) have recently investigated the Se contamination in the agricultural products and hair samples of local residents, and have found that the mean Se concentrations in rice and corn were 0.3–1.9 µg g^{-1}, and rice had accumulated more Se than corn. These high levels of Se obviously exceeded the Se safety limit of 0.3 µg g^{-1} for agricultural products set by the Chinese Health Care Center. Among the three studied locations (SangShuPo, BeiFengYa, and LaoXiongPo), a large variation in Se content in corn was observed because of soil Se variations. The dietary in Enshi was reported as follows (g day^{-1}): rice 400, tuber 30, bean 20, vegetable 400, meat 60, and egg 30 (Yi et al., 2006). The Se content (in µg g^{-1}) of tuber, bean, vegetable, meat, and egg was 1.54, 0.711, 2.00, 1.6, and 1.30, respectively. To be more specific, the daily Se intake in the three villages had been estimated as follows: 541.8 µg in BeiFengYa, 1065 µg in LaoXiongPo, and 410.2 µg in SangShuPo. All these daily Se intake values have

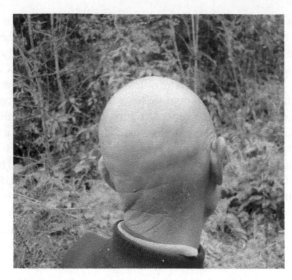

FIGURE 16.1 Hair loss in an Enshi resident due to Se toxicity.

been found to be higher than the toxic level (400 µg day⁻¹) for Chinese people stipulated by the Chinese Health Care Center. When excessive amounts of Se are taken up (such as 800 µg day⁻¹), the residents are likely to show Se toxic symptoms, such as hair loss (shown in Figure 16.1) or nail loss.

The average concentrations of Se in hair samples collected in BeiFengYa, SangShuPo, and LaoXiongPo were 2.89, 3.08, and 5.00 µg g⁻¹, respectively. Hair Se content mostly ranged from 1.0 to 4.0 µg g⁻¹ (Yin and Wang, in press). Generally, the hair Se content in females was higher than that in males, varying from 5.22 µg g⁻¹ in females to 2.71 µg g⁻¹ in males. This observation has suggested that females have a higher exposure to Se from daily coal-burning cooking. The mean hair Se concentrations both in females and males indicate that Se is still toxic to the residents originating from Se-loaded soils.

As Se is one of the few elements absorbed by plants in sufficient amounts that can be toxic to livestock, soils containing more than 0.5 mg Se kg⁻¹ are considered as seleniferous because the forages produced on such soils absorb Se more than the maximum permissible level suitable for animal consumption. Se binding onto soils and sediments depends upon the pH, Eh, Se species, competing anions, hydrous oxides of iron, and the type of clay minerals. Se in contaminated soils and water exists mainly as water-soluble selenate (SeO_4^{2-}) and selenite (SeO_3^{2-}).

In the Se-deficient areas of China, Finland, and New Zealand, the Keshan disease occurs because of a low Se intake from local food. To correct soil Se deficiency in Finland, the government initiated a national program to amend agricultural soils with Se to elevate the Se content in agricultural products. As a result, the daily dietary Se intake of residents in Finland increased from approximately 30 µg (a low value as compared to the recommended US National Research Council value of 55 µg) to 80 µg. Therefore, seleniferous soils considered as potentially Se-contaminated sites in one region of the globe become a valuable resource in Se-deficient areas in other places when the principles of green chemistry are applied.

16.4   PHYTOREMEDIATION STRATEGY

Soil Se accumulation associated with geochemical processes, mining, agricultural irrigation, and a variety of other industrial sources and processes frequently result in significant effects on animal health (Lemly, 1997). Although Se is an essential trace nutrient of importance to humans and animals as an antioxidant, its toxicity occurs at high concentrations due to its replacement of sulfur in amino acids eventually resulting in an incorrect folding of the protein, and dysfunction of proteins and enzymes.

One of the Se remediation strategies is to remove Se from the contaminated soils and waters through microbial and plant volatilization or phytoextraction (Wu et al., 1997). Most of the phytoremediation research and applications have been conducted on seleniferous soils in the United States, the United Kingdom, and India. Plants differ in their ability to accumulate Se in their tissues. Certain native plants are able to hyperaccumulate Se in their shoots when they are grown in seleniferous soils. These species are called Se accumulators and include *Astragalus, Stanleya, Morinda, Neptunia, Oonopsis*, and *Xylorhiza* (Terry et al., 2000). These accumulator species can accumulate Se up to hundreds or thousands of milligrams of Se per kg dry weight in their tissues. On the other hand, most forage and crop plants contain less than 25 mg Se kg^{-1} dry weight when grown in seleniferous soils, and those plants are defined as Se nonaccumulators. Crop plants grown in nonseleniferous soils typically have a Se concentration of 0.01–1.0 mg kg^{-1} dry weight. However, there are some plants, for example, the genus *Astragalus*, which contains both Se-accumulating species and nonaccumulating species and can grow next to each other on the same location. Trelease and Beath (1949) had observed that several species of plants growing in seleniferous soil of the Niobrara Formation had markedly different tissue Se concentrations with *Astragalus bisulcatus* containing 5530, *Stanleya pinnata* 1190, *Atriplex nuttallii* 300, and other grasses 23 mg Se kg^{-1} dry weight. Plants growing in soils of low-to-medium Se content and accumulating up to 1000 mg Se kg^{-1} dry weight are named as secondary Se accumulators.

Bañuelos et al. (1997, 2005) reduced the total Se inventory in the top 75 cm of soil by almost 50% over a period of 3 years in California's Central Valley. With their copious root systems, plants can scavenge large areas and volumes of soil, removing different chemical forms of Se through phytoextraction and volatilization processes. Once absorbed by plant roots, Se is translocated to shoots, where it may be harvested and removed from the site (Figure 16.2). However, one difficulty with phytoextraction is that Se is accumulated in plant tissues where it may become available to wildlife, especially birds. Many plant species have been evaluated for their efficacy in the phytoremediation of Se. The ideal plant species for phytoremediation of Se is one that can accumulate and volatilize large amounts of Se, grow rapidly and produce a large biomass on Se-contaminated soil, tolerate salinity and other toxic conditions, and provide a safe source of forage for Se-deficient livestock. For instance, Indian mustard (*Brassica juncea*), which typically contains 350 mg Se kg^{-1} dry

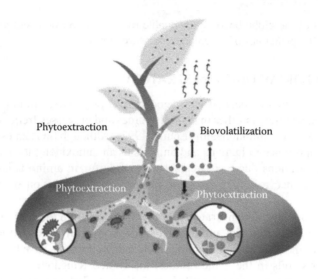

Phytoextraction

Biovolatilization

Phytoextraction

Phytoextraction

FIGURE 16.2   Phytoremediation of Se-laden soils.

weight, has most of the desired attributes. It is also easily genetically engineered by overexpressing the gene encoding ATP sulfurylase (Pilon-Smits and LeDuc, 2009). The transgenic plants exhibit much promise in the cleanup of Se from contaminated soils in that they accumulate two- to threefold more Se per plant than the wild-type plants (Bañuelos et al., 2005).

16.5   Se ACCUMULATION IN PLANTS AND CROPS

The translocation of Se in plants is dependent on the form of Se supplied. Selenate is transported much more easily than selenite and organic selenomethionine (SeMet). Zayed et al. (1998) showed that the shoot Se/root Se ratio ranged from 1.4 to 17.2 when selenate was supplied, but was only 0.6–1 for plants supplied with SeMet, and <0.5 with selenite. Arvy (1993) demonstrated that, within 3 h, 50% of the selenate taken up by bean roots moved to shoots, while most of the selenite Se remained in the root with a small fraction found in the shoot. The reason why selenite is poorly translocated to shoots maybe due to the rapid transformation of selenite to SeMet in roots (Zayed et al., 1998). The distribution of Se in plants differs from species, phases of development, and physiological conditions. In Se accumulators, Se is accumulated in young leaves during the early vegetative stage of growth; during the reproductive stage, high levels of Se are found in seeds while the Se content in leaves is drastically reduced. Nonaccumulating cereal crop plants often show about the same Se content in grain and in roots, but less in stems and leaves of a mature plant. Distribution of Se in plants also depends on the form and concentration of Se supplied to roots as well as the nature and concentrations of other substances, especially sulfur. Plants might absorb volatile Se from the atmosphere through the leaf surface as well. Using an isotope dilution method, Haygarth et al. (1995) assessed the importance of soil and atmospheric Se inputs to ryegrass. With a soil pH of 6.0, the percentage of Se in

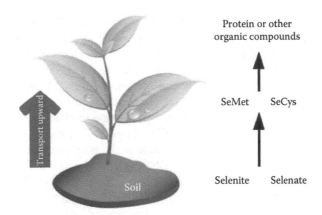

FIGURE 16.3 Se accumulation in plants.

the ryegrass leaves derived from the soil was 47, while with a pH 7.0 it was 70. It was assumed that the remainder of the Se came from the atmosphere.

Selenate readily competes with the uptake of sulfate and it has been proposed that both anions are taken up through a sulfate transporter in the root plasma membrane (Figure 16.3). Selenate uptake in other organisms, including *Escherichia coli* and yeast, is also mediated by a sulfate transporter. Research with yeast enabled the first sulfate transporter genes to be cloned from plants. The approach was to select yeast with resistance to high concentrations of selenate as a means of isolating mutants defective in sulfate transport (Terry et al., 2000).

The expression of the high-affinity sulfate transporter is regulated positively by oacetylserine, but negatively by sulfate and glutathione. Unlike selenate, there is no evidence that the uptake of selenite is mediated by membrane transporters. Selenite uptake was inhibited by 20% with the treatment of hydroxylamine (a respiratory inhibitor), while selenate uptake was inhibited by 80% in the same treatment. Asher et al. (1977) showed that, although the Se concentration in the xylem exudate of selenate-supplied detopped roots exceeded that of the external medium by 6–13 times, the Se concentration was always lower in the xylem exudate than that of the outside when selenite was supplied. Plants can also actively take up organic forms of Se (such as SeMet) through unknown uptake mechanisms. The Se/S discrimination coefficient (the ratio of the plant Se/S ratio to the solution Se/S ratio) indicated that Se accumulator species, such as *A. bisulcatus*, rice, and Indian mustard, are able to take up Se preferentially in the presence of a high sulfate supply. Other species such as alfalfa, wheat, ryegrass, barley, and broccoli had discrimination coefficient values of <1, ndicating that selenate uptake was significantly inhibited by increasing sulfate supply.

16.6 SELENIUM SPECIATION IN PLANTS

Recent studies have been conducted on Se health benefits, and several Se-enriched products were developed. Yin and his colleagues have focused on surveying the biological resource in high Se areas in Enshi in Hubei and Ziyang in Shanxi. The plants collected from Enshi included *Zea mays* (corn), *Glycine max* (soybean), *Medicago sativa* (alfalfa), *Polygonum multiflorum* (Chinese knotweed), *Cynanchum auriculatum*. Liu et al. (in press) analyzed the Se speciation of these samples by liquid chromatography–atomic force spectroscopy (LC–AFS). The results showed that there were large differences of Se concentrations among the samples collected in different regions, and the ratio of SeMet to the total Se ranged from 8.3% to 96.1%. Generally, SeMet was found in the greatest proportion in plants, at more than 90% in corn. Other Se speciation, such as SeCys, Se-methyl-seleno-cysteine, and selenomethionine, were very low or not detectable in the plant samples. Wang et al. (in press) compared the Se speciation in four Se-enriched corns from Enshi (provided by Setek Co., China) and Se-enriched yeast (provided Angel Co., China). The ratios of SeMet/total Se in the four corn and yeast samples were approximately 0.84, 1.31, 0.86, 0.91, and 0.47, respectively. SeCys in the corn was relatively lower, and their proportions were 0.175, 0.090, and 0.028, respectively (Figure 16.4). Similarly, in the Se-enriched yeast, SeCys only covered 1.4% of total Se. In these samples, most organic Se was present as SeMet. The Se-containing amino acids were mainly SeMet and their mass fractions were more than 80% in corn while only about 40% in the Se-enriched yeast. The proportions of SeMet among corns being relatively stable indicated that Se can be efficiently transformed from inorganic Se in soil to organic Se in corn. Hence, Se-biofortification (Lyons et al., 2003; Hawkesford and Zhao, 2007) from high Se areas can provide better forms of organic Se (such as SeMet) to people in Se-deficient areas to meet their daily dietary Se need.

Zhu et al. (in press) also reported concentrations of Se and enrichment factors of Se in plant and soil samples collected from Enshi. Concentrations of Se in the plant

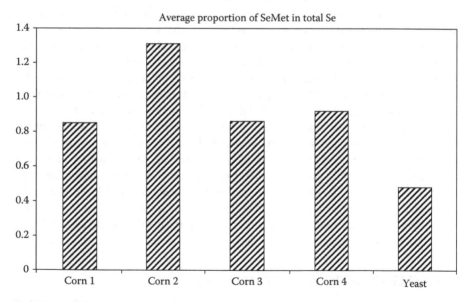

FIGURE 16.4   The proportions of SeMet in total Se of Se-enriched corns and yeast. (Data from Wang, W. et al. In press. *Selenium Deficiency, Toxicity, and Biofortification for Human Health.* USTC Press.)

samples ranged from 3.2 to 473.5 mg kg^{-1} (Table 16.1). *Siegesbeckia orientalis*, *Trifolium repens*, and *Medicago sativa* are Se accumulators showing Se transfer and enrichment factors greater than 1. Except *Miscanthus purpurascens* and *Elsholtzia splendens*, other plants are Se-tolerant plants. Generally, Se concentrations of the collected plants were higher than that of the rhizosphere soil, except for *Miscanthus sinensis*, *Miscanthus purpurascens*, *Polygonum hydropiper*, and *Rumex japonicus*.

The distribution of Se accumulated in plants is generally of the following order: leaves > stems > roots (Zhu et al., in press). Plant Se concentrations in Enshi were >25 mg kg^{-1} in most plants. *Siegesbeckia orientalis*, *Medicago sativa*, *Miscanthus purpurascens*, and *Sedum sarmentosum* contained Se concentrations >100 mg kg^{-1}. This finding indicated that high soil Se resulted in high Se accumulation in plants in Enshi, which is different from several previous reports by Bell et al. (1992) and Ellis and Salt (2003). Tea is one of the popular beverages; Se-enriched tea contains tea polyphenols and also more organic Se. Because tea from the natural Se-enriched areas has poor quality, low output, and large variation in Se content, the Se-biofortification of green tea by spraying selenite on leaves was performed (Hu et al., 2001). Huang et al. (in press) further used the Se fertilizer (available from Setek Co., China) to amend the soils of Biluochun, Kouhui Tea, and Huangshan Maofeng tea to investigate the uptake and accumulation of Se in tea to avoid higher residual of inorganic Se (Figure 16.5).

As a result, the Se content in leaves of Biluochun was elevated by approximately 50%. The highest tea Se concentration reached 291 µg kg^{-1}. Se contents in the tea tree tissues were as follows: leaf > bud > lateral branch > upper branch > lower branch.

TABLE 16.1
Concentrations and Bioconcentration Factors of Se in Dominant Plants and Rhizospheric Soils (mg kg^{-1}, dry weight)

Scientific Name of Species	Plant	Soil
Siegesbeckia orientalis	329.1 ± 3.6	45.33 ± 0.01
Trifolium repens	97.01 ± 1.39	45.62 ± 3.50
Medicago sativa	473.5 ± 14.4	42.63 ± 0.80
Miscanthus sinensis	64.87 ± 0.73	100.8 ± 9.5
Miscanthus purpurescens	212.1 ± 0.5	435.7 ± 5.8
Sedum sarmentosum	162.2 ± 2.8	138.9 ± 1.3
Polygonum hydropiper	48.91 ± 5.34	60.71 ± 4.62
Mosla dianthera	28.98 ± 0.88	28.11 ± 1.90
Erigeron annuus	46.42 ± 0.70	42.03 ± 1.34
Rumex japonicus	50.22 ± 1.04	79.08 ± 1.05
Elsholtzia splendens	31.70 ± 0.10	18.14 ± 0.12
Lycodium clavaturn	3.216 ± 0.042	3.187 ± 0.011
Artemisia lavandulaefolia	4.024 ± 0.073	3.761 ± 0.291

Source: Adapted from Zhu et al. In press. *Selenium Deficiency, Toxicity, and Biofortification for Human Health.* USTC Press.

FIGURE 16.5   Se fertilizing China green tea.

In Maofeng tea, the Se content in leaves reached 974.8 μg kg^{-1} with an increment of 916%, and 42.4 μg kg^{-1} in buds having increased only by 18%. The Se content in the tissues of the tea tree was as follows:

Leaf > Bud > Old branch > New branch

This suggested that the tea leaves have been a good sink of Se uptaken from the roots. Besides plants, Yin's group also carried out Se-biofortification in freshwater crab and chicken eggs. The feedstuff was supplemented with Se-enriched grain and other plants. After biofortification, the Se content in crab increased from 0.2 to 0.5 mg kg^{-1} and in egg from 0.1 to 0.3 mg kg^{-1}.

16.7   PLANT-BASED SELENIUM BIOFORTIFICATION FROM PHYTOREMEDIATION

As for the Se-enriched crops, the ability of absorption and accumulation of Se is the key point. The Se level in plants is:

Cross flower branch > Beans > Cereal

Se content in natural vegetation is:

Trees > Shrubs > Herbaceous, Coniferous > Broad-leaved,
Gymnosperms > Ferns > Dicotyledons > Onocotyledons

A large number of studies in Se-enriched crops cultivation have been reported. The black rice Jinlong No. 1 cultivated by the Jilin Academy of Agricultural Sciences reaches a Se content of 6.5 μg g^{-1}. Jiangsu Academy of Agricultural Sciences synthesized a Se-enriched hybrid variety of rice named Longqing No. 4 with the rice variety of Suzi No. 4 from Yunnan province and Crab rice. Shanxi Academy of Agricultural Sciences bred a new black wheat variety with a Se concentration of 112.8% higher than in ordinary ones. Research shows that breeding black rice, red rice, and other species enriched with Se can significantly increase the content of Se and other beneficial nutrients. However, there are exceptions. Other studies have shown that iron, zinc, calcium, and Se were higher in brown rice than in the milled rice.

The adoption of Se-enriched fertilizer in crop production to improve the content of Se in agricultural products is a fundamental strategy. The use of Se-enriched varieties of crops and their effects on the environment are closely related. The Se concentrations in the same black rice variety, which are planted in two different places can reach a difference of 2.87 times. Reports show that the main Se supplementation method is selenite application both at home and abroad. Soaking or spraying the crops with different concentrations of sodium selenite, so that Se can be absorbed by the crops and the agricultural products, can be achieved with certain Se contents.

One problem most often encountered in connection with Se-phytoremediation is how to dispose of the Se-rich vegetation. Without proper treatment, these Se-enriched plants may lead to secondary Se pollution. Considering that there are so many Se-deficient regions in the world, we may arrive at a conclusion of taking advantage of these plants as a Se supplement resource for animals and humans.

Bañuelos (2009) systematically stated this plant-based strategy. One option is to add the Se-containing plants to soils in Se-deficient areas as a source of bio-organic fertilizer supporting forage crops. Proper amounts of this bio-organic fertilizer can improve the condition of Se deficiency in the local soil as well as provide the crops with Se. Further, the crops cultivated with the fertilizer will improve the dietary intake of Se of the local animals and people (Figure 16.6). Second, Se is an essential trace element for proper nutrition and sound health in humans and animals. In this regard, one solution is to use seleniferous plant materials as forage for animals. Recently, Yin's group also began to utilize the Se-enriched plants from seleniferous soils as forage of animals in the deficient areas. As a result, Se-enriched plants were able to play an important role in improving the Se dietary intake of the animals in the Se-deficient areas. In addition, plants used for the phytoremediation of Se, may be used to generate heat by the combustion of dried plant material and bioenergy by fermentation to methane and/or ethanol. What is more, a large variety of chemical compounds (e.g., oil, sugars, fatty acids, proteins, pharmacological substances, and vitamins) are naturally produced by plants and may be useful as by-products of the phytoremediation process.

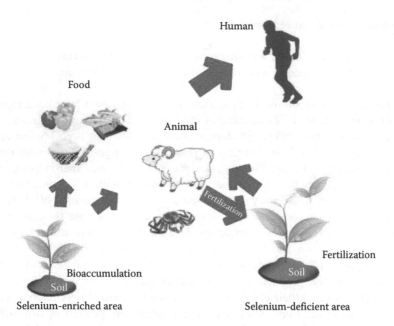

FIGURE 16.6   Conceptual model for plant-based Se biofortification.

The need of Se-biofortification has been demonstrated in Suzhou Industrial Park (SIP) in Suzhou, China, jointly constructed in 1998 by Singapore and China. First, Gao et al. (in press) determine the total Se contents in commonly consumed foods in SIP and estimated the Se daily intake by SIP residents based on their diet composition. According to the composition of residents' daily diet (Yuan et al., 2007) and the Se concentrations of the selected foods in SIP, Gao et al. (in press) considered the composition of the diets of the local residents and estimated the daily Se intake in SIP to be 42.9 μg day^{-1}. The most important source of daily Se intake for residents in SIP was found to be from meats that accounted for 60.0% (Figure 16.7). Cereals, a major staple food in the local diet accounted for 25.6%. Although vegetables and

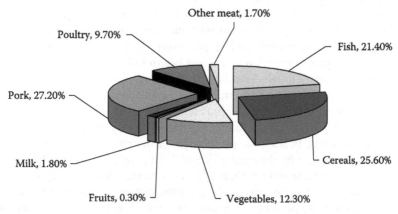

FIGURE 16.7   Contribution of food category to the daily Se intake.

TABLE 16.2

Se Content of Certain Foods in Different Regions in China ($\mu g \ kg^{-1}$)

Regions	Rice	Leafy Vegetables	Fish	Pork	Chicken
Keshan disease area	6	0.7	168	30	34
Beijing	49	20	283	97	—
Guangzhou	58.4	5.8	315.5	179.3	107.8
Taiyuan	37	11	399	61	174
SIP Suzhou	27.4	1.4	133.0	158.7	138.3

fruits were essential parts of the daily diet, each contributed only 12.3% and 0.33%, respectively, to the daily intake of Se.

Table 16.2 shows the comparison of food Se contents from different regions in China. Compared to other regions, SIP has a relatively low Se content in local foods. Concentrations of Se in leafy vegetables, fish, and rice were lower than the national averages, and close to those in the Keshan disease areas. By adding Na_2SeO_3 to animal feed, above-average high Se contents were observed in pork and chicken. In comparison with different countries (Table 16.3), the daily Se intake of residents in Suzhou Industrial Park is lower than that of some European countries, but similar to that observed in the United Kingdom. In the Keshan disease area, the daily Se intake is just about 17 $\mu g \ day^{-1}$, which is lower than the daily required amount of Se. The total human Se intake in China was reported by the China Nutrition Society (CNS) as 26–32 $\mu g \ day^{-1}$. Our study showed that the daily Se intake in SIP is about 2.5 times more than that in the Keshan disease area (slightly less than 55 $\mu g \ day^{-1}$), a value that the World Health Organization (WHO) recommends as the dietary value for adults (Dietary Reference Intakes (DRI), 2000). According to Chinese dietary culture, cereals and vegetables comprise a larger proportion of people's daily diet, while the intake of meat is less than it is in the western countries. Generally, the Se content of animal food is much higher than the plant food, which is likely the major reason why the daily Se intake of residents in most of the western

TABLE 16.3

Comparison of Daily Se Intake in Different Countries and Regions

Countries	Daily Intake (μg Se day^{-1})	References
SIP, China	42.9	This research group
Keshan, China	17.0	Yang et al. (1983)
UK	34	Barclay et al. (1995)
Netherlands	67.0	Foster and Sumar (1997)
Switzerland	70.0	Dumont et al. (2006)
Slovenia	87.0	Smrkolj et al. (2005)
Japan	129.0	Kazuko et al. (1996)
USA	60–160	Longnecker et al. (1991)

countries is higher than in China. To increase the Se intake, the SIP resident diet should contain more fish, shrimp, other seafood, or Se-biofortified food, such as meat and cereals.

16.8 CONCLUSIONS

Se in soil, plants, crops, and animal products can be utilized for the biofortification need. The concept of biofortification, inherently stemming from the concept of green chemistry (Michalak et al., 2009), embodies the transfer and delivery of high Se resources. Being toxic to human and animals in some special areas, these inorganic or organic Se compounds can be used to increase the Se content of the higher plants, animals, food in Se deficiency regions, and ultimately enhance the human Se intake. In order to ensure the concentration and speciation of Se in agricultural products to be within the safe or recommended amounts of human Se intake through biofortification, additional field studies should be conducted in the future.

ACKNOWLEDGMENT

This research work was financially supported by the National Natural Science Foundation of China (NSFC) (40601088).

REFERENCES

Arvy, M.P. 1993. Selenate and selenite uptake and translocation in bean plants (*Phaseolus vulgaris*). *Journal of Experimental Botany*, 44:1083–7.
Asher, C.J., Butler, G.W., and Peterson, P.J. 1977. Selenium transport in root systems of tomato. *Journal of Experimental Botany*, 28:279–91.
Bañuelos, G.S. 2009. Phytoremediation of selenium contaminated soil, and water produces biofortified products and new agricultural byproducts. In: Bañuelos, G. and Lin, Z.-Q. (Eds). 2008. *Development and Uses of Biofortified Agricultural Products*, pp. 57–70. Boca Raton, FL: Taylor & Francis Group.
Bañuelos, G.S., Ajwa, H.A., Mackey, M., Wu, L., Cook, C., Akohoue, S., and Zambruzuski, S. 1997. Evaluation of different plant species used for phytoremediation of high soil selenium. *Journal of Environmental Quality*, 26:639–46.
Bañuelos, G., Terry, N., LeDuc, D.L., Pilon-Smits, E.A.H., and Mackey, B. 2005. Field trial of transgenic Indian mustard plants shows enhanced phytoremediation of selenium-contaminated sediment. *Environmental Science and Technology*, 39:1771–7.
Barclay, M.N.I., Macpherson, A., and Dixon, J. 1995. Selenium content of a range of UK foods. *Journal of Food Composition and Analysis*, 8:307–18.
Bell, P.F., Parker, D.R., and Page, A.L. 1992. Contrasting selenate sulfate interactions in selenium accumulating and nonaccumulating plant species. *Soil Science Society of American Journal*, 56:1818–24.
Dietary Reference Intakes (DRI). 2000. *National Research Council*. Washington: National Academy Press.
Dumont, E., Vanhaecke, F., and Cornelis, R. 2006. Selenium speciation from food source to metabolites: A critical review. *Analytical and Bioanalytical Chemistry*, 385:1304–23.
Ellis, D.R. and Salt, D.E. 2003. Plants, selenium and human health. *Current Opinion Plant Biology*, 6:273–9.

Foster, L.H. and Sumar, S. 1997. Selenium in health and disease: A review. *Critical Reviews in Food Science and Nutrition*, 37:211–28.

Gao, J., Liu, Y., Lin, Z.-Q., Bañuelos, G.S., Lam, M., and Yin, X.B. In press. Daily dietary intake of selenium in Suzhou Industrial Park, China. *Selenium Deficiency, Toxicity, and Biofortification for Human Health*. USTC Press.

Hawkesford, M.J. and Zhao, F.-J. 2007. Strategies for increasing the selenium content of wheat. *Journal of Cereal Science*, 46:282–92.

Haygarth, P.M., Harrison, A.F., and Jones, K.C. 1995. Plant selenium from soil and the atmosphere. *Journal of Environmental Quality*, 24:768–71.

Hu, Q., Xu, J., and Pan, G. 2001. Effect of selenium spraying on green tea quality. *Journal of the Science of Food and Agriculture*, 81(4):1387–90.

Huang, Y., Li, F., and Yin, X.B. In press. The selenium uptake and accumulation by China green tea. *Selenium Deficiency, Toxicity, and Biofortification for Human Health*. USTC Press.

Kazuko, H., Katsuhiko, N., and Hiroshi, D. 1996. Selenium intake based on representative diets in Japan, 1957 to 1989. *Nutrition Research*, 16(9):1471–7.

Lemly, A.D. 1997. Environmental hazard of selenium in the Animas La Plata water development project. *Biomedical Environmental Science*, 10:92–6.

Lin, Z.-Q., Cervinka, V., Pickering, I.J., Zayed, A., and Terry, N. 2002. Managing selenium-contaminated agricultural drainage water by the integrated on-farm drainage management system: Role of selenium volatilization. *Water Research*, 12:3149–59.

Lin, Z.-Q., Hansen, D., Zayed, A., and Terry, N. 1999. Biological selenium volatilization: Method of measurement under field conditions. *Journal of Environmental Quality*, 28:309–15.

Lin, Z.-Q., Schemenauer, R.S., Cervinka, V., Zayed, A., Lee, A., and Terry, N. 2000. Selenium volatilization from the soil–plant system for the remediation of contaminated water and soil in the San Joaquin Valley. *Journal of Environmental Quality*, 29:1048–56.

Lin, Z.-Q. and Terry, N. 2003. Selenium removal by constructed wetlands. Quantitative importance of biological volatilization in the treatment of Se-laden agricultural drainage water. *Environmental Science and Technology*, 37:606–15.

Liu, H.Q., Zhu, Y.Y., Gao, J., and Yin, X.B. In press. A study on the determination of selenium speciation of high selenium plants in Enshi using LC-AFS. *Selenium Deficiency, Toxicity, and Biofortification for Human Health*. USTC Press.

Longnecker, M.P., Taylor, P.R., Levander, O.A., Howe, S.M., Veillon, C., Mcadam, P.A., Patterson, K.Y., Holden, J.M., Stampfer, M.J., Morris, J.S., and Willett, W.C. 1991. Selenium in diet, blood, and toenails in relation to human health in a seleniferous area. *American Journal of Clinical Nutrition*, 53:1288–94.

Lyons, G., Stangoulis, J., and Graham, R. 2003. High-selenium wheat: Biofortification for better health. *Nutrition Research Reviews*, 16:45–60.

Michalak, I., Chojnacka, K., and Glavič, P. 2009. The possibilities of the application of feed additives from macroalgae in sustainable mineral animal feeding. *American Journal of Applied Sciences*, 6:1458–66.

Pilon-Smits, E.A.H. and LeDuc, D.L. 2009. Phytoremediation of selenium using transgenic plants. *Current Opinion in Biotechnology*, 20:207–12.

Smrkolj, P., Pograjc, L., Hlastan-Ribič, C., and Stibilj, V. 2005. Selenium content in selected Slovenian foodstuffs and estimated daily intakes of selenium. *Food Chemistry*, 90:691–7.

Terry, N., Zayed, A.M., De Souza, M.P., and Tarun, A.S. 2000. Selenium in higher plants. *Annual Review of Plant Physiology and Plant Molecular Biology*, 51:401–32.

Trelease, S.F. and Beath, O.A. 1949. *Selenium, Its Geological Occurrence and its Biological Effects in Relation to Botany, Chemistry, Agriculture, Nutrition, and Medicine*. New York: Trelease and Beath.

Wang, W., Liu, H.Q., Zhu, Y.Y., Liu, S.F., Zhou, S.B., and Yin, X.B. In press. Determination of selenium speciation in selenium-enriched corn and yeast by HPLC-UV-HG-AFS. *Selenium Deficiency, Toxicity, and Biofortification for Human Health.* USTC Press.

Wu, L., Guo, X., and Bañuelos, G.S. 1997. Accumulation of seleno-amino acids in legume and grass plant species grown in selenium laden soil. *Journal of Environmental Toxicology and Chemistry,* 16(3):491–7.

Yang, G., Wang, S., Zhou, R., and Sun, S. 1983. Endemic selenium intoxication of humans in China. *The American Journal of Clinical Nutrition,* 37:872–81.

Yi, J.G. and Tang, X.Q. 2006. The changes in dietary pattern of Hubei residents from 1982 to 2002. *Acta Nutrimenta Sinica,* 28(3):195–7.

Yin, X.B. and Wang, Q.X. In press. Selenium toxicity and daily selenium intake in Enshi, China. *Selenium Deficiency, Toxicity, and Biofortification for Human health.* USTC Press.

Yuan, B.J., Pan, X.Q., Dai, Y., et al. 2007. Investigate the change of diet and nutrition of residents in Jiangsu Province. *Acta Nutrimenta Sinica,* 29:569–72.

Zayed, A., Lytle, C.M., and Terry, N. 1998. Accumulation and volatilization of different chemical species of Se by plants. *Planta,* 206:284–92.

Zhu, Y.Y., Liu, H.Q., Yin, X.B., and Zhou, S.B. In press. A preliminary investigation of selenium-accumulated plants in Enshi City, Hubei Province, China. *Selenium Deficiency, Toxicity, and Biofortification for Human Health.* USTC Press.

17 Biocatalysis
Green Transformations of Nitrile Function

Nicola D'Antona and Raffaele Morrone

CONTENTS

17.1 INTRODUCTION

Today, biocatalysis represents a well-established field of research at crossroads between chemistry, biology, chemical engineering, and bioengineering, and dealing with the application of biological systems such as microorganisms, enzymes, or catalytic antibodies to the synthesis of organic compounds. The essence of biocatalysis, as well as most of its current research, however, is captured in the interdisciplinary overlap between these individual areas. Biocatalysts do not operate by different scientific principles when compared with classic organic catalysts. The existence of a multitude of enzyme models proves that all the enzyme actions can be explained by rational chemical and physical principles. However, enzymes can create unusual and superior reaction conditions. They can be considered as catalysts that have been optimized through evolution to perform their physiological task upon which all forms of life depend. As a consequence it is not so surprising that enzymes are capable of performing a wide range of chemical reactions, most of which are extremely complex to carry out by chemical synthesis.

The fact that enzymes are active mostly at mild, near-ambient conditions of temperature and pH, and preferentially in aqueous media is often regarded as an advantage rather than a drawback nowadays. Anyway, the activity of many enzymes is compatible with the addition of organic cosolvents and others, such as lipases, can also work in anhydrous organic solvents. Additionally, well-developed immobilization techniques may protect the biocatalyst from the denaturing effects of organic solvents and cosolvents. In general, most biocatalysts can be satisfactorily used to realize chemo-, regio-, or stereoselective transformations. Goals for industrial processing such as "sustainable development," "green chemistry," or "environmentally benign manufacturing," can be achieved. Today, more mandatory conditions for industrial activities, would be much harder to attain without the availability of biocatalysts, which are able to work in the required green conditions. Two very strong and important driving forces, which are leading the chemical industry toward the future are:

1. Cost and margin pressure resulting from competition in an open market-oriented economy (even if the worldwide economical crisis of last years is creating doubts about the concept of free, and always wilder, global market)
2. Operation of the industry in a societal framework which puts emphasis on a clean (or at least less polluted) environment

Processing with a view toward this new set of conditions focuses on the development of production routes with fewer processing steps and higher yields on each step in order to save material and energy costs. Less waste is generated, and the treatment and disposal costs are lowered. In many cases, biocatalysis offers technological options and solutions that are not available through any other synthetic process; in such situations such as preparation of acrylamide, nicotinamide, or intermediates for antibiotics, biocatalysis, and more in general, biotechnology acts as "enabling technologies." In this context, the three terms afore mentioned are to a good extent synonymous even if developed in slightly different contexts:

Environmentally benign manufacturing is a movement toward manufacturing systems that aligns business needs with environmental needs.
Sustainable development is a worldwide chemical industry movement and represents a set of guidelines on how to manage resources such that

nonrenewables are minimized as much as possible, so that human needs can be met not only today, but also for the future generations.

Green chemistry is the design of chemical products and processes that reduce or eliminate the use and generation of hazardous substances.

Biocatalysis has the potential to combine the goals of all the three topics above. Amongst the enzymatic classes studied and synthetically exploited, nitrile-hydrolyzing enzymes are one of the most promising with a big potential for the preparation of bioactive compounds or of synthons useful as intermediates of molecules with strong biological properties. Even if this group of enzymes is not so "easy" and "flexible" to use if compared to other well-known catalysts such as lipases or proteases, significant efforts have been carried out and new important findings, both in the identification of new enzymatic systems and in the identification of new substrates, have been achieved. Nitrile compounds (organic molecules containing the –CN function) are widespread in the environment. In nature, they are mainly present as cyanoglycosides, which are produced by plants and animals (Conn, 1981). Plants also produce other nitrile compounds such as cyanolipids, ricinine, and phenylacetonitrile. Chemical industries make extensive use of various nitrile compounds for manufacturing a variety of polymers and other chemicals. For example, acrylonitrile and adiponitrile are required for the production of polyacrylonitrile and nylon-66 polymers. In general, different nitrile compounds are used as feedstock, solvents, extractants, in pharmaceuticals, for drug intermediates, and in pesticides (dichlobenil, bromoxynil, ioxynil, and buctril). Moreover, organic nitriles are used extensively in industries as precursors for the production of a wide variety of amides and acids by chemical synthesis. Conventional chemical hydrolysis of nitriles suffers from several disadvantages, including the requirement for highly acidic or basic reaction conditions, high-energy consumption, formation of undesirable by-products, low yields, and generation of large amounts of waste salts. Biocatalyzed hydrolysis, on the other hand, may be performed under mild conditions (lower temperatures and neutral pH) thus affording high conversion yields and selective conversion of the –CN functionality of compounds containing acidic or base labile groups (Banerjee et al., 2002). Furthermore, in sharp contrast with classical synthetic routes, biotransformations of nitriles into the corresponding carboxamides and carboxylic acids often proceed with excellent regio- and stereoselectivities.

But on the other hand, most of the nitriles are highly toxic, mutagenic, and carcinogenic (Pollak et al., 1991). The general toxicities of nitriles in humans are expressed as gastric diseases and vomiting (nausea), bronchial irritation, respiratory distress, convulsions, coma, and osteolathrysm that lead to lameness and skeletal deformities. Nitriles inactivate the respiratory system by tightly binding to cytochrome-c-oxidase (Solomonson and Spehar, 1981). Owing to the large quantities of cyanides, both inorganic and organic, currently used in industries, they are widely distributed in industrial wastewater. All used chemical methods for detoxification of cyanides such as alkaline chlorination, ozonization, and wet-air oxidation are expensive and require the use of hazardous chemicals such as chlorine and sodium hypochlorite. Again biotechnology and biocatalysis offer a convenient alternative: microbial degradation as an efficient, more acceptable, and environment-friendly way of removing highly toxic nitriles from the environment.

This chapter does not aim to give a complete treatise on the extensive literature on nitrile bioconversions, but rather aims at presenting an overview of enzymatic nitrile hydrolysis with a selection of recent and significant examples. Several reviews on the bioconversions of organic nitriles and their potential technological applications have been published namely by Banerjee et al. (2002), Mylerova and Martinkova (2003), Martinkova and Kren (2002).

17.2 NATURAL NITRILES

Cyanide is abundant in nature, present in more than 2000 plant species, and occurs both as inorganic cyanide (HCN) and as organic cyanide or nitriles (R-CN) mainly in the form of cyanogenic glycosides. This type of molecule is also found in the animal kingdom—among insects. It has been also demonstrated in other living organisms such as fungi, bacteria, sponges, and algae (Legras et al., 1990). Natural nitriles can be roughly divided into the following groups.

Cyanogenic glycosides that represent the largest group of natural compounds bearing the nitrile function. Approximately 50 compounds of this type have been isolated. Depending on the structure of the aglycon moiety and its probable origin, these compounds have been classified into six subgroups:

Cyanogenic glycosides originating from valine
Cyanogenic glycosides originating from leucine
Cyanogenic glycosides originating from phenylalanine
Cyanogenic glycosides originating from tyrosine
Cyanogenic glycosides with pentene structures
Other cyanogenic glycosides

Examples of compounds belonging to these subgroups are shown in Table 17.1. These nitriles release cyanide under mild conditions, and are used by plants as chemical protective agents against casual herbivores.

Cyanolipids, that is, esters of α-hydroxylated nitriles and different fatty acids are shown in Table 17.2.

Other natural nitriles fall into neither of the above groups. They also involve plant nitriles, but are mainly fungal nitriles and nitriles isolated from bacteria and the animal kingdom (e.g., 3-indolylacetonitrile (IAN), 4-amino-4-cyanobutyric acid, mandelonitrile—Table 17.2). IAN is an intermediate in the biosynthesis of the plant growth hormone (auxin) 5-indole-3-acetic acid (IAA) in the Brassicaceae (Ludwig-Muller and Hilgenberg, 1988; Muller and Weiler, 2000). The breakdown of glucosinolates (mustard oil glucosides) can also lead to the production of nitriles. Glucosinolates are not toxic, but breakdown of these compounds, which can occur on tissue damage, can release toxic compounds such as nitriles and isothiocyanates.

Both cyanogenic glycosides and glucosinolates are thought to play a part in the phytopathogen resistance (Vetter, 2000; Wittstock and Halkier, 2002).

TABLE 17.1

Representatives of Cyanogenic Glycosides Belonging to Different Subgroups Depending on Their Structure and Origin

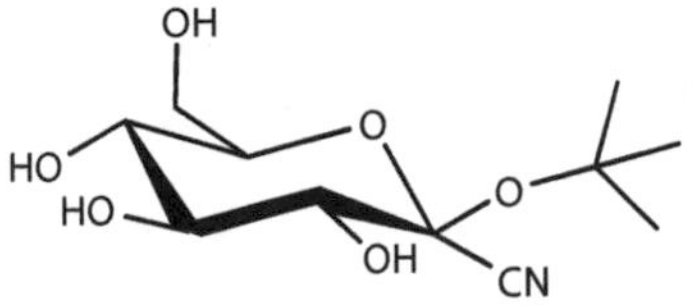

Linamarin

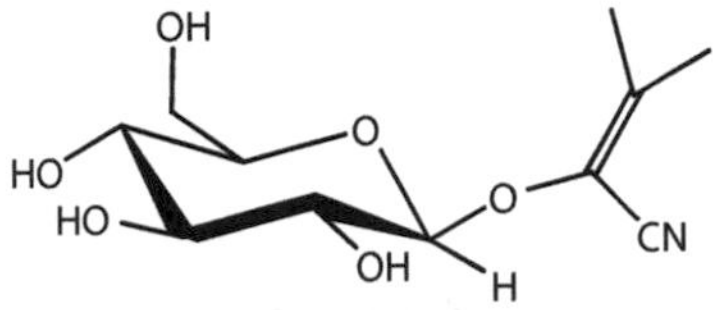

Acacipetalin

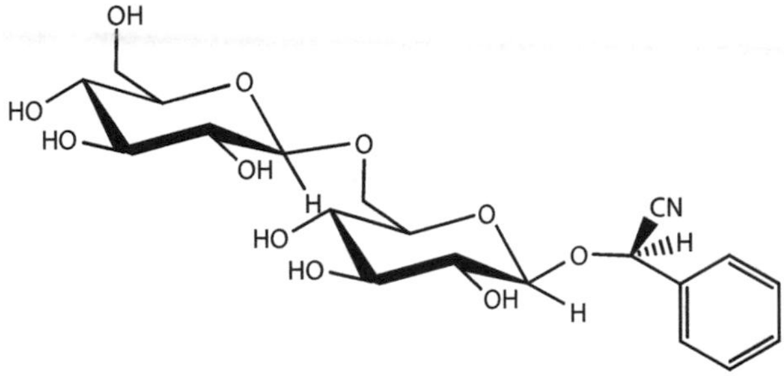

Amygdalin

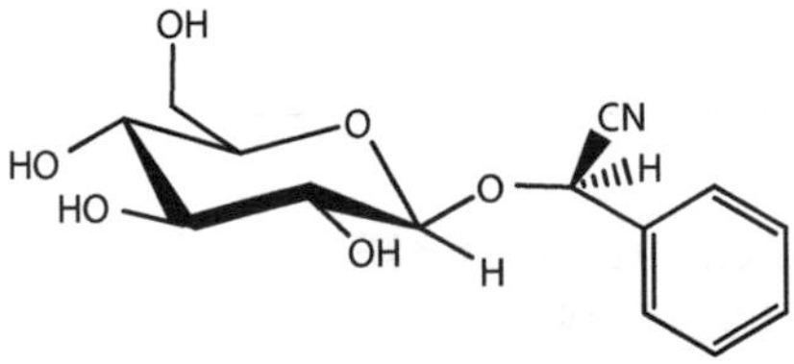

Taxiphyllin

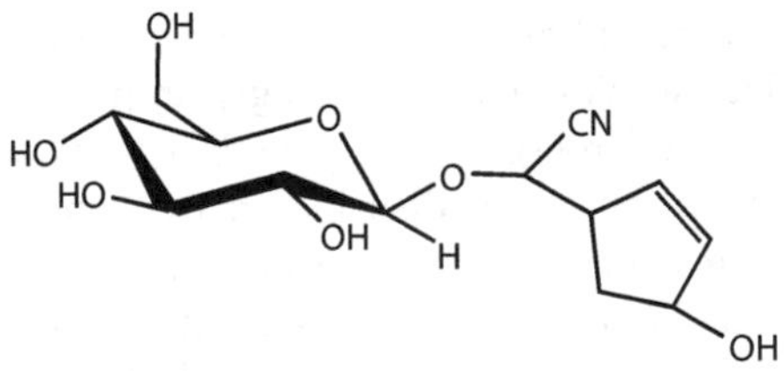

Tetraphyllin A

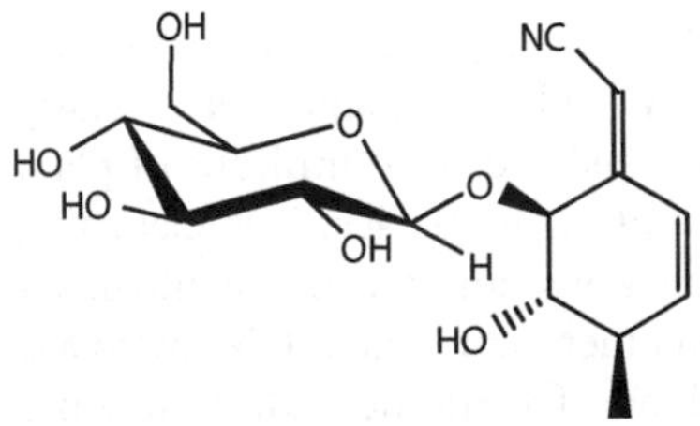

Dasycarponin

TABLE 17.2
Representatives of Cyanolipids and Other Natural Nitriles

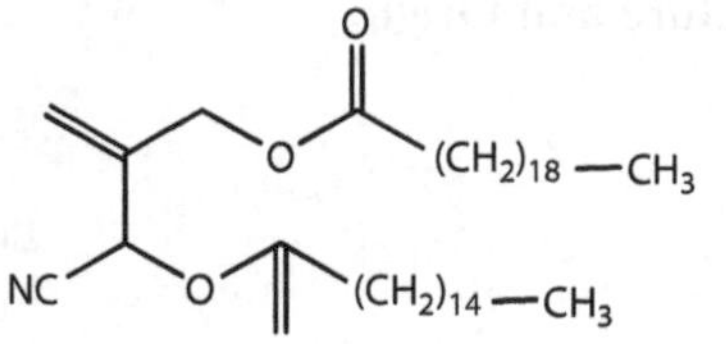

Cyanolipid

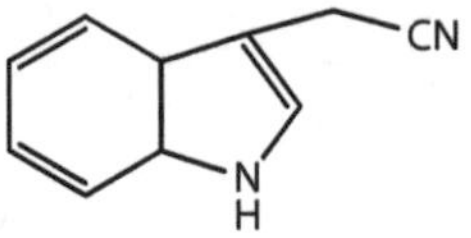

3-Indolyacetonitrile[a]

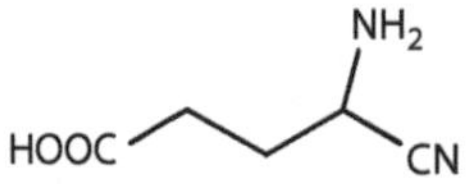

4-Amino-4-cyanobutyric acid[b]

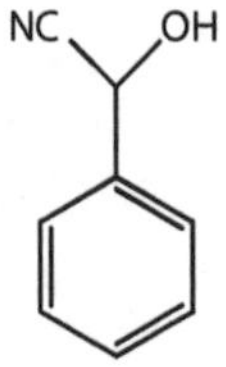

Mandelonitrile[c]

[a] Isolated from plants.
[b] Isolated from fungi.
[c] Isolated from the animal kingdom.

17.3 DISTRIBUTION AND PHYSIOLOGICAL ROLE OF NITRILE-DEGRADING ENZYME SYSTEMS

Nitrile-degrading activity, and especially cyanide and nitrile hydrolyzing activities, has been studied in a wide range of microbial species, and increasingly, in plants. Nitrile-hydrolyzing activity is very frequent in microorganisms, though without a more extensive screening it is uneasy to assess the actual distribution frequency. Nevertheless, a large number of bacteria (*Acinetobacter, Corynebacterium, Arthrobacter, Pseudomonas, Klebsiella, Nocardia*, and *Rhodococcus*) are known to metabolize nitriles as the sole source of carbon and nitrogen. Existence of nitrile-degrading enzyme activity has also been found in a number of fungal genera (*Fusarium, Aspergillus*, and *Penicillium*) (Harper, 1977). Today, with the complete and nearly complete genome sequence of several plant species and access to millions of plant EST and cDNA sequences covering a wide range of plant species, it is clear that enzymes such as nitrilases are ubiquitous in the plant kingdom. In contrast, nitrilases seem to be absent from algae, that anyway contain other members of CN-hydrolase superfamily (Piotrowsky, 2008). The physiological role of nitrile-degrading enzymes in microorganisms is still not clear even if the hypothesis that they participate in the defensive metabolic system is known. In plants, such activities are implicated in nutrient metabolism, particularly in the degradation of glucosinolates (Bestwick et al., 1993), in the synthesis of the growth hormones during plant germination and as constituents of a chemical defense system that guard plants from herbivores (Bartel and Fink, 1994). In some higher plants, nitrile-degrading activity is also required for cyanide detoxification (Miller and Conn, 1980). It has also been suggested that nitrile-degrading enzymes form components of complex pathways controlling both production and degradation of cyanogenic glycosides and related compounds where aldoximes are the key intermediates. While such a role is yet to be established in a microbial system, it is noted that some of the upstream enzyme activities, in particular the activity of aldoxime dehydratase, are responsible for the formation of nitriles from aldoximes (Kato et al., 2000), which further undergo hydrolysis, oxidation, and reduction by various enzymes.

17.4 METABOLISM OF NITRILES

Hydrolysis is the most common pathway for the microbial metabolism of nitriles. It can proceed by one of the two distinct mechanisms. The enzymes involved are nitrilases, or a nitrile hydratase–amidase system (Figure 17.1):

The single-step pathway is catalyzed by a nitrilase, which cleaves a nitrile into a carboxylic acid and ammonia.

The two-step pathway proceeds via an intermediate—amide—which is formed from a nitrile by the action of a nitrile hydratase. The amide is subsequently hydrolyzed into acid and ammonia by an amidase, which is usually produced in bacteria along with the nitrile hydratase.

These metabolic pathways enable microorganisms to utilize nitriles as sources of carbon or nitrogen or both. Most of the organisms have been shown to contain either a nitrilase or a nitrile hydratase/amidase bienzymatic system. Nevertheless, some microorganisms such as *Rhodococcus rhodochrous* LL100-21 (Dadd et al., 2000, 2001) or *Rhodococcus rhodochrous* J1 (Nagasawa et al., 1993; Kobayashi et al., 1994; Wieser et al., 1998) contain both nitrilase-catalyzed and nitrile hydratase/amidase-catalyzed degradation pathways. More specifically, *Rhodococcus rhodochrous* J1 contains two nitrile hydratases, one is a high molecular weight nitrile hydratase (H-NHase) that acts preferentially on aliphatic nitriles, and the other is a

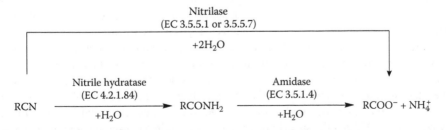

FIGURE 17.1 Different pathways of nitrile metabolism in microorganisms.

low molecular weight nitrile hydratase (L-NHase) with higher affinity for aromatic nitriles. Amazingly, the different enzymes can be induced selectively by the growth of the organism under different culture conditions.

Other hydrolytic pathways present in some microorganisms concern the HCN metabolism. HCN abiotically formed from various metal cyanides leads to various products through the action of different enzymes: cyanase produces CO_2 and NH_3 (Dorr and Knowles, 1989), cyanide hydratase leads to formamide (Wang et al., 1992), and cyanide dihydratase produces formic acid and NH_3 (Ingovorsen et al., 1991). Cyanide hydratase, although functionally different to nitrilases and cyanide dihydratases, has been shown to be closely related to these enzymes and shows no relationship to the more functionally similar nitrile hydratase (Wang et al., 1992; Cluness et al., 1993). Nitrilases, cyanide hydratase, and cyanide dihydratase have been classified into a single enzyme family, known as the nitrilase/cyanide hydratase family, which is part of a larger group of related proteins termed "nitrilase superfamily" (Pace and Brenner, 2001). Different nonhydrolytic pathways of nitrile degradation that will not be discussed further in this chapter include use of oxygenases by many plants and insects to oxidize some of the nitriles to cyanohydrins, which are further converted into an aldehyde and HCN by an oxynitrilase (hydroxynitrile lyase) (Johnson et al., 2000), and nitrogenases present in nitrogen-fixing organisms that are capable of reducing nitrile compounds to the corresponding hydrocarbons and NH_3 (Smith et al., 2004).

17.5　THE NITRILASE SUPERFAMILY

On the basis of global and structure-based sequence analysis, members of the nitrilase superfamily have been classified into 13 branches and the substrate specificity of members of nine branches can be anticipated (Table 17.3). Enzymes of the nitrilase superfamily have been found in all plants, animals, and fungi, and many of these organisms have multiple nitrilase-related proteins from more than one branch of the superfamily. Nitrilase-related sequences are also found in phylogenetically isolated prokaryotes that appear to have an ecological relationship to plants and animals. The nitrilase superfamily therefore probably emerged prior to the separation of plants, animals, and fungi, radiated into families, and then spread laterally to bacteria and archaea. Some branches of this superfamily are found only in prokaryotes; members of these branches could constitute rational antibiotic targets. It should be noted that the nitrilase branch of the nitrilase superfamily is probably the only branch that contains members able to perform nitrile hydrolysis (from a nitrile to the corresponding acid and ammonia); on the other hand, eight branches appear to be either amidases of various specificities or enzymes that condense acyl chains to amino groups. Nitrile hydratases are not members of the nitrilase superfamily. Moreover, despite the fact that most branches of the nitrilase superfamily are actually amidases, there are many amidases that are unrelated (Patricelli and Cravatt, 2000). Maybe in an anachronistical way, the "nitrilase" designation of this superfamily is retained because of the historical observation that aliphatic amidases are related to nitrilases (Bork and Koonin, 1994; Novo et al., 1995). All the superfamily members contain a conserved apparently catalytic triad of glutamate, lysine, and cysteine (Pace and Brenner, 2001).

TABLE 17.3
Branches of Nitrilase Superfamily with Their Substrates

Nitrilase Branch		Substrates
1	Nitrilase	Nitriles
2	Aliphatic-amidase	Amides
		Proteins
3	Amino-terminal amidase	Amides
		Proteins (asparagine and glutamine residues)
4	Biotinidase	Secondary amides
5	β-Ureidopropionase	Amides
		Carbamyles
6	Carbamylase	Amidasic
		Carbamyles
7	Prokaryote NAD synthetase	Predicted amides
8	Eukaryote NAD synthetase	Predicted amides
9	Apoliprotein N-acyltransferase	Amides (also reversal amidase activity) Proteins
10	Nit and Nitfhit	Unknown
11	NB11	Unknown
12	NB12	Predicted proteins
13	Nonfused outliers	Unknown

17.5.1 NITRILASE

The first nitrile-hydrolyzing enzyme discovered around 40 years ago was a nitrilase. It was known to convert 5-indole 3-acetonitrile (IAN) into 5-indole 3-acetic acid (IAA—an auxin). Since then a large number of bacteria and fungi with nitrilase activity was isolated, and their capability to metabolize various natural and synthetic nitriles were investigated (Podar et al., 2005). Nitrilases show a different range of biochemical characteristics and in particular the substrate specificity can vary widely, even if their activity is generally restricted to small molecules (Deigner et al., 1996). Initial investigations suggested them to be specific for aromatic nitriles. However, this distinction was reconsidered due to the growing information on this enzymatic class. At the present time, nitrilases are broadly classified into three different categories based on their substrate specificities, namely aliphatic, aromatic, and arylacetonitrilases. This classification has several drawbacks because substrate ranges of many nitrilases are much more blurred, and often these enzymes can accept both aliphatic and aromatic compounds even at different levels of activities, nevertheless bearing in mind these limitations, it can be a useful tool.

Cyanide hydratase and cyanide dihydratase belongs to the nitrilase branch of nitrilase superfamily, using HCN as the only efficient substrate and producing amide and acid products, respectively. Microorganisms appear in fact to have evolved separate metabolic pathways for the hydrolysis of inorganic cyanide. Thus, most nitrilases (as well as nitrile hydratases) till now investigated do not display activity toward cyanide. A reported exception is the nitrilase from *Pseudomonas* sp. S1 able to hydrolyze both potassium cyanide and organic nitriles (Dhillon and Shivaraman, 1999; Dhillon et al., 1999). The nitrilase of *Klebsiella pneumoniae* ssp. *Ozaenae* was isolated by selection on the herbicide bromoxynil (3,5-dibromo-4-hydroxybenzonitrile) as nitrogen source, and showed an almost complete substrate preference toward this compound (Stalker et al., 1998). In Table 17.4, a number of organisms with a well-characterized nitrilase activity, and their substrate classification is illustrated.

TABLE 17.4
Organisms with Characterized Nitrilase Activity and Substrate Classification

Organism	Substrate Class	References
Bacteria		
Acidovorax facilis 72W ATCC 55746	Aliphatic	Cooling et al. (2001), Gavagan et al. (1998, 1999)
Acinetobacter sp. (strain AK226)	Aliphatic	Yamamoto and Komatsu (1991)
Agrobacterium sp. DSM 6336	Heterocyclic nitriles	Wieser et al. (1997)
Alcaligenes faecalis DSM 6335	Arylacetonitrilases	Kiener et al. (1996)
Alcaligenes faecalis ATCC 8750	Arylacetonitrilases	Yamamoto et al. (1992)
Alcaligenes faecalis JM3	Arylacetonitrilases	Nagasawa et al. (1990)
Alcaligenes xylosoxidans ssp. *denitrificans* strain DF3	Cyanide dihydratase	Ingovorsen et al. (1991)
Arthrobacter sp. strain J1	Aromatic	Bandyopadhyay et al. (1986)
Arthrobacter nitroguajacolicus ZJUTB06-99	n.r.	Chen et al. (2009)
Bacillus pallidus Dac 521	Aromatic	Almatawah and Cowan (1999), Almatawah et al. (1999), Cramp and Cowan (1999), Graham et al. (2000)
Bacillus pumilus strains C1 and 8A3	Cyanide dihydratase	Jandhyala et al. (2003)
Bacillus subtilis ZJB-063	Arylacetonitrilases	Chen et al. (2008)
Bradyrhizobium japonicum USDA 110 (blr3397)	Aliphatic	Zhu et al. (2008)
Bradyrhizobium japonicum USDA 110 (blr6402)	Arylacetonitrilase	Zhu et al. (2007b)
Brassica napus	Aliphatic	Bestwick et al. (1993)
Comamonas testosteroni	Aliphatic	Levy-Schil et al. (1995)
Gordona terrae FERM BP-4535	Arylalkylnitriles	Hashimoto et al. (1996)
Klebsiella pneumoniae ssp. *ozaenae*	Bromoxynil-specific nitrilase	Stalker et al. (1998)
Microbacterium paraoxydans	n.r.	Kaul et al. (2004)
Microbacterium liquefaciens	n.r.	Kaul et al. (2004)
Nocardia globerula NHB-2	Aromatic	Sharma et al. (2006)

TABLE 17.4 (continued)
Organisms with Characterized Nitrilase Activity and Substrate Classification

Organism	Substrate Class	References
Nocardia (Rhodococcus) NCIMB 11215	Aromatic	Harper (1985)
Nocardia (Rhodococcus) NCIMB 11216	Aromatic	Tauber et al. (2000), Klempier et al. (1996)
Pseudomonas sp. S1	Aliphatic	Dhillon and Shivaraman (1999), Dhillon et al. (1999)
Pseudomonas fluorescens DSM 7155	Arylacetonitrilases	Layh et al. (1998)
Pseudomonas fluorescens EBC 191	Arylacetonitrilases	Kiziak et al. (2005)
Pseudomonas putida	Arylacetonitrilase	Banerjee et al. (2006)
Pseudomonas stutzeri AK61	Cyanide dihydratase	Sewell et al. (2003)
Rhodococcus rhodococcus K22	Aliphatic	Kobayashi et al. (1989)
Rhodococcus ATCC39484	Aromatic	Stevenson et al. (1992)
Rhodococcus rhodochrous LL100-21	Arylacetonitrilases	Dadd et al. (2000, 2001)
Rhodococcus rhodochrous J1 (FERM BP-1478)	Aromatic	Wieser et al. (1998), Nagasawa et al. (1993), Kobayashi et al. (1994)
Rhodococcus rhodochrous PA34	Aromatic	Bhalla et al. (1992)
Rhodococcus ruber	Acrylonitrile	Hughes et al. (1998)
Synechocystis sp. Strain PCC6803	Aliphatic	Heinemann et al. (2003)
Archaea		
Pyrococcus abyssi	Aliphatic	Mueller et al. (2006)
Fungi		
Aspergillus niger K10	Aromatic	Kaplan et al. (2006)
Cryptococcus sp. UFMG-Y28	Aromatic	Rezende et al. (2000)
Exophiala oligosperma R1		Shen et al. (2009)
Fusarium lateritium	Cyanide hydratase	Nolan et al. (2003)
Fusarium oxysporum N-10	Cyanide hydratase	Yanase et al. (2000)
Fusarium oxysporum f.sp. *melonis*	Aromatic	Goldlust and Bohak (1989)
Fusarium solani IMI196840	Aromatic	Harper (1977)
Fusarium solani O1	Aromatic	Vejvoda et al. (2008)
Gloeoecercospora sorghi	Cyanide hydratase	Wang and Van Etten (1992)
Leptosphaeria maculans	Cyanide hydratase	Sexton et al. (2000)
Penicillium multicolor	Aromatic	Kaplan et al. (2006)
Plants		
Arabidopsis thaliana	Aliphatic	Effenberger and Oswald (2001a, 2001b), Piotrowski et al. (2001)
Barley	Aromatic	Thimann and Mahadevan (1964)
Chinese cabbage	n.r.	Rausch and Hilgenberg (1980)

Nitrilases are generally inducible enzymes, and do not require metal cofactors or prosthetic groups for their activity. To date many nitrilases have been purified and their subunit molecular mass and primary structures have been determined. Most nitrilases consist of a single polypeptide having mass in the range 30–45 kDa, which aggregate to form the active holoenzyme under different conditions. The preferred form of this enzyme seems to be a large aggregate of 6–26 units. Most of the enzymes show substrate-dependent activation, but the presence of elevated concentrations of salt, organic solvents, pH, temperature, or even the enzyme itself may also trigger subunit association and therefore activation. The hydrophobic effect resulting from the above-mentioned conditions might change the conformation of the enzyme thereby exhibiting hydrophobic sites and enabling subunit associations and therefore activation (Nagasawa et al., 2000). Similar to other enzymes belonging to the nitrilase superfamily, a Cys residue has been shown to be essential for the catalytic activity of the nitrilases (Kobayashi et al., 1992a). This Cys is proposed to be part of the catalytic triad Cys-Glu-Lys. The three-dimensional view of the enzyme, its catalytic center, and genomic studies have been performed to understand its mechanistic details (Kobayashi et al., 1992a; Novo et al., 2002; Cilia et al., 2005; Kaplan et al., 2006). The Glu residue is thought to act as a general base by accepting the thiol hydrogen of the Cys residue. The activated Cys then carries out a nucleophilic attack on the CN carbon, forming a covalently bound enzyme–substrate complex, followed by attack of the first water molecule and formation of a tetrahedral thioimidate intermediate stabilized by the Lys residue (Nakai et al., 2000). Subsequent steps are the loss of ammonia, the attack of a second water molecule with the formation of an acyl tetrahedral intermediate, and the final break down of the enzyme–substrate complex to give the corresponding acid and the restored enzyme. This mechanism is graphically described in the pathway 1 of Figure 17.2.

Recently, the observation that the acyl intermediate would prevent access of a water molecule to the vicinity of the Glu active residue led to suggestions that the final hydrolysis could be catalyzed by a second glutamate residue whose side chain is in a suitable position to act as the general base catalyst (Kimani et al., 2007). Some significant amide concentration has been often detected in many nitrilases biocatalyzed reactions, suggesting that the tetrahedral thioimidate intermediate could evolve in some cases, toward the formation of an amidic compound like illustrated in pathway 2 of Figure 17.2. Recently, following experiments performed by using a recombinantly produced nitrilase from *Pseudomonas fluorescens* EBC 191 known to be prone to amide formation, it has been proposed that the charge distribution in the tetrahedral thioimidate intermediate, depending in turn on the sterochemical and electronic properties of the R group, acts as a mechanistic switch (Fernandes et al., 2006).

17.5.2 AMIDASE

The CO—NH amide bond is relatively energy-rich and can be hydrolyzed to free carboxylic acids and ammonia, by a variety of unrelated or distantly related enzymes, called amidases. Most of amidases are sulfhydryl enzymes like all members of the nitrilase superfamily, while other amidases such as those from *Pseudomonas*

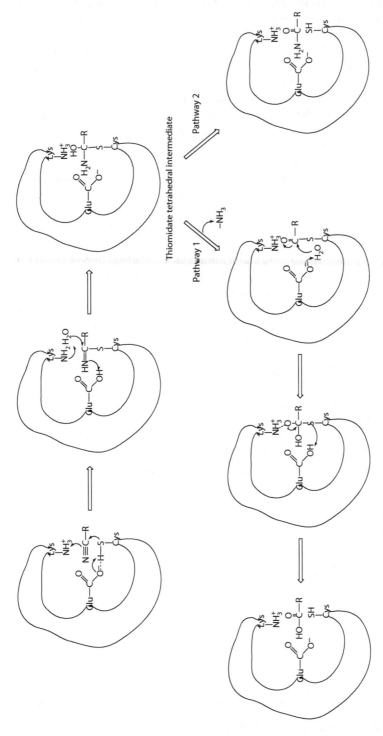

FIGURE 17.2 Reaction mechanim of nitrilase.

chlororaphis B23, *Rhodococcus* sp. N-774, or *R. rhodochrous* J1 belong to a different group of amidases containing a Gly-Gly-Ser-Ser signature in their amino acid sequence (Chebrou et al., 1996). Amazingly, the amidase of *R. rhodochrous* J1 has been found to catalyze the hydrolysis of the nitrile functional group to acid and ammonia, showing in this way an unexpected nitrilasic activity (Kobayashi et al., 1998b). Thus, the reaction mechanisms of both the nitrilase and amidase catalyzed reactions seem analogous, even if the active nucleophile present in both enzymes is different. Figure 17.3 shows the proposed reaction mechanism of amidase, involving also nitrile hydrolysis. The carbonyl group of amide undergoes a nucleophilic attack, resulting in the formation of a tetrahedral intermediate, which is converted into an acyl-enzyme with the removal of ammonia and subsequently hydrolyzed to acid.

Although there is no homology between the amidase and nitrilase groups, comparison of their reaction mechanisms can provide important insights for the development of catalysts required for the hydrolysis of both nitriles and amides. All the different amidases also exhibit acyl transfer activity in the presence of hydroxylamine (Fournand et al., 1998). The substrate specificity and biological functions of these enzymes vary widely including carbon/nitrogen metabolism in prokaryotes through hydrolysis of amides, the generation of properly charged tRNAGln in eubacteria through transfer of ammonia from glutamine (Curnow et al., 1997), and the degradation of neuromodulatory fatty acid amides in mammals (Cravatt et al., 1996). The mammalian enzyme belongs to "amidase signature family" and is defined by a conserved stretch of approximately 130 amino acids termed the "amidase signature sequence." Studies revealed that this group of amidases represents a large class of serine–lysine catalytic dyad hydrolases (Patricelli and Cravatt, 2000) and more specifically resemble serine hydrolases (Boger et al., 2000). While some amidases are specific for aliphatic substrates (Asano et al., 1982), others cleave amides of aromatic acids (Hirrlinger et al., 1996), and still others hydrolyze amides of α- or ω-amino acids (Stelkes-Ritter et al., 1995). Stereoselectivity has been reported to be generally associated with the amidases in the bienzymatic nitrile hydratase/amidase pathway (Mayaux et al., 1990), even if there are many examples of stereoselective nitrile hydratases. This is discussed later in the chapter.

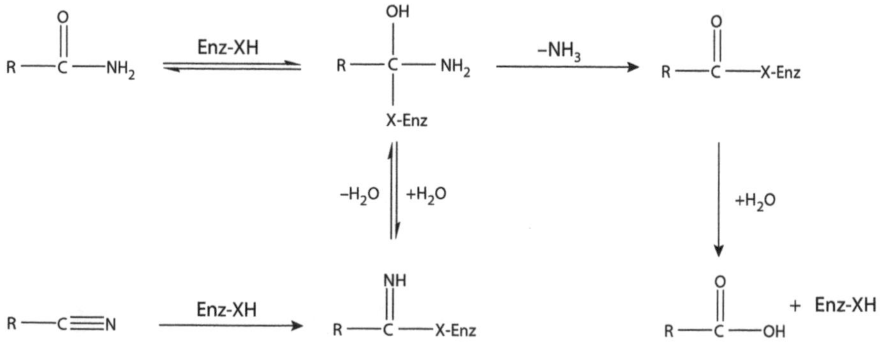

FIGURE 17.3   Reaction mechanism of amidase.

17.6 NITRILE HYDRATASE

Nitrile hydratase (NHase) catalyzes the conversion of nitriles into amides, which are further converted into a bienzymatic pathway to the corresponding acids by amidases. This class of enzymes can be broadly classified into two groups with respect to the kinds of metal ions composing the catalytic center: the Fe-type having a nonheme Fe^{3+} center (Endo et al., 2001; Nagashima, 1998), and the Co-type having a noncorrinoid Co^{3+} center at their active site (Kobayashi et al., 1999) (Table 17.5). Nitrile hydratase from *Agrobacterium tumefaciens* IAMB-261, which is involved in the biosynthesis of plant hormone indole-3-acetic acid has been found to contain both cobalt and ferric ions (Kobayashi et al., 1998). Recently, a few types of nitrile hydratases containing

TABLE 17.5
Organisms with Characterized Nitrile Hydratase Activity and Their Substrates

Organism	Substrates	References
Agrobacterium tumefaciens d3 (DSM 9674)	Arylnitriles, arylalkylnitriles	Bauer et al. (1998)
Agrobacterium tumefaciens IAMB-261	Indole-3-acetonitrile	Kobayashi et al. (1995)
Bacillus cereus	Acrylonitrile	Saroja et al. (2000)
Bacillus sp. BR449	Acrylonitrile	Kim and Oriel (2000), Padmakumar and Oriel (1999)
Bacillus sp. RAPc8	(Cyclo)alkylnitriles	Pereira et al. (1998)
Bacillus pallidus Dac 521	Alkylnitriles, heterocyclic nitriles, arylnitriles Alkylnitriles	Almatawah and Cowan (1999), Almatawah et al. (1999), Cramp and Cowan (1999), Graham et al. (2000)
Bacillus smithii SC-J05-1	Alkylnitriles	Takashima et al. (1998)
Brevibacterium imperialis CBS 489-74	Acrylonitrile	Cantarella et al. (1998a, 1998b), Alfani et al. (2001), Battistel et al. (1997)
Candida guilliermondii CCT 7207	Heterocyclic nitriles, (cyclo) alkylnitriles, arylnitriles	Dias et al. (2000)
Candida famata	Alkylnitriles	Linardi et al. (1996)
Cryptococcus sp. UFMG-Y28	Acetonitrile, proprionitrile	Rezende et al. (1999)
Cryptococcus flavus UFMG-Y61	Isobutyronitrile	Rezende et al. (1999)
Comamonas testosteroni ATCC 55744	Alkylnitriles	Gavagan et al. (1998)
Klebsiella oxytoca 38.1.2	Arylnitriles	Ewert et al. (2008)
Myrothecium verrucaria	Cyanamide	Marier-Greiner et al. (1991)
Pseudomonas chlororaphis B23 (FERM BP-187)	Alkylnitriles	Yamada and Kobayashi (1996)
Pseudomonas putida NRRL-B-18668	Arylalkylnitriles	Payne et al. (1997)

continued

TABLE 17.5 (continued)
Organisms with Characterized Nitrile Hydratase Activity and Their Substrates

Organism	Substrates	References
Pseudomonas putida	Acetonitrile, NaCN	Chapatwala et al. (1995)
Pseudonocardia thermophila JCM3095[a]	Acrylonitrile	Yamaki et al. (1997)
Rhodococcus sp. 409 (NOVO SP 409)	Heterocyclic nitriles, arylnitriles, arylalkylnitriles, (cyclo)alkylnitriles	Deigner et al. (1996)
Rhodococcus sp. C3II (DSM 9685)	Alkylnitriles, arylnitriles, arylalkylnitriles	Effenberger and Graef (1998)
Rhodococcus sp. N771	Alkylnitriles	Hann et al. (1999)
Rhodococcus sp. R312 (CBS 717.73)	Alkylnitriles, benzonitrile	Osprian et al. (1999), Cull et al. (2000, 2001)
Rhodococcus sp. YH3-3	Alkylnitriles, heterocyclic nitriles, arylnitriles, phenylacetonitrile	Kato et al. (1999)
Rhodococcus erythropolis	Acrylonitrile	Hughes et al. (1998)
Rhodococcus erythropolis BL1	Alkylnitriles, arylalkylnitriles	Langdahl et al. (1996)
Rhodococcus erythropolis FZB 53	Steroidal nitriles	Kaufmann et al. (1999)
Rhodococcus erythropolis MP50 (DSM 9675)	Alkylnitriles, arylnitriles, arylalkylnitriles,	Effenberger and Graef (1998)
Rhodococcus erythropolis NCIMB11540	Arylnitriles	Boesten and Cals. (1987),Vink et al. (2009)
Rhodococcus equi A4	Alkylnitriles, heterocyclic nitriles, arylnitriles, arylalkylnitriles, (cyclo) alkylnitriles	Prepechalova et al. (2001), Martinkova et al. (1998, 2001, 2002)
Rhodococcus rhodochrous LL100-21	Alkylnitriles, arylnitriles, arylalkylnitriles, 3-cyanopyridine,	Dadd et al. (2000, 2001)
Rhodococcus rhodochrous IFO 15564	Heterocyclic nitriles, arylalkylnitriles, (cyclo) alkylnitriles	Matoishi et al. (1998)
Rhodococcus rhodochrous NCIMB 11216	Alkylnitriles, heterocyclic nitriles, arylnitriles	Tauber et al. (2000), Klempier et al. (1996), Hoyle et al. (1998)
Rhodococcus rhodochrous AJ270	Heterocyclic nitriles, arylnitriles, arylalkylnitriles, (cyclo)alkylnitriles	Meth-Cohm and Wang (1997a, 1997b), Wang and Feng (2000), Wang et al. (2000a, 2000b, 2001a, 2001b)
Rhodococcus rhodochrous J1 (FERM BP-1478)	Alkylnitriles, heterocyclic nitriles, arylnitriles	Wieser et al. (1998), Nagasawa et al. (1993), Kobayashi et al. (1994)
Rhodotorula glutinis UFMG-Y5	Methacrylonitrile	Rezende et al. (1999)

other metal ions have been reported. NHase from the *Myrrothecium verrucaria* contain a Zn^{2+} ion in the active site (Marier-Greiner et al., 1991). The electron spin resonance (ESR) spectra of nitrile hydratase from *Bacillus pallidus* Dac521 have shown no unpaired electrons associated with a metal center suggesting that neither Fe^{3+} nor Co^{3+} atoms were cofactors. However, because ESR cannot detect Zn^{2+} or Ni^{2+}, the possible presence of these metal ions cannot be excluded (Cramp and Cowan, 1999).

The enzyme generally consists of α- and β-subunits and exists as α, β 46 kDa heterodimers or $(\alpha, \beta)_2$ 92 kDa tetramers (with one metal center), and the fragmentation into individual subunits has been reported to inactivate the enzyme. In addition to the function of the active center, metal ions may play a role in enhancing the folding or stabilization of the subunit polypeptides of the enzymes (Banerjee et al., 2002). Their primary sequences are well conserved among all known nitrile hydratases even if there is no apparent homology between the two subunits (Kobayashi et al., 1999). Since nitrile hydratases exhibit significant protein sequence homology, especially at their metal-binding domain where a highly conserved –C–S–L–C–S–C– motif is noted, it is believed that all NHases have the same structure and similar mechanism of catalysis. The relatively more robust Co-NHases have wider substrate specificity.

17.6.1   Fe-Type Nitrile Hydratase

Fe-type nitrile hydratase has been much more characterized than the Co-type. Endo et al. (1999) have reported the unique photoreactivity of the nitrile hydratase from *Rhodococcus* N771: *in vivo* the enzymatic activity of this NHase is lost during aerobic incubation in the dark (dark-inactivation), but it is almost completely recovered upon light irradiation (photoreactivation). However, the photoreactivated enzyme cannot be inactivated by darkness *in vitro*. Similar results have been observed in other Fe-type nitrile hydratases from *Rhodococcus* sp. N-774 (Endo and Watanabe, 1989) and R312 having identical amino acid sequences with *Rhodococcus* N-771. Spectra FT-IR (Noguchi et al., 1995) have clearly indicated that the cause of this phenomenon is the association of an endogenous nitric oxide (NO) molecule at the active center: binding of NO in dark results in deactivation of the enzyme, with the process reversed by loss of the same NO molecule upon light irradiation (Figure 17.4). The nitrile hydratase from *Comamonas testeroni* NI1, which has 50% sequence homology with *Rhodococcus* N771, and which is not reported to be photosensitive, shares the same property, namely inactivation with NO *in vitro* and photoactivation (Cerbelaud et al., 1995). This fact suggests that all Fe-type nitrile hydratases possess potentially the photoreactivity regulated by NO in spite of the fact that the microorganism produce or do not produce NO endogenously, and supports the hypothesis that in *Rhodococcus* N771 NHase, NO could be produced by a NO synthase present in the strain (Noguchi et al., 1995). This phenomenon is not observed with Co-NHases. The Fe-NHases are also inactivated by N^{3-}, CN^-, and by metal ions such as Hg^{2+} and Ag^+ (Nagasawa et al., 1987).

Again Endo et al. (1999), and previously Huang et al. (1997), have determined the structure of the NO-inactivated Fe-NHase from *Rhodococcus* sp. N-771: the Fe^{3+} center is located in a cleft situated at the interface of the two subunits of the heterodimer, and ligated to two deprotonated Cys and Ser carboxamido nitrogens from the peptide backbone and three Cys sulfur thiolates. Two of these metalcysteine sulfurs

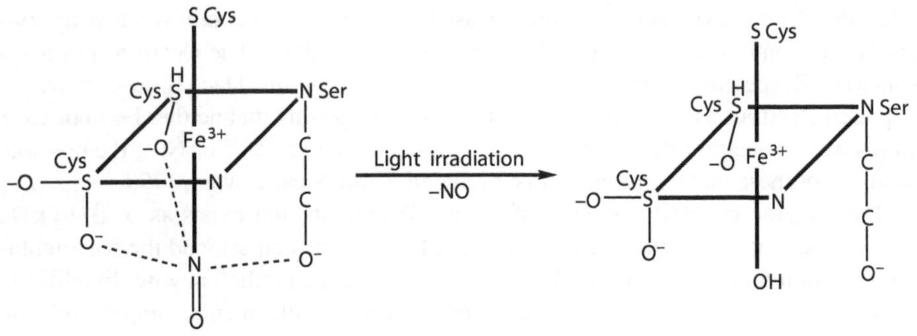

FIGURE 17.4 Mechanism of Fe-type nitrile hydratase photoactivation.

bound on the basal plane of the iron(III) center are post-translationally modified to the sulfinic (Cys-SO$_2$) and sulfenic (Cys-SO) groups and a molecule of NO is coordinated at the solvent exposed site of the metal center. In the photoactivated form of the enzyme, NO molecule is hypothesized to be substituted likely by an H$_2$O molecule (Cramp and Cowan, 1999) or by an OH$^-$ ion (Cerbelaud et al., 1995). Recent experiments have shown that modification of the Cys-residues is essential for the catalytic activity (Murakami et al., 2000). The mechanism of catalysis remains yet unresolved. However, two possible pathways have been proposed (Kobayashi and Shimitzu, 1998, 2000). In the first pathway, the nitrile substrate approaches a metal-bound hydroxide ion, which can act as a nucleophile attacking the nitrile carbon atom, while in a second possible mechanism, a metal-bound hydroxide ion acting as a general base activates a water molecule which then attacks the nitrile carbon resulting, in both the cases, in the formation of an imidate that by tautomerization gives the amide (Figure 17.5).

17.6.2 Co-Type Nitrile Hydratase

The Co-NHases belong to a small group of cobalt-dependent enzymes in nature. They are also the first examples of enzymes that incorporate a noncorrinoid Co^{3+} center in their structures. Although there is no report on the structure of a Co-NHase so far, the similarities of the pre-edge and extended x-ray absorption fine structure (EXAFS) spectra, and identical sequence homology at the metal-binding region suggests that the catalysis mechanism and the ligand environments of the metal ions in the cobalt and iron-containing NHases are similar. Cobalt nitrile hydratases from *Pseudomonas putida* and *Rhodococcus rhodochrous J1* have been found to have threonine in the −V−C−(T/S)−L−C−S−C− sequence at the active site, whereas ferric NHases have serine, so the difference in the metal cofactors could be ascribed to the different amino acid residues at this position (Payne et al., 1997).

17.7 IMPROVEMENTS IN BIOCATALYSTS

In general, a set of conditions are required for the successful development and scale up of a novel biocatalytic process: the availability of a suitable biocatalyst, methods

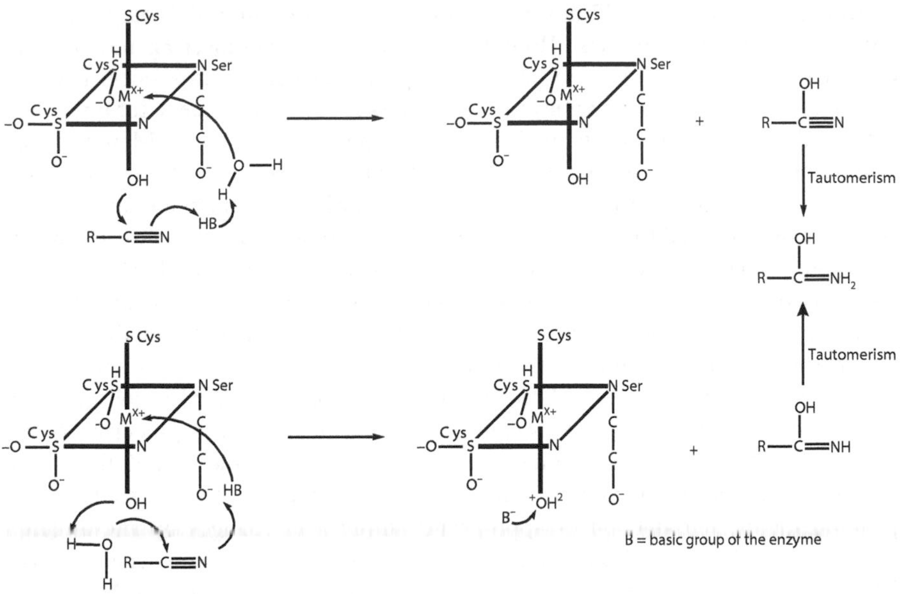

FIGURE 17.5 Mechanism of nitrile hydratase catalysis.

for enzyme stabilization to ease its application and reuse, and process engineering to deal with the specific conditions of an appropriate reaction system (aqueous or solvent system, batch or continuous, packed-bed or membrane reactor) and with upstream and downstream processing. In many cases, technologies to address these requirements are readily available needing only to be adapted to the process boundaries. However, obviously, the major precondition remains the availability of suitable enzymes for a given reaction. The traditional method to identify new enzymes such as nitrile hydrolyzing ones is based on screening of soil samples or strain collections by enrichment culture. "Enrichment culture" is a technique to isolate specific microorganisms by exploiting their characteristic to grow faster than others in media with limited nutrients, high temperatures, extreme pH values, or other well-defined physical–chemical parameters, thus becoming dominant after several transfers of the culture. Unluckily, common cultivation tools can give access only to a small fraction of the biodiversity, with a number of culturable microorganisms from a sample that has been estimated to be 0.001–1% of the overall, depending on their origin (Lorenz et al., 2003). Novel approaches to access biodiversity have been developed and comprise the metagenome approach. In the metagenome approach, the entire genomic DNA from uncultivated microbial consortia such as soil samples is directly extracted, cloned, and expressed in easily cultivable surrogate host cells such as *Escherichia coli*, that are subsequently subjected to screening or selection procedures to identify distinct enzymatic activities (Handelsman, 2004, 2005). The major advantage of this approach is the huge numbers of new biocatalysts that can be found, while the main disadvantages are that it is possible to find only those biocatalysts that can be expressed in the host organism and do not escape the activity tests. One amazing application of this kind of technique has been carried out by Diversa

Corporation discovering 137 novel nitrilases from more than 600 biotope-specific environmental DNA libraries (Robertson et al., 2004). Discovering new natural enzymes often is not enough to ensure an optimal biocatalytical process because almost all applications of enzymes in biocatalysis do not rely on the natural reaction catalyzed by them, but use synthetic or other nonnatural substrates. Moreover, the requirements of the reaction environment (solvent, concentration, pH, and temperature) can differ substantially from the conditions in which the enzymes have evolved in nature and are used to work. Thus, quite often, activity, stability, substrate specificity, and enantio-, regio-, or chemo-selectivity need to be improved. The classical route to overcome these limitations is the reaction engineering, which consist in the variations of the reaction parameters until conditions are found in which the biocatalyst meets the process requirements. Today, molecular biology provides a more efficient alternative by offering techniques to modify the amino acid sequence of an enzyme, thus improving its performance. In principle, genetic engineering offers two main strategies: rational protein design, which requires the availability of the three-dimensional structure (or a homology model) necessary to identify type and position for the introduction of appropriate amino acid changes by site-directed mutagenesis (Hellinga, 1997), or directed evolution techniques (Bornscheuer, 2001; Eijsink et al., 2005; Hibbert et al., 2005). Direct evolution comprises a first step of random mutagenesis of the gene(s) encoding the enzyme(s) followed by a second step in which desired biocatalyst variants within these mutant libraries are identified by screening or selection processes. Many methods have been developed to create mutant libraries and can be divided into two groups that use different approaches: a first set of techniques exploits nonrecombining mutagenesis tools, in which one parent gene is subjected to random mutagenesis leading to variants with point mutations. Conversely, a second group uses recombining methods (also referred to as sexual-mutagenesis) in which several parental genes (usually showing high sequence homology) are randomized, this eventually resulting in a library of chimeras rather than an accumulation of pointmutations. With the directed evolutionary technique, these mutation methods are repeatedly carried out to obtain desirable properties of the enzyme. As a consequence mutant libraries comprising $10^4–10^6$ enzymatic variants are commonly created and the main challenge remains is their accurate and fast identification. Two aforementioned different approaches can be applied: selection or screening (Martinkova et al., 2008). Selection methods are based on using a selective medium, which allows only the colonies with the desired enzyme to grow (the previously cited enrichment culture technique is an example), therefore they are more difficult to develop and provide only qualitative results, but have a very high throughput. Selection methods have been routinely used to obtain nitrile- or amide-degrading enzymes. On the other hand, screening methods are largely and readily available and are quantitative; however, every single mutant colony must be analyzed for enzyme activity. While the most frequent screening methods used are based on photometric and fluorimetric assays performed in microtiter-plate-based formats in combination with high-throughput robot assistance, not always the outcoming results furnish the desired enzyme for a real biocatalytic process as it is well known that "you get what you screen for." In fact, a typical biocatalytic process may include many steps such as fermentation for the enzyme production, biocatalytical conversion, product removal and purification, or enzyme recycling, and only screening methods able to mimic the entire process sequence can give a fully reliable answer (Hibbert et al., 2005). Finally, even if an enzyme is identified to be useful for a given reaction, its application is often hampered by its lack of long-term stability under process conditions, and also by difficulties in recovery and recycling when these steps are required because they need to be economically and environmentally advantageous. This problem can be overcome by immobilization, providing advantages such as enhanced stability, reduced costs due to the repeated or continuous use, easy separation from the reaction mixture, and possible modulation of the catalytic properties. Immobilization tools can be applied both to purified or semipurified enzymes and to whole cells depending on the specific technique used. The most frequently used immobilization techniques fall into two main categories: carrier-bound immobilization techniques comprising of noncovalent adsorption or deposition, covalent attachment and entrapment into a polymeric gel; and carrier-free immobilization techniques of cross-linking (Cao et al., 2003). All these approaches are always a compromise between maintaining high catalytic activity while achieving the advantages given before. Among the latter techniques, the cross-linked enzyme aggregate (CLEA) preparation of purified and semi-purified nitrilases or nitrile hydratases have been recently reported furnishing excellent results both in terms of retained activity and of stereorecognition properties (Van Pelt et al., 2009).

17.8   NITRILASE- AND NITRILE HYDRATASE-CATALYZED SYNTHETIC PROCESSES

Strategic importance of biocatalyzed synthetic transformations in terms of eco-compatibility and cheaper processes has been widely stressed previously. Among the developed biotransformations catalyzed by nitrilases or nitrile hydratases/amidases systems, a special interest is focused toward stereoselective reactions able to give access to molecules otherwise impossible to obtain by classical chemical routes. Hereby, selected examples aim to offer an overview of research in this direction. Examples of industrial processes using nitrile hydrolyzing biocatalysts are also illustrated.

17.8.1   ENANTIOSELECTIVE BIOTRANSFORMATIONS

Despite the fact that early experiments suggested low selectivity of nitrile-converting enzymes with respect to the substrate chirality (Faber, 1992), many recent works report the successful enantioselective bioconversion of nitriles catalyzed by nitrilases or nitrile hydratases, even if the stereoselectivity of nitrile hydratases remains often lower that that of coupled amidases.

17.8.1.1   Racemic Nitriles Bearing One α-Stereogenic Center

Optically active α-hydroxy carboxylic acids are useful intermediates in medicinal chemistry and asymmetric synthesis (Coppola and Schuster 1997). Enantioselective biotransformations of α-hydroxy nitriles (cyanohydrins) are important because they can lead to a dynamic kinetic resolution from readily available starting material.

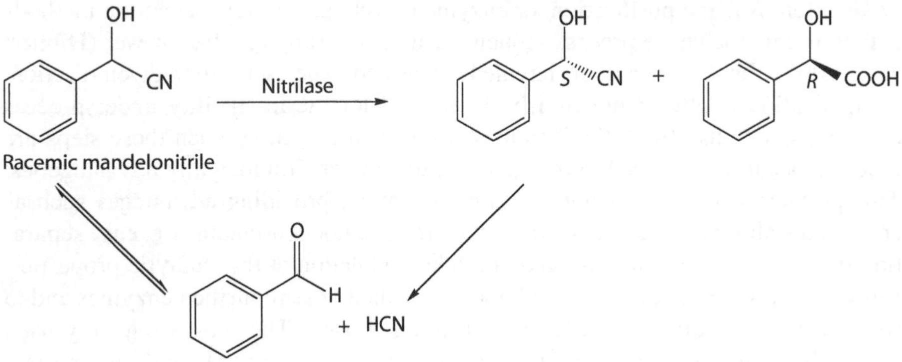

FIGURE 17.6 Nitrilase biocatalyzed transformation of racemic (*R/S*)-mandelonitrile for the production of (*R*)-mandelic acid.

Since cyanohydrins are known to racemize under basic conditions through reversible loss of HCN, the enzymes used are mainly nitrilases that are more tolerant toward the cyanide anion. (*R*)-Mandelic acid is an important key intermediate for the production of semi-synthetic cephalosporins (Terreni et al., 2001) and penicillins (Furlenmeier et al., 1976); it is also used as a chiral-resolving agent (Han et al., 2002) and chiral synthon for the production of antitumor (Surivet et al., 1998), and antiobesity agents (Mills et al., 1983). Whole cells of *Alcaligenes faecalis* ATCC8750 (Yamamoto et al., 1992), *Ps. putida* (Kaul et al., 2004), *Microbacterium paraoxydans* (Kaul et al., 2004), and *Microbacterium liquefaciens* (Kaul et al., 2004) have been reported to contain nitrilases able to catalyze the enantioselective production of optically enantiopure (*R*)-mandelic acid (Figure 17.6).

Kaul et al. (2004) also report the purification of the nitrilase from *Alcaligenes faecalis* ATCC8750, and the preparation of CLEAs to use for the preparation of (*R*) mandelic acid: this immobilization resulted in a 40% reduction of the reaction rates, but with a 5 times higher half-lives both under operational conditions (30°) and denaturing conditions (50°) (Kaul et al., 2007). Because no (*S*)-selective nitrilase for cyanohydrins has been yet identified, different bienzymatic routes have been proposed to have access to *S* enantiomer of mandelic acid. Mateo et al. (2006) used a (*S*)-selective oxynitrilase from *Manihot esculenta* coupled with a nonselective recombinant nitrilase from *Pseudomonas fluorescens* EBC 191 (Figure 17.7); this technique has been further developed by integrating and expressing both the genes encoding the arylacetonitrilase of *Ps. fluorescens* EBC 191 and the genes encoding the oxynitrilase of *Manihot esculenta* in the methylotrophic yeast *Pichia pastoris*

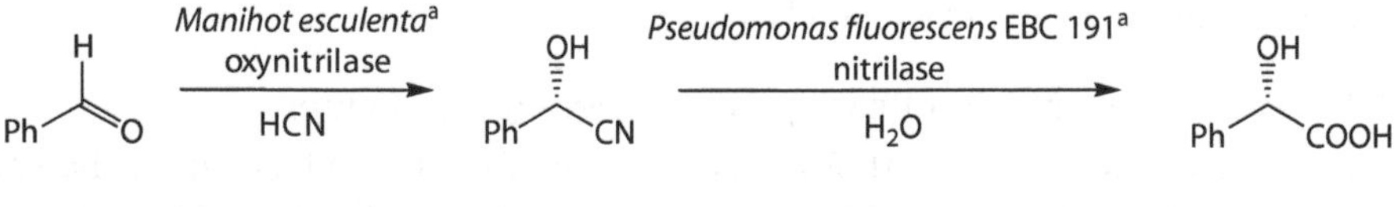

[a]CLEA immobilized

FIGURE 17.7 Bienzymatic oxynitrilase/nitrilase synthesis of (*S*)-mandelic acid.

known to act as an acidotolerant expression system (Rustler et al., 2008). Resting cells of this recombinant strain converted benzaldehyde and cyanide initially to (S) mandelonitrile and subsequently to (S) mandelic acid under acidic conditions.

Analogously, Van Pelt et al. (2009) have realized a one-pot bienzymatic cascade combining the (S)-selective oxynitrilase from *M. esculenta* and a purified nitrile hydratase from *Nitriliruptor alkaliphilus* for the synthesis of aliphatic (S)-α-hydroxycarboxylic amides (Table 17.6); both enzymes were immobilized as CLEAs to enhance their stability (Van Pelt et al., 2009).

These cascaded enzymatic routes theoretically can be adopted for the general synthesis of optically active (R)- or (S)-α-hydroxy acids and amides, simply by choosing an oxynitrilase of desired enantiopreference. Moreover, these routes are environmentally benign and compatible with a wide range of hydrolytically sensitive groups because of the mild reaction conditions. Very recently, Ewert et al. (2008) have reported the enantioselective biotransformation of mandelic acid catalyzed by the nitrile hydratase from *Klebsiella oxytoca*. By performing several biotransformations with different α-arylnitriles substrates, they have found that the isolated microorganism induces both an (S)-enantioselective nitrile hydratase and a putative (S)-selective amidase. Nevertheless, in the adopted work conditions, racemic mandelonitrile was stereoselectively converted into the (S)-amide obtained with a yield of 80% and an enantiomeric excess of 95%; no further conversion into mandelic acid occurred. Conversely, during the biotransformation of the racemic phenylglycine nitrile, the *Klebsiella oxytoca* nitrile hydratase/amidase system resulted in the generation of both (S)-phenylglycine (ee >99%) and (R)-phenylglicine amide (ee >99%). DeSantis et al. (2002) have screened a number of highly active and enantioselective nitrilases from an enzyme library. One of these nitrilases is able to catalyze the dynamic kinetic resolution of a wide spectrum of cyanohydrin substrates derived from substituted benzaldehydes, naphtaldehydes, and heterocyclic aldehydes to yield (R)-α-hydroxycarboxylic acids. Another screened

---

TABLE 17.6

Bienzymatic Oxynitrilase/Nitrile Hydratase Catalyzed Synthesis of (S)-mandelic acid

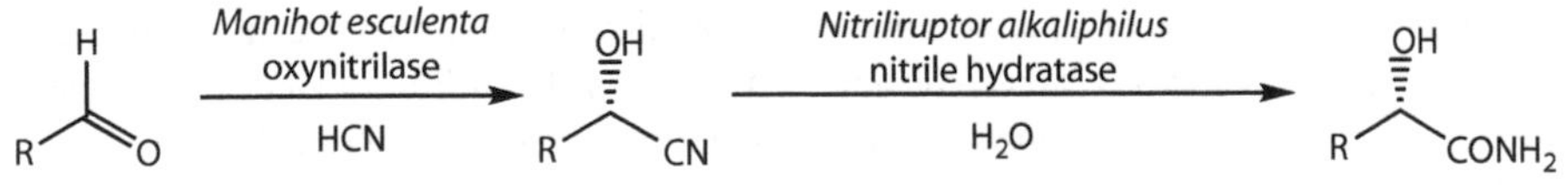

Entry	R	Amide	
		Produced (mM)	ee (%)–(S)
1	Et	44	86
2	*i*-Pr	35	88
3	Pr	41	88
4	Bu	42	90

Substrates concentration: 35–45 mM, pH 4.5, 21°C, maximum HCN concentration of 5 mM.

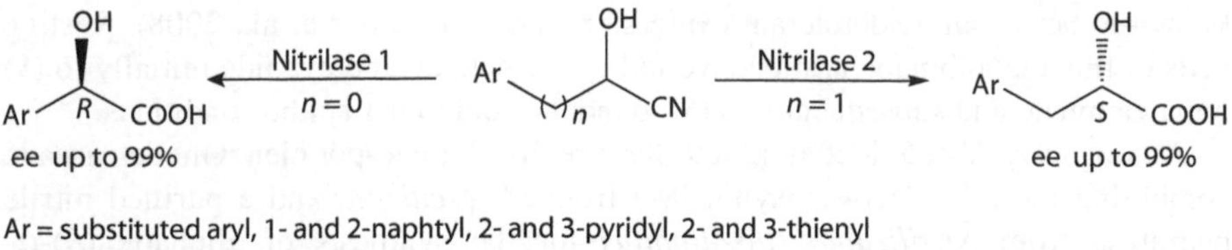

Ar = substituted aryl, 1- and 2-naphtyl, 2- and 3-pyridyl, 2- and 3-thienyl

FIGURE 17.8   Dynamic kinetic resolution of cyanohydrins catalyzed by nitrilases.

nitrilase has shown to convert a wide range of arylacetaldehyde-derived cyanohydrins into (*S*)-arylacetic acids in excellent yields with enantiomeric excess values ranging between 91% and 99% (DeSantis et al., 2002) (Figure 17.8).

Effenberger and Oswald (2001a) report the biotransformation of α-fluoroarylace-tonitriles catalyzed by nitrilase from the plant *Arabidopsis thaliana* that was over-expressed in *E. coli* (Effenberger and Oswald, 2001b). Unexpectedly, the main product is not the acid, but the corresponding amide obtained with yields ranging between 32% and 38%, ee between 88% and 92%, and *S* configuration. Only substrates with none or *meta*-substituents on aromatic ring are accepted by the enzyme (Figure 17.9).

Recently, Diversa Corporation focused their attention to identify alkalophilic nitrilases in their wide enzymatic library able to catalyze the hydrolysis of α-aminonitriles, for the synthesis of α-amino acids and more specifically of 4-fluoro-phenylglycine, an important chiral building block (Chaplin et al., 2004). They determined the conditions under which aromatic α-aminonitriles racemize and combining these conditions with the previously selected nitrilases, they were able to perform a dynamic kinetic asymmetric synthesis following two different approaches: hydrolysis of aromatic aminonitriles at high pH, and hydrolysis of *N*-acylaminonitrile at pH 8. Following the first approach, in presence of Nitrilase 5275 at pH 10.6, 4-fluorophenylglycinonitrile was converted into (*R*)-4-fluorophenylglycine with a yield of 79.5% and an ee of 96.3%. The second approach by using an N-protected aminonitrile, presented the advantage of milder racemization conditions when compared to conditions required by the nonpro-tected analogs. In such a way, working at a lower pH (pH 8) the aminonitrile substrate resulted in less labile and furnished higher yields retaining excellent enantioselectivity values: (*R*)-*N*-formyl-4-fluorophenylglycine in the presence of Nitrilase 5086 was obtained with 87% of yield and 98–99% ee. Benz et al. (2007) reported the nitrilase biocatalyzed synthesis of enantiopure 1,4-benzodioxane-2-carboxylic acid and 1,4-benzodioxane-6-formyl-2-carboxylic acid. The nitrilase from *Rhodococcus* R312 gave the best activity in terms of enantioselectivity and retained a good reaction rate. Both acids were recovered with enantiomer excesses in the range between 98% and 99%.

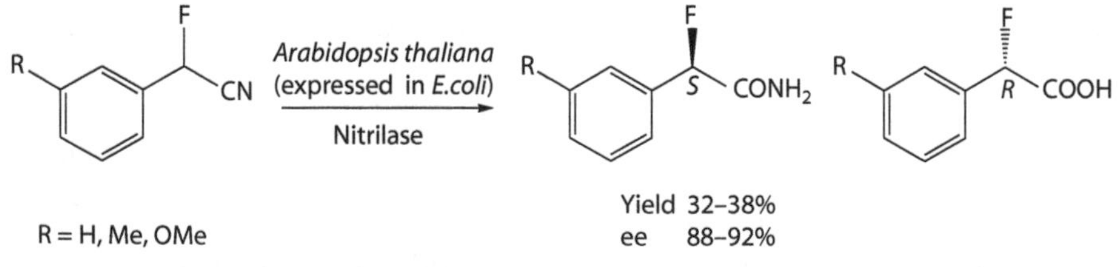

R = H, Me, OMe

Yield 32–38%
ee   88–92%

FIGURE 17.9   Nitrilase catalyzed transformation of α-fluoroarylacetonitriles.

TABLE 17.7
Nitrilase Biocatalyzed Resolution of 2-Cyano-1,4-Benzodioxane and 2-Cyano-6-Formyl-1,4-Benzodioxane

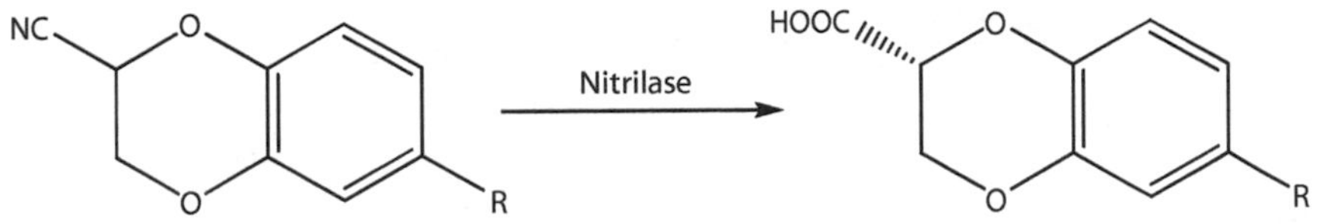

NC — structure → Nitrilase → HOOC — structure

Entry	Enzyme	R	% Cosolvent	Acid Yield %	ee %
1	*Alcaligenes faecalis* ATCC 8750	H	20% 2-Dodecanone	50	40 (S)
2	*Alcaligenes faecalis* ATCC 8750	H	3.1% Methanol	30	15 (S)
4	*Alcaligenes faecalis* ATCC 8750	Formyl	29% Toluene	50	41
5	*Alcaligenes faecalis* ATCC 8750	Formyl	4.3% Methanol	34	42
6	*Alcaligenes faecalis* ATCC 8750	Formyl	10% 2-Nonanone	40	42
7	*Pseudomonas fluorescens* DSM 7155	H	20% 2-Dodecanone	56	15 (S)
8	*Pseudomonas fluorescens* DSM 7153	H	20% 2-Dodecanone	50	9 (S)
9	*Pseudomonas fluorescens* DSM 7153	Formyl	4.3% Methanol	48	28
10	*Pseudomonas fluorescens* DSM 7153	Formyl	4.3% Methanol	26	35
11	*Pseudomonas fluorescens* DSM 7154	H	–	40	7 (S)
12	*Rhodococcus* R312	H	1% Methanol	25	98 (R)
13	*Rhodococcus* R312	Formyl	1% Methanol	25	99

Nitrilase from *Rhodococcus* R312 was found to be (R)-enantiospecific toward 2-cyano 1,4-benzodioxane, in contrast to other screened nitrilases that showed (S)-preferences; however, the authors do not report on the absolute configuration of the enantiopure synthesized 1,4-benzodioxane-6-formyl-2-carboxylic acid. An organic cosolvent was added to the reaction mixture to increase solubility of substrates (Table 17.7). A possible hypothesized application of enantiopure 1,4-benzodioxane-2-carboxylic acid is the synthesis of doxazosine methylate, member of the quinazoline family of drugs, and indicated for the treatment of hypertension.

17.8.1.2 Racemic Nitriles Bearing One β or Remote Stereogenic Center to Cyano Function

The use of nitrile hydrolyzing enzymes for the resolution of β-hydroxy nitriles has attracted attention by several groups. Optically active β-hydroxy nitriles are important precursors in the preparation of β-adrenergic blocking agents used in the treatment of cardiovascular and psychiatric disorders. On the other hand, the corresponding optically active acids find application as anti-inflammatory drugs (Abdel-Rahman and Hussein, 2006), and in the synthesis of α-amino acids (Thaisrivongs et al., 1987), β-lactames (Schostarez, 1988), β-lactones (Capozzi et al., 1993), and pheromones (Oertle et al., 1990). Hua and coworkers (Mukherjee et al., 2006) describe the synthetic applicability of the nitrilase from *Synechocystis* sp. Strain PCC 6803. This

TABLE 17.8

Nitrilase Catalyzed Biotransformation of β-Hydroxy Nitriles

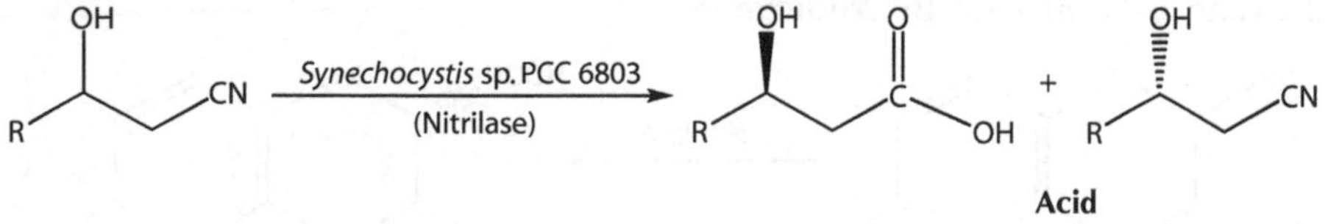

Entry	R	Yield (%)	ee (%)
1	Ph	51	46 (*S*)
2	*p*-F Ph	28	61 (*S*)
3	*p*-Cl Ph	30	65 (*R*)
4	*p*-Br Ph	50	23 (*S*)
5	*p*-Ac Ph	56	29 (*S*)
6	*p*-OMe Ph	48	32 (*S*)
7	*o*-naphtyl	93	2 (*S*)
8	*tert*-butyl	93	0 (*S*)

nitrilase was found to catalyze the enantioselective hydrolysis of β-hydroxy nitriles to give (*S*)-enriched β-hydroxy acids, although the enzyme was not enantioselective for the hydrolysis of β-hydroxy nitriles such as mandelonitrile (Table 17.8).

Reported data would indicate some effects exerted by the substituents at the *para* position of aromatic substrates on the enantioselectivity. These results are interesting because they are in contrast to the common observation that a chiral atom at the β-position to the reaction center is recognized with more difficulty than the one at the α-position. In general, it is believed that the movement of a stereocenter from the reactive site (α-position to the functional group) to a remote place gives rise to the decrease of enantioselectivity in asymmetric reactions. However, this notion may not be true for enzymatic reactions since the chiral recognition between the enzyme and a substrate might occur at a remote pocket from the reaction site. Again, Hua and coworkers (Kamila et al., 2006; Zhu et al., 2007a; Ankati et al., 2009) have developed a bienzymatic route for the stereospecific synthesis of β-hydroxy acids, using the corresponding β-ketonitriles as starting materials. By using two different carbonyl reductase from *Candida magnolia* (CMCR) and from *Saccharomyces cerevisiae* (Ymr226c) with opposite enantiopreference, it was possible to obtain both enantiomers of a number of aromatic β-hydroxy nitriles with high optical purity, that were further converted into β-hydroxy acids by less stereospecific nitrilases (from *Synechocystis* sp. strain PCC 6803 and from *Bradyrhizobium japonicum* strain USDA110–bll6402) (Table 17.9). Authors demonstrated the possibility to carry out the process in a "two-step-one-pot" fashion without the need to isolate intermediates with effective and environmental advantages.

Kinfe et al. (2009) report the kinetic resolution of various β-hydroxy nitriles, more specifically 3-hydroxy-4-aryloxybutanenitriles and 3-hydroxy-3-arylpropanenitriles,

TABLE 17.9
Bienzymatic Carbonyl Reductase/Nitrilase Catalyzed Route for the Stereospecific Synthesis of β-Hydroxy Acids

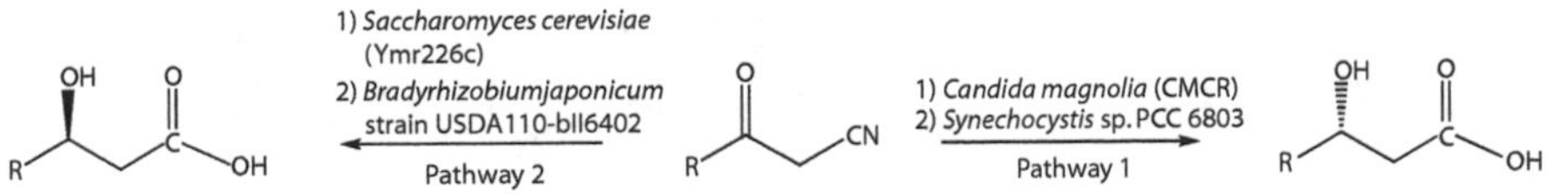

1) *Saccharomyces cerevisiae* (Ymr226c)
2) *Bradyrhizobiumjaponicum* strain USDA110-bll6402
Pathway 2

1) *Candida magnolia* (CMCR)
2) *Synechocystis* sp. PCC 6803
Pathway 1

		(R) Acid pathway 1		(S) Acid pathway 2	
Entry	R	Yield (%)	ee (%)	Yield (%)	ee (%)
1	Ph	72	99	75	99
2	p-F Ph	77	99		
3	p-Cl Ph	77	99	75	99
4	p-OMe Ph	80	98	82	99
5	o-F, p-F Ph			77	99

by a nitrile hydratase/amidase enzymatic system from *Rhodococcus rhodochrous* ATCC BAA-870 whole cells (Table 17.10). With all 3-hydroxy-4-aryloxybutanenitrile substrates, biotransformation gave access to (R)-amide and (S)-acid, with the exception of the substrate bearing a p-Cl-phenyl substituent where (S)-amide and (R)-acid were obtained. Stereochemical recognition was mainly due to a highly stereoselective amidase coupled with a nonselective nitrile hydratase. Although the rate of the reaction was not affected by the electron-withdrawing groups on the aromatic ring the selectivity decreased, while not significant effects were obtained by using electron-donating substituents. 3-Hydroxy-3-arylpropanenitriles gave lower reaction rates but enhanced stereoselectivity when substrates with substituents on the aromatic ring were used.

γ-Butyrolactones are components of natural flavors and fragrances, and building blocks for the synthesis of pharmaceuticals. Pollock et al. (2007) reported the one-pot enantioselective synthesis of a number of γ-butyrolactones by enzymatic hydrolysis of α- and β-hydroxynitriles to the corresponding hydroxyacid and subsequent lactonization. Nitrilase NIT 1003 showed (R) stereopreference and when R = $-C_5H_{11}$ group, the formed lactone was recovered with a yield of 30% and an ee of 88% (Table 17.11).

A method recently developed by Ma et al. (2008b) to enhance the enantioselectivity of nitrile hydratase/amidases enzymatic systems in the hydrolysis of β-hydroxy nitriles consists of the simple derivatization of the β-hydroxyl group. A simple protection of free hydroxyl by a benzyl (Ma et al., 2008a) or methyl group (Ma et al., 2008b) gave rise to a significant increase of the enantioselective action of *R. erythropolis* AJ270 whole cells (Figure 17.10).

Amazingly, reported data show opposite enantiopreferences when using substrates benzyl-protected and methyl-protected. This excellent stereorecognition stemmed again from the synergistic effect of a low enantioselective nitrile hydratase and a high enantioselective amidase. Ma's group used the same approach and obtained analogous results by using β-amino nitriles as substrates for the synthesis of valuable β-amino acids (Figure 17.11) (Ma et al., 2008a).

TABLE 17.10
Nitrile Hydratase/Amidase Biocatalyzed Transformation of 3-Hydroxy-4-Aryloxybutanenitriles and 3-Hydroxy-3-Arylpropanenitriles

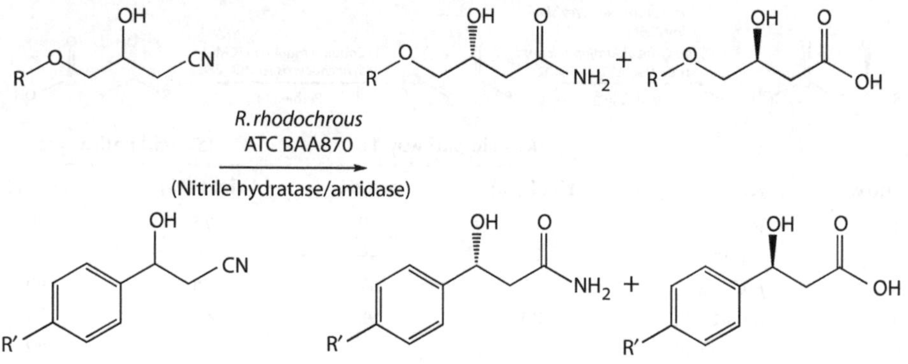

			Amide		Acid	
Entry	R	R′	Yield (%)	ee (%)	Yield (%)	ee (%)
1	Ph		32	>99 (R)	32	57 (S)
2	Bn		14	>99 (R)	37	60 (S)
3	p-Cl Ph		24	82 (S)	37	72 (R)
4	p-OMe Ph		14	96 (R)	34	70 (S)
5		H	8	3 (R)	46	0.5 (S)
6		Me	39	65 (R)	25	78 (S)

TABLE 17.11
Enantioselective Synthesis Nitrilase Biocatalyzed of γ-Butyrolactones

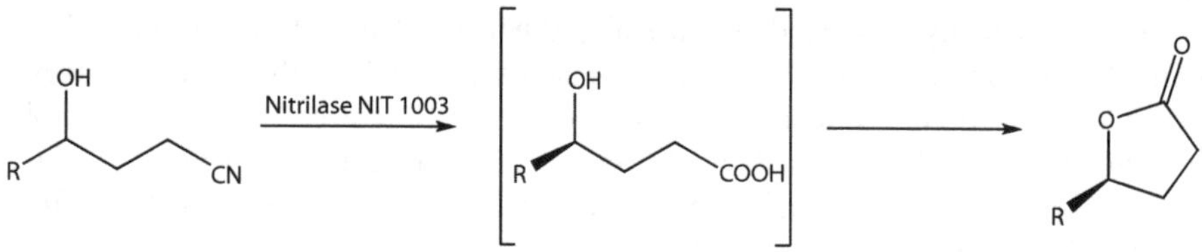

Entry	R	Yield (%)	ee (%)
1	Et	47	54
2	Bu	26	80
3	C_5H_{11}	30	88
4	C_8H_{17}	44	20

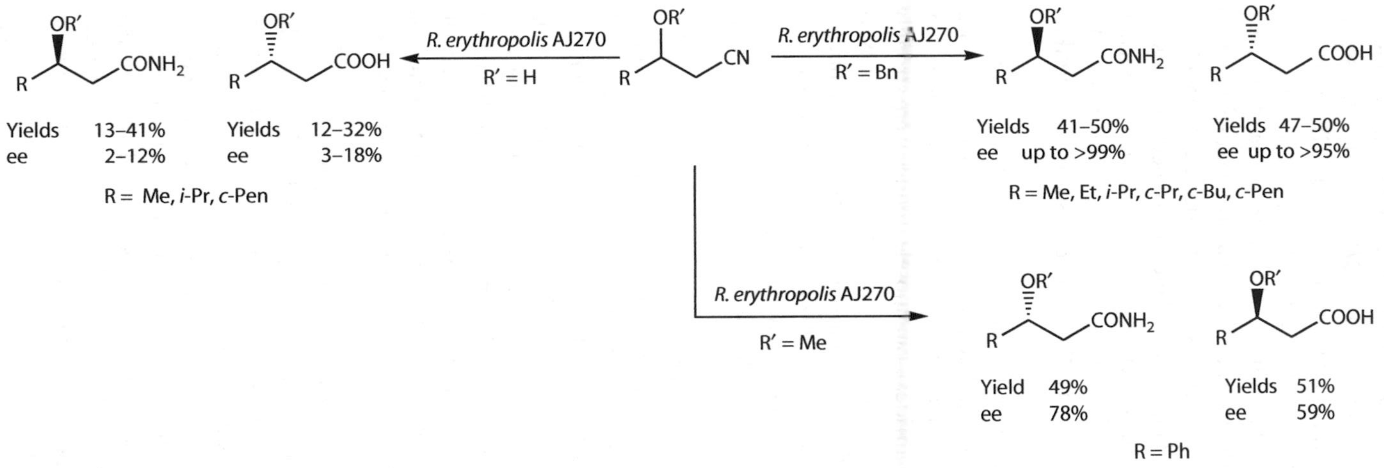

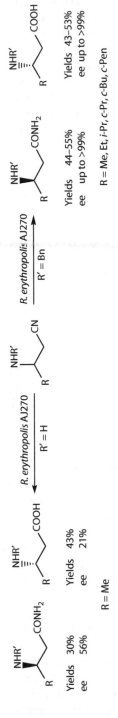

FIGURE 17.10 Nitrile hydratase/amidase catalyzed biotransformation of *O*-protected β-hydroxy nitriles.

FIGURE 17.11 Nitrile hydratase/amidase catalyzed biotransformation of *N*-protected β-amino nitriles.

β-Amino acids have gained considerable attention in the last years due to their antibiotic (Davies et al., 1999), antifungal (Theil and Ballschuh, 1996), and cytotoxic properties (Sone et al., 1993). An example is represented by (2R,3S)-3-phenylisoserine that is an important constituent of the antitumor agent paclitaxel (Denis et al., 1990). The replacement of β-amino acids in biologically active peptides by certain β-counterparts can have pronounced effects on their folding properties (Gellman, 1998), resulting in modified biological properties of the unnatural analogs (Giannis and Kolter, 1993). Klempier et al. (1996) have extensively investigated the synthesis and biotransformations of β-amino acids. They reported the stereoselective transformation of cis- and trans-N-protected-β-amino-cyclopentane/hexane nitriles biocatalyzed by nitrile hydrates/amidase bienzymatic systems from *Rhodococcus equi* A4, *Rhodococcus erythropolis* NCIMN 11540, and *Rhodococcus* sp. R312 (Winkler et al., 2005) (Table 17.12).

Biotransformations of five-membered alicyclic trans-N-protected-amino nitriles proceeded faster than in case of six-membered compounds. The products of the trans-N-protected-amino nitriles (amides and acids) were formed preferentially than the products of the cis-counterparts (only amides). Enantioselectivities were strongly dependent on the structure: the trans-five-membered substrates gave exclusively amides with excellent optical purity (94–99%), in contrast to the trans-six-membered substrates resulted in the formation of the acid with excellent enantiopurity (87–99%). The corresponding cis-compounds yielded much lower enantiomeric excesses. Nitrile precursor of α-methylene-β-amino acids was analogously investigated (Winkler et al., 2005) (Table 17.13).

Klempier and collaborators more recently reported the synthesis and biotransformations of β-amino acids, more specifically conformationally constrained γ-amino butyric acid (GABA) analogs, by using commercial nitrilases (Winkler et al., 2007a). GABA is the major inhibitory neurotransmitter in the mammalian central nervous system (Andersen et al., 2001). Recently, the application of conformationally restricted GABA mimics has contributed to a better understanding in GABA neuroreceptor research (Chebib and Johnston, 1999). Enzymes screened by Klempier and coworkers revealed a strong enantiopreference when cis-isomers were used (up to 99% ee), whereas trans-isomers gave ee up to 86%. Nitrilase NIT-106 proved to be highly R-selective in contrast to NIT-107 and others screened nitrilases that were S-selective (Table 17.14).

The same commercial enzymes have been used for the enantioselective synthesis of the nonproteinogenic heterocyclic N-protected amino acids pyrrolidine-3-carboxylic acid, nipecotic acid (piperidine-3-carboxylic acid), and isonipecotic acid (piperidine-4-carboxylic acid), starting from the corresponding N-protected nitriles (Winkler et al., 2007b). A pronounced difference in the catalytic activity was noticed depending on the ring size: five-membered pyrrolidine-3-carboxylic acids are completely transformed in less than 24 h, while all six-membered piperidine 3- and 4-carbonitriles required reaction times within days (Table 17.15). Nevertheless, the nitrilase worked with nearly unchanged activity over this period of time, permitting, for example, to obtain N-tosyl protected nipecotic acid with an ee of 93%.

Very recently, all substrates reported in Tables 17.14 and 17.15, were submitted to biotransformations in the presence of fungal nitrilases from *Fusarium solani* O1 and

TABLE 17.12
Nitrile Hydratase/Amidase Catalyzed Stereoselective Transformation of *cis*- and *trans*-N-Protected-β-amino-cyclopentane/hexane Nitriles

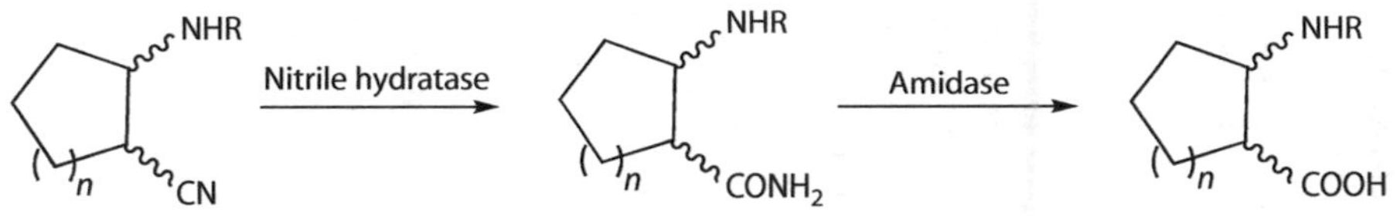

Entry	R	n	cis/trans	R. equi A4			R. erythropolis NCIMB 11540			R. sp. R312		
				Nitrile % (ee%)	Amide % (ee%)	Acid % (ee%)	Nitrile % (ee%)	Amide % (ee%)	Acid % (ee%)	Nitrile % (ee%)	Amide % (ee%)	Acid % (ee%)
1	Bz	1	*trans*	0	40 (94)	55 (75)	0	30 (>99)	63 (48)	0	7 (>99)	87 (15)
2	Bz	2	*trans*	38 (99)	22 (56)	36 (>95)	59 (44)	16 (67)	15 (>95)	61 (82)	14 (38)	7 (>95)
3	Ts	1	*trans*	40 (47)	14 (>99)	44 (2)	0	13 (>99)	86 (5)	46 (30)	10 (>99)	34 (14)
4	Ts	1	*cis*	71 (5)	14 (51)	0	50 (16)	49 (15)	0	11 (51)	75 (7)	0
5	Ts	2	*trans*	26 (78)	54 (65)	13 (>99)	24 (98)	56 (59)	15 (97)	33 (47)	42 (77)	16 (87)
6	Ts	2	*cis*	47 (8)	48 (6)	0	50 (10)	41 (8)	0	44 (10)	43 (4)	0

TABLE 17.13
Nitrile Hydratase/Amidase Catalyzed Stereoselective Transformation of N-Protected α-Methylene-β-Amino Nitrile

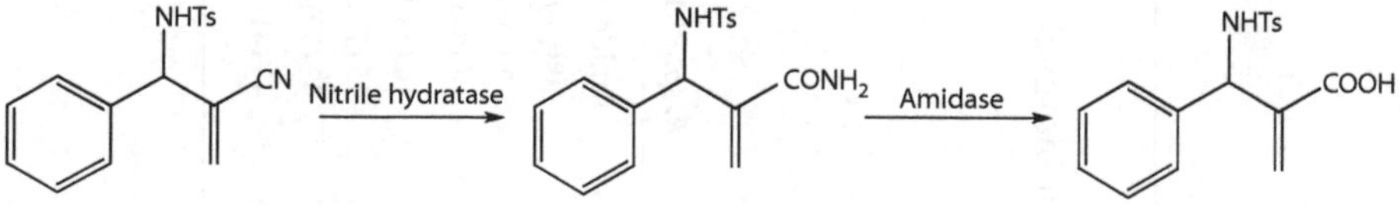

NHTs → (Nitrile hydratase) → NHTs → (Amidase) → NHTs
(Ph…CN) → (Ph…CONH₂) → (Ph…COOH)

	R. equi A4			R. erythropolis NCIMB 11540			R. sp. R312		
Entry	Nitrile % (ee%)	Amide % (ee%)	Acid % (ee%)	Nitrile % (ee%)	Amide % (ee%)	Acid % (ee%)	Nitrile % (ee%)	Amide % (ee%)	Acid % (ee%)
1	84 (0)	10 (11)	0	91 (1)	2 (32)	0	73 (0)	25 (6)	0

from *Aspergillus niger* K10. These enzymes showed broad substrate specificity and a strong diastereopreference for *cis*- versus *trans*-β-amino nitriles. However, enantioselectivities were rather low for all acids synthesized as well as their remaining nitriles (Winkler et al., 2009). An easy chemical route to synthesize optically active β-amino acids is represented by the reduction of the corresponding enantiopure azides. Despite the availability of preparative methods for organic azides, chiral organic azides represent yet a challenge and are mainly obtained from diasteroselective reactions of an azide ion with chiral substrates. Recently, Ma et al. (2006),

TABLE 17.14
Nitrilase Catalyzed Stereoselective Transformation of *cis*- and *trans*-N-Protected Conformationally Constrained γ-Amino Butyronitriles

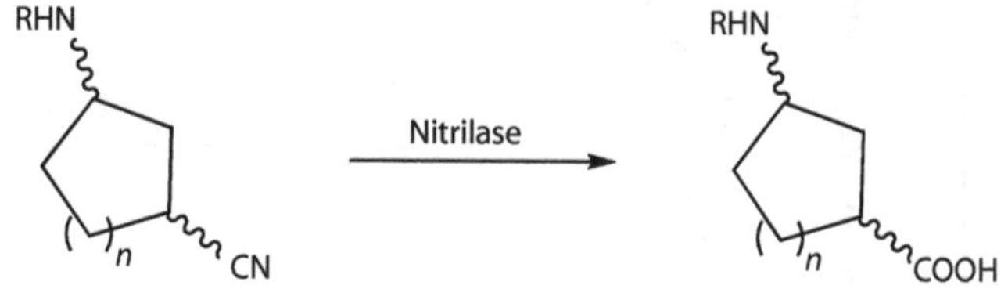

RHN …CN → (Nitrilase) → RHN …COOH

				NIT-106		NIT-107	
Entry	R	n	cis/ trans	Acid Yield (%)	Acid ee (%)	Acid Yield (%)	Acid ee (%)
1	Ts	1	cis	47	97 (1R,3S)		
2	Ts	1	trans	36	55 (1R,3R)		
3	Ts	2	cis	42	>99 (1R,3S)		
4	Ts	2	trans			46	86 (1S,3S)
5	Cbz	2	trans			41	74 (1S,3S)

TABLE 17.15

Nitrilase Catalyzed Enantioselective Synthesis of Nonproteinogenic Heterocyclic N-Protected Amino Acids

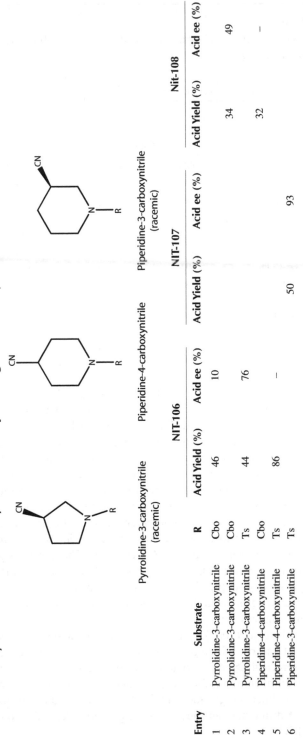

Entry	Substrate	R	NIT-106 Pyrrolidine-3-carboxynitrile (racemic)		NIT-107 Piperidine-4-carboxynitrile		Nit-108 Piperidine-3-carboxynitrile (racemic)	
			Acid Yield (%)	Acid ee (%)	Acid Yield (%)	Acid ee (%)	Acid Yield (%)	Acid ee (%)
1	Pyrrolidine-3-carboxynitrile	Cbo	46	10				
2	Pyrrolidine-3-carboxynitrile	Cbo					34	49
3	Pyrrolidine-3-carboxynitrile	Ts	44	76				
4	Piperidine-4-carboxynitrile	Cbo					32	–
5	Piperidine-4-carboxynitrile	Ts	86	–				
6	Piperidine-3-carboxynitrile	Ts			50	93		

TABLE 17.16

Enantioselective Biotransformations of Racemic α-Substituted β-Azidoproprionitriles

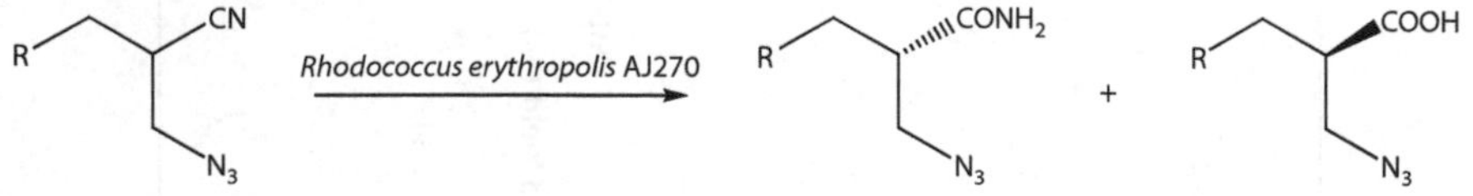

Entry	R	Amide		Acid	
		Yield (%)	ee (%) (S)	Yield (%)	ee (%) (R)
1	Ph	48.0	>99.5	49.0	96.2
2	p-MeO Ph	47.0	>99.5	48.0	94.0
3	p-Me Ph	48.0	97.2	49.0	>99.5
4	p-F Ph	48.0	>99.5	47.0	>99.5
5	p-Cl Ph	48.5	86.7	44.0	93.2
6	m-Cl Ph	50.0	>99.5	47.0	94.8
7	o-Cl Ph	48.5	97.3	48.0	>99.5
8	p-Br Ph	47	>99.5	46.5	93.6
9	Cyclohexyl	49	>99.5	49.0	95.5
10	Pr	41.0	83.0	58.0	58.6
11	Et	48	84.6	48.0	73.5

reported the first example of biocatalyzed resolution of a number of α-arylmethyl-β-azidoproprionitriles, illustrated in Table 17.16. Nitrile hydratase/amidase enzymatic system from *R. erythropolis* AJ270 was found to be highly (*R*)-enantioselective toward all arylic substrates, although the nature of the substituent and its substitution pattern on the aromatic ring played a part in determining the overall conversion rate. The high enantioselectivity of the process arises, as in many other cases here reported, by the combined effect of enantioselective nitrile hydratase and amidase with the latter playing a dominant role. Conversely, substrates containing alkylic substituents gave conversion rates comparable to those of aryl-substituted substrates, but only moderate enantioselectivities with the exception of the compound bearing a cyclohexylmethyl moiety (Table 17.16—entry 9). These data together with other insights coming from experiments here not illustrated, suggest that the presence of both an aryl unit or a six-membered cyclohexyl substituent and a methylene segment between the azido group and the stereogenic center in the structure is mandatory to obtain high enantioselectivities.

Kielbasiński et al. (2008) have recently demonstrated the first case of enzymatic recognition of a nitrile with a β chiral sulfur atom. Nine different commercial nitrilases were screened using cyanomethyl *p*-tolyl sulfoxide as substrate; all reactions occurred in a mixture of buffer phosphate and a co-solvent was used to dissolve the substrate. Very interestingly, they obtained both the acid and the amide as hydrolysis products with different ee and absolute configurations depending on the enzyme used (Table 17.17).

TABLE 17.17
Nitrilase Catalyzed Resolution of Cyanomethyl p-Tolyl Sulfoxide

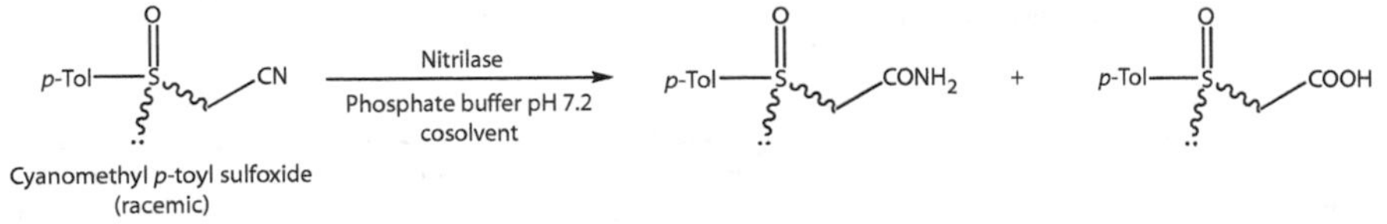

Cyanomethyl p-toyl sulfoxide
(racemic)

Entry	Nitrilase	Cosolvent	Nitrile		Amide		Acid	
			Yield (%)	ee (%)	Yield (%)	ee (%)	Yield (%)	ee (%)
1	NL 103	CHCl₃	40	37 (R)	30	21 (S)	30	35 (S)
2	NL 103	Acetone	46	33 (R)	27	20 (S)	19	31 (S)
3	NL 104	CHCl₃	10	44 (R)	32	71 (R)	57	27 (S)
4	NL 104	Acetone	26	26 (R)	33	62 (R)	36	24 (S)
5	NL 105	CHCl₃	72	0	0	—	25	1 (S)
6	NL 105	Acetone	80	0	0	—	9	1 (S)
7	NL 106	CHCl₃	0	—	10	37 (R)	90	7 (S)
8	NL 107	CHCl₃	0	—	36	13 (R)	60	17 (S)
9	NL 108	CHCl₃	63	0	0	—	30	2 (S)
10	NL 110	CHCl₃	100	0	0	—	0	—
11	NL 111	CHCl₃	100	0	0	—	0	—
12	NL 112	CHCl₃	0	—	62	45 (R)	38	83 (S)
13	NL 112	Acetone	0	—	35	42 (R)	58	76 (S)

The formation of amide in a nitrilase catalyzed reaction is not so uncommon and has been already discussed in this chapter together with hypothesis of the reaction mechanism. Nevertheless, all these explanations do not account for all the results obtained in this experiment, especially in the case in which the amide and the acid formed have opposite absolute configurations leading to the assumption that each reaction (formation of amide and formation of acid) may employ different pockets of the enzyme active site at the enantioselective water attack on the thiomidate intermediate; the fact that each enantiomer of the substrate produces preferentially one of the hydrolysis products might support this assumption. Another possible explanation hypothesized is that the nitrile is first converted into the amide in an active site large enough to enable the rotation of the amide within it; for the nitrile, the sulfoxide oxygen is likely to hydrogen bond to a residue in the active site, but for the amide it may be the carbonyl to selectively hydrogen bond to that residue, thus inverting the enantioselectivity of the hydrolysis to the acid. In other enzyme–substrate combinations such rotation might not be possible leading to a retention of configuration. It is clear enough that a better understanding of the nitrilase action mechanism is needed. Unluckily, it will have to await the determination of crystal structures of nitrilases, which so far do not exist.

17.8.1.3   Racemic Nitriles Containing a Three-Membered Ring

Nitriles and amides with a three-membered ring such as cyclopropane and epoxide are available from various synthetic methods, and represent useful molecules to act

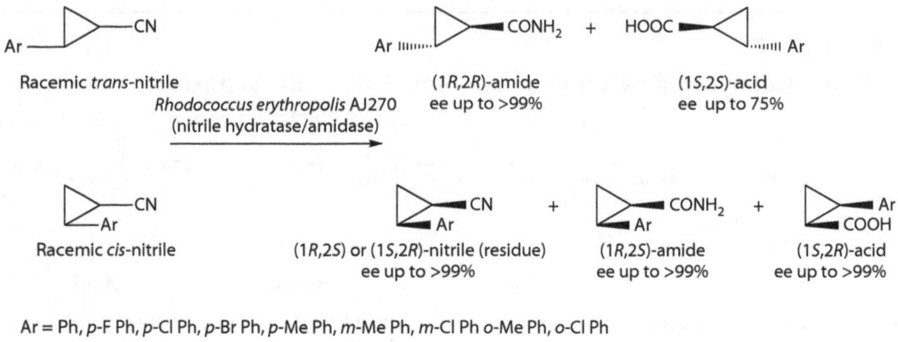

Racemic *trans*-nitrile
Rhodococcus erythropolis AJ270
(nitrile hydratase/amidase)

(1*R*,2*R*)-amide
ee up to >99%

(1*S*,2*S*)-acid
ee up to 75%

Racemic *cis*-nitrile

(1*R*,2*S*) or (1*S*,2*R*)-nitrile (residue)
ee up to >99%

(1*R*,2*S*)-amide
ee up to >99%

(1*S*,2*R*)-acid
ee up to >99%

Ar = Ph, *p*-F Ph, *p*-Cl Ph, *p*-Br Ph, *p*-Me Ph, *m*-Me Ph, *m*-Cl Ph *o*-Me Ph, *o*-Cl Ph

FIGURE 17.12   Enantioselective biotransformations of 2-arylcyclopropanecarbonitriles.

as probes in the study of the steric features of the active site of nitrile hydratases and amidases. Besides, it is still a challenging task to obtain optically active molecules such as cyclopropane, epoxide, or aziridine derivatives, especially polysubstituted and polyfunctionalized ones. Wang and collaborators have widely explored and studied the synthesis and biotransformations of such compounds, trying to create a predictive model of the enantioselective recognition of the nitrile hydratase/amidase enzymatic system from *Rhodococcus erythropolis* AJ270. *Rhodococcus* sp. AJ270 catalyzed the hydrolysis of racemic *trans*-2-arylcyclopropanecarbonitriles to produce optically active (1*R*,2*R*)-2-arylcyclopropancarboxamides and (1*S*,2*S*)-2-arylcyclopropanecarboxylic acids (Wang and Feng, 2000, 2002). Good and comparable enantiomeric excesses have been obtained for almost all substrates irrespective of the nature and the substitution pattern of the substituent, with the only exception of 4-methoxyphenyl-substituted nitrile, which gave very low enantioselectivity (Figure 17.12). All these data supported the conclusion that the nitrile hydratase involved in *Rhodococcus* sp. AJ270 is non enantioselective against the *trans*-2-arylcyclopropanecarbonitriles while the amidase is 1*S*-selective. On the other hand, both the nitrile hydratase and amidase showed much lower enzymatic activity but extraordinary high enantioselectivity against *cis*-configured 2-arylcyclopropanecarbonitriles furnishing in the case of the phenyl substituted substrate, enantiopure (1*R*,2*S*)-2-phenylcyclopropanecarboxamide, and (1*S*,2*R*)2-phenylcyclopropanecarboxylic acid in excellent yields (Wang and Wang, 2002). Introduction of a substituent larger than chlorine and methyl at the *para*-position of the benzene ring was found to lead to an even more sluggish hydration of nitriles and a complete inhibition of the hydrolysis of amides. It was also observed that the nitrile hydratase inversed its enantioselectivity from 1*S* to 1*R*, when the substituent on the benzene ring varied from others to a methoxy group (Wang and Feng, 2002).

These data and considerations on structural features of substrates have suggested the hypothesis that a readily reachable reactive site is embedded within a spacious pocket of the enantioselective nitrile hydratase, while the amidase comprises a relatively deep-buried and size-limited enantioselective active site. This hypothesis has been supported by the results obtained studying the biotransformations of differently configured 2,2-dimethyl-3-arylcyclopropanecarbonitriles catalyzed by

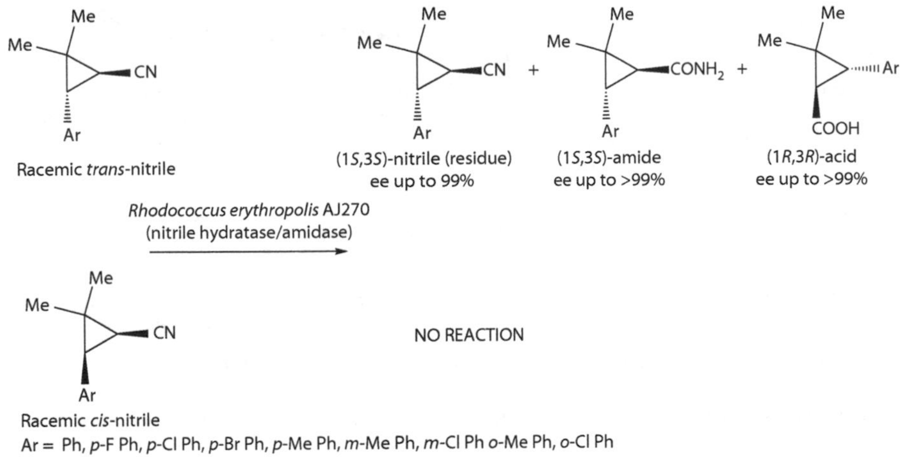

Me Me CN

Racemic *trans*-nitrile

Me Me CN

(1S,3S)-nitrile (residue)
ee up to 99%

Me Me CONH₂

(1S,3S)-amide
ee up to >99%

Me Me Ar
COOH

(1R,3R)-acid
ee up to >99%

Rhodococcus erythropolis AJ270
(nitrile hydratase/amidase)

Me Me CN

Racemic *cis*-nitrile

NO REACTION

Ar = Ph, *p*-F Ph, *p*-Cl Ph, *p*-Br Ph, *p*-Me Ph, *m*-Me Ph, *m*-Cl Ph *o*-Me Ph, *o*-Cl Ph

FIGURE 17.13 Enantioselective biotransformations of 2,2-dimethly-3-arylcyclopropane-carbonitriles.

the same *Rhodococcus* sp. AJ270 catalyst (Wang and Wang, 2002). Compared to *trans*-2-arylcyclopropanecarbonitriles and *trans*-2-arylcyclopropanecarboxamides, *trans*-2,2-dimethyl-3-arylcyclopropanecarbonitriles, and *trans*-2,2-dimethyl-3-arylcyclopropanecarboxamides have increased steric hindrance. Therefore, the interactions of these substrates with the active sites of the nitrile hydratase and amidase become slower but more stereoselective than their analogs. Conversely, in the case of *cis*-isomers, the introduction of geminally substituted dimethyl groups into the cyclopropane skeleton further increases the bulkiness of already sterically crowded molecules, and as a consequence the enzymatic reaction is completely inhibited (Wang, 2005). On the basis of these results it has been proposed that the reactivity and enantioselectivity of enzymatic hydrolysis of cyclopropanecarbonitriles and amides could be regulated by tuning the substituents either on the 2- or 3-position or on both positions of the three-membered ring (Figure 17.13).

Despite different chemical properties, *trans*-2,3-epoxy-3-arylpropanenitriles resemble *trans*-2-arylcyclopropanecarbonitriles to undergo an equally rapid *Rhodococcus* sp. AJ270-catalyzed hydrolysis. Almost the same substituent effects have been observed on the reaction rate, but enantiomeric excess values for amides, however, vary dramatically from >99% for *trans*-2,3-epoxy-3-phenyl- and 3-*para*—substituted-phenylpropanenitriles to 30–40% for 3-*ortho*-methylphenyl and 3-*meta*-chlorophenyl-substituted analogs. The corresponding arylglycidic acids are not isolated, however, because they undergo spontaneous decomposition under the reaction conditions. Being similar to *cis*-2-arylcyclopropanecarbonitriles, *cis*-2,3-epoxy-3-arylpropanenitriles undergo very sluggish reactions of nitrile hydration and subsequent amide hydrolysis. Amazingly, neither the nitriles are recovered nor amides formed are optically active, maybe due to the change of molecular polarity (Wang, 2005) (Figure 17.14).

The use of *trans*- and *cis*-3-aryl-2-methyloxyranecarbonitriles as substrates gave results that were in agreement with previous speculations (Wang et al., 2005).

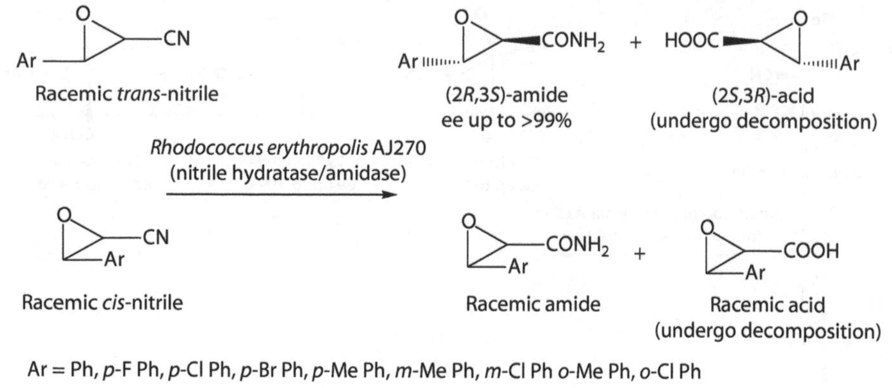

FIGURE 17.14 Enantioselective biotransformations of 2,3-epoxy-3-arylpropanecarbonitriles.

Trans-3-aryl-2-methyloxyranecarbonitriles gave fast reactions affording an optically active amide and an acid that could not be separated because they underwent decomposition analogously to the previous 2,3-epoxy-3-arylpropane carboxylic acids. The enantioselectivity of the amide hydrolysis was strongly dependent upon the structure of the substrate. More specifically, it is the substitution pattern rather than the nature of the substituent to play a crucial role in determining both the rate and the enantioselectivity. When the substituent is a phenyl or a *para*-derivative, the reaction is fast and gives amides with ee of 99–99.5%. Conversely, in the case of *meta* or *ortho*-phenyl substituents, long incubation times are necessary and amides obtained have low enantiopurity. *cis*-3-Aryl-2-methyloxyranecarbonitriles gave much more sluggish reactions, and after many days of incubation only small amounts of amides are recovered (up to 25%) with very poor ee (up to 28%) (Figure 17.15).

Recently, Wang et al. (2007) moved their investigations toward another group of three-membered ring containing nitriles, aziridines. The interest gained by this class

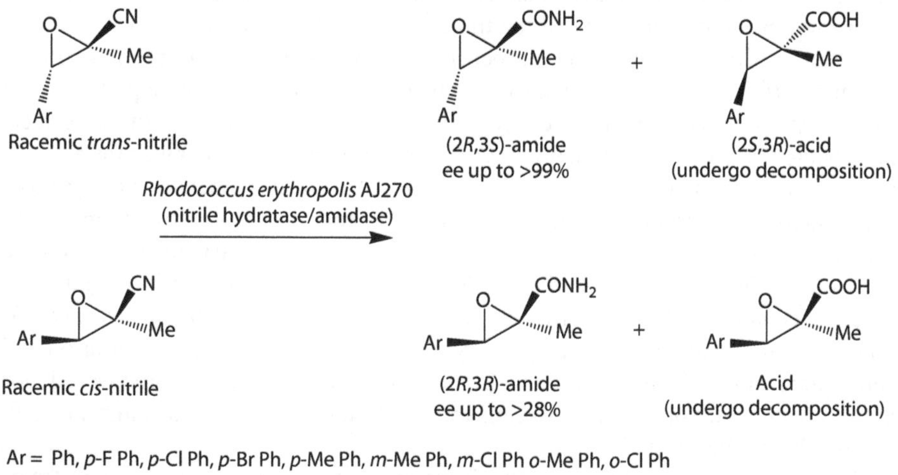

FIGURE 17.15 Enantioselective biotransformations of 3-aryl-2-methyloxyranecarbonitriles.

of compounds is rapidly increasing due to its intriguing and unique chemical and physical properties. Results obtained in 1-arylaziridine-2-carbonitriles biotransformation catalyzed by nitrilase from *Rhodococcus* sp. AJ270, are generally excellent and in agreement with all previous outcomes concerning cyclopropanecarbonitriles and oxiranecarbonitriles (Wang et al., 2007). It was found that all nitriles substrates underwent fast, even if with low enantioselectivity, hydration to amides, and the hydrolysis of these resulting amides catalyzed by the amidase was also efficient and enantioselective in most cases with yield ranging between 45% and 50% both for amides and acids. Enantiomeric excess values of obtained amides and acids were high (amides 99.5%, acids 87–99.5%) irrespective of the electronic nature of the substituent on the aromatic ring, with the exception of the *o*-methylphenyl-substituted aziridine-2-carbonitrile that furnished only optically inactive amide and acid products Figure 17.16.

17.8.2 DESYMMETRIZATION OF PROCHIRAL DINITRILES

3-Substituted glutaronitriles represent interesting prochiral dinitriles substrates in biocatalytic desymmetrization reaction because the resulting chiral products are key intermediates in organic synthesis. Wang and Feng (2000) report that nitrile hydratase/amidase-containing microorganism, *Rhodococcus* sp. AJ270 is able to hydrolyze 3-alkyl- and 3-arylglutaronitriles in a selective manner. Isolation of a (*S*)-monocyano acid as the sole product from the reaction indicated that the nitrile hydratase involved in this microbial cell acts as a regiospecific hydrating enzyme against dinitrile. The amidase, on the other hand, was highly efficient, converting rapidly and completely all monocyano amide into the acid, as demonstrated by the fact that in all cases no monocyano amide was obtained. Amazingly, the enantioselectivity of the overall hydrolysis seems to be derived from the stereospecific action of nitrile hydratase. The use of a cosolvent such as acetone was mandatory to obtain satisfactory enantiomeric excess values (from 32% to 95% depending on the nature of the 3-substituent) even if the role of this additive in the chiral recognition mechanism of nitrile hydratase is not yet fully understood (Table 17.18). Attempts to transform 3-methyl-3-phenylglutaronitrile, a quaternary carbon-containing prochiral dinitrile, into homochiral product gave no result after incubation with *Rhodococcus* sp. AJ270 cells for 6 days.

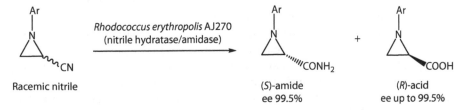

Ar = Ph, *p*-F Ph, *p*-Cl Ph, *p*-Br Ph, *p*-Me Ph, *m*-Me Ph, *o*-Me Ph, *p*-OMe Ph

FIGURE 17.16   Enantioselective biotransformations of 1-arylaziridine-2-carbonitriles.

TABLE 17.18

Enzymatic Desymmetrization of 3-Substituted Glutaronitriles

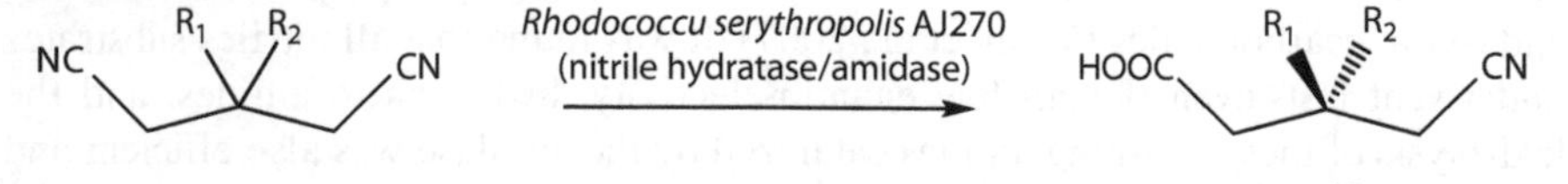

$$NC \overset{R_1 \; R_2}{\diagdown \diagup} CN \quad \xrightarrow[\text{(nitrile hydratase/amidase)}]{\textit{Rhodococcu serythropolis} \text{ AJ270}} \quad HOOC \overset{R_1 \; R_2}{\diagdown \diagup} CN$$

				Acid	
Entry	R_1	R_2	Cosolvent	Yield (%)	ee (%) (S)
1	Ph	H	—	88	39
2	Ph	H	Acetone	67	88
3	p-F Ph	H	—	81	25
4	p-F Ph	H	Acetone	16	76
5	p-Cl Ph	H	—	37	26
6	p-Cl Ph	H	Acetone	25	63
7	p-Me Ph	H	—	42	64
8	p-Me Ph	H	Acetone	25	95
9	p-MeO Ph	H	—	36	50
10	p-MeO Ph	H	Acetone	17	79
11	o-Me Ph	H	—	51	30
12	o-Me Ph	H	Acetone	40	35
13	Cyclohexyl	H	Acetone	63	31
14	Cyclohexyl	H	—	60	83
15	Bz	H	Acetone	90	29
16	Bz	H	—	62	32
17	Ph	Me	—	0	–

3-Hydroxyglutaronitrile (3-HGN) has been subject of independent investigations of many research groups. It is especially interesting because its optically active derivative ethyl (R)-4-cyano-3-hydroxybutyrate is a key intermediate in the synthesis of the statin Lipitor (Figure 17.17). Statins are HMG-CoA reductase inhibitors used for the treatment of hypocholesterolemia and atherosclerosis, and Lipitor represents actually the world's largest grossing drug with 2004 sales of 12 billion dollars. With the aim to develop highly enantioselective catalysts for the biotransformations of 3-HGN, DeSantis et al. (2003) have applied a directed evolution technique, the gene situ saturation mutagenesis (GSSM), to generate mutants with enhanced catalytical properties. Amino acid changes at 17 different residues led to enhanced enantioselectivity over the wild-type enzyme. Biotransformation of 3-HGN catalyzed by one of these mutants led to the formation of the enantiopure (R)-4-cyano-3-hydroxybutyric acid (yield 96%, 98.5% ee) (DeSantis et al., 2003). The same goal was achieved by Bergeron et al. (2006). By using nitrilase BD9570, they were able to synthesize (R)-4-cyano-3-hydroxybutyric acid acid with quantitative yield and an ee of 99% (Bergeron et al., 2006).

Vink et al. (2009) used an O-benzyl protected 3-hydroxyglutaronitrile in the presence of nitrilase from R. erythropolis NCIMB 11540 to produce the corre-

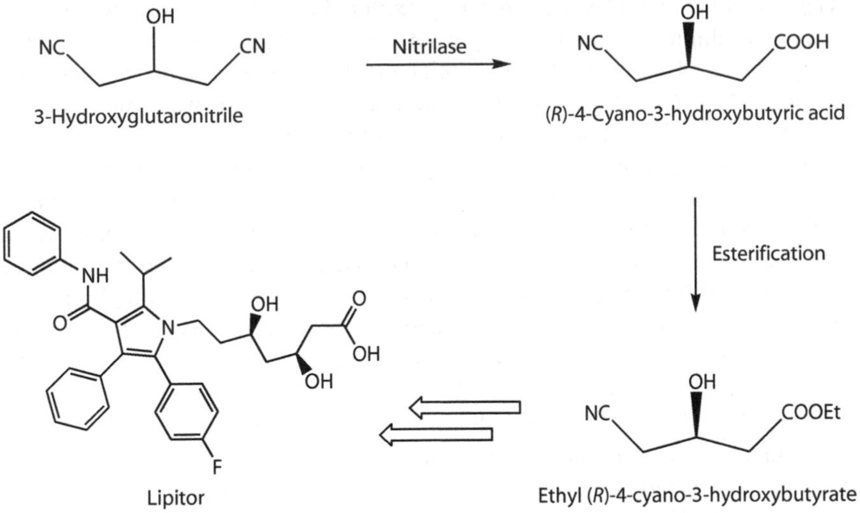

Biocatalysis 397

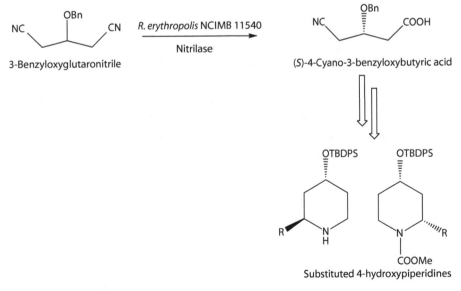

FIGURE 17.17 Enzymatic desymmetrization of 3-hydroxyglutaronitrile, key synthon for the production of Lipitor.

sponding enantiopure (S)-monoacid (ee 96%). Starting from these chiral synthons, they developed two different chemical routes to synthesize some optically active substituted 4-hydroxypiperidines, strategical intermediates in the preparation of important pharmacological molecules such as palinavir, a potent inhibitor of the HIV virus (Figure 17.18).

FIGURE 17.18 Enzymatic desymmetrization of 3-hydroxyglutaronitrile, synthon for the production of 4-hydroxypiperidines.

Kielbasiński et al. (2008) have recently explored the nitrilase catalyzed biotransformations of dinitriles containing a prostereogenic center located on heteroatoms. In the presence of a series of commercial nitrilases as well as of nitrilase from *R. erythropolis* NCIMB 11540, bis(cyanomethyl) sulfoxide was transformed in cyanomethylsulfinylacetamide, cyanomethylsulfinylacetic acid, and sulfinyldiacetic acid with ratios and enantioselectivities depending on used enzyme and illustrated in Table 17.19. These data show that the reaction does not occur in a chemoselective way since in several cases more than one product is formed. In cases of chiral products, the stereoselectivity vary from low to excellent depending on the catalyst, with the best results furnished by the Nitrilase-107 that afforded cyanomethylsulfinylacetic acid with an ee >99% and an (*R*)-stereopreference. The presence of an amide is imputable to the nitrile hydratasic activity of nitrilases as already discussed. It is however still of special interest that in case of Nitrilase-104, the amidic compound was obtained with a yield of 57% and an ee of 99% (Kielbasiński et al., 2007a).

The same enzymes have been used to catalyze the biotransformation of a dinitrile containing a phosphorous prostereogenic center, namely bis(cianomethyl)phenylphosphine oxide (Kielbasiński et al., 2007b). The results are shown in Table 17.20. Out of the five possible products only two, cyanomethylphenylphosphinacetamide and cyanomethylphenylphosphinylacetic acid were formed in various proportions and enantiopurities depending on the enzyme used. Enzymes were not always chemoselective, and nitrilases furnished in some experiments gave an amidic product. The stereoselectivity of the enzymes toward the phosphoryl group varied from low to very high.

TABLE 17.19
Enzymatic Desymmetrization of bis(Cyanomethyl) Sulfoxide

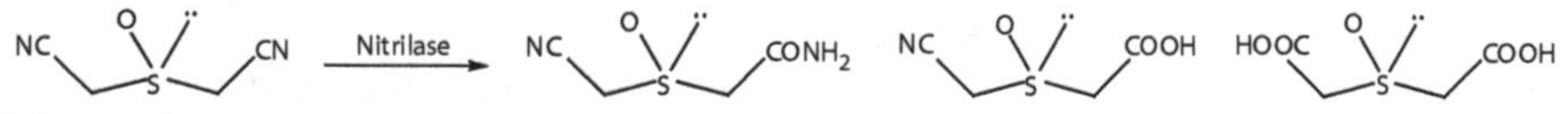

Bis(cyanomethyl) sulfoxide — Cyanomethylsulfinylacetic acid — Cyanomethylsulfinylacetamide — Sulfinyldiacetic acid

		Bis(cyanomethyl) Sulfoxide	Cyanomethylsulfi- nylacetamide		Cyanomethylsulfi- nylacetic Acid		Sulfinyldiacetic Acid
Entry	Nitrilase	Yield (%)	Yield (%)	ee (%)	Yield (%)	ee (%)	Yield (%)
1	NL 101	45.0	26.3	41 (*S*)	—	—	—
2	NL 102	—	—	—	—	—	55.8
3	NL 103	—	26.3	33 (*R*)	—	—	46.5
4	NL 104	—	61.4	>99 (*S*)	27.2	77 (*R*)	—
5	NL 105	—	8.8	82 (*S*)	43.5	52 (*S*)	23.2
6	NL 107	—	17.5	74 (*S*)	34.8	66 (*R*)	—
7	NL 108	50.0	21.9	41 (*S*)	4.4	66 (*S*)	—
8	*Rhodococcus erythropolis* NCIMB 11540	—	13.2	49 (S)	79.1	33 (*S*)	—

TABLE 17.20
Enzymatic Desymmetrization of bis(Cyanomethyl)Phenylphosphine Oxide

Bis(cyanomethyl)phenylphosphine oxide Cyanomethylphenylphosphinacetamide Cyanomethylphenylphosphinylacetic acid

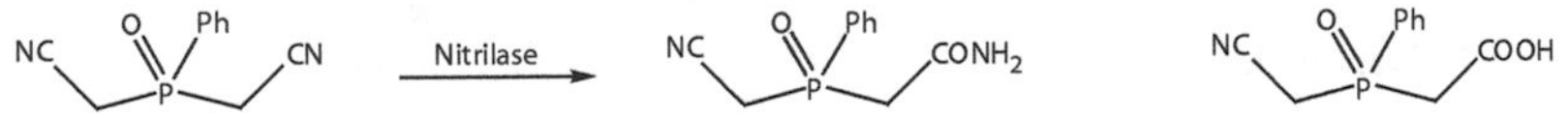

		Bis(cianomethyl) Phenylphosphine Oxide	Cyanomethylphe-nylphosphinacetamide		Cyanomethylphe-nylphosphinylacetic Acid	
Entry	**Nitrilase**	**Yield (%)**	**Yield (%)**	**ee (%)**	**Yield (%)**	**ee (%)**
1	NL 101	10.4	4.6	56 (*R*)	73.1	53 (*S*)
2	NL 102	8.2	9.6	49 (*R*)	82.2	16 (*S*)
3	NL 103	4.0	8.6	55 (*R*)	85.3	15 (S)
4	NL 105	80.0	7.4	36 (*R*)	—	—
5	NL 106	17.3	10.8	>99 (S)	51.0	70 (S)
6	NL 107	78.0	3.7	12 (*R*)	—	—
7	NL 108	16.0	18.7	96 (*R*)	70.7	60 (S)
8	*Rhodococcus erythropolis* NCIMB 11540	75.0	11.1	39 (*R*)	—	—

17.8.3 Industrial Applications

17.8.3.1 Nicotinamide

Nicotinic acid and its amide (nicotinamide), sometimes named niacin and niacinin-amide, are B-complex vitamins (vitamin B3) used in pharmaceutical formulations, and as additives in food and animal feed. This vitamin is also popularly known as pellagra-preventing (PP) factor as its deficiency in human beings causes pellagra. Nicotinamide is the physiologically active form of vitamin B3 in animals and it is a component of coenzymes NAD (nicotinamide adenine dinucleotide) and NADP (nicotinamide adenine dinucleotide phosphate). Nicotinamide derived from nicotinic acid is also used as a brightener in electroplating baths and stabilizer for pigmentation in cured meat. The world demand of nicotinamide and nicotinic acid is estimated to be 22,000 tons year^{-1} (Shimizu, 2001). The chemical processes for the industrial production of nicotinic acid involves, liquid-phase oxidation of either 2-methyl-5-ethyl pyridine with nitric acid under high temperature and pressure (Weissermel and Arpe, 1997), or 3-methylpyridine in combination with cobalt acetate, manganese acetate, and aqueous hydrobromic at 100 atm pressure and 210°C producing 32% of picoline and 19% of nicotinic acid (Hatanaka and Tanaka, 1993). Some electrochemical methods for oxidation of 3-picoline to nicotinic acid with a 70% yield have also been described (Toomey, 1984). These processes of course cannot be considered eco-friendly, but at the same time, cannot be considered cheap and economically convenient because they require high temperatures and pressures (and as a consequence large amounts of energy), expensive catalysts, harsh acidic or

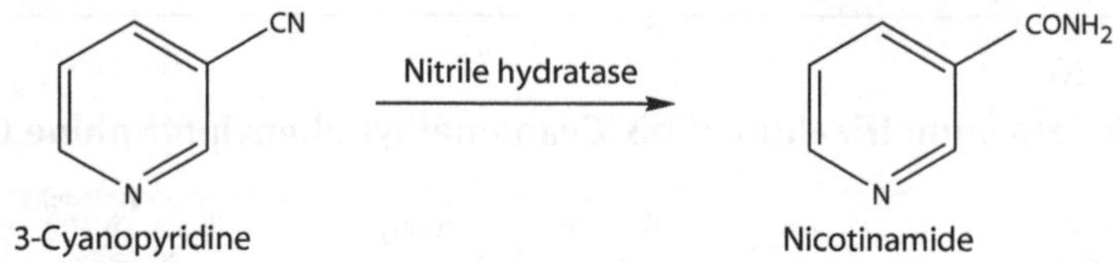

FIGURE 17.19   Nitrile hydratase-catalyzed production of nicotinamide.

basic conditions, and yielding undesired by-products such as nicotinic acid and HCN that increase dramatically the required costs of production. A proposed "greener" and cheaper alternative is the nitrile hydratase catalyzed biotransformation of 3-cyanopyridine (Figure 17.19). Both the inducible nitrile hydratases from *Rhodococcus rhodochrous* PA–34 (Prasad et al., 2007) and from *R. rhodochrous* J1 (Nagasawa et al., 1988) resting cells have been found to catalyze the process with quantitative yields (855 g of nicotinamide per liter of mixture in the former case, 1465 g L^{-1} in the latter case) and in very soft conditions (buffer phosphate pH 7.0– 8.0, temperatures 30–40°C). The nicotinamide produced in both the cases is stable and does not undergo any further conversion into nicotinic acid because of the very low amidases activity in these strains at the working conditions. Moreover, in large-scale preparations, there is not the need to add the substrate in small portions because the inhibitory effect of 3-cyanopyridine on the nitrile hydratase is small. The process biocatalyzed by *R. rhodochrous* J1 resting cells has been actually carried out at an industrial scale by Lonza Company.

An analogous biotransformation of 3-cyanopyridine catalyzed by a nitrilase has been proposed for the production of nicotinic acid. Nitrilases from *R. rhodochrous* J1 (Mathew et al., 1988) and *Nocardia globerula* NHB-2 (Sharma et al., 2006) resting cells have been found to be highly efficient in the production of nicotinic acid (Figure 17.20).

17.8.3.2   Acrylamide

Acrylamide with a demand of 200,000 tons year^{-1} is one of the most important commodities in the world. It is used for the preparation of coagulators, soil conditioners, stock additives for paper treatment, and in leather and textile industry as a component of synthetic fibers. Conventional chemical synthesis involving hydration of acrylonitrile with the use of copper salts as a catalyst has some disadvantages: rate of acrylic acid formation higher than acrylamide, by-products formation and polymerization, and high-energy inputs. To overcome these limits since 1985, the Japanese company "Nitto Chemical Industry" developed a biocatalyzed process to synthesize

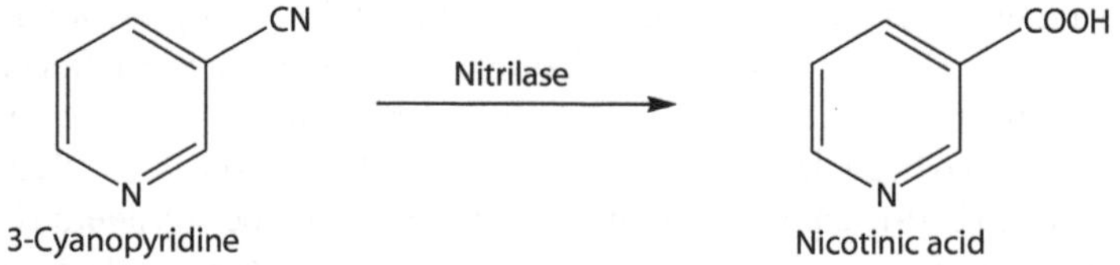

FIGURE 17.20   Nitrilase catalyzed production of nicotinic acid.

acrylamide. They found that the nitrile hydratase from *P. chlororaphis* B23 (induced by methacrylamide) was able to hydrolyze acrylonitrile to acrylamide with excellent results in terms of conversion (Ryuno et al., 1988). Compared to the classical chemical process, the recovery of unreacted acrylonitrile is no more necessary because in the biochemical process the conversion of the latter is more than 99.99%. Also, the removal of the copper from the product is no longer necessary. Overall, the enzymatic process is simpler and more economical, it is carried out at 10°C under mild reaction conditions, requires no special energy source, the immobilized cells are used repeatedly and a very pure product is obtained. Unfortunately, the process presented a secondary practical problem. The growth of cells, in the medium containing sucrose, was accompanied by the formation of mucilaginous polysaccharides that created serious difficulties during the phase of cells harvesting due to the high viscosity of the culture medium. Moreover, the accumulation of these polysaccharides also had a deleterious effect on aeration during the cultivation, resulting in a waste of the energy derived from sucrose as one of the carbon sources. Therefore, mucilage polysaccharide nonproducing mutants were isolated by the chemical mutagenesis method using N-methyl-N'-nitro-N-nitrosoguanidine. The resultant mutant was found to enhance considerably the productivity of acrylamide. Subsequently, a third generation of this process using a nitrile hydratase from *R. rhodochrous* J1 has been developed (Nagasawa et al., 1993). *R. rhodochrous* J1, maintaining all advantages of the previous catalyst, is more heat-stable and more resistant to a high concentration both of acrylonitrile and acrylamide. Moreover, the process uses a cheaper inductor, namely urea. More recently, Raj et al. (2006) have studied with excellent results the possibility to exploit a nitrile hydratase from *R. rhodochrous* PA-34 to catalyze the same reaction (Raj et al., 2006). Conversely, very recently Shen et al. (2009) have found a novel nitrilase-producing strain of *Arthrobacter nitroguajcolicus* ZJUTB06-99 able to catalyze the biotransformation of acrylonitrile to acrylic acid, a commodity chemical widely used in the production of many commercial products including paints, coatings, and paper (Shen et al., 2009).

17.9 BIOREMEDIATION OF NITRILE-CONTAMINATED MEDIA

As widely shown and illustrated previously, inorganic cyanides and nitrile compounds both of natural and synthetic origins are widely distributed on the Earth. In addition to the illustrated applications of synthetic nitriles, cyanide-containing compounds find a wide range of uses. For example, metal cyanides, chiefly potassium and sodium salts are employed in electroplating processes in the basic degreasing and the electroplating baths to control the concentration of metal ions (Smith, 2003). In the mining of precious metals, cyanide is widely used to leach gold and silver from the ore (Arslan et al., 2003). Another significant application of iron cyanide is as anticaking agents both in road salt and fire retardants; only in the United States, 700 tons of iron cyanide are used each year (Mudder and Botz, 2004). Similarly, the use of iron cyanide compounds in chemical retardants for firefighting results each year in 400 tons of cyanide derivatives being applied during forest fire control in the United States (Mudder and Botz, 2004). Because of the large volumes of effluents generated by these industrial activities, in total, 18 billion liters of cyanide containing wastes are estimated to be generated annually in the United States (ATSDR, 2004). Accidental spillages are quite common and are responsible for the contamination of soil and water. The most significant example in recent years occurred when the tailing dam of a mining operation in Romania was breached, releasing approximately 100,000 m^3 of cyanide and heavy metal-contaminated liquid waste into the Tizsa river system (UNEP, 2000) resulting in severe mortalities among aquatic organisms and animals living close to the contaminated rivers (Soldan et al., 2001). Another example is represented by the wide contamination occurred in India in 2001 when an earthquake led to the development of cracks in tanks storing acrylonitrile; neither volatilization nor natural attenuation significantly reduced the concentration of the pollutant in soil 8 months later. In this case, bioremediation through bioaugmentation and stimulation of indigenous soil microbes by nutrient supplementation, was required to reduce the acrylonitrile to below detectable levels (Deshkar et al., 2003).

Agriculture also contributes to the discharge of cyanides and nitriles into the environment, by the application of nitrile pesticides such as bromoxynil and chloro-thalonil (2,4,5,6-tetrachloroisopthalonitrile) with the latter used also as an antifouling agent on boat hulls (Thomas et al., 2003); significant concentrations have been found in the Mediterranean and in the United Kingdom coastal environments, leading to a concern that it could cause serious environmental damage impacting species other than the target fouling organisms (Sakkas et al., 2002). Investigations on the use of microorganisms with cyanides/nitriles hydrolyzing activities have been reported in last years and offer great promises in the field of bioremediation. Different from synthetic lab or industrial biotransformations in which the conditions of reaction can be fully or partly controlled, the success of a biodegradation depends upon the presence of microbes with the physiological and metabolic capabilities to degrade the pollutants in the contaminated environment, whose parameters cannot usually be modified. As a consequence, a number of factors comprising the concentrations of cyanides and nitriles within soil or water, the availability of nutrients, the presence of additional pollutants, and the physical nature of soils and sediments, can have a significant impact on biodegradation by influencing the indigenous microbial population, selecting for, or inhibiting, the growth of particular organisms (Kang and Park, 1997). Recently, it has been reported that the development of a stable mixed microbial population has been capable of degrading a highly toxic effluent generated by acrylonitrile production, containing also acrylonitrile, fumaronitrile, succino-nitrile, acrylic acid, acrylamide, acrolein, cyanopyridine, and maleimide (Wyatt and Knowles, 1995). This consortium has been capable of degrading acrylonitrile effluents from an acrylonitrile manufacturing plant in Texas. Muller and Gabriel (1999) reported the use of *Agrobacterium radiobacter* strains for the degradation of bromoxynil under nonsterile batch and in continuous conditions (Muller and Gabriel, 1999); bromoxynil-resistant transgenic plants resulting from the insertion of microbial bromoxynil-specific nitrilase genes into tomato and tobacco could be on the market in the near future (Freyssinet et al., 1996). Very recently, novel microorganisms haloalkaline-tolerant belonging to the *Bacillus* (Sorokin et al., 2007) and *Halomonas* (Chmura et al., 2008) species have been found in Siberian soda lakes and soda soils. These bacteria with nitrile hydrolyzing activity could find application both in synthetic processes in which cyanide is involved and in bioremediation of wastes with extreme haloalkaline characteristics. Analogously, Rustler and Stolz (2007) report the isolation of a novel nitrile hydrolyzing acid-tolerant black yeast, identified as *Exophiala oligosperma* R1 (Rustler et al., 2008) that could find application in the treatment of acidic soils, or wherever acid operative conditions are required.

17.10 CONCLUSIONS

The growing numbers of studies focused on the applications of nitrile-converting biocatalysts support the benefits of this relatively new group of enzymes now widely acknowledged by organic chemists. By virtue of their capability to eliminate highly toxic nitriles and cyanides, these classes of enzymes also play a significant role in protecting the environment. The advances in molecular biology and a better understanding of the structure and reaction mechanism of nitrile-metabolizing enzymes will provide a solution, in future, to new biocatalysts with improved properties such as higher activity, higher substrate- and product-tolerance, or higher acid-, alkaline-, and thermo-stability. The way to a full exploitation of their biotechnological potential is yet long, and further investigations and applications are required.

REFERENCES

Abdel-Rahman, H.M. and Hussein, M.A. 2006. Synthesis of β-hydroxypropanoic acid derivatives as potential antiinflammatory, analgesic and antimicrobial agents. *Archives in Pharmacology*, 339:378–87.

Alfani, F., Cantarella, M., Spera, A., and Viparelli, P. 2001. Operational stability of *Brevibacterium imperialis* CBS 489–74 nitrile hydratase. *Journal of Molecular Catalysis B-Enzymes*, 11:687–97.

Almatawah, Q.A. and Cowan, D.A. 1999. Thermostable nitrilase catalysed production of nicotinic acid from 3-cyanopyridine. *Enzyme Microbiology and Technology*, 25:718–24.

Almatawah, Q.A., Cramp, R., and Cowan, D.A. 1999. Characterization of an inducible nitrilase from a thermophilic bacillus. *Extremophiles*, 3:283–91.

Andersen, K.E., Sorensen, J.L., Lau, J., Lundt, B.F., Petersen, H., Hussfeldt, P.O., Suzdak, P.D., and Swetberg, M.D.B. 2001. Synthesis of novel γ-aminobutyric acid (GABA) uptake inhibitors. *Journal of Medicinal Chemistry*, 44:2152–63.

Ankati, H., Zhu, D., Yang, Y., Biehl, E.R., and Hua, L. 2009. Asymmetric synthesis of both antipodes of β-hydroxy nitriles and β-hydroxy carboxylic acids via enzymatic reduction or sequential reduction/hydrolysis. *Journal of Organic Chemistry*, 74:1658–62.

Arslan, F., Ozdamar, D.Y., and Muduroglu, M. 2003. Cyanidation of Turkish gold–silver ore and the use of hydrogen peroxide. *European Journal of Mineral Processing and Environmental Protection*, 3:309–15.

Asano, Y., Yasuda, T., Tani, Y., and Yamada, H. 1982. A new enzymatic method of acrylamide production. *Agricultural and Biological Chemistry*, 46:1183–9.

ATSDR (Agency for Toxic Substances and Disease Registry) 2004. U.S. Department of Health and Human Services, Atlanta, GA. http://www.atsdr.cdc.gov/toxprofiles

Bandyopadhyay, A.K., Nagasawa, T., Asano, Y., Fujishiro, K., Tani, Y., and Yamada, H. 1986. Purification and characterization of benzonitrilases from *Arthrobacter* sp. Strain J-1. *Applied and Environmental Microbiology*, 51:302–6.

Banerjee, A., Kaul, P., and Banerjee, U.C. 2006. Purification and characterization of an enantioselective arylacetonitrilase from *Pseudomonas putida*. *Archives of Microbiology*, 184:407–18.

Banerjee, A., Sharma, R., and Banerjee, U.C. 2002. The nitrile-degrading enzymes: Current status and future prospects. *Applied and Microbiology and Biotechnology*, 60:33–44.

Bartel, B. and Fink, G.R. 1994. Differential regulation of an auxin-producing nitrilase gene family in *Arabidopsis thaliana*. *Proceedings of the National Academy of Science USA*, 91:6649–53.

Battistel, E., Bernardi, A., and Maestri, P. 1997. Enzymatic decontamination of aqueous polymer emulsions containing acrylonitrile. *Biotechnology Letters*, 19:131–4.

Bauer, R., Knackmuss, H.J., and Stolz, A. 1998. Enantioselective hydration of 2-arylpropionitriles by a NHase from *Agrobacterium tumefaciens* strain d3. *Applied Microbiology and Biotechnology*, 49:89–95.

Benz, P., Muntwyler, R., and Wohlgemuth, R. 2007. Chemoenzymatic synthesis of chiral carboxylic acids via nitriles. *Journal of Chemical Technology & Biotechnology*, 82:1087–98.

Bergeron, S., Chaplin, D.A., Edwards, J.H., et al. 2006. Nitrilase-catalysed desymmetrization of 3-hydroxyglutaronitrile: Preparation of a statin side-chain intermediate. *Organic Process Reseach and Development*, 10:661–5.

Bestwick, L., Gronning, L., and James, D. 1993. Purification and characterization of a nitrilase from *Brassica napus*. *Physiologia Plantarum*, 89:811–6.

Bhalla, T.C., Miura, A., Wakamoto, A., et al. 1992. Asymmetric hydrolysis of α-aminonitriles to optically active amino acids by a nitrilase of *Rhodoccocus rhodocrous* PA-34. *Applied Microbiology and Biotechnology*, 37:184–90.

Boesten, W.H.J. and Cals, M.J.H. 1987. Process for the enzymatic hydrolysis of d-α-amino-acid amides. US Patent 4:705–52.

Boger, D.L., Fecik, R.A., Patterson, J.E., Miyachi, H., Patricelli, M.P., and Cravatt, B.F. 2000. Fatty acid amide hydrolase substrate specificity. *Bioorganic & Medicinal Chemistry Letters*, 10:2613–6.

Bork, K. and Koonin, E.V. 1994. A new family of carbon–nitrogen hydrolases. *Protein Science*, 2:1344–6.

Bornscheuer, U.T. 2001. Directed evolution of enzymes for biocatalytic applications. *Biocatalysis and Biotransformation*, 19:85–97.

Cantarella, M., Spera, A., and Alfani, F. 1998a. Characterization in UF membrane reactors of nitrile hydratase from *Brevibacterium imperialis* CBS 489–74 resting cells. *Annals of the New York Academy of Sciences*, 864:224–7.

Cantarella, M., Spera, A., Cantarella, L., and Alfani, F. 1998b. Acrylamide production in an ultrafiltration-membrane bioreactor using cells of *Brevibacterium imperialis* CBS 489–74. *Journal of Membrane Science*, 147:279–90.

Cao, L., Van Langen, L., and Sheldon, R. 2003. Immobilised enzymes: Carrier bound or carrier-free? *Current Opinion in Biotechnology*, 14:387–94.

Capozzi, G., Roelens, S., and Talami, S. 1993. A protocol for the efficient synthesis of enantiopure β-substituted β-lactones. *Journal of Organic Chemistry*, 58:7932–6.

Cerbelaud, E., Levy-Schil, S., Petre, D., et al. 1995. Nitrile hydratase and amidase from *Comamonas testoteroni*. 5-MGAM-4D PCT Int. Appl. WO 9504828 A1 950216.

Chapatwala, K.D., Babu, G.R.V., Armstead, E.R., White, E.M., and Wolfram, J.H. 1995. A kinetic study on the bioremediation of sodium cyanide and acetonitrile by free and immobilized cells of *Pseudomonas putida*. *Applied Biochemistry and Biotechnology*, 51/52:717–26.

Chaplin, J.A., Levin, M.D., Morgan, B., Farid, N., Li, J., Zhu, Z., McQuaid, J., Nicholson, L.W., Rand, C.A., and Burk, M.J. 2004. Chemoenzymatic approaches to the dynamic kinetic asymmetric synthesis of aromatic amino acids. *Tetrahedron: Asymmetry*, 15:2793–6.

Chebib, M. and Johnston, G.A.R. 1999. The "ABC" of GABA receptors: A brief review. *Clinical & Experimental Pharmacology & Physiology*, 26:937–40.

Chebrou, H., Bigey, F., Arnaud, A., et al. 1996. Study of the amidase signature group. *Biochimica and Biophysica Acta*, 1298:285–93.

Chen, J., Zheng, Y.G., and Shen, Y.C. 2008. Biosynthesis of *p*-methoxyphenylacetic acid from *p*-methoxyacetonitrile by immobilized *Bacillus subtilis* ZJB-063. *Process Biochemistry*, 43:978–83.

Chmura, A., Shapovalova A.A., Van Pelt, S., Van Rantwijk, F., Tourova, T.P., Muyzer, G., and Sorokin, D.Y. 2008. Utilization of arylaliphatic nitriles by haloalkaliphilic *Halomonas nitrilicus* sp. Nov. isolated from soda soils. *Applied Microbiology and Biotechnology*, 81:371–8.

Cilia, E., Fabbri, A., Uriani, M., Scialdone, G.G., and Ammendola, S. 2005. The signature amidase from *Sulfolobus solfataricus* belongs to the CX3C subgroup of enzymes cleaving both amides and nitriles: Ser195 and Cys145 are predicted to be the active site nucleophiles. *FEBS Journal*, 272:4716–24.

Cluness, M.J., Turner, P.D. Clements, E., Brown, D.T., and O'Reilly, C. 1993. Purification and properties of cyanide hydratase from *Fusarium lateritium* and analysis of the corresponding chy1 gene. *Journal of General Microbiology*, 139:1807–15.

Conn, E.E. 1981. In *Cyanide in Biology*. Vennesland, B., Conn, E.E., Knowles, C.J. et al., (eds.), pp. 183–96. London: Academic Press.

Cooling, F.B., Fager, S.K., Fallon, R.D., et al. 2001. Chemoenzymatic production of 1,5-dimethyl-2-piperidone. *Journal of Molecular Catalysis B-Enzyme*, 11:295–306.

Coppola, G. and Schuster, H. 1997. α-*Hydroxy Acids in Enantioselective Synthesis*. Weinheim, Germany: Wiley–VCH.

Cramp, R. and Cowan, D.A. 1999. Molecular characterization of a novel thermophilic nitrile hydratase *Biochimica and Biophysica Acta*, 1431:249–60.

Cravatt, B.F., Giang, D.K., Mayfield, S.P., Boger, D.L., Lerner, R.A., and Gilula, N.B. 1996. Molecular characterization of an enzyme that degrades neuromodulatory fatty acid amides. *Nature*, 384:83–7.

Cull, S.G., Holbrey, J.D., Vargas-Mora, V., Seddon, K.R., and Lye, G.J. 2000. Room-temperature ionic liquids as replacements for organic solvents in multiphase bioprocess operations. *Biotechnology and Bioengineering*, 69:227–33.

Cull, S.G., Woodley, J.M., and Lye, G.J. 2001. Process selection and characterisation for the biocatalytic hydration of poorly water soluble aromatic dinitriles. *Biocatalysis and Biotransformation*, 19:131–54.

Curnow, A.W., Hong, K., Yuan, R., Kim, S., Martins, O., Winkler, W., Henkin, T.M., and Söll, D. 1997. Glu-tRNAGln amidotransferase: A novel heterotrimeric enzyme required for correct decoding of glutamine codons during translation. *Proceedings of the National Academy of Science USA*, 94:11819–26.

Dadd, M.R., Claridge, T.D.W., Walton, R., Pettman, A.J., and Knowles, C.J. 2001. Regioselective biotransformation of the dinitrile compounds 2-, 3- and 4-(cyanomethyl) benzonitrile by the soil bacterium *Rhodococcus rhodochrous* LL100-21. *Enzyme Microbiology and Technology*, 29:20–7.

Dadd, M.R., Sharp, D.C.A., Pettman, A.J., and Knowles, C.J. 2000. Real-time monitoring of nitrile biotransformations by mid-infrared spectroscopy. *Journal of Microbiological Methods*, 41:69–75.

Davies, S.G., Smyth, G.D., and Chippindale, A.M. 1999. Syntheses of derivatives of L-daunosamine and its C-3 epimer employing as the key step the asymmetric conjugate addition of a homochiral lithium amide to tert-butyl (*E*,*E*)-hexa-2,4-dienoate. *Journal of the Chemical Society Perkin Transactions*, 21:3089–3104.

Deigner, H., Blencowe, C., and Freyberg, C.E. 1996. Prevalence of steric restrictions in enzymatic nitrile-hydrolysis of a preparation from *Rhodococcus* sp. 409. *Journal of Molecular Catalysis B-Enzyme*, 1:61–70.

Denis, J.N., Correa, A., and Greene, A.E. 1990. An improved synthesis of the taxol side chain and of RP 56976. *Journal of Organic Chemistry*, 55:1957–9.

DeSantis, G., Wong, K., Farwell, B., et al. 2003. Creation of a productive, highly enantioselective nitrilase through gene site saturation mutagenesis. *Journal of the American Chemical Society*, 125:11476–7.

DeSantis, G., Zhu, Z., Greenberg, W.A., et al. 2002. An enzyme library approach to biocatalysis: Development of nitrilases for enantioselective production of carboxylic acid derivatives. *Journal of the American Chemical Society*, 124:9024–5.

Deshkar, A., Dhamorikar, N., Godbole, S., Krishnamurthi, K., Saravanadevi, S., Vijay, R., Kaul, S., and Chakrabarti, T. 2003. Bioremediation of soil contaminated with organic compounds with special reference to acrylonitrile. *Annali di Chimica*, 93:729–37.

Dhillon, J.K. and Shivaraman, N. 1999. Biodegradation of cyanide compounds by a *Pseudomonas* species (S1). *Canadian Journal of Microbiology*, 45:201–8.

Dhillon, J.K., Chhatre, S., Shanker, R., and Shivaraman, N. 1999. Transformation of aliphatic and aromatic nitriles by a nitrilase from *Pseudomonas* sp. *Canadian Journal of Microbiology*, 45:811–5.

Dias, J.C.T., Rezende, R.P., Rosa, C.A., Lachance, M.-A., and Linardi, V.R. 2000. Enzymatic degradation of nitriles by a *Candida guilliermondii* UFMG-Y65. *Canadian Journal of Microbiology*, 46:525–31.

Dorr, P. and Knowles, C. 1989. Cyanide oxygenase and cyanase activities of *Pseudomonas fluorescens NCIMB 11764*. *FEMS Microbiology Letters*, 60:289–94.

Effenberger, F. and Graef, B.W. 1998. Chemo- and enantioselective hydrolysis of nitriles and acid amides, respectively, with resting cells of *Rhodococcus* sp. C3II and *Rhodococcus erythropolis* MP50. *Journal of Biotechnology*, 60:165–74.

Effenberger, F. and Oswald, S. 2001a. Enantioselective hydrolysis of (R,S)-2-fluoroarylacetonitriles using nitrilase from *Arabidopsis thaliana*. *Tetrahedron: Asymmetry*, 12: 279–85.

Effenberger, F. and Oswald, S. 2001b. (E)-Selective hydrolysis of (E,Z)-α,β-unsaturated nitriles by the recombinant nitrilase AtNIT1 from *Arabidopsis thaliana*. *Tetrahedron: Asymmetry*, 12:2581–7.

Eijsink, V.G.H., Gåseidnes, S., Borchert, T., and van der Burg, B. 2005. Directed evolution of enzyme stability. *Biomolecular Engineering*, 22:21–30.

Endo, I., Nojiri, M., Tsujimura, M., Nakasako, M., Nagashima, S., Yohda, M., and Odaka, M. 2001. Fe-type nitrile hydratase. *Journal of Inorganic Biochemistry*, 83:247–53.

Endo, I., Okada, M., and Yohda, M. 1999. An enzyme controlled by light: The molecular mechanism of photoreactivity in nitrile hydratase. *Trends in Biotechnology*, 17:244–8.

Endo, I. and Watanabe, I. 1989. Nitrile hydratase of *Rhodococcus* sp. N-774: Purification and characterization. *FEBS Letters*, 243:61–64.

Ewert, C., Lutz-Wahl, S., and Fisher Lutz. 2008. Enantioselective conversion of a-arylnitriles by *Klebsiella oxytoca*. *Tetrahedron: Asymmetry*, 19:2573–8.

Faber, K. 1992. *Biotransformations in Organic Chemistry*. Berlin: Springer-Verlag.

Fernandes, B.C.M., Mateo, C., Kiziak, C., Chmura, A., Wacker, J., van Rantwijk, F., Stolz, A., and Sheldon. R.A. 2006. Nitrile hydratase activity of a recombinant nitrilase. *Advanced Synthesis and Catalysis*, 348:2597–2603.

Fournand, D., Bigey, F., and Arnaud, A. 1998. Acyl transfer activity of an amidase from *Rhodococcus* sp. strain R312: Formation of a wide range of hydroxamic acids. *Applied and Environmental Microbiology*, 64:2844–52.

Freyssinet, G., Peleissier, B., Freyssinet, M., and Delon, R. 1996. Crops resistant to oxynils: From the laboratory to the market. *Field Crops Research*, 45:125–33.

Furlenmeier, A., Quitt, P., Vogler, K., et al. 1976. 6-Acyl derivatives of aminopenicillanic acid. US patent 3957758.

Gavagan, J.E., Di Cosimo, R., Eisenberg, A., et al. 1999. A Gram-negative bacterium producing a heat-stable nitrilase highly active on aliphatic dinitriles. *Applied Microbiology and Biotechnology*, 52:654–9.

Gavagan, J.E., Fager, S.K., Fallon, R.D., et al. 1998. Chemoenzymic production of lactams from aliphatic α,ω-dinitriles. *Journal of Organic Chemistry*, 63:4792–4801.

Gellman, S.H. 1998. Foldamers: A manifesto. *Accounts of Chemical Research*, 31:173–80.

Giannis, A. and Kolter, T. 1993. Peptidmimetica für Rezeptorliganden – Entdeckung, Entwicklung und medizinische Perspektiven. *Angewandte Chemie*, 105:1303–26.

Goldlust, A. and Bohak, Z. 1989. Induction, purification, and characterization of the nitrilase of *Fusarium oxysporum* f. sp. *melonis*. *Biotechnology and Applied Biochemistry*, 11:581–601.

Graham, D., Pereira, R.A., Barfield, D., and Cowan, D.A. 2000. Nitrile biotransformations using free and immobilized cells of a thermophilic *Bacillus* spp. *Enzyme Microbiology and Technology*, 26:368–73.

Han, Z., Krishnamurthy, D., Pflum, D., et al. 2002. First practical synthesis of enantiomerically pure (*R*)- and (*S*)-desmethylsibutramine (DMS) and unambiguous determination of their absolute configuration by single-crystal X-ray analysis *Tetrahedron: Asymmetry*, 13:107–9.

Handelsman, J. 2004. Metagenomics: Application of genomics to uncultured microorganisms. *Microbiology and Molecular Biology Reviews*, 68:669–85.

Handelsman, J. 2005. Sorting out metagenomes. *Nature Biotechnology*, 23:38–9.

Hann, E.C., Eisenberg, A., Fager, S.K., et al. 1999. 5-Cyanovaleramide production using immobilized *Pseudomonas chlororaphis* B23. *Bioorganic and Medicinal Chemistry*, 7:2239–45.

Harper, D.B. 1977. Fungal degradation of aromatic nitriles. Enzymology of C–N cleavage by *Fusarium solani*. *Biochemistry Journal*, 167:685–92.

Harper, D.B. 1985. Characterization of a nitrilase from *Nocardia* sp. (*Rhodochrous* group) NCIB 11215, using *p*-hydroxybenzonitrile as sole carbon source. *International Journal of Biochemistry*, 17:677–83.

Hashimoto, Y., Endo, T., Tamura, K., et al. 1996. Process for producing optically active a-hydroxycarboxylic acid having phenyl group using *Gordona terrae*. US 5 580 765.

Hatanaka, M. and Tanaka, N. 1993. Process for producing pyridinecarboxylic acid. World patent 93.0522A1.

Heinemann, U., Engels, D., Burger, S., et al. 2003. Cloning of a nitrilase gene from the *cyanobacterium Synechocystis* sp. Strain PCC6803 and heterologous expression and characterization of the encoded protein. *Applied and Environmental Microbiology*, 69: 4539–66.

Hellinga, H.W. 1997. Rational protein design: Combining theory and experiments. *Proceedings of the National Academy of Science USA*, 94:10015–7.

Hibbert, E.G., Baganz, F., Hailes, H.C., et al. 2005. Directed evolution of biocatalytic processes. *Biomolecular Engineering*, 22:11–19.

Hirrlinger, B., Stolz, A., and Knackmuss, H. 1996. Purification and characterization of an amidase from *Rhodococcus erythropolis* MP 50 which enantioselectively hydrolyzes 2-arylpropionamides. *Journal of Bacteriology*, 178:3501–7.

Hoyle, A.J., Bunch, A.W., and Knowles, C.J. 1998. The nitrilases of *Rhodococcus rhodochrous* NCIMB 11216. *Enzyme Microbiology and Technology*, 23:475–82.

Huang, W., Jia, J., Cummings, M., et al. 1997. Crystal structure of nitrile hydratase reveals a novel iron centre in a novel fold. *Structure*, 5:691–9.

Hughes, J., Armitage, Y.C., and Symes, K.C. 1998. Application of whole cell rhodococcal biocatalysts in acrylic polymer manufacture. *Antonie van Leeuwenhoek*, 74:107–18.

Ingovorsen, K., Hijer-Pedersen, B., and Godtfredsen, S. 1991. Novel cyanide-hydrolyzing anzyme from *Alcaligenes xylosoxidans* sub sp. *denitrificans*. *Applied and Environmental Microbiology*, 57:1783–9.

Jandhyala, D.M., Berman, M., Meyers, P.R., et al. 2003. Cyn D, the cyanide dihydratase from *Bacillus pumillus*: Gene cloning and structural studies. *Appied and Environmental Microbiology*, 69:4794–4805.

Johnson, D., Zabelinskaja-Mackova, A., and Griengl, H. 2000. Oxynitrilases for asymmetric C – C bond formation. *Current Opinion in Chemical Biology*, 4:103–9.

Kamila, S., Zhu, D., Biehl, E.R., et al. 2006. Unexpected stereorecognition in nitrilase-catalyzed hydrolysis of β-hydroxy nitriles. *Organic Letters*, 8:4429–31.

Kang, M.H. and Park, J.M. 1997. Sequential degradation of phenol and cyanide by a commensal interaction between two microorganisms. *Journal of Chemical Technology and Biotechnology*, 69:226–30.

Kaplan, O., Nikolaou, A., Pisvejcova, L., et al. 2006. Hydrolysis of nitriles and amides by filamentous fungi. *Enzyme Microbiology and Technology*, 38:260–4.

Kaplan, O., Vejvoda, V., Plihal, O., et al. 2006. Purification and characterization of a nitrilase from *Aspergillus niger* K10. *Applied Microbiology and Biotechnology*, 73:567–75.

Kato, Y., Ooi, R., and Asano, Y. 2000. Distribution of aldoxime dehydratase in microorganisms. *Applied and Environmental Microbiology*, 66:2290–6.

Kato, Y., Tsuda, T., and Asano, Y. 1999. Nitrile hydratase involved in aldoxime metabolism from *Rhodococcus* sp. strain YH3–3. *European Journal Biochemistry*, 263:662–70.

Kaufmann, G., Dautzenberg, H., Henkel, H., et al. 1999. Nitrile hydratase from *Rhodococcus erythropolis*: Metabolization of steroidal compounds with a nitrile group. *Steroids*, 64:535–40.

Kaul, P., Banerjee, A., Mayilraj, S., et al. 2004. Screening for enantioselective nitrilases: Kinetic resolution of racemic mandelonitrile to (*R*)-(–)-mandelic acid by new bacterial isolates. *Tetrahedron: Asymmtry*, 15:207–11.

Kaul, K., Stolz, A., and Banerjee, U.C. 2007. Cross-linked amorphous nitrilase aggregates for enantioselective hydrolysys. *Advances in Synthesis and Catalysis*, 349:2167–76.

Kielbasiński, P., Rachwalski, M., Mikolajczyk, M., et al. 2007a. Enzyme-promoted desymmetrization of prochiral bis(cyanomethyl) sulfoxide. *Advances in Synthesis and Catalysis*, 349:1387–92.

Kielbasiński, P., Rachwalski, M., Kwiatkowska, M., et al. 2007b. Enzyme-promoted desymmetrization of prochiral bis(cyanomethyl)phenylphosphine oxide. *Tetrahedron: Asymmetry*, 18:2108–12.

Kielbasiński, P., Rachwalski, M., Mikolajczyk, M., et al. 2008. Nitrilase-catalysed hydrolysis of cyanomethyl *p*-tolyl sulfoxide: Stereochemistry and mechanism. *Tetrahedron: Asymmetry*, 19:562–7.

Kiener, A., Roduit, J.P., and Gloeckler, R. 1996. Microbiological method for the synthesis of heteroaromatic caboxylic acids using microorganisms of the genus *Alcaligenes*. EP–B 0747486.

Kim, S.H. and Oriel, P. 2000. Cloning and expression of the nitrile hydratase and amidase genes from *Bacillus* sp. BR449 into *Escherichia coli*. *Enzyme Microbiology and Technology*, 27:492–501.

Kimani, S.L., Agarkar, V.B., Cowan, D.A., et al. 2007. Structure of an aliphatic amidase from *Geobacillus pallidus* RAP c8. *Acta Crystallographica D*, 63:1048.

Kinfe, H.H., Chhiba, V., Frederick, J., et al. 2009. Enantioselective hydrolysis of β-hydroxy nitriles using the whole cells biocatalyst *Rhodococcus rhodochrous* ATCC BAA–870. *Journal of Molecular Catalysis B Enzyme*, 59:231–6.

Kiziak, C., Conradt, D., Stolz, A., et al. 2005. Nitrilase from *Pseudomonas fluorescens* EBC191: Cloning and heterologous expression of the gene and biochemical characterization of the recombinant enzyme. *Microbiology*, 151:3639–48.

Klempier, N., Harter, G., de Raadt, A. et al. 1996. Chemoselective hydrolysis of nitriles by *Rhodococcus rhodochrous* NCIMB 11216. *Food Technology and Biotechnology*, 34:67–70.

Kobayashi, M., Goda, M., and Shimizu, S. 1998. Nitrilase catalyzes amide hydrolysis as well as nitrile hydrolysis. *Biochemical and Biophysical Research Communications*, 253: 662–6.

Kobayashi, M., Komeda, H., and Yanaka, N. 1992a. Nitrilase from *Rhodococcus rhodochrous* J1. Sequencing and overexpression of the gene and identification of an essential cysteine residue. *Journal of Biological Chemistry*, 267:20746–51.

Kobayashi, M., Nagasawa, H., and Yamada, H. 1992b. Enzymatic synthesis of acrylamide: A success story not yet over. *Trends in Biotechnology*, 10:402–8.

Kobayashi, M. and Shimizu, S. 1994. Versatile nitrilases: Nitrile-hydrolyzing enzymes. *FEMS Microbiology Letters*, 120:217–33.

Kobayashi, M. and Shimitzu, S. 1998. Metalloenzyme nitrile hydratase: Structure, regulation, and application to biotechnology. *Nature Biotechnology*, 16:733.

Kobayashi, M. and Shimitzu, S. 1999. Cobalt proteins. *European Journal of Biochemistry*, 261:1–9.

Kobayashi, M. and Shimitzu, S. 2000. Nitrile hydrolases. *Current Opinion in Chemical Biology*, 4:95–102.

Kobayashi, M., Suzuki, T., Fujita, T, et al. 1995. Occurrence of enzymes involved in biosynthesis of indole-3-acetic acid from indole-3-acetonitrile in plant-associated bacteria, *Agrobacterium* and *Rhizobium*. *Proceedings of the National Academy of Science USA*, 92:714–8.

Kobayashi, M., Yanaka, N., Nagasawa, T., et al. 1989. Purification and characterization of a novel nitrilase of *Rhodococcus rhodococcus* K22 that acts on aliphatic nitriles. *Journal of Bacteriology*, 172:4807–15.

Langdahl, B.R., Bisp, P., and Ingvorsen, K. 1996. Nitrile hydrolysis by *Rhodococcus erythropolis* BL1, an acetonitrile-tolerant strain isolated from a marine sediment. *Microbiology*, 142:145–54.

Layh, N., Parratt, J., and Willets, A. 1998. Characterization and partial purification of an enantioselective arylacetonitrilase from *Pseudomonas fluorescens* DSM 7155. *Journal of Molecular Catalysis B-Enzyme*, 5:467–74.

Legras, J.L., Chuzel, G., Arnaud, A., et al. 1990. Natural nitriles and their metabolism: Review. *World Journal of Microbiology and Biotechnology*, 6:83–108.

Levy-Schil, S. Soubrier, F., Crutz-LeCoq, A.M., et al. 1995. Aliphatic nitrilase from a soil isolated *Comamonas testosteroni* sp.: Gene cloning and overexpression, purification and primary structure. *Gene*, 161:15–20.

Linardi, V.R., Dias, J.C.T., and Rosa, C.A. 1996. Utilization of acetonitrile and other aliphatic nitriles by a *Candida famata* strain. *FEMS Microbiology Letters*, 144:67–71.

Lorenz, P., Liebeton, K., Niehaus, F., et al. 2003. The impact of non-cultivated biodiversity on enzyme discovery and evolution. *Biocatalysis and Biotransformation*, 21:87–91.

Ludwig-Muller, J. and Hilgenberg, W. 1988. A plasma membrane-bound enzyme oxidizes l-tryptophan to indole-3-acetaldoxime. *Physiologia Plantarum*, 74:240–50.

Ma, D.Y., Wang, D.X., Pan, J. et al. 2008a. Nitrile biotransformations for the synthesis of highly enantioenriched β-hydroxy and β-amino acid and amide derivatives: A general and simple but powerful and efficient benzyl protection strategy to increase enantioselectivity of the amidase. *Journal of Organic Chemistry*, 73:4087–4091.

Ma, D.Y., Wang, D.X., Pan, J., et al. 2008b. Nitrile biotransformations for the synthesis of enantiomerically enriched β²-, and β³-hydroxy and alkoxy acids and amides, a dramatic O-substituent effects of the substrates on enantioselectivity. *Tetrahedron: Asymmetry*, 19:322–29.

Ma, D.Y., Wang, D.X., Zheng, Q.Y., et al. 2006. Nitrile biotransformations for the practical synthesis of highly enantiopure azido carboxylic acids and amides, "click" to functionalized chiral triazoles and chiral β-amino acids. *Tetrahedron: Asymmetry*, 17:2366–76.

Marier-Greiner, U.H., Obermaier-Skrobranek, B.M.M., Estermaier, L.M., et al. 1991. Isolation and properties of nitrile hydratase from the soil fungus *Myrothecium verrucaria* that is highly specific for the fertilizer cyanamide and cloning of its gene. *Proceedings of the National Academy of Science USA*, 88:4260–4.

Martinkova, L., Klempier, N., Bardakji, J., et al. 2001. Biotransformation of 3-substituted methyl (*R,S*)-4-cyanobutanoates with nitrile- and amide-converting biocatalysts. *Journal of Molecular Catalysis B-Enzyme*, 14:95–99.

Martinkova, L., Klempier, N., Preiml, M., et al. 2002. Selective biotransformation of substituted alicyclic nitriles by *Rhodococcus equi* A4. *Canadain Journal of Chemistry*, 80:724–7.

Martinkova, L., Klempier, N., and Prepechalova, I. 1998. Chemoselective biotransformation of nitriles by *Rhodococcus equi* A4. *Biotechnology Letters*, 20:909–12.

Martinkova, L. and Kren, V. 2002. Nitrile- and amide-converting microbial enzymes: Stereo-, regio- and chemoselectivity. *Biocatalysis and Biotransformation*, 20:73–93.

Martinkova, L., Vejvoda, V., and Kren, V. 2008. Selection and screening for enzymes of nitrile metabolism. *Journal of Biotechnology*, 133:318–26.

Mateo, C., Chmura, A., Rustler, S., et al. 2006. Synthesis of enantiomerically pure (*S*)-mandelic acid using an oxynitrilase-nitrilase bienzymatic cascade: A nitrilase surprisingly shows nitrile hydratase activity. *Tetrahedron: Asymmetry*, 17:320–3.

Mathew, C.D., Nagasawa, T., Kobayashi, M., et al. 1988. Nitrilase-catalysed production of nicotinic acid from 3-cyanopyridine in *Rhodococcus rhodochrous* J1. *Applied and Environmental Microbiology*, 54:1030–2.

Matoishi, K., Sano, A., Imai, N., et al. 1998. *Rhodococcus rhodochrous* IFO 15564-mediated hydrolysis of alicyclic nitriles and amides: Stereoselectivity and use for kinetic resolution and asymmetrization. *Tetrahedron: Asymmetry*, 9:1097–1102.

Mayaux, J., Cerbelaud, E., Soubrier, F., et al. 1990. Purification, cloning, and primary structure of an enantiomer selective amidase from *Brevibacterium* sp. R312: Structural evidence for genetic coupling with nitrile hydratase. *Journal of Bacteriology*, 172:6764–73.

Meth-Cohm, O. and Wang, M.X. 1997a. Rationalisation of the regioselective hydrolysis of aliphatic dinitriles with *Rhodococcus rhodochrous* AJ270. *Chemical Communications*, 1041–2.

Meth-Cohm, O. and Wang, M.X. 1997b. An in-depth study of the biotransformation of nitriles into amides and/or acids using *Rhodococcus rhodochrous* AJ270. *Journal of the Chemical Society Perkin Transactions*, 1:1099–1104.

Miller, J.M. and Conn, E.E. 1980. Metabolism of hydrogen cyanide by higher plants. *Physiologia Plantarum*, 65:1199–1202.

Mills, J., Schmiegel, K.K., and Shaw, W.N. 1983. Phenethanolamines, compositions containing the same, and method for effecting weight control. US patent 4391826.

Mudder, T.I. and Botz, M.M. 2004. Cyanide and society: A critical review. *European Journal of Mineral Processing and Environmental Protection*, 4:62–74.

Mueller, P., Egarova, K., Vorgias, E.C., et al. 2006. Cloning, expression, and characterization of a thermoactive nitrilase from the hyperthermophile archaeon *Pyrococcus abyssi*. *Protein Expression and Purification*, 47:672–81.

Mukherjee, C., Zhu, D., Biehl, E.R., et al. 2006. Exploring the synthetic applicability of a cyanobacterium nitrilase as catalyst for nitrile hydrolysis. *European Journal of Organic Chemistry*, 23:5238–42.

Muller, D. and Gabriel, J. 1999. Bacterial degradation of the herbicide bromoxynil by *Agrobacterium radiobacter* in biofilm. *Folia Microbiologica*, 44:377–9.

Muller, A. and Weiler, E.W. 2000. IAA-synthase, an enzyme complex from *Arabidopsis thaliana* catalyzing the formation of indole-3-acetic acid from (*S*)-tryptophan. *Biological Chemistry*, 381:679–86.

Murakami, T., Nojiri, M., and Nakayama, H. 2000. Post-translational modification is essential for catalytic activity of nitrile hydratase. *Protein Science*, 9:1024–30.

Mylerova, V. and Martinkova, L. 2003. Synthetic applications of nitrile-converting enzymes. *Currrent Organic Chemistry*, 7:1–17.

Nagasawa, T., Mathew, C.D., Mauger, J., et al. 1988. Nitrile hydratase-catalysed production of nicotinamide from 3-cyanopyridine in *Rhodococcus rhodochrous* J1. *Applied and Environmental Microbiology*, 54:1766–9.

Nagasawa, T., Mauger, J., and Yamada, H. 1990. A novel nitrilase, arylacetonitrilase, of *Alcaligenes faecalis* JM3. Purification and characterization. *European Journal of Biochemistry*, 194:765–72.

Nagasawa, T., Nanba, H., Ryuno, K., et al. 1987. Nitrile hydratase of *Pseudomonas chlororaphis* B23. *European Journal of Biochemistry*, 162:691–8.

Nagasawa, T., Shimizu, H., and Yamada, H. 1993. The superiority of the third-generation catalyst, *Rhodococcus rhodochrous* J1 nitrile hydratase, for industrial production of acrylamide. *Applied Microbiology and Biotechnology*, 40:189–95.

Nagasawa, T., Wieser, M., Nakamura, T., et al. 2000. Nitrilase of *Rhodococcus rhodochrous* J1, conversion into the active form by subunit association. *European Journal of Biochemistry*, 267:138–44.

Nagashima, S., Nakasako, M., and Dohmaev, N. 1998. Novel non-heme iron center of nitrile hydratase with a claw setting of oxygen atoms. *Nature Structural & Molecular Biology*, 5:347–51.

Nakai, T., Hasegawa, T., Yamashita, E., et al. 2000. Crystal structure of N-carbamyl-D-amino acid amidohydrolase with a novel catalytic framework common to amidohydrolases. *Structure*, 8:729–37.

Noguchi, T., Honda, J., Nagamune, T., et al. 1995. Photosensitive nitrile hydratase intrinsically possesses nitric oxide bound to the non-heme iron center: Evidence by Fourier transform infrared spectroscopy. *FEBS Letters*, 358:9–12.

Nolan, L.M., Harnedy, P.A., Turner, P., et al. 2003. The cyanide hydratase enzyme of *Fusarium lateritium* also has nitrilase activity. *FEMS Microbiology Letters*, 221:161–5.

Novo, C., Farnaud, S., Tata, R., et al. 2002. Support for a three-dimensional structure predicting a Cys-Glu-Lys catalytic triad for *Pseudomonas aeruginosa* amidase comes from site-directed mutagenesis and mutations altering substrate specificity. *Biochemistry Journal*, 365:731–8.

Novo, C., Tata, R., Clemente, A., et al. 1995. *Pseudomonas aeruginosa* aliphatic amidase is related to the nitrilase/cyanide hydratase enzyme family and Cys[166] is predicted to be the active site nucleophile of the catalytic mechanism. *FEBS Letters*, 367:275–9.

Oertle, K., Beyeler, H., Duthaler, R.O., et al. 1990. A facile synthesis of optically pure (−)-(*S*)-Ipsenol using a chiral titanium complex. *Helvetica Chimica Acta*, 73:353–58.

Osprian, I., Jarret, C., Strauss, U., et al. 1999. Large-scale preparation of a nitrile-hydrolysing biocatalyst: *Rhodococcus* R 312 (CBS 717.73). *Journal of Molecular Catalysis B-Enzyme*, 6:555–60.

Pace, H.C. and Brenner, C. 2001. The nitrilase superfamily: Classification, structure and function. *Genome Biology*, 2:1–9.

Padmakumar, R. and Oriel, P. 1999. Bioconversion of acrylonitrile to acrylamide using a thermostable nitrile hydratase. *Applied Biochemistry and Biotechnology*, 79:671–9.

Patricelli, M.P. and Cravatt, B.F. 2000. Clarifying the catalytic roles of conserved residues in the amidase signature family. *Journal of Biological Chemistry*, 275:19177–84.

Payne, M.S., Wu, S., Fallon, R.D., et al. 1997. A stereoselective cobalt containing nitrile hydratase. *Biochemistry*, 36:5447–54.

Pereira, R.A., Graham, D., Rainey, F.A., et al. 1998. A novel thermostable nitrile hydratase. *Extremophiles*, 2:347–57.

Piotrowsky, M. 2008. Primary or secondary? Versatile nitrilases in plant metabolism. *Phytochemistry*, 69:2655–67.

Piotrowski, M., Schonfelder, S., and Weiler, W.E. 2001. The The *Arabidopsis thaliana* isogene *NIT4* and its orthologs in tobacco encode β-cyano-l-alanine hydratase/nitrilase. *Journal of Biological Chemistry*, 276:2616–21.

Podar, M., Eads, J.R., and Richardson, T.H. 2005. Evolution of a microbial nitrilase gene family: A comparative and environmental genomics study. *BMC Evolutionary Biology*, 5:42.

Pollak, P., Romender, G., Hagedorn, F., et al. 1991. Nitriles. In *Ullman's Encyclopedia of Industrial Chemistry*, 5th ed. Elvers, B., Hawkins, S., and Schulz, G. (Eds). Vol. A17, pp. 363–76. New York: Wiley.

Pollock, J.A., Clark, K.M., Martynowicz, B.J., et al. 2007. A mild biosynthesis of lactones via enantioselective hydrolysis of hydroxynitriles. *Tetrahedron: Asymmetry*, 18:1888–92.

Prasad, S., Raj, J., and Bhalla, T.C. 2007. Bench scale conversion of 3-cyanopyridine to nicotinamide using resting cells of *Rhodococcus rhodochrous* PA–34. *Indian Journal of Microbiology*, 47:34–41.

Prepechalova, I., Martinkova, L., Stolz, A., et al. 2001. Purification and characterization of the enantioselective nitrile hydratase from *Rhodococcus equi* A4. *Applied Microbiology and Biotechnology*, 55:150–6.

Raj, J., Shreenath, P., and Bhalla, T.C. 2006. *Rhodococcus rhodochrous* PA–34: A potential biocatalyst for acrylamide synthesis. *Process Biochemistry*, 41:1359–63.

Rausch, T. and Hilgenberg, W. 1980. Partial purification of nitrilase from Chinese cabbage. *Phytochemistry*, 19:747–50.

Rezende, R., Dias, J., Ferraz, V., et al. 2000. Metabolism of benzonitrile by *Cryptococcus* sp. UFMG–Y28. *Journal of Basic Microbiology*, 40:389–92.

Rezende, R.P., Dias, J.C.T., Rosa, C.A., et al. 1999. Utilisation of nitriles by yeasts strains isolated from a Brazilian gold mine. *The Journal of General and Applied Microbiology*, 45:185–92.

Robertson, D.E., Chaplin, J.A., DeSantis, G., et al. 2004. Exploring nitrilase sequence space for enantioselective catalysis. *Applied and Environmental Microbiology*, 70:2429–36.

Rustler, S., Motejadded, H., Altenbuchner, J., et al. 2008. Simultaneous expression of an arylacetonitrilase from *Pseudomonas fluorescens* and a (*S*)-oxynitrilase from *Manihot esculenta* in *Pichia pastoris* for the synthesis of (*S*)-mandelic acid. *Applied Microbiology and Biotechnology*, 80:87–97.

Rustler, S. and Stolz, A. 2007. Isolation and characterization of a nitrile hydrolysing acid tolerant black yeast—*Exophiala oligosperma* R1. *Applied Microbiology and Biotechnology*, 75:899–908.

Ryuno, K., Nagasawa, T., and Yamada, H. 1988. Isolation of advantageous mutants of *Pseudomonas chlororaphis* B23 for the enzymatic production of acrylamide. *Agricultural and Biological Chemistry*, 52:1813–6.

Sakkas, V.A., Lambropoulou, D.A., and Albanis, T.A. 2002. Study of chlorothalonil photodegradation in natural waters and in the presence of humic substances. *Chemosphere*, 48:939–45.

Saroja, N., Shamala, T.R., and Tharanathan, R.N. 2000. Biodegradation of starch-g-polyacrylonitrile, a packaging material, by *Bacillus cereus*. *Process Biochemistry*, 36:119–25.

Schostarez, H.J. 1988. A stereospecific synthesis of 3-aminodeoxystatine. *Journal of Organic Chemistry*, 53:3628–31.

Sewell, B.T., Berman, M.N., Meyers, P.R., et al. 2003. The cyanide degrading nitrilase from *Pseudomonas stutzeri* AK61 is a two-fold symmetric, 14-subunit spiral. *Structure*, 11:1–20.

Sexton, A.C. and Howlett, B.J. 2000. Characterization of a cyanide hydratase gene in the phytopathogenic fungus *Leptosphaeria maculans*. *Gene*, 263:463–470.

Sharma, N.N., Sharma, M., Kumar, H., et al. 2006. *Nocardia globerula* NHB-2: Bench scale production of nicotinic acid. *Process Biochemistry*, 41:2078–81.

Shen, M., Zheng, Y.G., and Shen, Y.C. 2009. Isolation and characterization of a novel *Arthrobacter nitroguajacolicus* ZJUTB06-99, capable of converting acrylonitrile to acrylic acid. *Process Biochemistry*, 44:781–5.

Shimizu, S. 2001. Vitamins and related compounds. In *Chemical Technology*, 24 Mark, H.F., Other, D.F., Overberger, C.G. et al. (Eds.). New York: Wiley.

Smith, G. 2003. Cyanides in metal finishing: Risks and alternatives. *Transactions of the Institute of Metal Finishing*, 81:B33–7.

Smith, B.E., Richards, R.L., and Newton, W.E. 2004. *Catalysts for Nitrogen Fixation—Nitrogenases, Relevant Chemical Models and Commercial Processes*. Dordrecht: Kluwer Academic Publishers.

Soldan, P., Pavonič, M., Bouček, J., et al. 2001. Baia Mare accident—brief ecotoxicological report of Czech experts. *Ecotoxicology and Environmental Safety*, 49:255–61.

Solomonson, L.P. and Spehar, A.M. 1981. *Cyanide as a Metabolic Inhibitor*. London: Academic Press.

Sone, H., Nemoto, T., Ishiwata, H., et al. 1993. Isolation, structure, and synthesis of dolastatin D, a cytotoxic cyclic depsipeptide from the sea gare *Dolabella auricularia*. *Tetrahedron Letters*, 34:8449–52.

Sorokin, D.Y., Van Pelt, S., Tourova, T.P., et al. 2007. Microbial isobutyronitrile utilization under haloalkaline conditions. *Applied and Environmental Microbiology*, 73:5574–9.

Stalker, D.M., Malyi, L.D., and McBride, K.E. 1998. Purification and properties of a nitrilase specific for the herbicide bromoxynil and corresponding nucleotide sequence analysis of the bxn gene. *Journal of Biological Chemistry*, 263:6310–4.

Stelkes-Ritter, U., Wyzgol, K., and Kula, M. 1995. Purification and characterization of a newly screened microbial peptide amidase. *Applied Microbiology and Biotechnology*, 44:393–8.

Stevenson, D.E., Feng, R., Dumas, F., et al. 1992. Mechanistic and structural studies on *Rhodococcus* ATCC 39484 nitrilase. *Biotechnology and Applied Biochemistry*, 15:283–302.

Surivet, J.P. and Valete, J.M. 1998. First total synthesis of (–)-8-*epi*-9-deoxygoniopypyrone. *Tetrahedron Letters*, 39:9681–2.

Takashima, Y., Yamaga, Y., and Mitsuda, S. 1998. Nitrile hydratase from a thermophilic bacillus. *Jounal of Industrial Microbiology and Biotechnology*, 20:220–6.

Tauber, M.M., Cavaco-Paulo, A., Robra, K.H. et al. 2000. Nitrile hydratase and amidase from *Rhodococcus rhodochrous* hydrolyze acrylic fibers and granular polyacrylonitriles. *Applied and Environmental Microbiology*, 66:1634–8.

Terreni, M., Pagani, G., Ubiali, D., et al. 2001. Modulation of penicillin acylase properties via immobilization techniques: One-pot chemoenzymatic synthesis of Cephamandole from Cephalosporin C. *Bioorganic and Medicinal Chemistry Letters*, 11:2429–32.

Thaisrivongs, S., Schostarez, H.J., Pals, D.T., et al. 1987. α,α-Difluoro-β-aminodeoxystatine-containing renin-inhibitory peptides. *Journal of Medicinal Chemistry*, 30:1837–42.

Theil, F. and Ballschuh, S. 1996. Chemoenzymatic synthesis of both enantiomers of cispentacin. *Tetrahedron: Asymmetry*, 7:3565–72.

Thomas, K.V., McHugh, M., Hilton M., et al. 2003. Increased persistence of antifouling paint biocides when associated with paint particles. *Environmental Pollution*, 123:153–61.

Toomey, J.J.E. 1984. Electrochemical oxidation of pyridine bases US patent 4482439. UNEP (United Nations Environment Programme). Office for the Coordination of Humanitarian Affairs 2000. *Cyanide Spill at Baia Mare Romania*. Report, Geneva, Switzerland.

UNEP/OCHA. 2000. Report on the cyanide spill at Baia Mare, Romania. The regional environmental center for central and eastern Europe, Szentendre.

Van Pelt, S., Van Rantwijk, F., and Sheldon, R.A. 2009. Synthesis of aliphatic (*S*)-a-hydroxycarboxylic amides using a one–pot bienzymatic cascade of immobilized oxynitrilase and nitrile hydratase. *Advances in Synthesis and Catalysis*, 351:397–404.

Vejvoda, V., Kaplan, O., Bezouska, K., et al. 2008. Purification and characterization of a nitrilase from *Fusarium solani* O1. *Journal of Molecular Catalysis B Enzyme*, 50:99–106.

Vetter, J. 2000. Plant cyanogenic glycosides. *Toxicon*, 38:11–36.

Vink, M.K.S., Schortinghuis, C.A., Luten, J., et al. 2009. A stereodivergent approach to substituted 4–hydroxypiperidines. *Journal of Organic Chemistry*, 67:7869–71.

Wang, M.X. 2005. Enantioselective biotransformations of nitriles in organic synthesis. *Topics in Catalysis*, 35:117–30.

Wang, M.X., Deng, G., Wang, D.X., et al. 2005. Nitrile biotransformation for highly enantioselective synthesis of oxiranecarboxamides with tertiary and quaternary sterocenters; efficient chemoenzymatic approaches to enantiopur a–methylated serine and isoserine derivatives. *Journal of Organic Chemistry*, 70:2439–44.

Wang, M.X. and Feng, G.Q. 2000. Enantioselective synthesis of chiral cyclopropane compounds through microbial transformations of *trans*–2–arylcyclopropanecarbonitriles. *Tetrahedron Letters*, 41:6501–5.

Wang, M.X. and Feng, G.Q. 2002. A novel approach to enantiopure cyclopropane compounds from biotransformation of nitriles. *New Journal of Chemistry*, 26:1575–83.

Wang, P., Matthews, D., and Van Etten, H. 1992. Purification and characterization of cyanide hydratase from the phytopathogenic fungus *Gloeocercospora sorghi. Archives in Biochemistry and Biophysics*, 298:569–75.

Wang, M.X., Liu, C.S., Li, J.S., et al. 2000a. Microbial desymmetrization of 3-arylglutaronitriles, an unusual enhancement of enantioselectivity in the presence of additives. *Tetrahedron Letters*, 41:8549–52.

Wang, M.X., Lu, G., Ji, G.J., et al. 2000b. Enantioselective biotransformations of racemic α-substituted phenylacetonitriles and phenylacetamides using *Rhodococcus* sp. AJ270 *Tetrahedron: Asymmetry*, 11:1123–35.

Wang, M.X., Liu, C.S., and Li, J.S. 2001a. Enzymatic desymmetrization of 3-alkyl- and 3-arylglutaronitriles, a simple and convenient approach to optically active 4-amino-3-phenylbutanoic acids. *Tetrahedron: Asymmetry*, 12:3367–73.

Wang, M.X., Li, J.J., Ji, G.J., et al. 2001b. Enantioselective biotransformations of racemic 2-aryl-3-methylbutyronitriles using *Rhodococcus* sp. AJ270. *Journal of Molecular Catalysis B-Enzyme*, 14:77–83.

Wang, P. and Van Etten, H.D. 1992. Cloning and properties of a cyanide hydratase gene from the phytopathogenic fungus *Gloeocercospora sorghi. Biochemical and Biophysical Research Communications*, 187:1048–54.

Wang, M.X. and Wang, G.Q. 2002. Enzymatic synthesis of optically active 2-methyl- and 2,2-dimethylcyclopropanecarboxylic acids and their derivatives. *Journal of Molecular Catalysis B-Enzyme*, 18:267–72.

Wang, M.X., Wang, D.X., Zheng, Q.Y. et al. 2007. Nitrile biotransformations for the efficient synthesis of highly enantiopure 1-arylaziridine-2-carboxylic acid derivatives and their stereoselective ring-opening reactions. *Journal of Organic Chemistry*, 72:2040–5.

Weissermel, K. and Arpe, H.J. 1997. *Industrial Organic Chemistry*, 3rd ed. Weinheim: VCH.

Wieser, M., Heinzmann, K., and Kiener, A. 1997. Bioconversion of 2-cyanopyrazine to 5-hydroxypyrazine-2-carboxylic acid with *Agrobacterium* sp. DSM 6336. *Applied Microbiology and Biotechnology*, 48:174–6.

Wieser, M., Takeuchi, K., Wada, Y. et al. 1998. Low-molecular-mass nitrile hydratase from *Rhodococcus rhodochrous* J1: Purification, substrate specificity and comparison with the analogous high molecular-mass enzyme. *FEMS Microbiology Letters*, 169:17–22.

Winkler, M., Martinkova, L., Knall, A.C., et al. 2005. Synthesis and microbial transformation of β-amino nitriles. *Tetrahedron*, 61:4249–60.

Winkler, M., Knall, A.C., Kulterer, M., et al. 2007a. Nitrilases catalyze key step to conformationally constrained GABA analogous γ-amino acids in high optical purity. *Journal of Organic Chemistry*, 72:7423–6.

Winkler, M., Meischler, D., and Klempier, N. 2007b. Nitrilase-catalyzed enantioselective synthesis of pyrrolidine- and piperidinecarboxylic acids. *Advances in Synthesis and Catalysis*, 349:1475–80.

Winkler, M., Kaplan, O., Vejvoda, V. et al. 2009. Biocatalytic application of nitrilases from *Fusarium solani* O1 and *Aspergillus niger* K10 *Journal of Molecular Catalysis B-Enzyme*, 59:243–7.

Wittstock, U. and Halkier, B. Glucosinolate research in the *Arabidopsis* era. 2002. *Trends in Plant Science*, 7:263–70.

Wyatt, J.M. and Knowles, C.J. 1995. Microbial degradation of acrylonitrile waste effluents: The degradation of effluents and condensates from the manufacture of acrylonitrile. *International Biodeterioration and Biodegradation*, 35:227–48.

Yamada, H. and Kobayashi, M. 1996. Nitrile hydratase and its application to industrial production of acrylamide. *Biosciences Biotechnology and Biochemistry*, 60:1391–1400.

Yamaki, T., Oikawa, T., Ito, K., et al. 1997. Cloning and sequencing of a nitrile hydratase gene from *Pseudonocardia thermophila* JCM3095. *Journal of Fermentation and Bioengineering*, 83:474–7.

Yamamoto, K. and Komatsu, K. 1991. Purification and characterization of nitrilase responsible for the enantioselective hydrolysis from *Acinetobacter* sp. AK 226. *Agricultural and Biological Chemistry Tokyo*, 55:1459–66.

Yamamoto, K., Fujimatsu, I., and Komatsu, K. 1992. Purification and characterization of the nitrilase from *Alcaligenes faecalis* ATCC8750 responsible for the enantioselective hydrolysis of mandelonitrile. *Journal of Fermentation and Bioengineering*, 73:425–30.

Yanase, H., Sakamoto, A., Okamoto, K., et al. 2000. Degradation of the metal-cyano complex tetracyanonickelate (II) by *Fusarium oxysporum* N–10. *Applied Microbiology and Biotechnology*, 53:328–34.

Zhu, D., Ankati, H., Mukherjee, C., et al. 2007a. Asymmetric reduction of β-ketonitriles with a recombinant carbonyl reductase and enzymatic transformations to optically pure β-hydroxy carboxylic acids. *Organic Letters*, 9:2561–3.

Zhu, D., Murkherjee, C., Biehl, E.R., et al. 2007b. Discovery of a mandelonitrile hydrolase from *Bradyrhizopium japonicum* USDA 110 by rational genome miming. *Journal of Biotechnology*, 129:645–50.

Zhu, D., Murkherjee, C., Yang, Y., et al. 2008. A new nitrilase from *Bradyrhizopium japonicum* USDA 110: Gene cloning, biochemical characterization and substrate specificity. *Journal of Biotechnology*, 133:327–33.

Index

A

4-amino-4-cyanobutyric acid, 362
AaP. *See Agrocybe aegerita* peroxidase (AaP)
ABEC. *See* Aqueous biphasic extraction
    chromatography (ABEC)
Acacipetalin, 361
ACC deaminase. *See* Aminocyclopropane-1-
    carboxylate (ACC) deaminase
Acetylation, 86–87
Acrylamide, 400
    grafting, 95, 97
    synthesis, 400–401
Activator generated by electron transfer
    (AGET), 287
Active pharmaceutical ingredients
    (APIs), 36
Adsorption, 81, 83. *See also* Biosorption
    protein, 101
AFM. *See* Atomic force microscopy
    (AFM)
Agency for Toxic Substances and Disease
    Registry (ATSDR), 226
AGET. *See* Activator generated by electron
    transfer (AGET)
Ag nanoparticles (Ag NPs), 16
    AFM study, 330–331
    formation and pH, 329–330
    microbes synthesizing, 318–320
    reduction reaction, 331, 332
    synthesis, 328, 329–332
    UV–vis spectra, 329
Ag NPs. *See* Ag nanoparticles (Ag NPs)
Agrobacterium tumefaciens, 371
Agrocybe aegerita peroxidase (AaP), 13
Air-filled porosity, 219, 220. *See also*
    Permeability
Alkylating agents, 186
Alkylation, 24, 186, 187, 190
    2-alkylated hydroquinones, 188, 189
    2-ethoxyphenol, 191
    C-alkylation, 187–189
    N-alkylation, 191–193
    N-alkylphthalimides, 191–192
    N-arylamines, 191–192
    O-alkylation, 190–191
    amino ethylated products, 189
    arenes, 188
    bicyclo[2.2.2]octanone, 189
    c–c bond, 187

cyclohexenes, 187
epoxypropoxyphenols, 190
Friedel–Crafts, 186
microwave irradiation, 186
nucleophilic alkylating agents, 186
principle types, 186–187
symmetrical ether, 190
Allylrethrone, 64
Amidases, 98, 368. *See also* Nitrilases
    microbial sources, 369–370
    reaction mechanism, 370
    substrates, 370
Aminocyclopropane-1-carboxylate 1-(ACC)
    deaminase, 140
Amygdalin, 361
ANN. *See* Artificial neural network (ANN)
Anthropocentrism, 111, 117
Anti-environment, 106
APIs. *See* Active pharmaceutical ingredients
    (APIs)
Aqueous biphasic extraction chromatography
    (ABEC), 288
Aromatization, 66
Arsenic (As), 128
    arsenate influx, 137
    AtACR2 RNAi knockdown, 141
    chelators, 139
    chronic exposure, 134
    concentrations, 137
    hyperaccumulation, 135–137, 139–140
    hyperaccumulators, 134–135
    metal chelators, 139
    phytoextraction, 137–139
    phytoremediation, 139–143
    pollution, 134
    properties, 133
    Pseudomonas, 142
    P. vittata, 136
    speciation, 133, 136
    toxicity, 133–134
    vascular effects, 134
Artificial neural network (ANN), 274, 282
As. *See* Arsenic (As)
As hyperaccumulators, 134–135
    BF, 135
    Chinese brake fern, 135
    mechanism, 135–137
    plants, 132
    P. vittata, 135
    TF, 135

Milton Keynes UK
Ingram Content Group UK Ltd.
UKHW020011071024
449327UK00031B/2739